AF616101

90 0779885 7

PROGRESS IN

Molecular Biology and Translational Science

Volume 108

PROGRESS IN

Molecular Biology and Translational Science

Recent Advances in Nutrigenetics and Nutrigenomics

edited by

C. Bouchard

John W. Barton, Sr. Chair in Genetics and Nutrition, Human Genomics Laboratory Pennington Biomedical Research Center, Baton Rouge, Louisiana, USA

J.M. Ordovas

U.S. Department of Agriculture, Human Nutrition Research Center, Tufts University Boston, Massachusetts, USA

Volume 108

ELSEVIER

AMSTERDAM • BOSTON • HEIDELBERG • LONDON
NEW YORK • OXFORD • PARIS • SAN DIEGO
SAN FRANCISCO • SINGAPORE • SYDNEY • TOKYO
Academic Press is an imprint of Elsevier

Academic Press is an imprint of Elsevier
32 Jamestown Road, London, NW1 7BY, UK
Radarweg 29, PO Box 211, 1000 AE Amsterdam, The Netherlands
225 Wyman Street, Waltham, MA 02451, USA
525 B Street, Suite 1900, San Diego, CA 92101-4495, USA

This book is printed on acid-free paper.

Library of Congress Cataloging-in-Publication Data
A catalog record for this book is available from the Library of Congress

British Library Cataloguing in Publication Data
A catalogue record for this book is available from the British Library

ISBN: 978-0-12-398397-8
ISSN: 1877-1173

For information on all Academic Press publications
visit our website at elsevierdirect.com

Printed and bound by CPI Group (UK) Ltd, Croydon, CR0 4YY

Transferred to digital print 2013

Contents

Contributors

Numbers in parentheses indicate the pages on which the authors' contributions begin.

Itziar Abete, Department of Nutrition, Food Science, Physiology and Toxicology, University of Navarra, Pamplona, Spain (323)

Michael Affolter, Department of BioAnalytical Sciences, Functional Genomics Group, Nestlé Research Center, Lausanne, Switzerland (51)

Smaragdi Antonopoulou, Department of Dietetics and Nutritional Science, Harokopio University of Athens, Athens, Greece (201)

Alaa Badawi, Office of Biotechnology, Genomics and Population Health, Public Health Agency of Canada, Toronto, Ontario, Canada (179)

Claude Bouchard, John W. Barton, Sr. Chair in Genetics and Nutrition, Human Genomics Laboratory, Pennington Biomedical Research Center, Baton Rouge, Louisiana, USA (1, 447)

Edith Brot-Laroche, Department of Physiology, Metabolism, Differentiation, Centre de Recherche des Cordeliers, Paris, France (113)

Chieh Jason Chou, Department of Gastrointestinal Health, Nestlé Institute of Health Sciences, Lausanne, Switzerland (51)

Karen D. Corbin, University of North Carolina at Chapel Hill, Nutrition Research Institute, Kannapolis, North Carolina, USA (159)

Dolores Corella, Genetic and Molecular Epidemiology Unit, School of Medicine, University of Valencia, Valencia, Spain, and CIBER Fisiopatología de la Obesidad y Nutrición, Instituto de Salud Carlos III, Madrid, Spain (261)

Marilyn C. Cornelis, Department of Nutrition, Harvard School of Public Health, Boston, Massachusetts, USA (293)

Laura A. Da Costa, Department of Nutritional Sciences, Faculty of Medicine, University of Toronto, Toronto, Ontario, Canada (179)

George Dedoussis, Department of Dietetics and Nutritional Science, Harokopio University of Athens, Athens, Greece (201)

Ahmed El-Sohemy, Department of Nutritional Sciences, Faculty of Medicine, University of Toronto, Toronto, Ontario, Canada (179)

María Mercedes Galindo, German Institute of Human Nutrition, Potsdam-Rehbrücke, Nuthetal, Germany (383)

Bibiana García-Bailo, Department of Nutritional Sciences, Faculty of Medicine, University of Toronto, and Office of Biotechnology, Genomics and Population Health, Public Health Agency of Canada, Toronto, Ontario, Canada (179)

Paul Haggarty, Lifelong Health, Rowett Institute of Nutrition and Health, University of Aberdeen, Aberdeen, UK (427)

Jiang He, Department of Epidemiology, Tulane University School of Public Health and Tropical Medicine, and Department of Medicine, Tulane University School of Medicine, New Orleans, Louisiana, USA (237)

Stavroula Kanoni, Department of Dietetics and Nutritional Science, Harokopio University of Athens, Athens, Greece, and Genetics of complex traits in humans (Team 147), Wellcome Trust Sanger Institute, Hinxton, Cambridge, UK (201)

Claudia Kappen, Developmental Biology Laboratory, Pennington Biomedical Research Center, Baton Rouge, Louisiana, USA (129)

Tanika N. Kelly, Department of Epidemiology, Tulane University School of Public Health and Tropical Medicine, New Orleans, Louisiana, USA (237)

Martin Kussmann, Proteomics and Metabonomics Core, Nestlé Institute of Health Sciences, Lausanne, Switzerland; and Faculty of Science, Aarhus University, Aarhus, Denmark (51)

Chao-Qiang Lai, Nutrition and Genomics Laboratory, Jean Meyer USDA Human Nutrition Research Center on Aging, Tufts University, Boston, Massachusetts, USA (461)

Maude Le Gall, Department of Physiology, Metabolism, Differentiation, Centre de Recherche des Cordeliers, Paris, France (113)

Armelle Leturque, Department of Physiology, Metabolism, Differentiation, Centre de Recherche des Cordeliers, Paris, France (113)

Amelia Marti, Department of Nutrition, Food Science, Physiology and Toxicology, University of Navarra, Pamplona, Spain (323)

J. Alfredo Martinez, Department of Nutrition, Food Science, Physiology and Toxicology, University of Navarra, Pamplona, Spain (323)

Wolfgang Meyerhof, German Institute of Human Nutrition, Potsdam-Rehbrücke, Nuthetal, Germany (383)

Santiago Navas-Carretero, Department of Nutrition, Food Science, Physiology and Toxicology, University of Navarra, Pamplona, Spain (323)

Tzortzis Nomikos, Department of Dietetics and Nutritional Science, Harokopio University of Athens, Athens, Greece (201)

Jose M. Ordovas, Nutrition and Genomics Laboratory, Jean Meyer USDA Human Nutrition Research Center on Aging, Tufts University, Boston, Massachusetts, USA (1)

George Papanikolaou, Department of Dietetics and Nutritional Science, Harokopio University of Athens, Athens, Greece (201)

Laurence D. Parnell, Computational Biologist, Nutritional Genomics Laboratory, JM-USDA Human Nutrition Research Center on Aging, Tufts University, Boston, Massachusetts, USA (17)

Louis Pérusse, Institute of Nutraceuticals and Functional Foods (INAF), and Department of Kinesiology, Laval University, Quebec, Canada (347)

Tuomo Rankinen, Human Genomics Laboratory, Pennington Biomedical Research Center, Baton Rouge, Louisiana, USA (447)

Iwona Rudkowska, Institute of Nutraceuticals and Functional Foods (INAF), Laval University, Quebec, Canada (347)

J. Michael Salbaum, Regulation of Gene Expression Laboratory, Pennington Biomedical Research Center, Baton Rouge, Louisiana, USA (129)

Nanette Yvette Schneider, German Institute of Human Nutrition, Potsdam-Rehbrücke, Nuthetal, Germany (383)

Frauke Stähler, German Institute of Human Nutrition, Potsdam-Rehbrücke, Nuthetal, Germany (383)

Maria G. Stathopoulou, Department of Dietetics and Nutritional Science, Harokopio University of Athens, Athens, Greece (201)

Jonas Töle, German Institute of Human Nutrition, Potsdam-Rehbrücke, Nuthetal, Germany (383)

John P. Vanden Heuvel, Department of Veterinary and Biomedical Sciences and Center for Excellence in Nutrigenomics, Penn State University, University Park, Pennsylvania, USA, and Indigo Biosciences, Inc., State College, Pennsylvania, USA (75)

Steven H. Zeisel, University of North Carolina at Chapel Hill, Nutrition Research Institute, Kannapolis, North Carolina, USA (159)

Preface

This volume entirely devoted to the recent scientific advances in nutrigenetics and nutrigenomics comes at the right time for these ever-expanding fields of research. It is also highly relevant to the serial *Progress in Molecular Biology and Translational Science* (PMBTS). As editors, we are extremely pleased by the distinguished panel of authors that we have been able to assemble for this volume. Forty authors from eight countries have contributed to the publication. We are very grateful for their willingness to participate in this effort. We would like to express our gratitude to them not only for their outstanding science but also for the timely delivery of their contributions.

The leadership of the PMBTS publication series has been a delight to work with. We would like to express our thanks to Dr. Michael Conn, editor of the PMBTS serial, who proposed to one of us that a volume on nutrigenetics and nutrigenomics would be a timely addition to the PMBTS banner. We also benefited greatly from the support of Lisa Tickner and Mary Ann Zimmerman, acquisition editors, and Sarah Latham, editorial project manager, all at Elsevier Inc. They were all very supportive at various stages of the development of the publication, and we would like to express our warmest thanks to them.

Finally, we would not have been able to undertake the task of serving as editors for this volume without the outstanding and competent support of Allison Templet at the Pennington Biomedical Research Center. Allison worked diligently with each author in order to ensure that the manuscript was complete and met all the requirements of the publisher. Her expertise in editing complex scientific material and her dedication to excellence in scientific publication have made a substantial difference in our ability to deliver a high-quality volume. We both feel greatly indebted to her.

Claude Bouchard and Jose Ordovas
February 2012

Fundamentals of Nutrigenetics and Nutrigenomics

Claude Bouchard* and
Jose M. Ordovas†

*John W. Barton, Sr. Chair in Genetics and Nutrition, Human Genomics Laboratory, Pennington Biomedical Research Center, Baton Rouge, Louisiana, USA

†Nutrition and Genomics Laboratory, Jean Meyer USDA Human Nutrition Research Center on Aging, Tufts University, Boston, Massachusetts, USA

This volume of *Progress in Molecular Biology and Translational Science* is devoted to the exciting and promising field of nutrigenetics and nutrigenomics. The introductory chapter defines the basic concepts necessary for the interpretation of the material covered in the remainder of the volume. Emphasis is on the concept of personalized nutrition and its likely role in public health and disease prevention, as well as in therapeutics. Nutrigenetics refers to the role of DNA sequence variation in the responses to nutrients, whereas nutrigenomics is the study of the role of nutrients in gene expression. This research is predicated on the assumption that there are individual differences in responsiveness to acute or repeated exposures to a given nutrient or combination of nutrients. Throughout human history, diet has affected the expression of genes, resulting in phenotypes that are able to successfully respond to environmental challenges and that allow better exploitation of food resources. These adaptations have been key to human growth and development. Technological advances have made it possible to investigate not only specific genes but also to explore in unbiased designs the whole genome-wide complement of DNA sequence variants or transcriptome. These advances provide an opportunity to establish the foundation for incorporating biological individuality into dietary recommendations, with significant therapeutic potential.

Progress in Molecular Biology
and Translational Science, Vol. 108
DOI: 10.1016/B978-0-12-398397-8.00001-0

I. Definitions of Basic Concepts

The study of the interactions between genes and nutrients has been a topic of interest in the past few decades. As in other fields of biology, the first experiments focused on one or only a few genes primarily because of technological limitations at that time. More recently, the technological barriers have been lifted, and it has become possible to apprehend a large number of genes and, at times, the whole genome in a single study. It is in that context that the concepts of nutrigenetics and nutrigenomics have evolved. Evolution in the definitions of these concepts has been discussed elsewhere.[1]

In the past few years, with the creation of the International Society of Nutrigenetics/Nutrigenomics and the launching of the *Journal of Nutrigenetics and Nutrigenomics* in 2008, these concepts have become more widely used in a consistent way by the community of active scientists. Although all conceptual issues have not been resolved, enough progress has been made to be able to use both terms in ways that improve the clarity of communication.

Simply put, nutrigenetics refers to the role of DNA sequence variation in the responses to nutrients, whereas nutrigenomics is the study of the role of nutrients in gene expression.

Nutrigenetics focuses on the potential effects of single-nucleotide polymorphisms, copy number variants, epigenetic marks, and other genomic markers on the biological and behavioral responses to micronutrients, macronutrients, and calories. A substantial body of data has accumulated over the years, with an emphasis on cardiovascular disease (CVD), cancer, and other common diseases or associated risk factors.[1] These studies were typically observational and focused on one or a few genetic markers in a candidate gene.

Two major issues need to be addressed for nutrigenetics to move forward, and these issues are being progressively dealt with, as this volume reveals. The first is the incorporation of high-throughput genotyping and sequencing technologies so that genome-wide explorations of DNA variant–nutrient interactions can be undertaken. The sample sizes required are well beyond the scope of any individual study. This has been addressed through large consortia such as CHARGE (http://web.chargeconsortium.com), CARe (http://public.nhlbi.nih.gov/GeneticsGenomics/home/care.aspx), and GIANT (http://www.broadinstitute.org/collaboration/giant/index.php/GIANT_consortium), to name but a few. A shift from candidate genes to genome-wide tests will translate into new unbiased targets that have the potential to alter our current understanding of the underlying biology and generate new pathways and mechanisms of interest. However, much of this optimism is predicated on the assumption that we can obtain accurate and reliable dietary information. Despite decades of efforts, the current instruments and methods in this area remain inadequate. The second issue represents an even greater challenge. We need to augment the quality of not only observational studies but also controlled trials and experimental designs. Cross-sectional studies are useful for the identification of potentially interesting targets, but they can also be impacted by confounders that limit their usefulness, particularly when dealing with DNA markers and gene–nutrient interactions whose effect sizes tend to be small.

Nutrigenomics has evolved to signify the field concerned by the investigation of the effects of nutrients on gene expression and related downstream molecular and biological events. Nutrigenomics will increasingly incorporate transcriptomics, proteomics, and metabolomics, as well as more dynamic approaches brought up by fluxomic technologies in the quest to understand the specific effects of nutrients at the molecular as well as at the tissue and organ levels.[1–3] Nutrigenomics science should eventually be taken one step further with designs that take advantage of experimentally derived gene–nutrient interaction effects. For instance, experiments in which subjects are randomized by genotype shown to participate in a gene–nutrient interaction phenomenon would allow even more informative studies regarding the role of specific nutrients on cellular responses.

Even though the concepts of "nutriepigenetics" and "nutriepigenomics" do not appear in the title of this volume, they are always present in the background. The body of knowledge is still admittedly limited on the role of specific nutrients in the modulation of DNA methylation, histone modifications, and noncoding RNAs (e.g., microRNAs), which can potentially influence gene expression and cellular biology. However, there are already striking examples of such actions. Since some of these epigenetic events have the potential to attenuate or enhance gene–nutrient interaction effects, nutriepigenetics will undoubtedly contribute significantly in the future. Likewise, as epigenetic events are known to influence gene expression under some conditions, nutriepigenomics is poised to become more prominent.

II. Historical Background

The origins of the nutrigenetics/nutrigenomics concepts are lost in time, but historically, the well-known Greek aphorism "What is food for one, is to others bitter poison" (Lucretius) is one of the first written references alluding to the interindividual variability in response to dietary factors and its effects on human health. In modern times, one of the first reports describing an interaction between genes and diet can be traced back to 1945.[4] The investigators demonstrated that a diet of whole wheat and whole dried milk was able to promote a higher survival rate among W-Swiss mice subjected to *S. enteritidis* infection than a "synthetic" diet. However, the ability of diet to condition natural resistance was found to depend upon the genetic constitution of the mice employed. The term nutrigenetics appeared much later, and one of the first appearances is found in the title of a 1975 book by Brennan and Mulligan.[5] However, it was not until the 1980s that we acquired the technical capacity to interrogate specific genes and identify variants associated with variability in response to dietary factors. Since then, the field has been incorporating new technological advances, and we are currently witnessing the use of genome-wide association studies (GWAS) combined with dietary information to identify the genetic basis of differential dietary response.

III. Individual Differences and Nutrition

Nutrigenetics and nutrigenomics research is predicated on the assumption that there are individual differences in responsiveness to acute or repeated exposures to a given nutrient or a combination of nutrients. Although indirect evidence for interindividual differences in response to exposures can be obtained from observational studies,[1,6] causal relationships cannot be established from such studies. Of course, observational studies can be large, even extraordinarily so, but this cannot compensate for the fact that many confounders can contribute to the heterogeneity of the trait of interest, with the end result of augmenting or decreasing the magnitude of the gene–diet interaction effect. More powerful evidence for a role of genomic characteristics on dietary response can be derived from intervention studies, preferably randomized controlled trials.

There is a whole body of research that has documented beyond a shadow of a doubt that there are considerable interindividual differences in the response of key indicators of metabolism to a given nutrient dose or dietary regimen. The issue has been studied extensively, particularly for plasma lipids and lipoproteins.[7] From this body of data, the notion that there were hypo-responders and hyper-responders to dietary lipids and dietary cholesterol and that these traits were reproducible was proposed more than 25 years ago.[8,9] Human heterogeneity in response to a controlled overfeeding protocol has also been documented.[10] Exposure to hypocaloric diets for 2 months was shown to result in large individual differences in plasma leptin level changes.[11] Similar patterns were observed for blood pressure response to dietary sodium and potassium contents.[12]

Interestingly, the phenomenon is not limited to dietary exposure. For instance, individual differences in the lipid-lowering response to statins were observed in a study of about 4000 subjects from three cohorts.[13] A range of responses was also found for platelet aggregation in response to aspirin therapy.[14] Similarly, considerable human heterogeneity is found in the responses of metabolic and cardiovascular traits to regular exercise.[15,16] To make matters even more complicated, it has been observed that the extent of lipoprotein responses to defined dietary regimens was influenced by the exercise level of subjects.[17]

As suggested by the above, and as documented throughout this volume, the assumption that there are substantial interindividual differences in the response of morphological and metabolic traits to dietary exposures is a reasonable one. The central question, from our point of view, then becomes: what is the role of genetics and genomics factors in the variance in responsiveness?

IV. Evolution and Diversification

Evolution, diversification, and positive selection have been crucial in driving the changes that have led to the present situation of humankind. Throughout human history, nutrients have been interacting with genes in a "two-way interaction." On one hand, diet affects the expression of genes, resulting in phenotypes that can successfully respond to environmental challenges, such as those resulting from the dispersal to new habitats. On the other hand, diet also provides metabolic support for the development of functions that in turn allow better exploitation of food resources in the new habitat.[18] These adaptations have been key to human development, especially in relation to cognitive abilities. Compared to other primates and mammals, we allocate a much larger share of our energy intake to feed our brain. In order to accommodate the high energy demands of our large and complex brain, we must consume diets that are more dense in energy and fat than those of our closest relatives, the nonhuman primates[19–21] and, specifically, much higher levels of essential long-chain polyunsaturated fatty acids (PUFAs) that are critical to brain development.[22,23] Moreover, humans have higher body fatness than other primates, especially early in life. The need for an energy-rich diet may have also shaped our ability to detect and metabolize high-fat foods. We show strong preferences for lipid-rich foods that are based on the smell, texture, and taste of fatty foods,[24–26] and our brain has the ability to estimate the energy content of foods.[27]

We have acquired an enhanced capacity to digest and metabolize higher fat diets. Our gastrointestinal tract, with its expanded small intestine and reduced colon, is quite different from those of chimpanzees and gorillas and is consistent with the consumption of a high-quality diet with large amounts of animal food,[28] driven by the evolution of key "meat-adaptive" genes.[29] Therefore, a shift from principally carbohydrate-based to fish- and meat-based eating habits provided sufficient fuel and building blocks to facilitate encephalization and attain the current human brain size and structure. The dramatic expansion in brain size and function is likely related to specific mutations and other genetic changes, which were maintained across generations with critical contributions from dietary factors. The search for these specific mutations has revealed interesting results, both in terms of what was found as well as what was not found. The initial surveys of protein-coding sequences for evidence of positive selection in humans or chimpanzees identified only a few genes known to function in neural or nutritional processes, despite pronounced differences between humans and chimpanzees in behavior, cognition, and diet. Conversely, Haygood *et al.*[30] hypothesized that most such differences are due to changes in gene regulation rather than protein structure. These researchers conducted the first survey of promoter (5′-flanking) regions, which are rich in *cis*-regulatory sequences, for evidence of positive selection in humans. Their data suggested that positive selection has targeted the regulation of many genes known to be involved in neural development and function in the brain and elsewhere in the nervous system, as well as in nutritional metabolism, particularly in glucose metabolism.

Following the initial gene–diet interactions that shaped brain size and function of the *Homo sapiens*, there have been more recent examples of continuous adaptation to different ecological niches. They were driven primarily by the transition from food collection to food production (FP) which dramatically modified the nature of selective pressures, and several studies illustrated that genetic adaptation to a changing lifestyle has occurred in humans since the agricultural revolution. This is exemplified by the high levels of genetic variation at *CYP2D6*, a locus coding for a detoxifying enzyme of the cytochrome P450 complex.[31] Comparisons of DNA sequences and predicted levels of enzyme activity were undertaken across 10 African, Asian, and European populations, 6 of which currently rely on hunting and gathering (HG) and 4 on FP. Both HG and FP populations showed similar levels of *CYP2D6* diversity but displayed different substitution patterns at coding DNA sites, possibly related to selective differences. The differences between HG and FP populations suggest that new lifestyle and dietary habits acquired in the transition to agriculture affected the variation pattern at *CYP2D6*, leading to an increase in FP populations of the frequency of alleles that are associated with a slower rate of metabolism. These alleles reached a balanced coexistence with other previously selected variants.

Lactase persistence is one of the best examples of recent genetic adaptation driven by nutrition in humans. Lactase is the enzyme responsible for the digestion of the milk sugar lactose, and its production decreases after the weaning phase in most mammals, including our human ancestors. However, more recently, some humans have continued to produce lactase throughout adulthood, a trait known as "lactase persistence." In European populations, a single mutation (−13910*T) explains the distribution of the phenotype, whereas several mutations are associated with it in Africa and the Middle East. Current estimates for the age of lactase persistence-associated alleles bracket those for the origins of animal domestication and the culturally transmitted practice of dairying around 8000 years ago.[32]

Another striking example relates to starch consumption, which is a prominent characteristic of agricultural societies and hunter-gatherers in arid environments. In contrast, rainforest and circum-arctic hunter-gatherers and some pastoralists consume much less starch. This behavioral variation raises the possibility that different selective pressures have acted on amylase, the enzyme responsible for starch hydrolysis. Perry *et al.*[33] found that the copy number of the salivary amylase gene (*AMY1*) is correlated positively with salivary amylase protein level and that individuals from populations with high-starch diets have, on average, more *AMY1* copies than those with traditionally low-starch diets. This example of positive selection on the number of copies of a gene is one of the first discovered in the human genome. Higher *AMY1* copy numbers and protein levels probably improve the digestion of starchy foods and may buffer against the fitness-reducing effects of intestinal disease.

Against this evolutionary background, an unprecedented epidemic of obesity has evolved over the past few decades. But the roots of this epidemic can be traced back to the time in which our enhanced cognitive capacities enabled the control of fire and the manufacturing of tools, which increased energy yield from food even further and made it easier to defend against predators. The latter development relieved the selective pressure to maintain a normal level of body weight (driven by predation of overweight individuals). Since then, random mutations allowing body weight to increase have spread in the human gene pool by genetic drift and other forces. Also, (seasonal) food insecurity in hunter-gatherer societies spurred the increase in the prevalence of genes that maximize nutrient intake and energy storage when food is available. The agricultural and industrial revolutions rapidly changed our habitat: virtually unlimited stocks of (refined) foodstuffs and mechanical substitutes of physical efforts tilted the energy balance, particularly in those who are still biologically suited for former environmental conditions (i.e., those who carry genes favoring high energy intake and lack [genetic] protection against weight gain). Intrauterine epigenetic mechanisms potentially reinforce the impact of these genes on the propensity to become obese.[34]

Another example of coevolution and gene–environment interactions is provided by the gut microbiome.[35–37] The three-way interactions between human genetics, diet, and the microbiota fundamentally shaped modern populations and continue to affect health globally. Demonstrating the importance of the gut microbiota in human health and well-being represents a major transformational task in both medical and nutritional research. Owing to high-throughput -omics methodologies, the complexity, evolution with age, and individual nature of the gut microflora have recently been more thoroughly investigated. The balance between the complex community of gut bacteria, food nutrients, and intestinal genomic and physiological milieu is increasingly recognized as a major contributor to human health and disease.

V. Personalized Nutrition and Disease Prevention

Nutritional science has a long tradition of recommending specific diets to people based on well-defined characteristics that translate into different nutrient and energy needs. For instance, dietary recommendations have been developed for infants, children, and adolescents; pregnant women; adult men and women; older individuals; athletes engaged in specific training programs and athletic events; diabetic patients; and persons with dyslipoproteinemia, hypertension, and other conditions. These dietary guidelines are based on broad definitions of nutrient and energy needs of individuals belonging to defined segments of the population. While this approach recognizes that there are global differences in nutritional requirements, it does not specifically take into account the wide interindividual differences observed in response to given doses of nutrients and energy intakes.

The observation that there is human variation in the responses to acute and chronic exposures to a given nutrient dose offers an extraordinary opportunity to match dietary guidelines to the biology of the individual for optimal growth, disease prevention, and successful aging, to name but a few. While the paradigm is very attractive, it is also extraordinarily complex. The effort will have to be grounded in the genomic sequence of the individual. But knowing the exact genotype at significant genomic sites will not be sufficient in most cases. Adequately capturing biological individuality is likely to require information not only on the genotype but also on the epigenome, transcriptome, proteome, and metabolome in relevant tissues and body fluids. Indeed, in the end, matching biology to nutrition will necessitate more complex models than the simple relationship between genomic information and dietary nutrients. However, with the progress made on the genomics front, an opportunity is offered to take the first significant steps toward the goal of incorporating biological individuality into dietary recommendations.

A proof of concept of the critical importance of genomics evidence is provided by specifically designed dietary practices that alleviate some of the deleterious effects of monogenic disorders such as phenylketonuria, lactose intolerance, and other inherited metabolic diseases.[38] Beyond inborn errors of metabolism, advances in nutritional sciences have progressively led to an expanding and increasingly sophisticated segmentation of the consumer population based on life stages, cultural preferences, existing disease states, and physiological events such as digestive discomfort, body weight control, and energy level.[2] Although these advances constitute progress in the effort to match the "individual" to a set of dietary advices, they still fall short of meeting the ambitious goals of personalized preventive nutrition.

The average response to a given dose of a nutrient is tacitly considered as the "normal" response. However, as discussed previously, the observation that there are large interindividual differences in adaptation to a given nutrient dose imposes on us a paradigm shift, one that forces us to prioritize biological individuality over the notion of "normal." To make this transition successfully, access to comprehensive genomic information is a prerequisite, and fortunately this is becoming more of a reality with every passing day. The availability of relevant genomic information is, and is likely to remain, the cornerstone of personalized preventive nutrition. Other layers of complex -omics information will undoubtedly complement what can be learned about gene–nutrient interaction effects, but having the basic DNA sequence with the panel of nutrient-relevant variants will provide the initial foundation for the development of personalized preventive nutrition.

The interest in personalized nutrition is strong at this time, as evidenced by numerous publications directed to this topic (e.g., Refs. 3, 39–41). Although much remains to be done, and despite the fact that we are at a very early stage in this journey, one can anticipate that personalized genomic nutrition, and more generally personalized preventive nutrition, will be predictive of individual vulnerabilities to diseases, will offer evidence-based disease prevention guidelines, will provide opportunities to attenuate disease progression, and could globally result in more precise and safer dosages (see, for instance, the Preface in Ref. 42).

VI. Personalized Nutrition and Therapeutic Applications

Personalized nutrition has been used for decades to palliate the effects of rare inborn metabolic errors. Most current nutrigenetics efforts aim to bring similar approaches to the general population as a preventive or therapeutic tool to deal with common chronic diseases. The initial findings in this area and some of the promises and claims have spurred interest in developing and marketing tests designed for personalized nutrition. The goal of personalized nutrition is that, based on the genetic information regarding a person's health risk profile, it should be possible to prescribe individualized nutrition recommendations to reduce disease risk. The current information suggests that consumers have a positive attitude toward the testing of their genetic profile to be used in nutritional advice. A recent paper by Roosen *et al.*[43] found that about 45% of their sample population would agree to such a test and would like to obtain personalized advice on nutrition. These and other results show that the concept of personalized nutrition and its therapeutic potential are promising. However, the science needed to provide these recommendations is not mature yet, and the early launching of products promising too much and delivering too little could seriously compromise the trust of the population and slow the pace of growth of the field.

VII. Scope of Topics Covered

This volume is devoted to the latest advances in nutrigenetics and nutrigenomics, with an emphasis on current peer-reviewed literature as well on the most promising research designs and technologies. These topics are covered in some depth through 17 chapters written by colleagues who are intimately involved in the development of scientific basis of the field. Here is a brief overview of the scope of each of these chapters.

Chapter 2 by LD Parnell focuses on technologies and study designs. Downstream effects of the availability of the sequence of the human genome include in-depth probing of the molecular mechanisms underlying the response to micronutrients and macronutrients. Changes in gene expression resulting from genetic perturbations or DNA sequence variation are a critical component of nutrigenetics and nutrigenomics research.

Chapter 3 by CJ Chou, M Affolter, and M Kussmann deals with nutrigenomics and protein intake. They first review readouts of protein intake assessed by -omics technologies, including gene expression, proteomics, and metabolite profiling. Then protein benefits are addressed beyond macronutrient supply, with an emphasis on how to generate, analyze, and leverage bioactive peptides. Finally, protein turnover as evidenced by proteomics tools is discussed.

Chapter 4 targets omega-3 PUFAs and is authored by JP Vanden Heuvel. Four nuclear receptor subfamilies that respond to dietary and endogenous ligands are emphasized: (1) peroxisome proliferator-activated receptors, (2) retinoid X receptors, (3) liver X receptors, and (4) farnesoid X receptor. In addition to the different responses elicited by varying structures of fatty acids, responses may vary because of genetic variation in enzymes that metabolize omega-3 and omega-6 PUFAs or that respond to them. In particular, polymorphisms in the fatty acid desaturases and the aforementioned nuclear receptors contribute to the complexity of nutritional effects seen with omega-3 PUFAs. Understanding the nutrigenomics and nutrigenetics of dietary fatty acids is key to understanding critically important human diseases such as CVD and cancer.

Chapter 5 by A Leturque, E Brot-Laroche, and M Le Gall focuses on carbohydrates. Sugar intake is regulated by several factors, and gene polymorphisms are involved in determining sugar preference. Nutrigenomic adaptations to carbohydrate availability have been evidenced in the persistence of lactose digestion and *AMY* copy number. In addition, dietary oligosaccharides, fermentable by gut flora, can modulate microbiotal diversity to benefit the host. Genetic diseases linked to mutations in disaccharidase and transporter genes impact carbohydrate intake. Carbohydrate intolerance is revealed upon exposure to the offending sugar, and withdrawal of this sugar from the diet typically prevents disease symptoms.

Chapter 6 by JM Salbaum and C Kappen deals with folate. The emphasis of the chapter is on the evolutionary aspects of folate transport and metabolism, the consequences of genetic defects in folate genes in mice and humans, and the implications of the nutritional methyl-donor supply for epigenomics of health and disease. It is argued that dietary folate provides the maintenance and stability of the genome, as well as regulation of gene expression via the methylation of DNA and histones.

Chapter 7 authored by KD Corbin and SH Zeisel addresses the issue of the dietary requirements for choline. Choline is an essential nutrient with a wide range of biological functions. Gender, single-nucleotide polymorphisms, estrogen status, and gut microbiome composition have been shown to influence its optimal intake level. The chapter discusses the challenges we face in developing individualized nutrition recommendations, using the biological function of choline and the consequences of inadequate choline nutrition to illustrate the concepts.

Chapter 8 by LA Da Costa, B Garcia-Bailo, A Badawi, and A El-Sohemy deals with dietary antioxidants. Modulating oxidative stress by dietary antioxidant micronutrients such as vitamins C and E or phytochemicals such as carotenoids may help prevent or delay the development of certain diseases. However, research on antioxidant supplementation and disease has yielded inconsistent findings, which may be due, in part, to interindividual genetic variation. DNA sequence variants in genes encoding host antioxidant enzymes or proteins responsible for the absorption, transport, distribution, or metabolism of dietary antioxidants affect antioxidant status and response to supplementation.

Chapter 9 is authored by G Dedoussis and collaborators from the University of Athens. It focuses on the nutrigenetics and nutrigenomics of dietary minerals, elements playing key roles in the regulation of metabolism. Although defining optimal dietary requirements of minerals across geographic locations and cultural practices is a challenge, adequate intake is essential for physiological homeostasis, cell protection, functionality, and overall health. Mineral intake deficiencies are associated with well-characterized illnesses. Calcium, copper, iron, selenium, and zinc dietary intakes are emphasized because of their roles in a variety of biological processes.

Chapter 10 is coauthored by TN Kelly and J He. It focuses on salt intake, with an emphasis on blood pressure regulation. Monogenic blood pressure disorders have illuminated the role of renal salt handling in blood pressure regulation. They have been instrumental in implicating genes and pathways related to salt sensitivity. Candidate gene studies have evidenced the contribution of genes and variants in the renin–angiotensin–aldosterone system, renal sodium channels, and transporters modulating the blood pressure response to salt intake. Advances based on the latest results from GWAS complete this review.

Chapter 11 is written by D Corella and deals with alcohol intake. Alcohol consumption is a major research topic in nutritional genomics. This chapter reviews findings on alcohol consumption trends, methodological limitations in the analysis of alcohol consumption, and genes and polymorphisms that have been associated with alcohol intake. The surprisingly inconsistent results from GWAS are discussed. The effects of alcohol consumption on CVD and cancer and the interactions between genomic markers and alcohol in disease-related traits are highlighted.

Chapter 12 deals with the nutrigenetics and nutrigenomics of coffee intake and is authored by MC Cornelis. The extraordinary popularity and availability of coffee beverages have fueled concerns over the potential side effects and health consequences. Thus far, studies of coffee intake and health outcomes have been inconsistent. One possible reason has to do with the exact chemical components of coffee potentially involved. Another reason for the discrepancies in the results may be human heterogeneity in sensitivity to caffeinated beverages. The chapter details the variability in physiological effects of coffee intake.

Chapter 13 addresses the issue of caloric restriction and is authored by JA Martinez and collaborators from the University of Navarra. Responses to caloric restriction programs show a wide range of interindividual differences. The expression patterns and nutritional regulation of several obesity-related genes are discussed. The association between genomic markers and the response to caloric restriction is also reviewed.

Chapter 14 complements the preceding chapter and addresses the issue of individualized body weight management. It is authored by I Rudkowska and L Perusse. Gene–environment interactions are important in determining a predisposition to excess weight, but they can also influence the outcome of weight loss programs and weight management strategies. The chapter reviews relevant gene expression studies and the effects of various obesity candidate gene polymorphisms on the response of body weight traits to weight loss programs.

Chapter 15 focuses on taste preferences and is authored by W Meyerhof and colleagues from the German Institute of Human Nutrition, Potsdam. Taste has an important input into food preference. It plays a critical role in differentiating nutritive and harmful substances and in the selection of nutrients. Gustatory stimuli make contact with specialized receptors and channels expressed in taste buds in the oral cavity. Gustatory information is then conveyed via afferent nerves to the central nervous system, where the gustatory information results in stimulus recognition, integration with metabolic needs, and control of eating behavior. The role of genetic variability is highlighted.

Chapter 16 is authored by P Haggarty and deals with nutrition and the epigenome. There is an increasing awareness of the importance of epigenetic events in health and disease. Epigenetic mechanisms include DNA methylation, histone modifications, and regulation by noncoding RNAs. There is strong evidence for nutritional effects on the epigenome. The dietary substrates for epigenetic reactions include acetyl and methyl groups. Current challenges in nutritional epigenetics include the question of tissue-specific epigenetic events and the issue of the heterogeneity of response at specific epigenetic loci.

Chapter 17 addresses the specialized topic of exercise genomics and is authored by T Rankinen and C Bouchard. The advances in our knowledge of the role of DNA sequence variants in physical activity level and indicators of health-related fitness are reviewed. Public health agencies around the world recommend that people be active most days in order to enjoy a higher quality of life and prevent morbidities and premature death. The ability to respond favorably to regular exercise is influenced by genetic individuality, and the evidence to that effect is discussed.

Chapter 18 deals with the complex topic of genetic variation and population differences. It is authored by CQ Lai. The human genome has been shaped by demographic forces and adaptation to local environments, including regional climate, landscape, food sources, culture, and pathogens. Individuals from different populations sharing the same environments can exhibit differences in disease risk, as do individuals from the same population living in various regions of the globe. The chapter discusses the methodology employed in examining adaptive genetic variation across populations, the importance of adaptive genetic variation to human health, and the implications for nutrigenetics and nutrigenomics research, as well as disease prevention.

The various chapters of this volume provide numerous examples supporting the often-made argument that technological advances and new instrumentations are key drivers of scientific progress. Genome-wide technologies have already transformed nutrigenetics and nutrigenomics. The belief that personalized dietary recommendations are within our reach is rather pervasive. It is already widely recognized that one diet does not fit all. The question has now become: when will our scientific foundation be such that it will become the ethical standard practice to match diet to biological individuality?

References

1. Ordovas JM, Corella D. Nutritional genomics. *Annu Rev Genomics Hum Genet* 2004;**5**:71–118.
2. Kussmann M, Fay LB. Nutrigenomics and personalized nutrition: science and concept. *Pers Med* 2008;**5**:447–55.
3. Kaput J, Rodriguez RL. Nutritional genomics: the next frontier in the postgenomic era. *Physiol Genomics* 2004;**16**:166–77.

4. Schneider HA, Webster LT. Nutrition of the host and natural resistance to infection. 1. The effect of diet on the response of several genotypes of Mus musculus to Salmonella-enteritidis infection. *J Exp Med* 1945;**81**:359–84.
5. Brennan RO, Mulligan WC. *Nutrigenetics: new concepts for relieving hypoglycemia.* New York and Philadelphia: M. Evans; 1975.
6. Marti A, Goyenechea E, Martinez JA. Nutrigenetics: a tool to provide personalized nutritional therapy to the obese. *J Nutrigenet Nutrigenomics* 2010;**3**:157–69.
7. Masson LF, McNeill G, Avenell A. Genetic variation and the lipid response to dietary intervention: a systematic review. *Am J Clin Nutr* 2003;**77**:1098–111.
8. Beynen AC, Katan MB, Van Zutphen LF. Hypo- and hyperresponders: individual differences in the response of serum cholesterol concentration to changes in diet. *Adv Lipid Res* 1987; **22**:115–71.
9. Katan MB, Beynen AC, de Vries JH, Nobels A. Existence of consistent hypo- and hyperresponders to dietary cholesterol in man. *Am J Epidemiol* 1986;**123**:221–34.
10. Bouchard C, Tremblay A, Despres JP, Nadeau A, Lupien PJ, Theriault G, et al. The response to long-term overfeeding in identical twins. *N Engl J Med* 1990;**322**:1477–82.
11. de Luis DA, Aller R, Izaola O, Sagrado MG, Conde R. Influence of Lys656Asn polymorphism of leptin receptor gene on leptin response secondary to two hypocaloric diets: a randomized clinical trial. *Ann Nutr Metab* 2008;**52**:209–14.
12. The GenSalt Collaborative Research Group . GenSalt: rationale, design, methods and baseline characteristics of study participants. *J Hum Hypertens* 2007;**21**:639–46.
13. Barber MJ, Mangravite LM, Hyde CL, Chasman DI, Smith JD, McCarty CA, et al. Genome-wide association of lipid-lowering response to statins in combined study populations. *PLoS One* 2010;**5**:e9763.
14. Mitchell BD, McArdle PF, Shen H, Rampersaud E, Pollin TI, Bielak LF, et al. The genetic response to short-term interventions affecting cardiovascular function: rationale and design of the Heredity and Phenotype Intervention (HAPI) Heart Study. *Am Heart J* 2008;**155**:823–8.
15. Bouchard C, Rankinen T. Individual differences in response to regular physical activity. *Med Sci Sports Exerc* 2001;**33**:S446–51 discussion S452–S453.
16. Rankinen T, Bouchard C. Gene-physical activity interactions: overview of human studies. *Obesity (Silver Spring)* 2008;**16**(Suppl. 3):S47–50.
17. Williams PT, Blanche PJ, Rawlings R, Krauss RM. Concordant lipoprotein and weight responses to dietary fat change in identical twins with divergent exercise levels 1. *Am J Clin Nutr* 2005;**82**:181–7.
18. Rotilio G, Marchese E. Nutritional factors in human dispersals. *Ann Hum Biol* 2010; **37**:312–24.
19. Leonard WR, Robertson ML. Nutritional requirements and human evolution: a bioenergetics model. *Am J Hum Biol* 1992;**4**:179–95.
20. Leonard WR, Robertson ML. Evolutionary perspectives on human nutrition: the influence of brain and body size on diet and metabolism. *Am J Hum Biol* 1994;**6**:77–88.
21. Popovich DG, Jenkins DJ, Kendall CW, Dierenfeld ES, Carroll RW, Tariq N, et al. The western lowland gorilla diet has implications for the health of humans and other hominoids. *J Nutr* 1997;**127**:2000–5.
22. Cordain L, Watkins BA, Mann NJ. Fatty acid composition and energy density of foods available to African hominids. Evolutionary implications for human brain development. *World Rev Nutr Diet* 2001;**90**:144–61.
23. Crawford MA, Bloom M, Broadhurst CL, Schmidt WF, Cunnane SC, Galli C, et al. Evidence for the unique function of docosahexaenoic acid during the evolution of the modern hominid brain. *Lipids* 1999;**34**(Suppl.):S39–47.
24. Gaillard D, Passilly-Degrace P, Besnard P. Molecular mechanisms of fat preference and overeating. *Ann N Y Acad Sci* 2008;**1141**:163–75.

25. le Coutre J, Schmitt JA. Food ingredients and cognitive performance. *Curr Opin Clin Nutr Metab Care* 2008;**11**:706–10.
26. Sclafani A. Psychobiology of food preferences. *Int J Obes Relat Metab Disord* 2001;**25**(Suppl. 5): S13–6.
27. Toepel U, Knebel JF, Hudry J, le Coutre J, Murray MM. The brain tracks the energetic value in food images. *Neuroimage* 2009;**44**:967–74.
28. Milton K. Primate diets and gut morphology: implications for hominid evolution. In: Harris M, Ross EB, editors. *Food and evolution: toward a theory of human food habits*. Philadelphia, PA: Temple University Press; 1987. pp. 93–116.
29. Finch CE, Stanford CB. Meat-adaptive genes and the evolution of slower aging in humans. *Q Rev Biol* 2004;**79**:3–50.
30. Haygood R, Fedrigo O, Hanson B, Yokoyama KD, Wray GA. Promoter regions of many neural- and nutrition-related genes have experienced positive selection during human evolution. *Nat Genet* 2007;**39**:1140–4.
31. Fuselli S, de Filippo C, Mona S, Sistonen J, Fariselli P, Destro-Bisol G, et al. Evolution of detoxifying systems: the role of environment and population history in shaping genetic diversity at human CYP2D6 locus. *Pharmacogenet Genomics* 2010;**20**:485–99.
32. Gerbault P, Liebert A, Itan Y, Powell A, Currat M, Burger J, et al. Evolution of lactase persistence: an example of human niche construction. *Philos Trans R Soc Lond B Biol Sci* 2011;**366**:863–77.
33. Perry GH, Dominy NJ, Claw KG, Lee AS, Fiegler H, Redon R, et al. Diet and the evolution of human amylase gene copy number variation. *Nat Genet* 2007;**39**:1256–60.
34. Pijl H. Obesity: evolution of a symptom of affluence. *Neth J Med* 2011;**69**:159–66.
35. Dimitrov DV. The human gutome: nutrigenomics of the host-microbiome interactions. *OMICS* 2011;**15**:419–30.
36. Guerzoni ME. Human food chain and microorganisms: a case of co-evolution. *Front Microbiol* 2010;**1**:106.
37. Walter J, Ley R. The human gut microbiome: ecology and recent evolutionary changes. *Annu Rev Microbiol* 2011;**65**:411–29.
38. Valle D, Beaudet AL, Vogelstein B, Kinzler KW, et al. (Eds.) *Online metabolic and molecular bases of inherited disease*. www.ommbid.com. Published 2006. Updated 28 March 2011. Accessed 11 May 2011..
39. Chadwick R. Nutrigenomics, individualism and public health. *Proc Nutr Soc* 2004;**63**:161–6.
40. Ordovas JM, Corella D. Nutrition and diet in the era of genomics. In: Willard HF, Ginsburg GS, editors. *Genomic and personalized medicine*, vol. 2. Oxford, UK: Elsevier; 2009. pp. 1204–20 2 vols.
41. Simopoulos AP, Milner JA, editors. *Personalized nutrition: translating nutrigenetic/nutrigenomic research into dietary guidelines. World Rev Nutr Diet*, vol. 101. Basel: Karger; 2010.
42. Willard HF, Ginsburg GS, editors. *Genomic and personalized medicine*. Oxford, UK: Elsevier; 2009 2 vols.
43. Roosen J, Bruhn M, Mecking RA, Drescher LS. Consumer demand for personalized nutrition and functional food. *Int J Vitam Nutr Res* 2008;**78**:269–74.

Advances in Technologies and Study Design

LAURENCE D. PARNELL

Computational Biologist, Nutritional Genomics Laboratory, JM-USDA Human Nutrition Research Center on Aging, Tufts University, Boston, Massachusetts, USA

Progress in Molecular Biology
and Translational Science, Vol. 108
DOI: 10.1016/B978-0-12-398397-8.00002-2

1877-1173/12 $35.00

The initial draft sequence of the human genome was the proving ground for significant technological advancements, and its completion has ushered in increasingly sophisticated tools and ever-increasing amounts of data. Often, this combination has multiplicative effects such as stimulating research groups to consider subsequent experiments of at least equal if not greater complexity or employ advanced technologies. As applied to the fields of nutrigenetics and nutrigenomics, these advances in technology and experimental design allow researchers to probe the biological, biochemical, and physiological mechanisms underpinning the response to micro- and macronutrients, along with downstream health effects. It is becoming ever more apparent that effects on gene expression as a consequence of genetic variation and perturbations to cellular and physiological systems are an important cornerstone of nutrigenomics and nutrigenetics research. A critical, near-term objective, however, must be to determine where and how nutrients and their metabolites augment or disrupt the genetic variation–gene expression axis. Downstream effects on protein and metabolite measures are also seen with growing regularity as vital components to this research. Thus, this chapter reviews the scope of recent progress and innovation in genomics and associated technologies as well as study designs as applied to nutrigenomics and nutrigenetics research and provides concrete examples of the application of those advancements in genomics-oriented nutrition research.

I. DNA Sequencing

A. Implications of Advances in Sequencing Technology

There is little doubt that DNA sequencing has seen great leaps in terms of technology and decreased costs in the 11 years since the initial publication of a complete human genome sequence.[1,2] Technological advances are being applied to the sequencing of genomes from more individuals, targeted sequencing of exons or of the pool of expressed genes, and identification of organisms comprising the oral and intestinal flora.

The past few years in particular have seen substantial advances in sequencing technology. Technologies such as reversible dye terminators, sequencing by ligation, and pyrosequencing are all broadly classified as next-generation sequencing and are used to gain information on whole-genome data, including single-nucleotide polymorphisms (SNPs) and copy number variants, gene expression, and characterization of an individual's microbiome. Notably, the cost of sequencing is approaching \$1000 per human genome, which implies that nutrition-based research can now realistically consider whole-genome sequence data as a viable component to the research plan. However, a consideration not to be overlooked is the ripple effect in terms of the data-generating capacity of the sequencing machines, data compression and storage with computational manipulation, and, finally, interpretation of the data. These ancillary developments are profound and demand constant attention, yet no easy, ready-made solutions exist.

As sequencing technology advances and costs decrease, attempts to find rare variants and to ascertain their association with the risk of common disease no longer seem far fetched. Similarly, as of early 2011, the cost of assaying messenger RNA (mRNA) levels with array technology was about half that of sequencing the entire sample's worth of mRNA, so-called RNA-seq (see Section I.D). It is expected that the competing technological approaches of sequencing and array applications will be about equal by late 2012. However, some believe that array-based measures of gene expression remain a good complement to RNA-seq and one should not supplant the other.[3] Although less than biological variation, technical variation remains large enough that it cannot be disregarded, principally as this is greatest at low coverage levels.[4]

Another recent improvement in sequencing technology applies to the number of samples that can be processed simultaneously. Researchers at the Royal Institute of Technology in Solna, Sweden, have devised a method whereby upwards of 5000 samples can be run at the same time and at the same price.[5] This novel strategy uses a combination of two tags to tie sequencing reads back to their sources from a pool of samples with these tags incorporated in two steps. This dramatic reduction in cost has significant implications for DNA or

RNA sequencing studies. Also of note is a product available from Illumina®, Inc., which detects degraded mRNA species, as often occurs in formalin-fixed, paraffin-embedded (FFPE) samples. The Whole-Genome DASL HT Assay can detect more than 29,000 gene targets in FFPE samples from as little as 50 ng total RNA. Thus, archived samples could more readily serve as source material in basic research.

B. Personal Genomes

Yet another area in which the fields of nutrigenomics and nutrigenetics are experiencing evolution is with regard to personal genomes. There are three major aspects to these changes.

- First, the 1000 Genomes Project[6] is deciphering the genetic sequence of a large number of humans from diverse ethnic and ancestral backgrounds with the goal of fully cataloging human genetic variation. While phenotypes on these subjects are scant, the effort is uncovering a large number of genetic variants and adding to our knowledge of genetic variation and genome architecture.
- Second, many companies offer services where for a fee an individual can submit a sample from which genomic DNA is extracted and genotyped for a wide array of genetic variants. Such direct-to-consumer (DTC) genetic testing firms often provide health advice or put the alleles into a health context based on the genetic profile, sometimes in conjunction with gender and ethnicity. At this time, however, such reports on the customer's genetic profile with respect to health risk do not include gene–environment or gene–diet interactions.
- Third, although just a trickle now, there are individuals who have had a large number, typically 1 million, common genetic variants genotyped or have had their entire genome sequenced and who also have made those data available to the public. The Personal Genomes Project (http://www.personalgenomes.org/) is one portal to these data.

Although such efforts draw attention to issues of genome patentability, destiny not being genetically programmed, and a person's right to obtain his or her genomic data independent of a physician's prior approval, these actions do bring into sharper view the intersection of genome, health, and lifestyle—especially diet and exercise.

C. Exome Sequencing

The exome is defined as the collection of exons, that is, protein-coding segments plus their affiliated 5'- and 3'-untranslated regions. Exome sequencing resolves the genomic sequence of the exome and in a way is a natural intermediary between genome-wide association studies (GWAS) and whole-genome sequencing. Many studies apply exome sequencing to the discovery of rare genetic variants mapping within protein-coding regions that may lie at the root of both common and, especially, rare disorders. Pertinent to the transition of a research project from variant discovery to association of that variant with a disease phenotype or disease risk is the ability to actually detect the genetic variants in the sequence data, especially in the midst of sequencing errors (not a true genetic variant) and missing data. A prime objective of resequencing or exome-sequencing efforts is to detect with confidence variants in a population that can then be used to conduct an association study. Methods to correctly call variants from such data have been reviewed recently.[7] Association studies that use resequencing or exome data are best undertaken with a genome continuum model partnered with principal components in order to detect with accuracy and confidence the genetic associations.[8] Several exome-sequencing–genetic association studies pertinent to nutrigenetics are under way, particularly those examining metabolic syndrome and heart disease as outcomes, but results are not yet available as data collection and analysis of those initial studies are ongoing.

D. RNA Sequencing

High-throughput-sequencing technology is often used in the study of gene expression. Twenty years ago, what was then called "high-throughput sequencing" was applied to the sequencing of partial-length DNA copies of "expressed sequence tags," or ESTs, in order to generate a fractional view of the repertoire of expressed genes in a sample.[9] Today, the same is done, albeit on a much grander scale and under the new term of RNA-seq, which is essentially sequence data representing those genes that are active in a given biological sample. Libraries of expressed genes, as mRNA copied into cDNA, are rapidly and deeply sequenced. Here, the depth to which a given exon is represented in the data is indicative of the degree to which that sequence element is expressed as mRNA. Thus, RNA-seq is a means to assess gene expression levels without *a priori* knowledge of which segments of the genome are expressed. In other words, there is no design at the heart of the experiment, such as an array of oligonucleotide probes, that is subject to selection bias, whether an element is transcribed into RNA or even which strand is transcribed.

II. Epigenetics and Dietary Influences on Epigenetic Marks

A. Types of Epigenetic Marks

Epigenetics can be defined as on-top-of genetics, meaning inherited characteristics, phenotypes, and chemical entities that are superimposed on the genetic code and do not follow basic Mendelian laws. Collectively, the catalog of the presence or absence of epigenetic marks is called "the epigenome," and the makeup of the epigenome differs between cell types. These changes to the DNA occur continually throughout life, and so each person dies with an epigenome that is quite different from that with which he or she was born. Differences in these epigenetic marks, either from an earlier sampling for an individual or from the population mean, can be important contributors to altered regulation of gene expression and disease status.

Epigenetic marks can occur either as methyl (CH_3–) tags on the DNA itself or as methyl, acetyl, phosphate, ribosyl, ubiquitin, citrulline conversion via deamination, or glucose tags on specific amino acid residues, typically lysine, on certain histone proteins that comprise the nucleosomes and the DNA superstructure. These are chemical marks, often added to the amino-terminal "tails" of histones or certain deoxycytosine residues in the genomic sequence, and so epigenetics is really a chemical modification. Standard technologies employed in the measurement of epigenetic marks now include chromatin immunoprecipitation sequencing (ChIP-seq), methyl-seq, and genome-wide DNase hypersensitivity assays. Resulting data give information on chromatin modifications and structure. In general, the epigenetic marks are signposts signifying the closed or accessible state of a genome segment, a topic that has been reviewed recently.[10]

B. DNA Methylation at CpG Islands

A cytosine–guanine dinucleotide pair in the DNA sequence is known as CpG, and longer iterations or high frequencies of this dinucleotide are often referred to as "CpG islands." A high degree of methylation, or hypermethylation, across a gene region corresponds to reduced levels of gene expression, while low levels of methylation, or hypomethylation, at CpG islands are indicative of active gene expression. Single-base variants within a CpG island, such as a transversion from a C to a T residue, which cannot be methylated, could affect the methylation status. Others would disagree saying that loss of a single methylation site within a larger CpG island would not invoke a gross or functionally consequential change to the overall methylation status. In this instance, insertions or deletions that encompass all or a significant portion of the CpG island are much more likely to exert a phenotypic effect. Discerning the links between genetic variation, differential CpG methylation status, and gene expression has begun.[11] While this work demands that all data be evaluated for influential differences in methylation status, a developing trend suggests that more attention is directed at the DNA regions just neighboring CpG islands, the so-called CpG shores, in order to demarcate the boundaries between hyper- and hypomethylated genomic regions.

C. Dietary Sources of Methyl Donors

The study of epigenetic changes in the context of nutrigenomics investigations is exciting because of the important role of diet in providing the basic substrate. The principal methyl donors from the diet include folate or folic acid, choline, and vitamins B6 and B12. Dietary folate in the United States is mainly from fortified grain products, but it is also found in abundance in lentils, spinach, and many varieties of beans. Choline is most abundant in eggs and liver. Vitamins B6 and B12 are cofactors that regulate the pathways whereby the methyl group is made available to DNA (cytosine-5-) methyltransferases, or DNMT enzymes, for incorporation into the DNA. Thus, there is a rather strong nutritional genetic alignment formed by dietary methyl donors, epigenetic modifications, and disease condition.[12] For example, a comparative analysis of whole-genome methylation status between benign and malignant peripheral nerve sheath tumors found over 100,000 cancer-associated differentially methylated regions with significant enrichment in CpG island shores and gene promoters but not within CpG islands.[13] Further studies have shown that several nutritionally relevant genes have demonstrated differential DNA methylation status, including the glucocorticoid receptor nuclear receptor subfamily 3, group C, member 1 (*NR3C1*), leptin (*LEP*), and peroxisome proliferator-activated receptor alpha (*PPARA*).[14a]

D. Diet and Histone Modification

Diet has been shown to affect histone methylation and acetylation in important ways. In human colon cancer cells, sulforaphane, an isothiocyanate derived from a glucoraphanin abundant in broccoli, induced cell cycle arrest but did not alter histone deacetylase (HDAC) activity, unlike in controls.[14b] A likely partner to the most sensitive HDAC (HDAC3) was found to be peptidyl-prolyl *cis/trans* isomerase NIMA-interacting 1. There has been substantial work showing that chronic alcohol intake in animal models detrimentally affects histone modification, acting primarily through euchromatic histone-lysine *N*-methyltransferase 2 (histone H3-K9 methyltransferase 3) and leading to stress within the endoplasmic reticulum.[14c] Lastly, life span extension in human cells brought about by glucose restriction has been found to have epigenetic and genetic mechanisms mediated in part by sirtuin 1 (*SIRT1*).[14d]

E. Issues Concerning Sample Preparation and Data Interpretation

A salient aspect of epigenetics research performed with tissue samples, whether in the context of nutrigenomics studies or another discipline, is the detected signals represent an average pattern as measured from the mixture of different cell types within that tissue sample. In fact, data for gene expression, protein or peptide abundance, or metabolite levels are also dependent on the composition of different cell types within the tissue sample or the particular state of activation or differentiation of those cells. Also in this regard, methylation patterns correlate with gene silencing but also with gene activation, and this, too, depends on where and when the observation took place—by cell type and differentiation timeline. These are all reminders that experimental design and data analysis must be undertaken with care.

III. MicroRNAs, Gene Regulation, and Nutrition

A. MicroRNAs and mRNA Decay

MicroRNAs (miRs) are small RNA molecules, in humans typically 21-24 ribonucleotides long when fully processed. The miRs are encoded by stand-alone genes, often in a polycistronic message, or are cleaved from an intron that has been spliced during maturation of the mRNA of a protein-coding gene. A miR targets an mRNA, typically within the 3'-untranslated region at what is known as the "seed site." The double-stranded nature of the miR–mRNA interaction then elicits a decay process whereby the mRNA is degraded. This serves as a chief mode of gene regulation—mRNAs are cleared and are not available for translation to protein by ribosomes. Three vignettes follow to illustrate the roles of miRs in gene regulation in the context of nutrition and nutrition-relevant diseases.

1. *MIR33* and Cholesterol Efflux

MIR33A is an intronic miR located within the gene encoding sterol regulatory element-binding transcription factor 2 and this miR inhibits the expression of ATP-binding cassette, subfamily A, member 1 (*ABCA1*), with the effect of easing cholesterol efflux to apolipoprotein A-1.[15] The ABCA1 protein functions as a cholesterol efflux pump. When *MIR33* is silenced *in vivo*, *ABCA1* expression in liver and plasma levels of HDL cholesterol both increase.[15] These results strongly imply that *MIR33* regulates both HDL biogenesis in the liver and cellular cholesterol efflux. *MIR33A* and *MIR33B* target *ABCA1* and the Niemann–Pick disease, type C1 gene (*NPC1*), which mediates intracellular cholesterol trafficking, and both *ABCA1* and *NPC1* are involved in cholesterol homeostasis.[16]

2. *PLIN4*, *MIR522*, and Obesity

The perilipin 4 (*PLIN4*) gene encodes a lipid droplet surface protein. In an investigation of whether genetic variation in the human *PLIN4* gene modulates anthropometric measures of obesity in response to dietary polyunsaturated fatty acids, the common SNP rs8887 was identified in the 3'-untranslated region of *PLIN4*.[17] This genetic variant associates with a set of obesity measures, and importantly, those associations are modulated by dietary *n*-3 polyunsaturated fatty acid intake. Interestingly, the less common A allele of rs8887 creates a sequence motif in the *PLIN4* mRNA that exhibits perfect complementarity to the seed site for *MIR522*. Thus, individuals with the A allele will see downregulation of the *PLIN4* mRNA in adipose tissue via this miR–mRNA interaction, while the more common G allele does not promote such an interaction. This is the first example of a human genetic variant creating a miR target site resulting in the modulation by diet of obesity-related traits.

3. *IRGM*, *MIR196*, and Crohn's Disease

Allele-specific interactions between the *MIR196* family and the mRNA for the gene encoding the immunity-related GTPase family, M (*IRGM*) have been reported as causal for Crohn's disease.[18] Although the *IRGM* exonic SNP c.313C>T (rs10065172) is in perfect linkage disequilibrium ($r^2 = 1.0$) with a 20-kbp deletion polymorphism mapping upstream of the *IRGM* gene and the deletion had been associated strongly with Crohn's disease in several European populations or those with European ancestry,[19–22] it is the functional consequences of this SNP that provide the details of causation. That the c.313C>T variant calls for leucine at codon 105 irrespective of allele suggested that there could be allele-specific consequences to protein expression over protein function. It was observed that predicted binding between *MIR196* and *IRGM* mRNA was affected by the variation at SNP c.313C>T.[18] Importantly, it was demonstrated that not only was the *MIR196–IRGM* interaction real, but that expression of *MIR196* was elevated in inflammatory epithelia from Crohn's sufferers. Although the report did not present original GWAS results, it did build on the results of four previous studies[19–22] in an important way, offering a mechanism whereby a synonymous variant leads to the disease condition.

B. Analysis Tools

miR genes, with a promoter, possibly driven by enhancer elements, controlling transcription, are not so dissimilar to protein-coding genes. Genetic variation within the promoter of a miR gene could be a source of allele-specific transcription rates of the downstream miR. The recent publication of the dPORE database provides an examination of the role of genetic variation in transcription factor binding sites (TFBSs) in the promoter regions of miR genes. dPORE (available at http://cbrc.kaust.edu.sa/dpore/) integrates information from promoter regions of human miR genes, SNPs, and predicted TFBSs in those promoter regions.[23] This tool allows the exploration of the effect of SNPs on the regulation of transcription of miR genes. Here, the focus is on those particular SNPs that alter TFBSs or lead to the creation of a new

TFBS. This gives a tool not cluttered by all known or predicted sites of transcription factor binding to DNA, but only those changed by genetic variation. Other software tools such as miRanda,[24] PicTar,[25] and TargetScan[26] provide miR–mRNA interactions, both predicted and validated. Gene expression analysis tools offered by Ingenuity® Systems now include analysis of miR genes. Expression data for both miR and miR target genes are merged to assess the biological effects of altered expression.

C. MicroRNA Expression and Diet

Although there are several reports of the effects of certain diets or dietary components on the expression of specific miR genes, the landscape at the intersection of miR expression and diet is sparse. Evidence is especially lacking on the consequences of bioactive food components in affecting the physical aspects of the miR–mRNA interaction. Of great interest to nutrigenomics are the effects of the miR–food interaction on the miR–mRNA interaction and rates of translation into protein. In other words, does compound X, normally found in the diet, disrupt a specific miR–mRNA interaction whereby more, and how much more, of that mRNA is translated? While such questions remain unanswered for now, evidence has been reported to suggest that miRNAs are key metabolic regulators.

In the adipose tissue of mice, expression of miRNAs was shown to be sensitive to conjugated linoleic acid in the diet.[27] Rats fed a diet of corn oil/fish oil with pectin/cellulose and in which colonic tumors were induced exhibited a number of miRNAs, including *MIR16*, *MIR19b*, *MIR21*, *MIR26b*, *MIR27b*, *MIR93*, and *MIR203*, with altered expression that was linked to oncogenic signaling pathways.[28] Also in rats, downregulation in liver of three miRNAs (*MIR27*, *MIR122*, and *MIR451*) and upregulation of *MIR200a*, *MIR200b*, and *MIR429* were noted after feeding either a high-fat or high-fructose diet with consequences of diet-induced nonalcoholic fatty liver disease.[29] In mice, pregnant and lactating dams fed a high-fat diet displayed reduced expression of *MIR26a*, *MIR122*, *MIR192*, *MIR194*, *MIR709*, and the *Mirlet-7* family, with a common predicted target of methyl-CpG binding protein 2.[30]

A comparison of miR expression profiles in subcutaneous adipose of women highlighted 11 miRNAs as significantly deregulated in obese human subjects with and without type 2 diabetes.[31] Many of the same miRNAs also showed significant deregulation during adipocyte differentiation. The role of diet in regulating miR expression in prostate cancer,[32a] as well as the interplay between phytochemicals and miRs,[32b] has been reviewed.[32a] The farnesoid X receptor/small heterodimer partner (FXR/SHP) signaling cascade regulates *MIR34a* and its target *SIRT1*,[33] which likely functions as either a regulator of epigenetic gene silencing or an intracellular regulatory protein with mono-ADP-ribosyltransferase activity. Using a mouse diet-induced obesity model, it was shown that hepatic expression of *MIR107* decreases while its target gene, fatty acid synthase (*Fasn*), increases.[34] In summary, there is a growing body of evidence to strongly implicate miRNAs as having significant functions in regulating the metabolic-based response of many cell types.

D. MicroRNAs Present in Plasma

That miR species have been detected in plasma[35,36] suggests a type of hormone-like role for these small interfering RNAs—produced in one cell type, secreted and acting at a distance on a different cell type to elicit a specific response. While interest in miRs is rapidly developing within many disciplines of biomedical research, of some value for nutrigenomics is the finding that miRs have been detected as constituents of the HDL particle and that HDL then functions as a vehicle to deliver miRs to cells that take up HDL.[37] It is noteworthy that levels of several of the miRs most abundant in the HDL particle were different between normolipidemic individuals and those suffering from familial combined hyperlipidemia.[37] In addition, an extensive role for miRs in cardiovascular function and disease has been described.[38] Taken together, these reports indicate that the regulation of gene expression is far more complex than earlier hypothesized and that the contributions of miRs certainly will be defined much more richly as research into nutrition and disease progresses.

IV. Metagenomics and the Bacteria Living on and Within Us

A. Microbiome and Health

Increasing attention is being directed toward the intestinal microbiota or gut flora, its role in metabolic health, and its overall metabolic capacity. Fast and large-scale sequencing efforts combine effectiveness and speed in cataloging the genera and species present in the gut, what is known as "defining the enterotype." Although the application of principles of ecosystems suggests that the intestinal microbial community progresses toward and maintains a certain stable composition, analysis and interpretation of the sequencing data nonetheless are bottlenecks to understanding the total metabolic capacity of the gut flora and the health consequences for the individual. Progress is being made by considering the total metabolic capacity of the mixture of species detected and not by thinking about biochemical pathways of species in isolation (Fig. 1). In fact, animal studies have demonstrated that intestinal microbial communities have the ability to affect the efficiency of energy extraction from the diet, with subsequent impact on susceptibility to obesity.[39] In mice, it has been shown that colonocytes rely on the output of bacteria that utilize butyrate as an energy

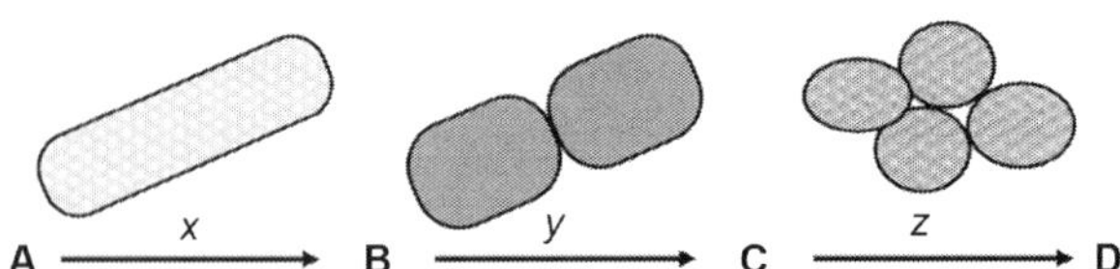

FIG. 1. Conversion of compound A to compound D in the human gut in three successive steps. Each step in this hypothetical conversion process is catalyzed by a different bacterial species, which supplies a unique enzyme (*x*, *y*, or *z*). It is the sum metabolic capacity of the microflora that supports the conversion of A into D.

source over glucose.[40] Other studies have proposed roles of the gut flora in insulin resistance and nonalcoholic fatty liver disease.[41,42] Such studies, along with others that show support for important metabolic functions and health consequences of the microbiome in humans, have led to discussion of therapeutic opportunities.[43] The goals of treating the microbiome could be slightly altering the mix of organisms within the colon or providing new sought-after functions to the intestinal community that add to its total metabolic capacity, such as altering the capacity of the gut flora to metabolize a dietary component. However, it remains unclear how easily and stably the enterotype of an individual can be changed.

B. Cardiovascular Disease

Of late, a search for metabolites associated with cardiovascular disease (CVD) pinpointed significant differences in serum concentrations of choline, betaine, and trimethylamine *N*-oxide (TMAO).[44] These are metabolites of a lipid found in the diet known as phosphatidylcholine or lecithin. Choline is liberated from phosphatidylcholine by phospholipases, and some species comprising the intestinal flora then metabolize it into trimethylamine (TMA), which is then absorbed and transported by the blood. In the liver, TMA is oxidized by flavin containing monooxygenase 3 (FMO3), to yield TMAO. The by-product TMAO has been shown to promote atherosclerotic plaque formation and increased accumulation of cholesterol in macrophages, in addition to enhancing the formation of foam cells—all hallmarks of CVD.[44] Accordingly, the future of human nutrigenomics research very likely will encompass greater doses of bacteriology and attempts to unravel the intricacies of the human–microbiome symbiosis.

C. Gut Enterotypes and Oral Flora

Through a meta-analysis based on published gut metagenomics data for dozens of individuals, combined with new metagenome sequence information for 22 more individuals from a total of six countries, three gut microbial communities or enterotypes were defined.[45] The enterotypes are dominated by bacteria from the *Bacteroides*, *Prevotella*, or *Ruminococcus* genera but this finding has raised objections. Because genes and functional markers found in these communities seem to coincide with certain host attributes, characterizing individuals' enterotypes and the cadre of bacterial genes present in the gut might eventually benefit our understanding of responses to diet and drug treatment. Concurrently, research is progressing in identifying oral flora and the relationships between population types and health outcomes. This is particularly relevant for periodontal diseases and associations with CVD and myocardial infarction.[46,47]

D. *H. pylori* and Epigenetic Changes

In the stomach, the bacterial species with the greatest health impact is *Helicobacter pylori*, as it is the strongest known risk factor for developing gastric cancer. Although gastric cancer is a heterogeneous condition, people with Lynch syndrome, marked by difficulty in repairing DNA mismatch mutations, develop gastric cancer at an elevated rate. Infection with *H. pylori* causes a similar effect: cells exposed to *H. pylori* also have diminished ability to repair DNA mismatches. Recently, this defect has been shown partially to arise from promoter methylation at mutL homolog 1, colon cancer, nonpolyposis type 2 (*Escherichia coli*) (*MLH1*), a critical gene in the mismatch repair pathway.[48] Methylation reduces *MLH1* expression and impairs the ability to repair mismatches. Emerging work is attempting to map across the genome of the human host the DNA methylation patterns associated with *H. pylori* infection (Gu J and Wang T, personal communication).

E. Databases

Because data repositories and information exchange will be vital as efforts in metagenomics of the human organism proceed, the National Institutes of Health has built a Web portal (http://www.ncbi.nlm.nih.gov/genomeprj/43021?report=HMP) to its Human Microbiome Roadmap Project. This includes data from samples taken from the digestive tract, mouth, nose, and many locations from the skin and female urogenital tract of human volunteers. Reference genomes for over 600 microbes are available.

F. From Sequence Data to Reliable Gene Models

Lastly, a bacterial genome sequence is only as useful as the calculated gene models and ascribed function of the encoded proteins. There are many algorithms designed to identify open reading frames in bacterial genome sequence, but the Glimmer-MG software (http://www.cbcb.umd.edu/software/glimmer-mg/) is specifically designed to assess accurately the functional capacity of a metagenomic sample. Glimmer-MG uses data from a sample where many sequences cover only a portion of a gene and originate from species that are rare, common, or abundant in that sample. Specifically, phylogenetic classifications (as opposed to percent G + C content), sequence clustering, and modeling of both insertion/deletion and substitution sequencing errors altogether yield highly accurate gene predictions.[92]

V. Advances in Metabolomics

A. Genetics and Differences in Metabolite Levels

Metabolomics is the study of small molecules found in a biological sample. Data on various small-molecule metabolites present in a clinical sample—typically blood serum or urine—are collected by mass spectrometry (electrospray ionization or triple quadrupole), nuclear magnetic resonance, or isotope labeling. These data yield information that describes metabolic abundances, which can then be used to determine enzyme activities and to describe metabolic fluxes. The practice of metabolomics is well established for the screening of newborns for a number of metabolic deficiencies which if untreated can result in serious health problems. It should be emphasized that the course of treatment often involves adherence to a specific diet.

Interest in metabolomics is growing, and advances in the technology used to collect data as well as in data analysis continue to push this discipline into ever closer union with other omics or high-throughput data. For example, the use of metabolomic data from blood plasma and urine in GWAS has led to new findings on the roles of specific phosphatidylcholine species in lipid homeostasis and heart disease[49] and of β-aminoisobutyrate in detoxification capacity and hypertension, congestive heart failure, progression of chronic kidney disease, and atherosclerosis via a polymorphism in the alanine–glyoxylate aminotransferase 2 gene (*AGXT2*).[50] In such studies, the ratios of reactant to product for certain biochemical reactions are used as a phenotype in order to demonstrate a functional link between the metabolite pairs.

B. New Opportunities for Metabolomics in Coronary Artery Disease

Recent reports describing new genetic loci associating with coronary artery disease (CAD) indicate that new pathways are likely involved in disease onset, progression, and overall risk.[51,52] In one report, 10 of 13 new loci carrying common variants associated with CAD were not associated with traditional CAD clinical measures such as HDL-, LDL-, and total cholesterol, body mass index, type 2 diabetes, hypertension, and smoking.[51]This is rather noteworthy because the aforementioned phenotypes of blood lipids, blood pressure, and anthropometrics are strongly influenced by diet and other lifestyle factors. Furthermore, and perhaps of greater importance, is the supposition that such results open new avenues of research into other pathways, processes, or lifestyle factors that may contribute to CAD in ways not before imagined. For example, metabolomics could be employed to identify compounds that were not previously considered as associated with CAD but which do associate with these newly described loci. The genes and their pathways can then be targeted either by dietary or by pharmaceutical intervention for disease prevention or amelioration. Thus, the combined application of genetic variation, gene expression, and proteomic and metabolomic data with respect to human intervention studies is an area of continued growth in nutrigenomics, a topic that has been reviewed recently[53] and is further explored in Section XII.

C. Databases

An extensive literature review combined with implementation of new data capture and analysis methods has produced a comprehensive set of metabolites regularly observed in human serum.[54] This information was compiled into data tables containing 4229 confirmed and highly probable human serum compounds with information on concentration, literature references, and known associations to disease. Many of these entries are directly applicable to nutrigenomics research because they are compounds found in the typical human diet, are products of biochemical reactions, interact with nutrition-relevant proteins, or are pertinent to a metabolic or nutrition-based disease. See http://www.serummetabolome.ca to access these data. An interesting feature of this database, in addition to assignments of the metabolite to biochemical, physiological, and some disease pathways, is inclusion of a large number of hand-drawn pathways applicable to metabolites observed in human serum. Furthermore, based on the National Center for Biotechnology Information's Online Mendelian Inheritance in Man® database,[55] numerous metabolites have been assigned to human diseases. As a whole, the human serum metabolome database is an important resource for investigations of small-molecule metabolites and their relationships to health within a nutrition context.

VI. Lipidomics

A. Application of Lipidomics to Nutrition Studies

Another nutrition-related concern is the health consequences of dietary fat—too little, too much, specific types. Dietary fat is composed of any of several types of lipid and fatty acid, and the broad study of all the moieties present in a biological sample is termed "lipidomics." These molecules, including sphingolipids, phospholipids, glycerolipids, prenols, sterols, and fatty acids, have been classified into a comprehensive system to assist the experimentation in lipid biology.[56]

Issues of solubility as well as a wide range of molecular sizes and weights have complicated lipidomics research. Nonetheless, the ingestion of lipids and fats and their measures in biological samples have obvious relevance to nutrigenomics studies. For instance, 24 different lipid moieties were described as specific to or highly concentrated in atherosclerotic plaques found in carotid and femoral arteries by an analysis of triple quadrupole mass spectrometer data in which over 150 different lipids were examined.[57] Another study showed positive association between consumption of coffee and two classes of sphingomyelins, *N*-hydroxylacyloylsphingosyl-phosphocholine and *N*-hydroxyldicarboacyloylsphingosyl-phosphocholine, as well as negative correlations between coffee intake and medium- and long-chain acylcarnitines.[58] Issues of lipotoxicity were addressed by positron emission tomography coupled to computed tomography to determine postprandial organ-specific deposition of different fatty acids in humans.[59a] Lastly, lipidomics analyses of human adipose tissue identified specific compositional differences in adipose tissue membranes between lean and obese discordant twins.[59b]

B. Databases, Tools, and Projects

Although preliminary and demonstrative of a methodology, such results indicate that application of lipidomics has the potential to identify novel biomarkers, especially if these or related lipids show aberrant levels in serum prior to atherosclerotic plaque detection. Diet modification could then serve as the initial course of therapy. In this area of nutrigenomics research, essential databases, tools, and projects include the work of the Kansas Lipidomics Research Center (http://www.k-state.edu/lipid/lipidomics/), the lipidomics gateway at the journal *Nature* (http://www.lipidmaps.org/), the European Lipidomics Initiative (http://www.lipidomics.net/), and lastly, a strategic partnership formed between Agilent Technologies, Inc., and the National University of Singapore in order to develop workflow solutions that will provide a better picture of the role of lipids in disease states.

VII. Systems Biology and Deep Phenotyping

A. Nodes, Edges, and Networks

Integrated analysis of different data types, leading to the description—either mathematically with a series of differential equations or biologically in terms of interaction networks—of a biological process can be termed “systems biology.” Types of biological processes described in this manner include signal transduction and gene expression cascades, models of disease progression, and interactions between genes and phenotypes. The networks consist of nodes and edges, where nodes are biological entities and edges are the connections—either uni- or bidirectional—between nodes.

Changes to the environment or polymorphic genetic markers were long thought to have direct impact solely on nodes. That thought has been challenged by the notion that environmental and genetic variations can affect the relationships between, or the edges joining, the nodes. Hence, "edgetic" perturbations are a means to describe altered relationships within a biological network.[60] Considered differently, the node or biological entity is not so much altered in activity or function as are the relationships (i.e., the edges) that entity has with other components of the network. This perspective and analysis approach toward systems involving genes, genetic variation, and human disease are particularly well suited to nutrition.

B. Nutrition Systems Biology and Deep Phenotyping

In the context of nutrition research, systems biology as applied to nutrigenomics and nutrigenetics data seeks to describe fully how dietary components affect gene expression and protein and metabolite levels and how genetic variation alters those effects.[53] Exhaustive phenotyping providing an extensive list of measurements taken during the course of an experiment or trial has been called "a key element" for the full application of systems biology to nutrition research.[61a] Such deep phenotyping can take two forms. First, the collection of measurements can be broadened and subcategories defined. This was done for the Framingham Heart Study and cholesterol particle measures, for example, where HDL- and LDL-cholesterol particles were divided into different subclasses as a means to better differentiate individuals and their postprandial responses. Second, as time is often a critical parameter in nutrition-based phenotypes, particularly time after a meal, thorough phenotyping may entail the taking of more measures over a given time period. The postprandial response of serum triglyceride levels to a high-fat meal is a prime example. When measures are taken at 0, 3, and 6 h after the meal, the course of the response and its peak value may be misinterpreted. Instead, measures should be extended to 11 h, particularly in a laboratory setting, by which time many blood lipids return to fasting levels,[61b] as area under the curve for postprandial responses is an informative measure.

C. Phenotyping Exhaled Breath

For many years, assayed bodily fluids have included blood plasma and urine, with saliva tested in some applications. The implementation of new technologies now also permits the chemical analysis of exhaled breath. Although these data are complex, representing a wide array of exposures, they nonetheless can be incorporated via a systems biology approach into the study of gene–environment interactions.[62] Not only does analysis of the breath give information on health status, but also the biomarker patterns are constructive in reviewing public health risks from environmental exposures.

D. Development of Data Standards with a Nutrition Focus

A significant challenge sitting at the intersection of today's nutrition and health research is determining how to devise and execute food-based strategies that stabilize and then can return an individual to metabolic homeostasis and a healthy condition.[63] Because this research involves numerous bioactive compounds that exert effects on a multitude of networks of interacting processes, the basics of this research are multifaceted.[63] Additionally, nutrition research demands that specific analytical and bioinformatics procedures be put into practice in order to better interpret the flood of high-throughput data.[53] These procedures and the research they support would benefit from efforts to develop standards and organize data and databases with a nutrition–health focus. With these goals in mind, a gene–environment interaction database for metabolic traits was assembled.[64a]

VIII. Cancer Genomics as a Technological Model for Nutrigenomics

A. Advances in Cancer Genomics Applicable to Nutrition Research

In genomics research, many of the technological breakthroughs and advances in both procedures and data analysis arise from and are being applied to the study of the genetic origins of cancer. A brief list of some of the most important applications of genomics technology to cancer research includes rapid sequencing of normal and tumor tissue to identify altered genes, applying those sequencing results to identify critical signaling pathways disrupted or exploited during oncogenesis, and analyzing formalin-embedded tissue or tumor samples to identify genetic differences and altered gene expression.[64b,c] The identification of altered pathways or gene–protein modules widens the perspective of the researcher to consider that a gene or its encoded protein is really just a part of a module, which leads to the consideration of pathways as functional units instead of genes.

B. *TP53* and Nutrient Reprogramming During Oncogenesis

From the standpoint of nutrigenomics, one key recent finding is a description of mechanisms leading to nutrient reprogramming during oncogenesis, particularly with respect to the critical oncogene *TP53*, commonly known as *p53*, which is a transcriptional regulator that reacts to diverse cellular stresses. Frequently mutated in human tumors, cellular tumor antigen p53 (TP53) binds to glucose-6-phosphate dehydrogenase and inhibits the pentose phosphate pathway, thereby suppressing glucose consumption, NADPH production, and biosynthesis.[65] The metabolic functions of TP53 can block the shift to glycolysis typically observed in cancer cells while promoting energy production by mitochondrial oxidative phosphorylation.[66] Curiously, isothiocyanates, such as those found in broccoli, selectively deplete mutant TP53 in cancer cells to the exclusion of normal TP53.[67] While such findings do not in themselves demonstrate advances in technology or study designs, they do provide an opportunity to glimpse what kinds of experiments could be performed within a nutrigenetics or nutrigenomics group when the resources, technologies, and tools of cancer biology are brought to bear on nutrition research.

IX. Biomarkers of Disease Onset or Progression and Diet Adherence

A. Developments in Analysis of Genome-Wide Association Data

GWAS reached a well-recognized level of maturity in terms of design, reproducibility, and acceptance around 2009–2010. While more numerous and more diverse phenotypes have been tested and larger cohorts, some over 100,000 participants, have been employed, the basic GWAS approach remains largely unchanged. There are, however, three notable developments in GWAS data analysis that deserve mention. One, pathway analysis seeks to uncover the genetic contributors to phenotypic differences that exist across a defined set of genes whose encoded products function in a concerted manner.[68,69] This approach is not unlike dissection of the genetic basis of cancer. Two, GWAS of metabolomic data give information on the genetic basis for differences in pathway fluxes such as energy homeostasis, detoxification, and metabolic aberrations (see Section V). Three, the integration of GWAS signals with other high-throughput data can yield deeper insight into the genetic basis of the observed phenotypic differences. For example, a recent study showed that numerous regions of the human genome that harbor tissue-specific epigenetic marks of methylated and acetylated histones also carry genetic variants identified by GWAS for phenotypes relevant to the tissue where the differential histone marks were noted.[70] This finding emphasizes the links between regulation of gene expression and disease.

B. Adherence to a Dietary Regimen

Protein or metabolite biomarkers can measure organ dysfunction or adherence to a given dietary regimen. The example of urinary glucose measures and type 2 diabetes is classic. The usefulness of prostate-specific antigen levels and prostate cancer status has diminished. Thus, descriptions of the genes and proteins that function within a particular cell type and the metabolites that enter and exit that cell, as well as the influences of genetic differences, all must become more complete before one can make confident, individualized assessments of disease progression and diet adherence based on measurements of biomarkers.

C. Environment-Wide Association Studies

Environment-wide association studies (EWAS) aim to identify environmental factors associated with disease on a broad scale. This type of study is gaining attention and was applied recently to type 2 diabetes.[71] In the past, epidemiological studies used a similar approach, but with EWAS, using cross-sectional epidemiological data, participants were asked about health status and a subset of subjects received a battery of clinical and laboratory tests. Environmental attributes that were assayed include allergens, bacterial/viral organisms, chemical toxins, nutrients, and pollutants.

EWAS are comprised of two methodological steps, both of which possess counterparts in GWAS. First, a panel of many unique environmental assays is measured in cases and controls. These can be considered environmental "loci" and can yield several environmental factors with significant associations with the disease state while controlling for multiple hypotheses. Second, validation of the associations is undertaken in other cohorts, which is analogous to validation and replication in GWAS. As applied to the National Health and Nutrition Examination Survey cohorts from the years 1999 to 2006, EWAS-significant associations to type 2 diabetes were noted for heptachlor epoxide (a pesticide derivative), γ-tocopherol, and elevated levels of polychlorinated biphenyls such as PCB170.[71] A protective effect of β-carotenes was also noted.

X. Gene–Gene and Gene–Environment Interactions

A. Gene–Gene and Gene–Environment Interactions and Disease Risk

There is a growing body of evidence to show that the genetic basis of disease risk arises in part from altered or differential responses in gene expression.[72] As this evidence accumulates, it follows that environmental factors that modulate or amplify the disease risk or pairs of genes that function in concert will be observed as affecting gene regulation and protein–protein interaction networks. In addition, many nutrition-based phenotypes can be thought of as complex traits, with many markers contributing to both baseline and "response" values. Furthermore, even more complex are afflictions such as CVD, type 2 diabetes, complications of the vascular system, kidney ailments, and osteoporosis, among others, all of which can arise from poor or improper nutrition. These diseases involve an array of genetic factors, physiological and cellular processes, and interacting lifestyle choices or inputs. Thus, an added, significant layer of complexity key to nutrigenetics studies arises from gene–gene and gene–environment interactions.[64a,73,74]

A gene–gene (G × G) interaction can be described as an association of a phenotype with two genetic factors, or alleles, both of which must be present for the association to be observed. This is known as "epistasis" (Fig. 2). A gene–environment (G × E) interaction is an association between a genetic factor, or allele, and a phenotype, in which that association is modified by some factor in the environment. Environmental factors, which may include dietary fat as a percent of energy intake, level of daily physical activity, average daily alcohol consumption, and tobacco smoking, represent an instance where disease risk can be modified in an environmentally specific manner. An example G × E interaction is depicted in Fig. 3.

The G × G and G × E interactions are situations where a favorable allele can compensate for one allele with risk potential or where the response to a stimulus is buffered or enhanced. G × E interactions are points of entry

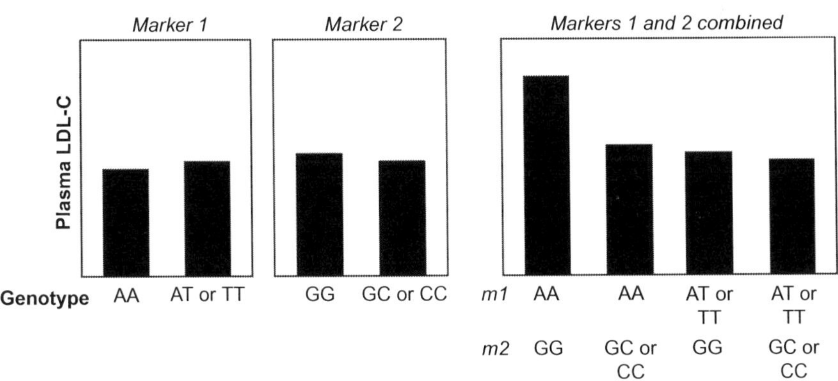

FIG. 2. An example of epistasis. In this hypothetical example of two genetic markers or variants associating with plasma LDL-cholesterol (LDL-C) levels, neither marker 1 nor marker 2 alone shows a significant association. When the two markers are analyzed together, individuals who are both AA genotype at marker 1 and GG at marker 2 show a significantly elevated LDL-C.

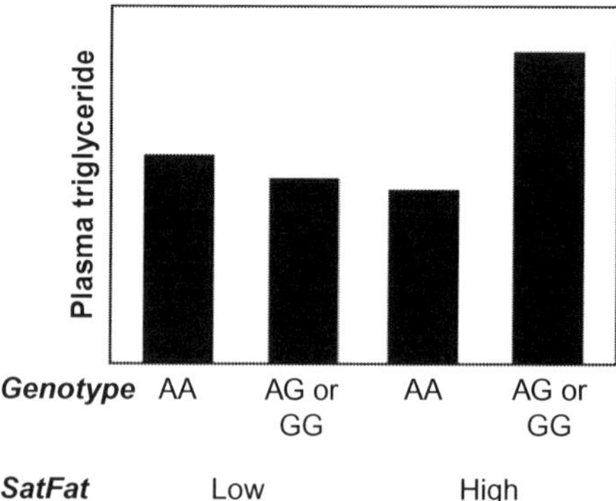

FIG. 3. An example of gene–environment interaction. Carriers of the G allele at a particular genetic locus show a significant association to plasma triglyceride levels only when the diet is high in saturated fat. In many such cases, the association of genetic marker and phenotype is not detected until the study population is categorized by a specific environmental or dietary factor.

for personalized medicine or nutrition. In particular, the environmental component of the G × E interaction is often modifiable—one can alter the intake of saturated fat, or stop smoking, or exercise more— and with such modifications to lifestyle come altered and usually reduced risks of disease onset or progression. Thus, as both epistasis and G × E interactions gain wider acceptance as factors in determining disease risk, some researchers have sought to develop computational methods to predict these genetic interactions. This is of keen interest to those who wish to garner more information from GWAS, as it is much easier to reanalyze available data when the phenotypes are accurate than to recruit new study volunteers.

B. Limitations in Detecting Gene–Environment Interactions

Importantly, it has been proposed via computer simulation of G × E interactions in family-based studies that, if the population structure includes the exposure, an interaction involving a gene × exposure term is susceptible to inflated Type I error rates.[75a] This occurs because Type I errors arise through subtle dependencies which researchers often fail to appreciate. An offered solution suggests that, in a study of affected children and their healthy parents, the addition of a sibling who is not genotyped but assayed for exposures adds robustness to the study of both genetic and exposure effects, as well as G × E interactions.[75a]

XI. Lifestyle Measures

Clinical and dietary measurements are integral components of disease risk calculations. Power to interpret these measures or translate them into disease risk grows when studies are combined, but doing so can bring to light interoperability problems, that is, situations where different systems cannot operate synchronously or where data exist in different, incompatible formats. Such is the realm of the Open Geospatial Consortium (http://www.opengeospatial.org/ogc) and PhenX, the latter seeking to establish a set of accepted, high-priority, and broadly applicable phenotypic measures for each of several research domains.[75b] The PhenX Toolkit offers large-scale genomic studies with well-established measures that are of high quality and low burden—all in order to facilitate the integration of like data. Similar issues of data storage and sharing relevant to nutrition systems biology have been addressed.[63]

A. Clinical Data

Accurate, quick, and cost-effective collection of data from volunteer subjects participating in an association or epidemiological study remains a challenge. The volunteer cannot be burdened with too many questions or poorly worded or irrelevant inquiries. A facilitator should be used only sparingly in order to keep costs low and prevent unnecessarily long interviews. A remarkable technological trend that may increase the accuracy of interviews and reduce cost and time expenditures is the use of devices that interface directly with the study volunteer and capture data in a digital format.

REDCap, or Research Electronic Data Capture, is exclusively designed as a secure, Web-based application that collects data for clinical and translational research studies.[76] Such a system offers intuitive data entry with validation, audit trails, options to export data, protocols to import data, and features such as branching logic and calculated fields. For the Boston Puerto Rican Health Study, facilitators are using portable computers running REDCap in both English and Spanish to collect data in volunteer subjects' homes. A particularly useful feature is the ability to download data into Access or Excel (both from Microsoft®), R (an open-source statistical package), SAS®, SPSS®, and Stata®. Of equal importance is the ability of REDCap to accept prespecified minimum and maximum values so that when a value beyond the predetermined accepted range is entered, the program prompts the facilitator to verify the value as a double check. Experience with paper-based forms to collect volunteer information in the Boston Puerto Rican Health Study[77] has shown how valuable a system can be when data on sleep, migration history, and self-rating of health status are collected.

B. Dietary Data

Two methods in widespread use for recording the dietary intake of subject volunteers are the food diary and the food frequency questionnaire (FFQ). Both are prone to many different errors, including not understanding the FFQ question, forgetting to record an item, over- or, especially, underreporting of amounts, and disregard for recipes or methods of food preparation.[74] The result is the correlation between known, verified intakes of specific food items or components of consumed food and the information garnered from a standard food diary or FFQ is typically moderate at best. This presents a problem when attempting to uncover connections between a dietary component and measures of health status.

Recently, advances have been made in the adaptation of new technology to better capture the information on what an individual has actually consumed. For example, VioFFQ from Viocare®, Inc. is a Web-based system that permits research subjects to self-administer the FFQ from a standard computer connected to the Internet. A special feature of this method uses improved branching logic, which intuitively omits questions considered immaterial based on answers to previous questions. For example, a respondent who declares zero coffee consumption will not then be asked questions regarding milk, cream, or sugar added to the coffee. This improves the quality of dietary assessment. Research is also seeking to build a device that recognizes from a photograph, such as from a cell phone, what foods and how much of each is on a plate both before and after the meal. Once proven reliable, such a system could supersede all formats of the FFQ.

C. Measuring Physical Activity

Accelerometers provide accurate, real-time data on sedentary, inactive, and physically active time periods and are another example of technological improvements germane to nutrigenetics research and population studies. The term "inactive" describes a person who does not engage in enough physical activity or who does not meet physical activity guidelines. In contrast, "sedentary" describes the act of sitting itself. The distinction is necessary, as accumulating evidence suggests that sitting for too long does not have the same physiological impact as exercising too little, and thus researchers need to be aware to treat the two issues as separate behaviors.[78,79]

XII. Study Design

A. Types of Study Designs

Epidemiological studies form the backbone of much nutrigenetics research and are classified by the degree of experimentation, as well as by characteristics such as the number and type of subjects, how those subjects were selected, exclusion criteria, and the level of follow-up.[73] Thus, there are observational studies that can be subcategorized as case–control, cross-sectional, cohort, and ecological, and experimental studies, which involve clinical and community trials, often with an intervention arm. The simultaneous assessment of exposures and disease status in subjects selected from a defined population constitutes a key feature of the cross-sectional design. A case–control study will collect data on dietary intakes, for example, from participants with a particular disease and from those who are disease free, but because of the retrospective nature of this study design, errors are often higher. Therefore, there is a real need for these studies to incorporate the technological advances described in Section XI. This need also applies to prospective cohort studies in which nutrigenetics researchers assess exposure, past or present, to dietary variables and other risk factors, such as tobacco and alcohol use, and the study subjects are revisited over a period of time in order to evaluate changes in disease status and assign, when possible, links between an exposure and that disease.[80,81]

While experimental studies grapple with the cost of the intervention and accurate measures of adherence to the study protocol, observational studies must contend with the control population not becoming future members of the case cohort, the cross-section as an adequate representation of a larger population, and confounding or correlative issues. Using more specific phenotypes can assist the discovery of genetic variants associating with increased risk in the latter situations. Lastly, large sample sizes are required in order to detect G × E interactions but are confounded by environmental variation within the sampled population in observational studies. These issues have been reviewed elsewhere.[73,74]

B. Studies Generating or Testing Hypotheses

Initial studies and many ongoing investigations into the genetic basis of disease risk employ hypothesis-testing approaches where prior knowledge about a gene justifies its inclusion in the test for disease association. Such knowledge includes protein function, association to phenotypes in humans or model organisms or diseases highly similar to the currently studied disease, and pathway assignment. The quick and wide adoption of GWAS to studies of disease risk has at its foundation a simple hypothesis that there are genetic factors affecting risk. GWAS, however, offer no preconception that genes with membership in a specific pathway must or cannot contribute to disease risk. Additionally, it must be stressed that a significant *P*-value for association does not confer causality, but rather proof is required and effect size must be evaluated. A common form of this proof is consistency in results across different studies, which acts as replication.[82]

In this regard, it is important to keep in mind that an observed phenotype may be the result of nontraditional elements—genetic factors, metabolites, cell types—some of which may be functionally or physically distant from that observed phenotype. In other words, the cascade leading to disease risk may involve factors farther upstream or outside the orbit of traditional factors. Furthermore, association studies that use genotyping data or lifestyle measures in isolation will not be able to describe completely the factors leading to disease risk. Studies that recognize the value of both genetics and lifestyle measures in combination would be able to capitalize fully on the breadth of $G \times E$ interactions that modulate risk of metabolic diseases, cancer, and many other chronic ailments.

C. Intervention Studies and High-Throughput Data

Intervention studies in nutrition have been making use of high-throughput data with greater regularity.[53] These data include measures of gene transcription and protein and metabolite levels. Some studies now divide the population on the basis of genotype prior to the intervention in order to assess genotype-specific responses. Such studies, however, by design employ hypothesis-testing approaches because a single specific gene variant is assessed. Other studies recruit volunteers within a narrow age range. While that is an obvious limitation of a study, it also reduces the variability in many of the measured values, something that could be important as the number of measured phenotypes rises.

D. Studies Using Monozygotic Twins

Studies of monozygotic twins provide greater statistical power to detect smaller effect sizes and to discover the impact of nutrient intake while controlling for genetic variation. However, while these twins have essentially identical genome sequences at the nucleotide level, their lifetime exposures are distinct and so their epigenetic patterns will be as well, notably at nutritionally relevant genes.[14a] Their microbiome constitutions may also differ. Thus, an exhaustive catalog of exposures is warranted.

E. Mendelian Randomization

The combining of genetics and classical epidemiological analysis to deduce causality of an environmental exposure is used in Mendelian randomization.[83a] Importantly, Mendelian randomization is based on the laws of independent assortment and sets genetic variants as proxies for environmental exposures. Thus, the resulting associations are less prone to confounding and reverse causation. Studies on dietary and lifestyle factors and biomarkers applicable to nutrition-based diseases that have taken a Mendelian randomization approach have been reviewed.[83a]

F. Epigenome-Wide Association Studies

The contributions to phenotypic variation attributable to epigenetic variation remain for the most part untested and unknown. From a technological standpoint, certainly with respect to DNA methylation, it is feasible to design an epigenome-wide association study. Although similar to GWAS, an epigenome-wide association study must contend with specific design considerations, as outlined.[83b] Two important concerns are (1) the tissue- and developmental stage-specific nature of DNA methylation patterns, which themselves change over the life of the individual and (2) the associations between phenotype and epigenetic mark, which can be causal or consequential for the phenotype under study.[83b] Lastly, for such studies to succeed, thorough collection of data on environmental exposures is a must, as these may confound associations.

XIII. In the Physician's Office

For reasons of ancestry, health risk, or pure curiosity, growing numbers of patients are interested in their own genetic makeup, and many also wish to have an informed and frank discussion with a health care provider about how their genes affect their health. Furthermore, nutrigenetics has shown that many genetic variants assayed by DTC genotyping outlets pertain to genes or health conditions that are modifiable to some degree by diet. Challenges remain, however, in fully incorporating genetic data and lifestyle activities of the patient into the conversation between patient, physician, and nutritionist for the benefit of that patient's health. Nonetheless, as more patients enter the waiting room with DTC genotyping data in hand, pressure will increase to engage in that conversation. Thus, while researchers are challenged with discovering the relationships of genetics, environment, and disease risk, physicians must be fully trained in these very topics.[84–88]

Perhaps an important but seemingly underappreciated key recent advance in nutrigenomics research is the deeper understanding that a wide array of factors are at play in complex diseases and conditions. Obesity risk, for example, is influenced by genetics, energy intake, food production, activity level, potential for activity given the environment, psychological factors, and physiology. The interactions and influences of these various factors are of course complex and difficult for the physician to consider when making recommendations, but drawing up such a list is a start. At some point, elements of design and interactivity will be used as complex data invade mainstream conversations, such as those in a physician's office or under the guise of public health policy. Accordingly, taking stock of the environmental and behavioral factors that contribute to excessive and prolonged weight gain will better inform health professionals.[89]

Certain technological advances, such as AnatOnMe from Microsoft®, a handheld, projection-enabled mobile device, are designed to improve patient–doctor communication in a medical setting. The device is designed to project directly onto the patient's skin an image of musculoskeletal elements—bones, tendons, and muscles—involved in an injury. At this point, applicability to metabolic diseases or conditions affected by diet is not present, but one can imagine a similar device that incorporates personal genetic and dietary information to demonstrate health effects.

XIV. Social Media

Social media outlets such as Twitter®, Facebook®, and LinkedIn® are virtual gathering places where scientific professionals can exchange ideas and learn about important new publications or news releases.[90] Joining a network allows an individual to sample the wisdom of the crowd in a direct and timely manner. For this writing assignment, Twitter® was used to solicit ideas from nutrigenomics and nutrigenetics professionals, thus showing the permeation of social media. However, acceptance of social media still has a long way to go, as only a small minority of scientists is connected in this way.[91] Many of the same principles and advantages of larger social media sites can be applied to sites such as BioStar (www.biostars.org), which allows scientists to solicit advice for solving problems or overcoming obstacles in computational and bioinformatics aspects of genomics research.[93]

XV. Summary

Advances in genomics and other high-throughput technologies and the increasing sophistication of study designs and phenotype collection have provided a solid foundation for nutrigenomics and nutrigenetics research while continuing to stimulate novel discoveries. Greater ease of obtaining DNA sequence and epigenetics data and adoption of systems biology—all following paths explored by cancer genome biologists—in conjunction with studies of the microbiome offer great potential to nutrigenomics and nutrigenetics. These and other data could be crystallized into concise information that leads to a better understanding of the scientific basis of functional foods and how exactly those foods or food components trigger a biological response leading to enhanced health. Such information would facilitate better informed choices when applying nutrition to the modulation of disease risk. Health care professionals, particularly physicians and nutritionists, will need to understand how high-throughput data are collected, what those data likely entail in terms of good health and disease risk, and how to communicate the information to patients.

At the same time, researchers must keep in mind that as different types of data are integrated to identify the entities of a nutrition-based disease, the affected tissue may not always be the tissue with the defect that causes the observed phenotype. Taken together, the recent achievements described here set the stage for numerous and unrestrained opportunities that should bring further successes to the maturing disciplines of nutrigenetics and nutrigenomics.

Acknowledgments

Mention of trade names or commercial products in this publication is solely for the purpose of providing specific information and does not imply recommendation or endorsement by the U.S. Department of Agriculture (USDA). The USDA is an equal opportunity provider and employer.

References

1. International Human Genome Sequencing Consortium . Initial sequencing and analysis of the human genome. *Nature* 2001;**409**:860–921.
2. Human genome at ten: the sequence explosion. *Nature* 2010;**464**:670–1.
3. Malone JH, Oliver B. Microarrays, deep sequencing and the true measure of the transcriptome. *BMC Biol* 2011;**9**:34.
4. McIntyre L, Lopiano K, Morse A, Amin V, Oberg A, Young L, et al. RNA-seq: technical variability and sampling. *BMC Genomics* 2011;**12**:293.
5. Neiman M, Lundin S, Savolainen P, Ahmadian A. Decoding a substantial set of samples in parallel by massive sequencing. *PLoS One* 2011;**6**:e17785.
6. The 1000 Genomes Project Consortium . A map of human genome variation from population-scale sequencing. *Nature* 2010;**467**:1061–73.
7. Nielsen R, Paul JS, Albrechtsen A, Song YS. Genotype and SNP calling from next-generation sequencing data. *Nat Rev Genet* 2011;**12**:443–51.
8. Luo L, Boerwinkle E, Xiong M. Association studies for next-generation sequencing. *Genome Res* 2011;**21**:1099–1108 doi:10.1101/gr.115998.110.
9. Adams MD, Dubnick M, Kerlavage AR, Moreno R, Kelley JM, Utterback TR, et al. Sequence identification of 2,375 human brain genes. *Nature* 1992;**355**:632–4.
10. Chahwan R, Wontakal SN, Roa S. The multidimensional nature of epigenetic information and its role in disease. *Discov Med* 2011;**11**:233–43.
11. Gibbs JR, van der Brug MP, Hernandez DG, Traynor BJ, Nalls MA, Lai SL, et al. Abundant quantitative trait loci exist for DNA methylation and gene expression in human brain. *PLoS Genet* 2010;**6**:e1000952.
12. Li CCY, Cropley JE, Cowley MJ, Preiss T, Martin DIK, Suter CM. A sustained dietary change increases epigenetic variation in isogenic mice. *PLoS Genet* 2011;**7**:e1001380.
13. Feber A, Wilson GA, Zhang L, Presneau N, Idowu B, Down TA, et al. Comparative methylome analysis of benign and malignant peripheral nerve sheath tumors. *Genome Res* 2011;**21**:515–24.
14. (a)Campión J, Milagro FI, Martínez JA. Individuality and epigenetics in obesity. *Obes Rev* 2009;**10**:383–92; (b)Rajendran P, Delage B, Dashwood WM, Yu TW, Wuth B, Williams DE, et al. Histone deacetylase turnover and recovery in sulforaphane-treated colon cancer cells: competing actions of 14-3-3 and Pin1 in HDAC3/SMRT corepressor complex dissociation/reassembly. *Mol Cancer* 2011;**10**:68; (c)Esfandiari F, Medici V, Wong DH, Jose S,

Dolatshahi M, Quinlivan E, et al. Epigenetic regulation of hepatic endoplasmic reticulum stress pathways in the ethanol-fed cystathionine beta synthase-deficient mouse. *Hepatology* 2010;**51**:932–41; (d)Li Y, Tollefsbol TO. p16(INK4a) suppression by glucose restriction contributes to human cellular lifespan extension through SIRT1-mediated epigenetic and genetic mechanisms. *PLoS One* 2011;**6**:e17421.

15. Rayner KJ, Suárez Y, Dávalos A, Parathath S, Fitzgerald ML, Tamehiro N, et al. MiR-33 contributes to the regulation of cholesterol homeostasis. *Science* 2010;**328**:1570–3.
16. Najafi-Shoushtari SH, Kristo F, Li Y, Shioda T, Cohen DE, Gerszten RE, et al. MicroRNA-33 and the SREBP host genes cooperate to control cholesterol homeostasis. *Science* 2010;**328**:1566–9.
17. Richardson K, Louie-Gao Q, Arnett DK, Parnell LD, Lai CQ, Davalos A, et al. The PLIN4 variant rs8887 modulates obesity related phenotypes in humans through creation of a novel miR-522 seed site. *PLoS One* 2011;**6**:e17944.
18. Brest P, Lapaquette P, Souidi M, Lebrigand K, Cesaro A, Vouret-Craviari V, et al. A synonymous variant in IRGM alters a binding site for miR-196 and causes deregulation of IRGM-dependent xenophagy in Crohn's disease. *Nat Genet* 2011;**43**:242–5.
19. Parkes M, Barrett JC, Prescott NJ, Tremelling M, Anderson CA, Fisher SA, et al. Sequence variants in the autophagy gene IRGM and multiple other replicating loci contribute to Crohn's disease susceptibility. *Nat Genet* 2007;**39**:830–2.
20. Wellcome Trust Case Control Consortium . Genome-wide association study of 14,000 cases of seven common diseases and 3,000 shared controls. *Nature* 2007;**447**:661–78.
21. Barrett JC, Hansoul S, Nicolae DL, Cho JH, Duerr RH, Rioux JD, et al. Genome-wide association defines more than 30 distinct susceptibility loci for Crohn's disease. *Nat Genet* 2008;**40**:955–62.
22. Franke A, McGovern DP, Barrett JC, Wang K, Radford-Smith GL, Ahmad T, et al. Genome-wide meta-analysis increases to 71 the number of confirmed Crohn's disease susceptibility loci. *Nat Genet* 2010;**42**:1118–25.
23. Schmeier S, Schaefer U, MacPherson CR, Bajic VB. dPORE-miRNA: polymorphic regulation of microRNA genes. *PLoS One* 2011;**6**:e16657.
24. Betel D, Wilson M, Gabow A, Marks DS, Sander C. The microRNA.org resource: targets and expression. *Nucleic Acids Res* 2008;**36**:D149–53.
25. Chen K, Rajewsky N. Natural selection on human microRNA binding sites inferred from SNP data. *Nat Genet* 2006;**38**:1452–6.
26. Friedman RC, Farh KKH, Burge CB, Bartel DP. Most mammalian mRNAs are conserved targets of microRNAs. *Genome Res* 2009;**19**:92–105.
27. Parra P, Serra F, Palou A. Expression of adipose microRNAs is sensitive to dietary conjugated linoleic acid treatment in mice. *PLoS One* 2010;**5**:e13005.
28. Shah MS, Schwartz SL, Zhao C, Davidson LA, Zhou B, Lupton JR, et al. Integrated microRNA and mRNA expression profiling in a rat colon carcinogenesis model: effect of a chemo-protective diet. *Physiol Genomics* 2011;**43**:640–54.
29. Alisi A, Da Sacco L, Bruscalupi G, Piemonte F, Panera N, De Vito R, et al. Mirnome analysis reveals novel molecular determinants in the pathogenesis of diet-induced nonalcoholic fatty liver disease. *Lab Invest* 2011;**91**:283–93.
30. Zhang J, Zhang F, Didelot X, Bruce KD, Cagampang FR, Vatish M, et al. Maternal high fat diet during pregnancy and lactation alters hepatic expression of insulin like growth factor-2 and key microRNAs in the adult offspring. *BMC Genomics* 2009;**10**:478.
31. Ortega FJ, Moreno-Navarrete JM, Pardo G, Sabater M, Hummel M, Ferrer A, et al. MiRNA expression profile of human subcutaneous adipose and during adipocyte differentiation. *PLoS One* 2010;**5**:e9022.
32. (a)Saini S, Majid S, Dahiya R. Diet, microRNAs and prostate cancer. *Pharm Res* 2010;**27**:1014–26; (b)Parasramka MA, Ho E, Williams DE, Dashwood RH. MicroRNAs,

diet, and cancer: new mechanistic insights on the epigenetic actions of phytochemicals. *Mol Carcinog* 2011;**51**:213–30 doi:10.1002/mc.20822.

33. Lee J, Kemper JK. Controlling SIRT1 expression by microRNAs in health and metabolic disease. *Aging (Albany NY)* 2010;**2**:527–34.
34. Park JH, Ahn J, Kim S, Kwon DY, Ha TY. Murine hepatic miRNAs expression and regulation of gene expression in diet-induced obese mice. *Mol Cells* 2011;**31**:33–8.
35. Turchinovich A, Weiz L, Langheinz A, Burwinkel B. Characterization of extracellular circulating microRNA. *Nucleic Acids Res* 2011;**39**:7223–33 doi:10.1093/nar/gkr254.
36. Cortez MA, Bueso-Ramos C, Ferdin J, Lopez-Berestein G, Sood AK, Calin GA. MicroRNAs in body fluids-the mix of hormones and biomarkers. *Nat Rev Clin Oncol* 2011;**8**:467–77 doi:10.1038/nrclinonc.2011.76.
37. Vickers KC, Palmisano BT, Shoucri BM, Shamburek RD, Remaley AT. MicroRNAs are transported in plasma and delivered to recipient cells by high-density lipoproteins. *Nat Cell Biol* 2011;**13**:423–33.
38. Small EM, Olson EN. Pervasive roles of microRNAs in cardiovascular biology. *Nature* 2011;**469**:336–42.
39. Turnbaugh PJ, Ley RE, Mahowald MA, Magrini V, Mardis ER, Gordon JI. An obesity-associated gut microbiome with increased capacity for energy harvest. *Nature* 2006;**444**:1027–31.
40. Donohoe DR, Garge N, Zhang X, Sun W, O'Connell TM, Bunger MK, et al. The microbiome and butyrate regulate energy metabolism and autophagy in the mammalian colon. *Cell Metab* 2011;**13**:517–26.
41. Bäckhed F, Ley RE, Sonnenburg JL, Peterson DA, Gordon JI. Host-bacterial mutualism in the human intestine. *Science* 2005;**307**:1915–20.
42. Dumas ME, Barton RH, Toye A, Cloarec O, Blancher C, Rothwell A, et al. Metabolic profiling reveals a contribution of gut microbiota to fatty liver phenotype in insulin-resistant mice. *Proc Natl Acad Sci USA* 2006;**103**:12511–6.
43. Sonnenburg JL, Fischbach MA. Community health care: therapeutic opportunities in the human microbiome. *Sci Transl Med* 2011;**3**:78ps12.
44. Wang Z, Klipfell E, Bennett BJ, Koeth R, Levison BS, DuGar B, et al. Gut flora metabolism of phosphatidylcholine promotes cardiovascular disease. *Nature* 2011;**472**:57–63.
45. Arumugam M, Raes J, Pelletier E, Le Paslier D, Yamada T, Mende DR, et al. *Enterotypes of the human gut microbiome*. *Nature* 2011;**473**:174–80 doi:10.1038/nature09944.
46. Kweider M, Lowe GD, Murray GD, Kinane DF, McGowan DA. Dental disease, fibrinogen and white cell count; links with myocardial infarction? *Scott Med J* 1993;**38**:73–4.
47. Li X, Tse HF, Jin LJ. Novel endothelial biomarkers: implications for periodontal disease and CVD. *J Dent Res* 2011;**90**:1062–9 doi:10.1177/0022034510397194.
48. Yao Y, Tao H, Park DI, Sepulveda JL, Sepulveda AR. Demonstration and characterization of mutations induced by Helicobacter pylori organisms in gastric epithelial cells. *Helicobacter* 2006;**11**:272–86.
49. Gieger C, Geistlinger L, Altmaier E, Hrabé de Angelis M, Kronenberg F, Meitinger T, et al. Genetics meets metabolomics: a genome-wide association study of metabolite profiles in human serum. *PLoS Genet* 2008;**4**:e1000282.
50. Suhre K, Wallaschofski H, Raffler J, Friedrich N, Haring R, Michael K, et al. A genome-wide association study of metabolic traits in human urine. *Nat Genet* 2011;**43**:565–9.
51. Schunkert H, König IR, Kathiresan S, Reilly MP, Assimes TL, Holm H, et al. Large-scale association analysis identifies 13 new susceptibility loci for coronary artery disease. *Nat Genet* 2011;**43**:333–8.
52. Coronary Artery Disease (C4D) Genetics Consortium . A genome-wide association study in Europeans and South Asians identifies five new loci for coronary artery disease. *Nat Genet* 2011;**43**:339–44.

53. Wittwer J, Rubio-Aliaga I, Hoeft B, Bendik I, Weber P, Daniel H. Nutrigenomics in human intervention studies: current status, lessons learned and future perspectives. *Mol Nutr Food Res* 2011;**55**:341–58.
54. Psychogios N, Hau DD, Peng J, Guo AC, Mandal R, Bouatra S, et al. The human serum metabolome. *PLoS One* 2011;**6**:e16957.
55. Online Mendelian Inheritance in Man, OMIM (TM). McKusick-Nathans Institute of Genetic Medicine, Johns Hopkins University (Baltimore, MD) and National Center for Biotechnology Information, National Library of Medicine (Bethesda, MD);22 April 2011. World Wide Web URL: http://www.ncbi.nlm.nih.gov/omim/..
56. Fahy E, Subramaniam S, Brown HA, Glass CK, Merrill Jr. AH, Murphy RC, et al. A comprehensive classification system for lipids. *J Lipid Res* 2005;**46**:839–61.
57. Stegemann C, Drozdov I, Shalhoub J, Humphries J, Ladroue C, Didangelos A, et al. Comparative lipidomics profiling of human atherosclerotic plaques. *Circ Cardiovasc Genet* 2011;**4**:232–42.
58. Altmaier E, Kastenmüller G, Römisch-Margl W, Thorand B, Weinberger KM, Adamski J, et al. Variation in the human lipidome associated with coffee consumption as revealed by quantitative targeted metabolomics. *Mol Nutr Food Res* 2009;**53**:1357–65.
59. (a)Labbé SM, Grenier-Larouche T, Croteau E, Normand-Lauzière F, Frisch F, Ouellet R, et al. Organ-specific dietary fatty acid uptake in humans using positron emission tomography coupled to computed tomography. *Am J Physiol Endocrinol Metab* 2011;**300**:E445–53; (b) Pietiläinen KH, Róg T, Seppänen-Laakso T, Virtue S, Gopalacharyulu P, Tang J, et al. Association of lipidome remodeling in the adipocyte membrane with acquired obesity in humans. *PLoS Biol* 2011;**9**:e1000623.
60. Zhong Q, Simonis N, Li QR, Charloteaux B, Heuze F, Klitgord N, et al. Edgetic perturbation models of human inherited disorders. *Mol Syst Biol* 2009;**5**:321.
61. (a)de Graaf AA, Freidig AP, De Roos B, Jamshidi N, Heinemann M, Rullmann JA, et al. Nutritional systems biology modeling: from molecular mechanisms to physiology. *PLoS Comput Biol* 2009;**5**:e1000554; (b)Delgado-Lista J, Perez-Martinez P, Perez-Jimenez F, Garcia-Rios A, Fuentes F, Marin C, et al. ABCA1 gene variants regulate postprandial lipid metabolism in healthy men. *Arterioscler Thromb Vasc Biol* 2010;**30**:1051–7.
62. Pleil JD, Stiegel MA, Sobus JR, Liu Q, Madden MC. Observing the human exposome as reflected in breath biomarkers: heat map data interpretation for environmental and intelligence research. *J Breath Res* 2011;**5**:037104.
63. van Ommen B, Bouwman J, Dragsted LO, Drevon CA, Elliott R, de Groot P, et al. Challenges of molecular nutrition research 6: the nutritional phenotype database to store, share and evaluate nutritional systems biology studies. *Genes Nutr* 2010;**5**:189–203.
64. (a)Lee YC, Lai CQ, Ordovas JM, Parnell LD. A database of gene-environment interactions pertaining to blood lipid traits, cardiovascular disease and type 2 diabetes. *J Data Mining Genomics Proteomics* 2011;**2**:106; (b)Moreno-Sánchez R, Saavedra E, Rodríguez-Enríquez S, Gallardo-Pérez JC, Quezada H, Westerhoff HV. Metabolic control analysis indicates a change of strategy in the treatment of cancer. *Mitochondrion* 2010;**10**:626–39; (c)MacConaill LE, Garraway LA. Clinical implications of the cancer genome. *J Clin Oncol* 2010;**28**:5219–28.
65. Jiang P, Du W, Wang X, Mancuso A, Gao X, Wu M, et al. p53 regulates biosynthesis through direct inactivation of glucose-6-phosphate dehydrogenase. *Nat Cell Biol* 2011;**13**:310–6.
66. Ide T, Chu K, Aaronson SA, Lee SW. GAMT joins the p53 network: branching into metabolism. *Cell Cycle* 2010;**9**:1706–10.
67. Wang X, Di Pasqua AJ, Govind S, McCracken E, Hong C, Mi L, et al. Selective depletion of mutant p53 by cancer chemopreventive isothiocyanates and their structure-activity relationships. *J Med Chem* 2011;**54**:809–16.
68. Braun R, Buetow K. Pathways of distinction analysis: a new technique for multi-SNP analysis of GWAS data. *PLoS Genet* 2011;**7**:e1002101.

69. Kim YA, Wuchty S, Przytycka TM. Identifying causal genes and dysregulated pathways in complex diseases. *PLoS Comput Biol* 2011;**7**:e1001095.
70. Ernst J, Kheradpour P, Mikkelsen TS, Shoresh N, Ward LD, Epstein CB, et al. Mapping and analysis of chromatin state dynamics in nine human cell types. *Nature* 2011;**473**:43–9.
71. Patel CJ, Bhattacharya J, Butte AJ. An environment-wide association study (EWAS) on type 2 diabetes mellitus. *PLoS One* 2010;**5**:e10746.
72. Kim J, Gibson G. Insights from GWAS into the quantitative genetics of transcription in humans. *Genet Res (Camb)* 2010;**92**:361–9.
73. Ordovas JM, Corella D. Nutritional genomics. *Annu Rev Genomics Hum Genet* 2004;**5**:71–118.
74. Tucker KL. Assessment of usual dietary intake in population studies of gene-diet interaction. *Nutr Metab Cardiovasc Dis* 2007;**17**:74–81.
75. (a)Shi M, Umbach DM, Weinberg CR. Family-based gene-by-environment interaction studies: revelations and remedies. *Epidemiology* 2011;**22**:400–7; (b)Hamilton CM, Strader LC, Pratt JG, Maiese D, Hendershot T, Kwok RK, et al. The PhenX Toolkit: get the most from your measures. *Am J Epidemiol* 2011;**174**:253–60 doi:10.1093/aje/kwr193.
76. Harris PA, Taylor R, Thielke R, Payne J, Gonzalez N, Conde JG. Research electronic data capture (REDCap)—a metadata-driven methodology and workflow process for providing translational research informatics support. *J Biomed Inform* 2009;**42**:377–81.
77. Tucker KL, Mattei J, Noel SE, Collado BM, Mendez J, Nelson J, et al. The Boston Puerto Rican Health Study, a longitudinal cohort study on health disparities in Puerto Rican adults: challenges and opportunities. *BMC Public Health* 2010;**10**:107.
78. Katzmarzyk PT, Church TS, Craig CL, Bouchard C. Sitting time and mortality from all causes, cardiovascular disease and cancer. *Med Sci Sports Exerc* 2009;**41**:998–1005.
79. Tremblay MS, Colley RC, Saunders TJ, Healy GN, Owen N. Physiological and health implications of a sedentary lifestyle. *Appl Physiol Nutr Metab* 2010;**35**:725–40.
80. Freudenheim JL. Study design and hypothesis testing: issues in the evaluation of evidence from research in nutritional epidemiology. *Am J Clin Nutr* 1999;**69**:1315–21.
81. Willett WC. Nutritional epidemiology issues in chronic disease at the turn of the century. *Epidemiol Rev* 2000;**22**:82–6.
82. Weed DL. Interpreting epidemiological evidence: how meta-analysis and causal inference methods are related. *Int J Epidemiol* 2000;**29**:387–90.
83. (a)Qi L. Mendelian randomization in nutritional epidemiology. *Nutr Rev* 2009;**67**:439–50; (b) Rakyan VK, Down TA, Balding DJ, Beck S. Epigenome-wide association studies for common human diseases. *Nat Rev Genet* 2011;**12**:529–41.
84. Scientific foundations for future physicians . Report of the AAMC-HHMI Committee. In: . pp. 1–46.
85. Feero WG. *Creating a blueprint for genomic medical training.* ACP Internist; 2009 sep 2009.
86. Genetics education and training . Report of the Secretary's Advisory Committee on Genetics. febIn: USA.: Health, and Society—Department of Health and Human Services; 2011. pp. 1–211.
87. Walt DR, Kuhlik A, Epstein SK, Demmer LA, Knight M, Chelmow D, et al. Lessons learned from the introduction of personalized genotyping into a medical school curriculum. *Genet Med* 2011;**13**:63–6.
88. Friend SH, Ideker T. POINT: are we prepared for the future doctor visit? *Nat Biotechnol* 2011;**29**:215–8.
89. Giskes K, van Lenthe F, Avendano-Pabon M, Brug J. A systematic review of environmental factors and obesogenic dietary intakes among adults: are we getting closer to understanding obesogenic environments? *Obes Rev* 2011;**12**:e95–e106.
90. Maxmen A. Science networking gets serious. *Cell* 2010;**141**:387–9.
91. Bonetta L. Should you be tweeting? *Cell* 2009;**139**:452–3.

92. Kelley DR, Liu B, Delcher AL, Pop M, Salzberg SL. Gene prediction with Glimmer for metagenomic sequences augmented by classification and clustering. *Nucleic Acids Res* 2012;**40**:e9.
93. Parnell LD, Lindenbaum P, Shameer K, Dall'Olio GM, Swan DC, Jensen LJ, et al. BioStar: an online question & answer resource for the bioinformatics community. *PLoS Comput Biol* 2011;**7**:e1002216.

A Nutrigenomics View of Protein Intake: Macronutrient, Bioactive Peptides, and Protein Turnover

Chieh Jason Chou,[*] Michael Affolter,[†] and Martin Kussmann[‡,§]

[*]*Department of Gastrointestinal Health, Nestlé Institute of Health Sciences, Lausanne, Switzerland*

[†]*Department of BioAnalytical Sciences, Functional Genomics Group, Nestlé Research Center, Lausanne, Switzerland*

[‡]*Proteomics and Metabonomics Core, Nestlé Institute of Health Sciences, Lausanne, Switzerland*

[§]*Faculty of Science, Aarhus University, Aarhus, Denmark*

Proteins are needed for the development and sustainability of life. They are the molecular machines and building blocks in the human body that drive or exert most biological functions and confer structure and function to cell and tissue architecture. Dietary proteins provide essential amino acids and complement lipid and carbohydrate as a major source of energy. Therefore, humans must consume a sufficient amount and quality of proteins to stay healthy and avoid deficiencies. Even with a reasonable amount of intake,

Progress in Molecular Biology
and Translational Science, Vol. 108
DOI: 10.1016/B978-0-12-398397-8.00003-4

1877-1173/12 $35.00

example, A/J mice preferred eating bedding materials and their own feces over a semipurified low-fat diet.[3] In another study, Gelegen *et al.* tested the behavior of A/J and C57BL/6J mice in a restricted feeding schedule[4]; during the food-restriction period, C57BL/6J mice developed food-anticipatory activity prior to scheduled feedings and showed reduced dark-phase running wheel activity compared to the baseline values. In contrast, A/J mice did not develop food-anticipatory activity and showed increased dark-phase running wheel activity relative to baseline. These data suggest that A/J mice lack the motivation to eat, which is similar to patients suffering anorexia nervosa. Unique phenotypes of A/J but not of other strains suggest that the genetic component is a contributing factor to the modifications of eating behavior. However, precise information about the responsible genes for the aberrant food intake behavior of A/J mice is not available.

Habitual consumption of proteins has limited impact on obesity. Summerbell *et al.* reviewed the food frequency questionnaire data in several longitudinal studies and reported that in six out of eight prospective cohort studies baseline protein intake did not have any association with increased body weight or BMI in follow-up visits.[5] On the other hand, increasing the amount of protein in a calorie-restricted diet is an effective nutritional intervention to reduce body weight. Proteins have been shown to affect energy balance by enhancing satiety, diet-induced thermogenesis, and lean body mass, factors that favor body weight control.[6–10]

To better understand how the amount and the quality of proteins impact on weight management, the pan-European study DiOGenes (diet, obesity, and genes) was conducted from 2005 to 2009. The aims of the study were to (a) identify diets that are effective in preventing weight regain after a period of weight loss with very low-caloric feeding, (b) better comprehend why only certain subjects benefit from certain diets,[11] and (c) develop candidate markers for prediction of successful weight management by dietary interventions. In 773 participants assigned to one of five weight maintenance diets, weight loss was better maintained with the diet featuring a moderate increase in protein content and a modest reduction of glycemic index of food.[12] Within the same cohort of subjects, more than 30 plasma proteins in the categories of vascular factors, adipokines, insulin and related hormones, immunoproteins, growth factors, satiety hormones, and some steroid hormones were analyzed.[13] Interestingly, a greater reduction of angiotensin-converting enzyme (ACE) during the 8 weeks of hypocaloric feeding was the most important predictor for later weight loss in the subsequent 6-month weight maintenance stage.[13]

Circulating level and activity of ACE are highly influenced by genetic polymorphisms of the gene.[14] While the role of *ACE* in diet-mediated weight loss is not clear, it is listed as one of the candidate genes for obesity in "The Human Obesity Gene Map: The 2005 Update".[15] A related study examining 13

ACE gene polymorphisms in three black populations living in Nigeria, Jamaica, and the United States concluded that the haplotype ACE1–ACE5 TACAT, located in the promoter region, was significantly overtransmitted from parents to obese offspring in the U.S. and Nigerian subjects but not in the Jamaican population.[16] In another study, Strazzullo *et al.* reported that the *ACE* I/D genotype was a significant predictor of overweight and abdominal adiposity in men, and DD homozygosity was associated with larger increases in body weight and blood pressure in aging individuals, as well as with higher incidence of overweight.[17] A significant association between BMI and the transmission of the *ACE* D allele has also been found in a Chinese population.[18] However, the role of dietary protein as a modulating factor of ACE-related body weight regulation remains to be elucidated, and it would be interesting to find out whether such a relationship exists in a large longitudinal cohort study such as DiOGenes.

Proteomic and metabonomic responses to dietary factors and supplements in general were reviewed by Astle and colleagues[19] on the occasion of the related symposium within Experimental Biology 2007, Washington, DC, USA. Advancement of targeted metabonomics has provided new insights into metabolic regulation within individuals. With such tools, new studies have discovered an interesting role of amino acids in insulin resistance. Newgard *et al.* used tandem mass spectrometry (MS/MS) to measure the plasma amino acid and acylcarnitine profiles of 73 obese and 67 lean subjects[20]; eight out of sixteen measured amino acids were more abundant in the obese than in the lean samples. Using principal components analysis, lean and obese subjects could be discriminated on the basis of a combination of branched-chain amino acids (BCAAs), methionine, glutamate/glutamine, phenylalanine and tyrosine, and C3 and C5 acylcarnitines. In addition, BCAA-related metabolites showed a positive and linear correlation to homeostatic model assessment, an insulin resistance index. Association between insulin resistance and elevated plasma BCAAs was also reported in two other studies.[21,22] Based on the evidence, it appears plausible that in a condition with excessive fat intake, BCAAs contribute to the development of obesity-associated insulin resistance. Clinically, the most effective treatment for morbid obesity is gastric bypass surgery. In addition to rapid reduction of body weight, rapid improvement of insulin sensitivity has been found with this surgery. Interestingly, BCAA concentrations in the plasma were more reduced after the gastric bypass surgery than after an equivalent 10 kg weight loss with dietary intervention.[23]

A link between plasma BCAAs and diabetes was also reported by Wang *et al.*, who investigated whether metabonomic profiles could be used to predict the later onset of diabetes.[24] In the study, samples were selected from 2422 nondiabetic subjects in the Framingham Offspring Study, and fasting metabolite profiles at baseline were measured in 189 individuals who developed diabetes during the 12-year follow-up and in 189 matched control subjects who remained healthy otherwise. Their results show that high concentrations of five BCAAs (isoleucine, leucine, valine, tyrosine, and phenylalanine) were significantly associated with the development of diabetes during the follow-up. Compared to individuals in the lowest quartile of amino acid score, which is a composite index based on isoleucine, phenylalanine, and tyrosine, individuals in the top quartile had a five- to sevenfold higher risk of developing diabetes. Other reports in animals and humans showing increased insulin resistance after BCAA supplementation further raise the concern of high BCAA intake in an overnutrition condition.[20,25] One possible mechanism by which high amounts of plasma BCAAs increase the risk of diabetes is by affecting insulin signaling pathways. Insulin receptor substrate 1 serine phosphorylation has been shown as a consequence of long-term stimulation of the MTOR/S6K1 pathway by chronic exposure to leucine.[26] Nevertheless, the role of BCAAs in the development of insulin resistance and diabetes remains controversial. Given the benefits of proteins in diet-induced thermogenesis, enhancement of satiety, and increase of lean body mass, the long-term consequences of consuming proteins rich in BCAAs remain to be elucidated.

B. Omics-Level Views of Dietary Protein Effects

Several studies have applied omics technologies to examine the effect of protein intake on metabolism. Endo *et al.* tested whether hepatic gene expression would be affected when proteins are removed from a diet.[27] In comparison with the rats fed a 12% casein diet, 281 genes were differentially regulated by at least twofold in the liver of the protein-free group. Among them, 97 genes were upregulated and 184 genes were downregulated. It is worth noting that 11 genes in the cholesterol metabolism were downregulated, and none of the genes in this pathway was upregulated. In the same study, liver transcriptomic profiles were also compared between two other diets, namely, one made of gluten and the other based on casein; a lower number of differentially regulated genes was observed (61 upregulated and 50 downregulated in the gluten group). Similar to the comparison between protein-free and casein diets, cholesterol metabolism was highly sensitive to the source of proteins, and all 15 differentially regulated genes in this pathway were unilaterally increased in the gluten group as compared to the casein group. Interestingly, despite the opposite expression patterns of genes in hepatic cholesterol metabolism, both protein-free and gluten feeding resulted in a significant reduction of serum total cholesterol and HDL cholesterol.

The hypocholesterolemic effect of proteins, especially soy protein, has been extensively examined since the 1960s; a meta-analysis of 38 controlled clinical trials confirmed the beneficial effect of dietary soy protein on reducing serum total and LDL-cholesterol concentrations.[28] Consistent results from these studies led to U.S. Food and Drug Administration approval of the health claim that consumption of soy protein (25 g/day) as a part of a diet low in saturated fat and cholesterol may reduce the risk of coronary heart disease by lowering blood cholesterol levels. In an attempt to reveal the mechanism by which soy protein or soy protein isolate improves blood cholesterol levels, plasma parameters and hepatic transcriptomic profiles were measured in animals fed a soy protein isolate or a casein diet.[29] Results of the study confirmed the general observations and showed a reduced plasma total cholesterol level in rats fed a soy protein isolate diet. Transcriptomic analysis indicated that 115 genes were differentially expressed, with 61 upregulated and 54 downregulated in the soy group. Among the upregulated genes, nearly 20% are involved in steroid metabolism. Concurrent upregulation of genes in steroid metabolism and reduction of serum cholesterol by soy protein suggest that a well-coordinated transcriptional regulation is involved. In fact, overabundance of sterol regulatory element-binding protein 2 (SREBP2)[30] and upregulation of 3-hydroxy-3-methylglutaryl-coenzyme A reductase (HMGCR), which is both an SREBP2 target gene and the rate-limiting enzyme for cholesterol biosynthesis,[31] were reported in rodents fed a soy protein diet.

Isoflavones in soy protein isolate probably play an important role in SREBP2-mediated transcriptional activation of cholesterol metabolism genes. The argument is supported by the following evidence: (1) soy isoflavones can independently activate SREBP2, the master transcriptional factor of cholesterol biosynthesis genes, and its target genes in HepG2 cells[32] and (2) reduction of serum cholesterol and upregulation of *Srebp2* and *Hmgcr* mRNAs were absent in rats fed an isoflavone-poor soy protein diet.[33] However, another feeding study reported contradicting results and showed that a significant reduction of serum cholesterol was achieved in rats fed a diet made of isoflavone-poor soy protein isolate as compared that made of casein.[34]

Inconsistent findings among different experiments are likely due to factors such as the remaining amount of isoflavone in the soy protein isolates, the source of soy proteins, and the chemical composition of soy protein products. To illustrate such inconsistent quality of commercially available soy protein isolates, one study compared the composition of soy isolates by proteomics[35]; proteins were first separated by two-dimensional gel electrophoresis, and the identities of proteins were then determined by matrix-assisted laser desorption/ionization mass spectrometry (MALDI-MS). Total soybean proteins extracted from defatted soy flour contained α, α', and β subunits of β-conglycinin (7S globulin) and heavy and light subunits of glycinin (11S globulin). However, the main constituents of a commercial soy concentrate (Croksoy) were breakdown products of β-conglycinin and intact glycinin. In another soy protein isolate (SUPRO), only some light-chain and intact heavy subunits of glycinin were found, and none of the fragments corresponding to β-conglycinin was detected. Since administration of soy β-conglycinin has been shown to ameliorate atherosclerosis in apolipoprotein E knockout and LDL receptor knockout mice,[36] the amount of β-conglycinin in various preparations of soy protein products, together with different amounts of isoflavone, might explain different experimental outcomes. This problem calls for a better standardization of dietary interventions, not only at the level of overall amounts of protein and macronutrients, but also at the level of composition and quality, if we would like to use soy protein isolates as a means to manage hypercholesterolemia.

The biological functions of isoflavone-containing soy protein isolate have also been assessed in adipose tissue[30,37]; rats fed a soy protein isolate diet had a reduced adipocyte area and lowered triglyceride concentrations in the adipose tissue compared to those fed a casein diet.[37] Microarray analyses of rat adipose tissues indicated that feeding soy protein isolate led to differential expression of genes that are overrepresented in the renin–angiotensin system, lipid metabolism, signaling and carbohydrate metabolism, and adipokines and cytokines. Pathway analyses revealed networks of many genes centered at leptin (*Lep*). Although the biological relevance of *Lep* expression in soy protein-fed rats remains to be elucidated, generation of the hypothesis would not have been possible without omics-level study readouts.

III. Protein Beyond Macronutrient

A. Bioactive Food-Derived Peptides

A broad range of animal and plant food proteomes have been characterized.[38–40] However, it is noteworthy that food-derived proteins and peptides deliver more than the bulk macronutrient complement and the building blocks for protein synthesis. Food-derived proteins and peptides are becoming increasingly appreciated because of their bioactive properties and functions, including serving as growth factors, antimicrobials, antihypertensives, immune regulators, or modifiers of food intake. Bioactive peptides and proteins were reviewed by Moller *et al.*,[41] and a database of bioactive peptides has been created for classification and bioactivity of food proteins.[42] The so-termed biologically active motifs in polypeptide chains remain inactive as long as they "reside" in their precursor sequences, but upon release by proteolytic enzymes, they may interact with receptors and exert bioactivity.[43,44] Bioactive peptides may be released by the host or microbial enzymes during digestion.[45] They can also be generated during food processing (industrial processing) or ripening (natural processing). In order to properly investigate bioavailability and bioefficacy at a systems level (i.e., in the blood) and at an organ level (e.g., in the stomach and gut), bioactive peptides and proteins need to be identified and quantified from the food matrix to the target tissues in the body.

B. Bioactive Peptides in Milk and Dairy Products

Milk has coevolved with humans to ideally support neonatal healthy growth and development and to favor the maturation and maintenance of a balanced immune system.[46] Mammalian milk bioactive peptides stem from the protein/peptide,[47] lipid,[48] and oligosaccharide complement.[49] More information can be found in a review that we recently published on omics approaches to unravel the protein/peptide, lipid, and carbohydrate complement of human and animal milk.[50]

Human milk proteins comprise caseins and whey, with a 50:50 w/w ratio.[47] Caseins function as precursors of various bioactive peptides, and whey proteins exhibit activity in immune modulation and defense.[51] Together with the caseins, α-lactalbumin, lactoferrin, albumin, and various immunoglobulins as parts of whey proteins account for >99% of the total human milk protein mass. Lactoferrin is an abundant mammalian iron-binding milk glycoprotein that is found in whey and profoundly impacts the host defense system; it has been shown to prevent microbial growth by direct interaction with the membrane of Gram-negative bacteria.[52] Lactoferrin peptides derived from the full-length protein influence cytokine production in cell cultures, thereby becoming a candidate for bioactive peptides capable of modulating immune and inflammatory actions of the body.[53] Notably, the remaining <1% of the human milk protein complement encompasses a complex mixture of bioactive proteins and peptides, which is far from being completely characterized and leveraged.

Human breast milk is the gold standard for neonate and infant nutrition. Secretory immunoglobulins, lysozyme, interferon, and growth factors are known as the immunological "assets" of breast milk. Breast milk partly promotes the inhibition of pathogens and favors the growth of a protective colonic microbiota.[54] Apart from delivering basic nutrition to the newborn, milk also protects the neonate and the mammary gland against infection; breast-fed newborns typically experience less gastrointestinal infections and inflammatory, respiratory, and allergic disorders. These benefits have been ascribed to diverse protective factors in breast milk. One such specific bioactive protein in mother's milk is the soluble monocyte differentiation antigen CD14 (sCD14)[55,56]; one study revealed a key role of sCD14 during bacterial colonization of the gut and proposed sCD14 to be implicated in modulating local innate and adaptive immune responses, thereby controlling homeostasis of the neonatal intestine. Another related study unraveled an interaction between soluble toll-like receptor 2 and sCD14 in plasma and milk, suggesting a novel innate immune mechanism that regulates bacteria-induced toll-like receptor signaling.[57]

Biologically active peptides derived from cow and human milk have been shown to exert both functional and physiological roles *in vitro* and *in vivo*, including immunomodulatory, antibacterial, antihypertensive, and opioid-like properties.[58] Fermentation of milk proteins from lactic acid bacterial proteolysis can generate functional products that are enriched in bioactive peptides, and this strategy is being technologically exploited. For example, cell envelope proteinases (CEPs) of the lactobacilli are central to bacterial nutrition, and their activities can generate bioactive and health-beneficial peptides from milk proteins, thereby contributing to organoleptic properties of fermented milk products. Hebert *et al.*[59] studied the breakdown products of α(s1)- and β-casein by a CEP from *Lactobacillus delbrueckii* (subsp. *lactis* CRL 581) and the concomitant bioactive peptide release, including the influence of peptide supply, carbohydrate source, and osmolites on CEP activity. Mass spectrometric screening of the main HPLC-isolated peptide peaks identified 33 and 32 peptides in the α(s1)- and β-casein hydrolysates, respectively. A pattern of α(s1)- and β-casein breakdown was established with a series of potentially bioactive peptides (antihypertensives and phosphopeptides).

Bioactive milk proteins have also been researched for their benefits in dental health,[60] with caseins and their hydrolysates being investigated for the ability to prevent caries through inhibition of plaque-forming bacteria, inhibition of tooth enamel demineralization, and subsequent enamel remineralization. Caseinophosphopeptides (CPPs) and glycomacropeptides (GMPs) have been shown to inhibit growth of *Streptococcus* mutants and other species. Moreover, CPP forms nanoclusters with amorphous calcium phosphate at the tooth surface, thereby providing a calcium and phosphate ion reservoir. Glycosidic structures attached to GMP are the key for its bioactivity, including anticariogenic effects. Like CPP, GMP inhibits tooth enamel demineralization and promotes enamel remineralization.

C. Bioactive Peptide Discovery

In silico and *in vitro* methods can be used to discover and identify bioactive peptides in food sources. Figure 1 shows the workflows for the traditional hydrolysis and screening approach (Fig. 1A) and the bioinformatics/genomics-driven strategy (Fig. 1B).

The classical *in vitro* approach (Fig. 1A) starts with the generation of a functional fraction or extract, potentially containing bioactive substances. These peptide mixtures are typically tested in *in vitro* assays (e.g., examination of binding capacity to a targeted receptor). Once an interesting fraction or extract has been identified, it undergoes further fractionation and subsequent proteomic and peptidomic analysis in order to narrow down to a few candidates, which then can be further evaluated. The application of typically MS-based peptidomics (the analogue of proteomics for the lower molecular weight, i.e., peptide range) for bioactive peptide identification and quantification must consider the complexity and heterogeneity of full-length, native, bioactive peptides; these peptides differ from the typically tryptic peptides

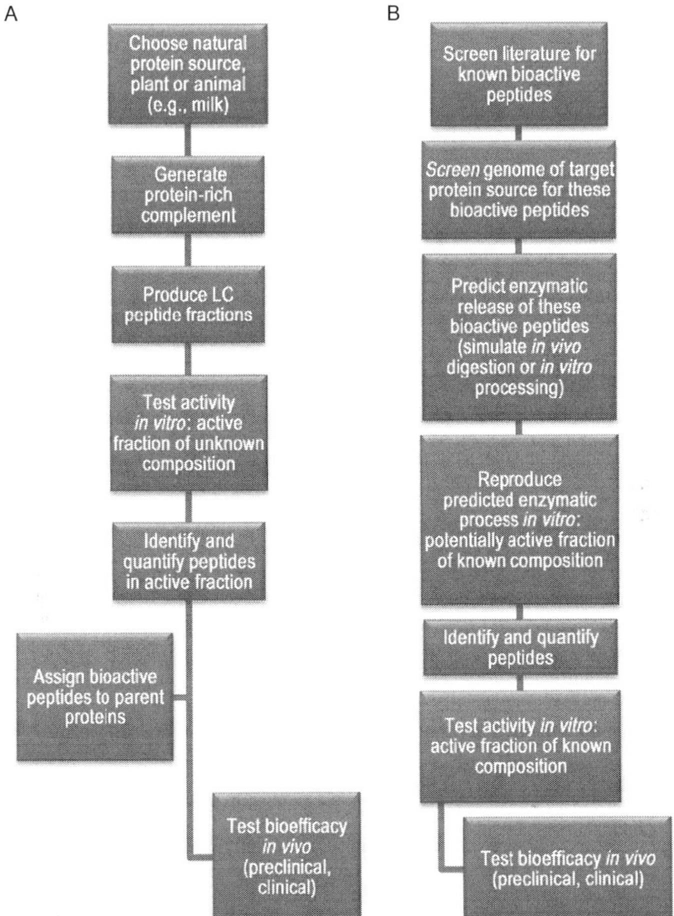

FIG. 1. Analytical workflows for discovery and validation of bioactive peptides: (A) classical hydrolysis/fractionation and *in vitro* screening approach and (B) reverse-genome engineering with *in silico* prediction of peptide sequences and their release, plus subsequent *in vitro* validation. (For color version of this figure, the reader is referred to the online version of this chapter.)

that are generated for protein biomarker identification and are more similar in size and C-terminal amino acids. We have reviewed proteomics and peptidomics tools for the discovery and characterization of bioactive substances and how these techniques differ from the classical protein biomarker discovery.[61,62]

We and others increasingly follow a complementary and bioinformatics-driven strategy to discover and leverage bioactive peptides (Fig. 1B). The approach has been coined "reverse-genome engineering" and enables *in silico* discovery and prediction of bioactive peptides.[45] First, known bioactive peptides and their functions are identified in public databases; these sequences are then mapped onto suitable food genomes from plant or animal sources where these peptides may be residing in parent protein sequences. In a further *in silico* analysis, either human digestive or food processing-related conditions are mimicked to reveal the potential release of the bioactive peptides from their parent sequences. The result of this bioinformatic, sequence-based approach is a set of peptide sequences that can potentially be released from a food source under a given condition. Then, *in vitro* and *in vivo* experiments can follow to validate the desired bioactive properties. Clearly, this top-down approach can significantly reduce the number of bioactivity tests to be performed and help filter the nutritionally relevant and feasible candidates.

Ideally, one would like to expand from this literature-dependent approach (based on peptides with reported or suggested health benefits) to peptide sequences in general, but the prediction of bioactivity based on amino acid sequence alone is not (yet) feasible.

IV. Protein and Proteome Turnover

A. Protein Turnover and Its Classical Assessment

Protein turnover is the net result of continuous synthesis and breakdown of body proteins and ensures maintenance of optimally functioning proteins.[63] Figure 2 depicts the different states of protein turnover, namely, steady-state, pool expansion, and pool contraction. In the steady state, protein synthesis equals degradation, and neither the related transcriptome nor metabolome is affected. Protein pool expansion can be achieved through either increased synthesis or decreased degradation with the related transcriptome or metabolome being modulated, respectively. By contrast, the protein pool is contracted when synthesis decreases (transcriptome changes accordingly) or degradation increases (metabolome changes).

Protein turnover is a fundamental biological process in all living organisms, and therefore, scientists have tried for decades to quantify turnover rates.[65] Sprinson and Ritterberg[66] introduced the "end-product method" in 1949 for quantitative measurements of protein turnover in humans; they used a constant infusion of a tracer and measured the excretion of labeled urea and ammonia in the urine.[65] Initially, radioactive isotope-labeled amino acids were used in combination with scintillation counting, but because of the health concerns of radioactivity, these tracers became restricted for human studies and were eventually replaced by stable-isotope tracer amino acids labeled with ^{2}H, ^{13}C, or ^{15}N isotopes for which mass spectrometric detection

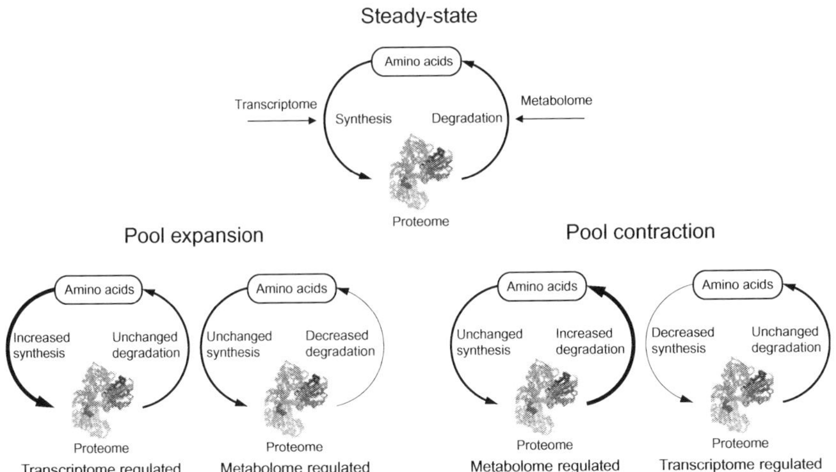

FIG. 2. States of protein turnover. Steady state: protein synthesis equaling degradation with neither the related transcriptome nor the metabolome affected. Protein pool expansion: increased synthesis or decreased degradation with the related transcriptome or metabolome being modulated, respectively. Protein pool contraction: decreased synthesis (transcriptome changed) or increased degradation (metabolome changed). Adapted from Ref. 64. (For color version of this figure, the reader is referred to the online version of this chapter.)

methods have been developed. Currently, the most commonly used tracers in whole-body protein turnover studies are [^{15}N]glycine,[67] L-[^{13}C]leucine,[68] and L-[$^{2}H_5$] phenylalanine.[69]

Although easily utilized, the end-product method has been largely replaced by the "precursor" approach. Today, the precursor method is considered the gold standard for measuring whole-body protein turnover in humans. "Constant infusion" and "flooding dose" are two strategies used to measure protein synthesis in humans.[70,71] Constant infusion requiring a prolonged labeling period has the advantage over concurrent determination of whole-body and tissue-specific turnover rates[72] and is well suited for the measurement of proteins with slow turnover rates, such as muscle proteins. However, the synthesis rate of high-turnover proteins may be underestimated because of recycling of the tracer or possible secretion of newly synthesized proteins from the target organ.

The flooding dose approach involves the injection of a large amount of unlabeled amino acid (tracee) along with the isotope-labeled amino acid (tracer). The method is based on the assumption of equilibration between the precursor pools,[73] such as the extracellular, intracellular, and aminoacyl-transfer RNA pool.[74] This strategy rapidly increases the labeling of the intracellular amino acid pools, thereby favoring measurements of proteins with a fast turnover rate, as typically observed for liver proteins.

Although the flooding dose method offers practical advantages, its validity is still under debate. Some studies showed differences between flooding dose and constant infusion methods, which could be due to a stimulation of protein synthesis by a large dose of essential amino acids,[75] whereas others reported similar synthesis rates obtained by the two approaches.[76] It seems that these concerns are less or not evident when short labeling periods of less than 30 min are studied. Whole-body protein turnover, or more specifically the fractional synthesis rate of proteins, can be estimated either by the continuous infusion or by the flooding dose method.

A considerable number of human nutritional studies investigated different health aspects of protein turnover in animals and humans. Although by far most of them deployed classical turnover assessments (i.e., bulk body or tissue protein synthesis/degradation rather than proteome turnover at single-protein resolution), only a few of them will be briefly mentioned here. The intimate interplay between protein intake, exercise, and muscular health has been reviewed, in particular with regard to aging (Walrand and Boirie[77]), lean body mass loss (sarcopenia) (Evans[78]), and compromised physical mobility and recovery (Koopman and van Loon[79]), respectively. Typical clinical nutrition interventions administered stable-isotope-labeled amino acids[80,81] and always compared body composition changes by bioelectrical impedance analysis, dual-energy X-ray absorptiometry,[82] or whole-body magnetic resonance imaging.[83]

B. Mass Spectrometric Protein Turnover Analysis

The study of dynamic changes in the protein and amino acid pool requires methods to measure the incorporation or loss of a tracer. Historically, the tracer was an unstable (radioactive) isotope determined by scintillation counting. Over time, it has been replaced or complemented by stable (nonradioactive) isotopes measured by MS.

MS is considered today as clearly the best technique to measure isotopic ratios and enrichments because of its precision, sensitivity, and accuracy.[84] Two classes of MS systems are most suited to determine very small differences in isotope ratios: (1) isotope ratio MS (IRMS) and (2) atmospheric pressure MS, both commonly operated with gas chromatography but more recently also with liquid chromatography. Considering carbon as one of the most commonly used elements, the precision of measurement for $^{13}C/^{12}C$ ratios is in the range of 0.0002% for GC–IRMS and 0.05% for GC–MS.[84] Typically, the IRMS value is expressed relative to a standard in ‰ ($\delta^{13}C$). Another difference between the two MS devices is the type of species measured; IRMS determines the isotopic ratio after conversion of the organic molecules into ionized CO_2, whereas lower precision MS measures isotopomer ratios of ionized molecules. Baseline resolution of analytes is critical for IRMS, as it distinguishes only CO_2, but less decisive for MS as coelution of compounds can be discriminated in the analyzer.

The precision of IRMS is ideally suited for the determination of natural abundance variations and protein turnover studies using stable-isotope-enriched tracers, whereas the lower resolution MS is complementary for the measurement of samples with higher isotope enrichment.[85]

C. From Single-Protein to Proteome-Scale Turnover

Total protein turnover has been measured in a variety of organisms and tissues, either by the continuous infusion or by the flooding dose method. Information on whole-body turnover is valuable, especially in clinical application, but yields only a global view of the total protein pool either at the whole-tissue or organism level. Until the emergence of proteomics, limited effort was invested to study protein synthesis and degradation at the level of individual proteins.[64] More recently, new methods have been developed to address the technical gap by applying multiplexed mass spectrometric protein analysis in the same labeling experiment.

Initially, proteomics technologies were developed for simultaneous (qualitative) identification of proteins in mixtures, but these have evolved more recently to (relative and absolute) quantitative characterization of whole proteomes.[61,86–88] Nevertheless, the classical approaches did not address the dynamics of the proteome in a steady state, for example, synthesis and degradation of individual protein species; it is the net balance of synthesis and degradation that determines the concentration of any protein at a steady state or a given moment in time (proteomics snapshot). Expanding from proteomics to transcriptomics, some of the discrepancy observed between gene expression and protein abundance data[89] might in fact be explained by the underlying variations of protein synthesis/degradation rates that are not captured by the traditional proteomics snapshots.[90]

Addressing protein turnover dynamics at single-protein resolution but proteome-wide scale presents significant technical challenges for analysis. Whereas efficient and complete stable-isotope labeling of cells and tissue cultures has been shown to be feasible,[90,91] the high isotope enrichment needed for proteome-wide studies is still difficult to achieve in more complex organisms such as animals. Modern proteomics MS systems readily deliver a mass resolution in the low ppm range, meaning that mass differences of a few ppm can still be seen; however, resolution is not everything in protein turnover studies because the natural isotope envelope of the unlabeled peptide can overlap with the profile of the labeled peptide. Therefore, mass differences of 4–10 Da introduced by a labeled precursor are ideal for distinguishing labeled from unlabeled peptide forms.

Several considerations have to be addressed in practice in order to select the "ideal" labeling precursor[92]:

- Metabolic isolation (labeled atoms are specific per amino acid and do not distribute through metabolic reactions into any other amino acid)
- Cell that is auxotrophic for the tracer amino acid (no dilution of precursor pool through *de novo* synthesis; for human studies, this means that only essential amino acids should be labeled)
- Metabolically active precursor pool (rapid change in precursor isotope pool)
- Abundant amino acid (high incorporation probability; stable pool size)
- Sufficient number of heavy atom centers (4–10 Da mass offset between labeled and unlabeled amino acid)

Proteome dynamics was measured in chicken fed with a semisynthetic diet containing [$^{2}H_{8}$]-valine at a calculated relative isotope abundance of 0.5.[93] The relative isotope abundance was stable over an extended labeling window and enabled calculation of the rates of synthesis and degradation of individual proteins. By calculating the partition between newly synthesized (new) and preexisting (old) components and by factoring in the total pool size as well as assumptions about tissue expansion, a detailed time course for the replacement of eight individual muscle proteins, reflecting muscle growth, was generated. This study demonstrated for the first time the feasibility of analyzing the turnover of individual proteins in whole animals.

Turnover of a human cellular proteome has been reported more recently using dynamic incorporation of stable isotopes in cultured cells with amino acids (dynamic stable-isotope labeling of amino acids in cell culture, dynamic SILAC).[94] Almost 600 proteins from human adenocarcinoma cells were characterized for time-dependent changes by examining the incorporation of [$^{13}C_{6}$]-arginine in a classical pulse-chase experiment. Although a large number of proteins were analyzed and turnover rates deduced, the data reflect the protein turnover only in cultured cells and thus exclude proteome-wide assessment *in vivo*. Nevertheless, it demonstrates the power of modern proteomics for determining synthesis and degradation rates of individual proteins in the proteome.

All aforementioned approaches require high enrichment levels of labeled proteins, which can be achieved in animals by feeding highly enriched isotopes for a long period. By contrast, this strategy is not feasible in humans, and therefore the measurement of fractional synthesis rates of multiple plasma proteins after a meal with intrinsically labeled milk proteins was established as an alternative.[95] In order to provide labeled proteins for metabolic studies, Boirie and coworkers developed 15 years ago a method to produce a large amount of milk proteins intrinsically labeled with [^{13}C]-leucine.[96] The enrichment achieved in the milk proteins (casein and whey protein fraction) was between 10% and 20% ([^{13}C]-leucine atom percent excess), which is sufficient for studying human protein metabolism.

Jaleel and colleagues[95] used the labeled milk and measured the incorporation rate of amino acids liberated during digestion into plasma proteins in humans. Their approach combined proteomics technologies (plasma protein prepurification, one-dimensional gel separation, and LC–MS/MS-based protein identification) with traditional GC–MS analysis for isotope enrichment ($[^{13}C_6]$-phenylalanine) analysis; 29 individual plasma proteins were identified, and their corresponding postprandial fractional synthesis rates were calculated based on the rate of $[^{13}C_6]$-phenylalanine incorporation, showing a 30-fold range in synthesis rates.

In a more recent study, the same group reported a methodology for measuring synthesis rates of multiple muscle mitochondrial proteins in rats.[97] Skeletal muscle mitochondrial dysfunction is observed in conditions such as aging and insulin resistance, but the underlying molecular pathways remain unclear. Jaleel and coworkers measured the synthesis rates of 68 mitochondrial and 23 nonmitochondrial proteins isolated from a skeletal muscle mitochondria fraction; these rates varied 10-fold between the lowest and highest; the lowest rate was found for a structural protein such as myosin heavy chain ($0.16 \pm 0.04\%$/h) and the highest for a mitochondrial protein such as that encoded by the dihydrolipoamide branched-chain transacylase E2 gene ($1.5 \pm 0.42\%$/h). This approach corroborates the transcriptomic analysis and offers a great opportunity to better understand the complex regulatory steps in protein turnover.

Stable-isotope metabolic labeling applied to proteome turnover measurements in animals has been recently reported by the Turck group.[98] Applying isotopic tracers to live animals requires an assessment of relative isotope abundance in a precursor pool. In addition to this challenge, data analysis becomes difficult in cases of low label incorporation, resulting in a complex convolution of mass spectrometric signals from labeled and unlabeled peptides. Turck *et al.* administered a ^{15}N-labeled diet as an isotopic tracer in mice for limited time periods. The resulting partially labeled proteins were digested and analyzed by LC–MS/MS. The group interpreted the mass spectrometric data with its ProTurnyzer software, which facilitates the determination of protein fractional synthesis rates in the absence of precursor relative isotope abundance information. ProTurnyzer results were validated with *Escherichia coli* protein data, and the method was applied to mouse brain and plasma proteomes for automated turnover studies.[98]

D. Protein Intake and Turnover—Today's Knowledge and Future Implications

Modern advances in studying protein synthesis, breakdown, and metabolism in the body are intimately coupled to the progress in stable-isotope analysis by MS. These methods have impressively improved over the past few decades; modern GC-combustion–MS systems enable accurate and sensitive measurements of isotope enrichment, whereas IRMS analysis gives ultimate precision for determination of natural abundance variations and for studies using stable-isotope-labeled tracers. Both MS technologies represent cornerstones for studies of protein metabolism.

With the development of proteomics technologies for qualitative and quantitative analysis of entire proteomes, new high-resolution and high-mass-accuracy MS instruments have increasingly been deployed to assess isotope enrichment in labeled proteins. Although current types of mass spectrometers used in proteomics may not yet be sufficiently sensitive for measuring less than a few percent of a stable-isotope-labeled variant, higher enrichment levels already meet readily the analytical performance levels of modern high-resolution instruments. This sets the stage for an expected breakthrough in proteome-wide turnover studies, addressing protein metabolism at individual protein level. The ideal and comprehensive approach should combine tracer studies (constant infusion or flooding dose) and proteomics techniques for protein isolation, purification, and characterization. Together with GC–MS and IRMS analysis for tracer enrichment quantification, this would enable the largest coverage of protein identities with concurrent synthesis and degradation data for individual proteins. This information will become even more important in the future because systems biology requires integration and correlation of transcriptome, proteome, and metabolome data; with protein synthesis and breakdown information at hand, transcriptome–proteome correlations can be refined because turnover data allow protein abundance changes arising from modulated transcription to be distinguished from those due to altered synthesis/degradation.

V. Conclusions

Genome-wide readouts of protein intake effects have delivered mechanistic insights into the health effects of dietary protein(s). Classical, "food pyramid-like" recommendations for macronutrient proportions are suitable for general populations but fall short in addressing the needs of individuals. Nutrigenomic analysis of protein consumption has begun to shed more light on interindividual differences in protein requirements at the level of quantity or quality, setting a foundation for personalized nutrition. Genomics, and particularly proteomics, views of animal and plant protein compositions have furthermore yielded a more detailed analysis of protein quality depending on the source and origin. This information should increasingly be integrated into nutritional interventions in order to improve protein content standardization and comparability.

Proteomics has greatly refined the appreciation of protein as a nutrient beyond bulk energy and amino acid supply; protein-specific analysis of food matters at proteome-wide scale, combined with bioinformatic analysis of the food genomes, enable a more targeted and top-down approach to the dietary delivery of bioactive peptides. This new approach offers an alternative or complement to the cumbersome classical empirical strategy, which requires hydrolyzing food proteins, screening fractions for activity, and predicting and confirming bioactive peptides embedded in food proteins.

Protein turnover can now be assessed globally, on a per-protein basis. This is a paradigm shift compared to the bulk turnover analyses per body compartment or tissue; incorporation of labeled amino acids into specific proteins followed by proteomics *ex vivo* analysis delivers protein-specific synthesis and degradation information. Apart from deeper insights into dietary modulation of protein turnover, protein-specific turnover is a great addition to gene expression and protein abundance data, typically derived from transcriptomics and proteomics experiments, because mRNA levels often do not correlate with protein abundances. With protein turnover information in hand, one can now distinguish pretranslational from posttranslational events and benefit from a more dynamic view of protein abundance changes, beyond the static transcriptomics and proteomics snapshots.

Overall, today's available and emerging nutrigenomics tools further pave the way toward increasingly mechanism-based nutritional assessments both on the food and host side, to eventually deliver personalized dietary solutions at protein level and beyond.

REFERENCES

1. Lindon JC, Nicholson JK. Spectroscopic and statistical techniques for information recovery in metabonomics and metabolomics. *Annu Rev Anal Chem* 2008;**1**:45–69.
2. Smith BK, Andrews PK, West DB. Macronutrient diet selection in thirteen mouse strains. *Am J Physiol Regul Integr Comp Physiol* 2000;**278**:R797–805.
3. Parekh PI, Petro AE, Tiller JM, Feinglos MN, Surwit RS. Reversal of diet-induced obesity and diabetes in C57BL/6J mice. *Metabolism* 1998;**47**:1089–96.
4. Gelegen C, van den Heuvel J, Collier DA, Campbell IC, Oppelaar H, Hessel E. Dopaminergic and brain-derived neurotrophic factor signalling in inbred mice exposed to a restricted feeding schedule. *Genes Brain Behav* 2008;**7**:552–9.
5. Summerbell CD, Douthwaite W, Whittaker V, Ells LJ, Hillier F, Smith S, et al. The association between diet and physical activity and subsequent excess weight gain and obesity assessed at 5 years of age or older: a systematic review of the epidemiological evidence. *Int J Obes* 2009;**33**: S1–S92.
6. Westerterp-Plantenga MS, Lejeune MP, Nijs I, van OM, Kovacs EM. High protein intake sustains weight maintenance after body weight loss in humans. *Int J Obes Relat Metab Disord* 2004;**28**:57–64.

7. Westerterp-Plantenga MS, Rolland V, Wilson SA, Westerterp KR. Satiety related to 24 h diet-induced thermogenesis during high protein/carbohydrate vs high fat diets measured in a respiration chamber. *Eur J Clin Nutr* 1999;**53**:495–502.
8. Raben A, Agerholm-Larsen L, Flint A, Holst JJ, Astrup A. Meals with similar energy densities but rich in protein, fat, carbohydrate, or alcohol have different effects on energy expenditure and substrate metabolism but not on appetite and energy intake. *Am J Clin Nutr* 2003;**77**:91–100.
9. Lejeune MP, Kovacs EM, Westerterp-Plantenga MS. Additional protein intake limits weight regain after weight loss in humans. *Br J Nutr* 2005;**93**:281–9.
10. Layman DK, Boileau RA, Erickson DJ, Painter JE, Shiue H, Sather C, et al. A reduced ratio of dietary carbohydrate to protein improves body composition and blood lipid profiles during weight loss in adult women. *J Nutr* 2003;**133**:411–7.
11. Rubio Aliaga I, Marvin-Guy LF, Wang P, Wagnière S, Mansourian R, Fuerholz A, et al. Mechanisms of weight maintenance under high- and low-protein, low-glycaemic index diets. *Mol Nutr Food Res* 2011;**55**:1603–12.
12. Larsen TM, Dalskov SM, van BM, Jebb SA, Papadaki A, Pfeiffer AF, et al. Diets with high or low protein content and glycemic index for weight-loss maintenance. *N Engl J Med* 2010;**363**:2102–13.
13. Wang P, Holst C, Andersen MR, Astrup A, Bouwman FG, van OS, et al. Blood profile of proteins and steroid hormones predicts weight change after weight loss with interactions of dietary protein level and glycemic index. *PLoS One* 2011;**6**:e16773.
14. Rigat B, Hubert C, Alhenc-Gelas F, Cambien F, Corvol P, Soubrier F. An insertion/deletion polymorphism in the angiotensin I-converting enzyme gene accounting for half the variance of serum enzyme levels. *J Clin Invest* 1990;**86**:1343–6.
15. Rankinen T, Zuberi A, Chagnon YC, Weisnagel SJ, Argyropoulos G, Walts B, et al. The human obesity gene map: the 2005 update. *Obesity (Silver Spring)* 2006;**14**:529–644.
16. Kramer H, Wu X, Kan D, Luke A, Zhu X, Adeyemo A, et al. Angiotensin-converting enzyme gene polymorphisms and obesity: an examination of three black populations. *Obes Res* 2005;**13**:823–8.
17. Strazzullo P, Iacone R, Iacoviello L, Russo O, Barba G, Russo P, et al. Genetic variation in the renin-angiotensin system and abdominal adiposity in men: the Olivetti Prospective Heart Study. *Ann Intern Med* 2003;**138**:17–23.
18. Wang JG, He X, Wang GL, Li Y, Zhou HF, Zhang WZ, et al. Family-based associations between the angiotensin-converting enzyme insertion/deletion polymorphism and multiple cardiovascular risk factors in Chinese. *J Hypertens* 2004;**22**:487–91.
19. Astle J, Ferguson JT, German JB, Harrigan GG, Kelleher NL, Kodadek T, et al. Characterization of proteomic and metabolomic responses to dietary factors and supplements. *J Nutr* 2007;**137**:2787–93.
20. Newgard CB, An J, Bain JR, Muehlbauer MJ, Stevens RD, Lien LF, et al. A branched-chain amino acid-related metabolic signature that differentiates obese and lean humans and contributes to insulin resistance. *Cell Metab* 2009;**9**:311–26.
21. Huffman KM, Shah SH, Stevens RD, Bain JR, Muehlbauer M, Slentz CA, et al. Relationships between circulating metabolic intermediates and insulin action in overweight to obese, inactive men and women. *Diabetes Care* 2009;**32**:1678–83.
22. Tai ES, Tan ML, Stevens RD, Low YL, Muehlbauer MJ, Goh DL, et al. Insulin resistance is associated with a metabolic profile of altered protein metabolism in Chinese and Asian-Indian men. *Diabetologia* 2010;**53**:757–67.
23. Laferrere B, Reilly D, Arias S, Swerdlow N, Gorroochurn P, Bawa B, et al. Differential metabolic impact of gastric bypass surgery versus dietary intervention in obese diabetic subjects despite identical weight loss. *Sci Transl Med* 2011;**3**:80re2.

24. Wang TJ, Larson MG, Vasan RS, Cheng S, Rhee EP, McCabe E, et al. Metabolite profiles and the risk of developing diabetes. *Nat Med* 2011;**17**:448–53.
25. Krebs M, Krssak M, Bernroider E, Anderwald C, Brehm A, Meyerspeer M, et al. Mechanism of amino acid-induced skeletal muscle insulin resistance in humans. *Diabetes* 2002; **51**:599–605.
26. Tremblay F, Lavigne C, Jacques H, Marette A. Role of dietary proteins and amino acids in the pathogenesis of insulin resistance. *Annu Rev Nutr* 2007;**27**:293–310.
27. Endo Y, Fu Z, Abe K, Arai S, Kato H. Dietary protein quantity and quality affect rat hepatic gene expression. *J Nutr* 2002;**132**:3632–7.
28. Anderson JW, Johnstone BM, Cook-Newell ME. Meta-analysis of the effects of soy protein intake on serum lipids. *N Engl J Med* 1995;**333**:276–82.
29. Tachibana N, Matsumoto I, Fukui K, Arai S, Kato H, Abe K, et al. Intake of soy protein isolate alters hepatic gene expression in rats. *J Agric Food Chem* 2005;**53**:4253–7.
30. Torre-Villalvazo I, Tovar AR, Ramos-Barragan VE, Cerbon-Cervantes MA, Torres N. Soy protein ameliorates metabolic abnormalities in liver and adipose tissue of rats fed a high fat diet. *J Nutr* 2008;**138**:462–8.
31. Ascencio C, Torres N, Isoard-Acosta F, Gomez-Perez FJ, Hernandez-Pando R, Tovar AR. Soy protein affects serum insulin and hepatic SREBP-1 mRNA and reduces fatty liver in rats. *J Nutr* 2004;**134**:522–9.
32. Mullen E, Brown RM, Osborne TF, Shay NF. Soy isoflavones affect sterol regulatory element binding proteins (SREBPs) and SREBP-regulated genes in HepG2 cells. *J Nutr* 2004; **134**:2942–7.
33. Shukla A, Brandsch C, Bettzieche A, Hirche F, Stangl GI, Eder K. Isoflavone-poor soy protein alters the lipid metabolism of rats by SREBP-mediated down-regulation of hepatic genes. *J Nutr Biochem* 2007;**18**:313–21.
34. Takahashi Y, Ide T. Effects of soy protein and isoflavone on hepatic fatty acid synthesis and oxidation and mRNA expression of uncoupling proteins and peroxisome proliferator-activated receptor gamma in adipose tissues of rats. *J Nutr Biochem* 2008;**19**:682–93.
35. Gianazza E, Eberini I, Arnoldi A, Wait R, Sirtori CR. A proteomic investigation of isolated soy proteins with variable effects in experimental and clinical studies. *J Nutr* 2003;**133**:9–14.
36. Adams MR, Golden DL, Franke AA, Potter SM, Smith HS, Anthony MS. Dietary soy beta-conglycinin (7S globulin) inhibits atherosclerosis in mice. *J Nutr* 2004;**134**:511–6.
37. Frigolet ME, Torres N, Uribe-Figueroa L, Rangel C, Jimenez-Sanchez G, Tovar AR. White adipose tissue genome wide-expression profiling and adipocyte metabolic functions after soy protein consumption in rats. *J Nutr Biochem* 2011;**22**:118–29.
38. Fong BY, Norris CS, Palmano KP. Fractionation of bovine whey proteins and characterisation by proteomic techniques. *Int Dairy J* 2008;**18**:23–46.
39. Gao L, Wang A, Li X, Dong K, Wang K, Appels R, et al. Wheat quality related differential expressions of albumins and globulins revealed by two-dimensional difference gel electrophoresis (2-D DIGE). *J Proteomics* 2009;**73**:279–96.
40. Sakata K, Ohyanagi H, Nobori H, Nakamura T, Hashiguchi A, Nanjo Y, et al. Soybean proteome database: a data resource for plant differential omics. *J Proteome Res* 2009;**8**:3539–48.
41. Moller NP, Scholz-Ahrens KE, Roos N, Schrezenmeir J. Bioactive peptides and proteins from foods: indication for health effects. *Eur J Nutr* 2008;**47**:171–82.
42. Minkiewicz P, Dziuba J, Iwaniak A, Dziuba M, Darewicz M. BIOPEP database and other programs for processing bioactive peptide sequences. *J AOAC Int* 2008;**91**:965–80.
43. Schlimme E, Meisel H. Bioactive peptides derived from milk proteins. Structural, physiological and analytical aspects. *Nahrung* 1995;**39**:1–20.
44. Meisel H, Bockelmann W. Bioactive peptides encrypted in milk proteins: proteolytic activation and thropho-functional properties. *Antonie Van Leeuwenhoek* 1999;**76**:207–15.

45. Grigorov MG, van Bladeren PJ. Functional peptides by genome reverse engineering. *Curr Opin Drug Discov Devel* 2007;**10**:341–6.
46. German JB, Morgan CJ, Ward RE. Milk: a model for nutrition in the 21st century. *Aust J Dairy Technol* 2003;**58**:49–54.
47. Severin S, Xia W. Milk biologically active components as nutraceuticals: review. *Crit Rev Food Sci Nutr* 2005;**45**:645–56.
48. German JB, Dillard CJ. Composition, structure and absorption of milk lipids: a source of energy, fat-soluble nutrients and bioactive molecules. *Crit Rev Food Sci Nutr* 2006;**46**:57–92.
49. Ninonuevo MR, Youmie P, Hongfeng Y, Jinhua Z, Ward RE, Clowers BH, et al. A strategy for annotating the human milk glycome. *J Agric Food Chem* 2006;**54**:7471–80.
50. Casado B, Affolter M, Kussmann M. OMICS-rooted studies of milk proteins, oligosaccharides and lipids. *J Proteomics* 2009;**73**:196–208.
51. Madureira AR, Pereira CI, Gomes A-MP, Pintado ME, Malcata FX. Bovine whey proteins—overview on their main biological properties. *Food Res Int* 2007;**40**:1197–211.
52. Farnaud S, Evans RW. Lactoferrin—a multifunctional protein with antimicrobial properties. *Mol Immunol* 2003;**40**:395–405.
53. Crouch SP, Slater KJ, Fletcher J. Regulation of cytokine release from mononuclear cells by the iron-binding protein lactoferrin. *Blood* 1992;**80**:235–40.
54. Levy J. Immunonutrition: the pediatric experience. *Nutrition* 1998;**14**:641–7.
55. Labeta MO, Vidal K, Nores JE, Arias M, Vita N, Morgan BP, et al. Innate recognition of bacteria in human milk is mediated by a milk-derived highly expressed pattern recognition receptor, soluble CD14. *J Exp Med* 2000;**191**:1807–12.
56. Vidal K, Labeta MO, Schiffrin EJ, Donnet-Hughes A. Soluble CD14 in human breast milk and its role in innate immune responses. *Acta Odontol Scand* 2001;**59**:330–4.
57. LeBouder E, Rey-Nores JE, Rushmere NK, Grigorov M, Lawn SD, Affolter M, et al. Soluble forms of Toll-like receptor (TLR)2 capable of modulating TLR2 signaling are present in human plasma and breast milk. *J Immunol* 2003;**171**:6680–9.
58. Hayes M, Stanton C, Fitzgerald GF, Ross RP. Putting microbes to work: dairy fermentation, cell factories and bioactive peptides. Part II: bioactive peptide functions. *Biotechnol J* 2007;**2**:435–49.
59. Hebert EM, Mamone G, Picariello G, Raya RR, Savoy G, Ferranti P, et al. Characterization of the pattern of alphas1- and beta-casein breakdown and release of a bioactive peptide by a cell envelope proteinase from Lactobacillus delbrueckii subsp. lactis CRL 581. *Appl Environ Microbiol* 2008;**74**:3682–9.
60. Aimutis WR. Bioactive properties of milk proteins with particular focus on anticariogenesis. *J Nutr* 2004;**134**:989S–95S.
61. Panchaud A, Affolter M, Moreillon P, Kussmann M. Experimental and computational approaches to quantitative proteomics: status quo and outlook. *J Proteomics* 2008;**71**:19–33.
62. Kussmann M, Panchaud A, Affolter M. Proteomics in nutrition: status quo and outlook for biomarkers and bioactives. *J Proteome Res* 2010;**9**:4876–87.
63. Waterlow JC. *Protein turnover.* Cambridge: CABI; 2006.
64. Doherty MK, Beynon RJ. Protein turnover on the scale of the proteome. *Expert Rev Proteomics* 2006;**3**:97–110.
65. Duggleby SL, Waterlow JC. The end-product method of measuring whole-body protein turnover: a review of published results and a comparison with those obtained by leucine infusion. *Br J Nutr* 2005;**94**:141–53.
66. Sprinson DB, Rittenberg D. The rate of interaction of the amino acids of the diet with the tissue proteins. *J Biol Chem* 1949;**180**:715–26.
67. Fern EB, Garlick PJ, McNurlan MA, Waterlow JC. The excretion of isotope in urea and ammonia for estimating protein turnover in man with [15N]glycine. *Clin Sci (Lond)* 1981;**61**:217–28.

68. Matthews DE, Motil KJ, Rohrbaugh DK, Burke JF, Young VR, Bier DM. Measurement of leucine metabolism in man from a primed, continuous infusion of L-[1-13C]leucine. *Am J Physiol* 1980;**238**:E473–9.
69. Thompson GN, Pacy PJ, Merritt H, Ford GC, Read MA, Cheng KN, et al. Rapid measurement of whole body and forearm protein turnover using a [2H5]phenylalanine model. *Am J Physiol* 1989;**256**:E631–9.
70. Hasten DL, Pak-Loduca J, Obert KA, Yarasheski KE. Resistance exercise acutely increases MHC and mixed muscle protein synthesis rates in 78-84 and 23-32 yr olds. *Am J Physiol Endocrinol Metab* 2000;**278**:E620–6.
71. Garlick PJ, McNurlan MA, Preedy VR. A rapid and convenient technique for measuring the rate of protein synthesis in tissues by injection of [3H]phenylalanine. *Biochem J* 1980; **192**:719–23.
72. Davis TA, Reeds PJ. Of flux and flooding: the advantages and problems of different isotopic methods for quantifying protein turnover in vivo: II. Methods based on the incorporation of a tracer. *Curr Opin Clin Nutr Metab Care* 2001;**4**:51–6.
73. Davis TA, Fiorotto ML, Nguyen HV, Burrin DG. Aminoacyl-tRNA and tissue free amino acid pools are equilibrated after a flooding dose of phenylalanine. *Am J Physiol Endocrinol Metab* 1999;**277**:E103–9.
74. Rennie MJ. An introduction to the use of tracers in nutrition and metabolism. *Proc Nutr Soc* 1999;**58**:935–44.
75. Smith K, Reynolds N, Downie S, Patel A, Rennie MJ. Effects of flooding amino acids on incorporation of labeled amino acids into human muscle protein. *Am J Physiol* 1998;**275**:E73–8.
76. Southorn BG, Kelly JM, McBride BW. Phenylalanine flooding dose procedure is effective in measuring intestinal and liver protein synthesis in sheep. *J Nutr* 1992;**122**:2398–407.
77. Walrand S, Boirie Y. Optimizing protein intake in aging. *Curr Opin Clin Nutr Metab Care* 2005;**8**:89–94.
78. Evans WJ. Protein nutrition, exercise and aging. *J Am Coll Nutr* 2004;**23**:601S–9S.
79. Koopman R, van Loon LJ. Aging, exercise, and muscle protein metabolism. *J Appl Physiol* 2009;**106**:2040–8.
80. Murphy C, Miller BF. Protein consumption following aerobic exercise increases whole-body protein turnover in older adults. *Appl Physiol Nutr Metab* 2010;**35**:583–90.
81. Mansoor O, Breuille D, Bechereau F, Buffiere C, Pouyet C, Beaufrere B, et al. Effect of an enteral diet supplemented with a specific blend of amino acid on plasma and muscle protein synthesis in ICU patients. *Clin Nutr* 2007;**26**:30–40.
82. Baier S, Johannsen D, Abumrad N, Rathmacher JA, Nissen S, Flakoll P. Year-long changes in protein metabolism in elderly men and women supplemented with a nutrition cocktail of beta-hydroxy-beta-methylbutyrate (HMB), L-arginine, and L-lysine. *JPEN J Parenter Enteral Nutr* 2009;**33**:71–82.
83. Morais JA, Ross R, Gougeon R, Pencharz PB, Jones PJ, Marliss EB. Distribution of protein turnover changes with age in humans as assessed by whole-body magnetic resonance image analysis to quantify tissue volumes. *J Nutr* 2000;**130**:784–91.
84. Godin JP, Fay LB, Hopfgartner G. Liquid chromatography combined with mass spectrometry for 13C isotopic analysis in life science research. *Mass Spectrom Rev* 2007;**26**:751–74.
85. Montigon F, Boza JJ, Fay LB. Determination of 13C- and 15N-enrichment of glutamine by gas chromatography/mass spectrometry and gas chromatography/combustion/isotope ratio mass spectrometry after N(O, S)-ethoxycarbonyl ethyl ester derivatisation. *Rapid Commun Mass Spectrom* 2001;**15**:116–23.
86. Beynon RJ, Doherty MK, Pratt JM, Gaskell SJ. Multiplexed absolute quantification in proteomics using artificial QCAT proteins of concatenated signature peptides. *Nat Methods* 2005; **2**:587–9.

87. Gygi SP, Rist B, Gerber SA, Turecek F, Gelb MH, Aebersold R. Quantitative analysis of complex protein mixtures using isotope-coded affinity tags. *Nat Biotechnol* 1999;**17**:994–9.
88. Gerber SA, Rush J, Stemman O, Kirschner MW, Gygi SP. Absolute quantification of proteins and phosphoproteins from cell lysates by tandem MS. *Proc Natl Acad Sci USA* 2003; **100**:6940–5.
89. Gygi SP, Rochon Y, Franza BR, Aebersold R. Correlation between protein and mRNA abundance in yeast. *Mol Cell Biol* 1999;**19**:1720–30.
90. Pratt JM, Petty J, Riba-Garcia I, Robertson DH, Gaskell SJ, Oliver SG, et al. Dynamics of protein turnover, a missing dimension in proteomics. *Mol Cell Proteomics* 2002;**1**:579–91.
91. Cargile BJ, Bundy JL, Grunden AM, Stephenson Jr. JL. Synthesis/degradation ratio mass spectrometry for measuring relative dynamic protein turnover. *Anal Chem* 2004;**76**:86–97.
92. Beynon RJ, Pratt JM. Metabolic labeling of proteins for proteomics. *Mol Cell Proteomics* 2005; **4**:857–72.
93. Doherty MK, Whitehead C, McCormack H, Gaskell SJ, Beynon RJ. Proteome dynamics in complex organisms: using stable isotopes to monitor individual protein turnover rates. *Proteomics* 2005;**5**:522–33.
94. Doherty MK, Hammond DE, Clague MJ, Gaskell SJ, Beynon RJ. Turnover of the human proteome: determination of protein intracellular stability by dynamic SILAC. *J Proteome Res* 2009;**8**:104–12.
95. Jaleel A, Nehra V, Persson XM, Boirie Y, Bigelow M, Nair KS. In vivo measurement of synthesis rate of multiple plasma proteins in humans. *Am J Physiol Endocrinol Metab* 2006;**291**:E190–7.
96. Boirie Y, Fauquant J, Rulquin H, Maubois JL, Beaufrere B. Production of large amounts of [13C]leucine-enriched milk proteins by lactating cows. *J Nutr* 1995;**125**:92–8.
97. Jaleel A, Short KR, Asmann YW, Klaus KA, Morse DM, Ford GC, et al. In vivo measurement of synthesis rate of individual skeletal muscle mitochondrial proteins. *Am J Physiol Endocrinol Metab* 2008;**295**:E1255–68.
98. Zhang Y, Reckow S, Webhofer C, Boehme M, Gormanns P, Egge-Jacobsen WM, et al. Proteome scale turnover analysis in live animals using stable isotope metabolic labeling. *Anal Chem* 2011;**83**:665–1672.

Nutrigenomics and Nutrigenetics of ω3 Polyunsaturated Fatty Acids

John P. Vanden Heuvel

Department of Veterinary and Biomedical Sciences and Center for Excellence in Nutrigenomics, Penn State University, University Park, Pennsylvania, USA

Indigo Biosciences, Inc., State College, Pennsylvania, USA

Diets rich in ω3 polyunsaturated fatty acids (ω3-PUFAs) such as alpha-linolenic acid, eicosapentaenoic acid, and docosahexaenoic acid are associated with decreased incidence and severity of several chronic diseases including cardiovascular disease (CVD) and cancer. At least some of the beneficial effects of these dietary fatty acids are via metabolites such as prostaglandins, leukotrienes, thromboxanes, and resolvins. The effects of ω3-PUFAs are in contrast to those of fatty acids with virtually identical structures, such as the ω6-PUFAs linoleic acid and arachidonic acid, and their corresponding metabolites. The purpose of this chapter is to discuss both the nutrigenomics (nutrient–gene interactions) and nutrigenetics (genetic variation in nutrition) of dietary fatty acids with a focus on the ω3-PUFAs (Gebauer *et al.*, 2007[1]). Important in the biological response for these fatty acids or their metabolites are cognate receptors that are able to regulate gene expression and coordinately affect metabolic or signaling pathways associated with CVD and cancer. Four nuclear receptor (NR) subfamilies will be emphasized as receptors that respond to dietary and endogenous ligands: (1) peroxisome proliferator-activated receptors, (2) retinoid X receptors, (3) liver X receptors, and (4) farnesoid X receptor. In

Progress in Molecular Biology
and Translational Science, Vol. 108
DOI: 10.1016/B978-0-12-398397-8.00004-6

1877-1173/12 $35.00

addition to the different responses elicited by varying structures of fatty acids, responses may vary because of genetic variation in enzymes that metabolize ω3- and ω6 fatty acids or that respond to them. In particular, polymorphisms in the fatty acid desaturases and the aforementioned NRs contribute to the complexity of nutritional effects seen with ω3-PUFAs.

Following a brief introduction to the health benefits of ω3-PUFAs, the regulation of gene expression by these dietary fatty acids via NRs will be characterized. Subsequently, the effects of single-nucleotide polymorphisms (SNPs) in key enzymes involved in the metabolism and response to ω3-PUFAs will be described. An outline of the events to be explored is shown in Fig. 1. Understanding the nutrigenomics and nutrigenetics of dietary fatty acids is key to understanding the etiology, as well as prevention, of critically important human diseases including CVD and cancer.

I. Dietary Fatty Acids and Health

Polyunsaturated fatty acids (PUFAs) that cannot be made in the body are known as essential fatty acids, and two of these are linoleic acid (LA, ω6) and alpha-linolenic acid (ALA, ω3) (Fig. 2). In some conditions, such as LA deficiency, arachidonic acid (AA) is also considered essential. Once in the body, LA and ALA may be converted to other PUFAs such as AA, eicosapentaenoic acid (EPA), and docosahexaenoic acid (DHA) (see Fig. 2). Although some fats have been associated with increased risk of disease (saturated and *trans*-fatty acids), EPA and DHA have been associated with a variety of beneficial health effects as described in this chapter. An important question is, why are some PUFAs, in particular ω3-PUFAs (ALA, EPA, DHA), associated with reduced risk of disease, while the closely related ω6 (LA, AA) and saturated fats (palmitic acid) are either not as effective in reducing risk or are detrimental to heart health? One explanation may be that ω3-PUFAs are metabolized to products that have more favorable effects than the corresponding ω6-PUFAs, in addition to the fact that the ω3-PUFAs can interfere with the metabolism of ω6-PUFAs. Thus, diets rich in ω3-PUFAs would support lower levels of metabolites associated with platelet aggregation, inflammation, and vasoconstriction (leukotriene B_4 [LTB_4], prostaglandin I_2 [PGI_2], thromboxane A_2 [TXA_2]) at the expense of metabolites with anti-aggregation, anti-inflammation, and anti-vasoconstriction properties (LTB5, PGI_3, TXA_3). Another explanation, and the option explored herein, is that some cognate receptors preferentially respond to a particular structure of fatty acid or its metabolites.[2,3] These specific "lipid sensors" would affect gene expression in a tissue-, sex-, and developmentally specific manner and

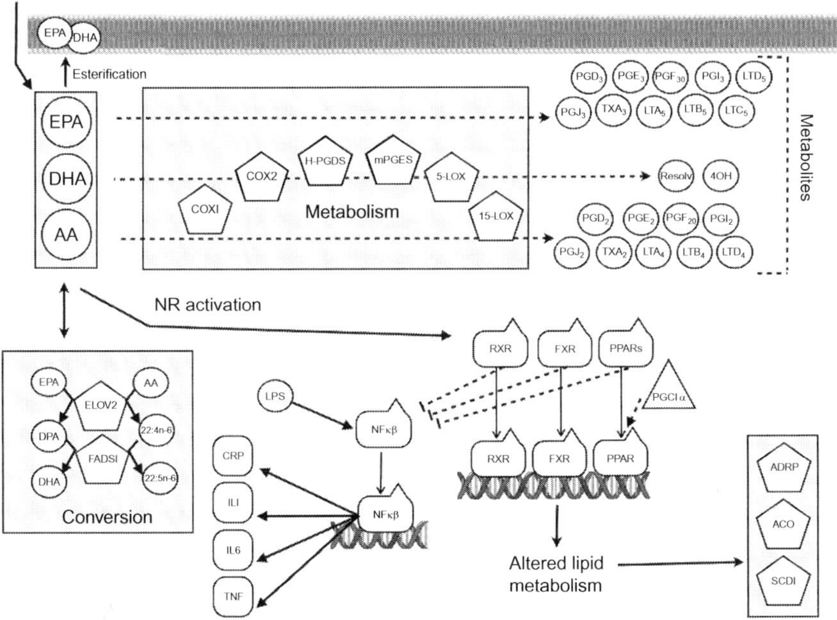

FIG. 1. Nutrigenomics and nutrigenetics of dietary fatty acids. Dietary fatty acids and their metabolites alter gene expression, in part through activation of nuclear receptors, ultimately resulting in altered lipid metabolism and decreased inflammation. Many of the enzymes involved in metabolism of dietary fatty acids, the nuclear receptors that affect responsiveness, and the molecules that elicit the beneficial effects are genetically polymorphic in the human population. *Abbreviations*: 4OH, 4-hydroxy DHA; AA, arachidonic acid; ACO, acyl-CoA oxidase; ADRP, adipose differentiation related protein; COX, cyclooxygenase; CRP, C-reactive protein; DHA, docosahexaenoic acid; DPA, docosapentaenoic acid; ELOV, elongase; EPA, eicosapentaenoic acid; FADS, fatty acid desaturase; FXR, farnesoid X receptor; IL1, interleukin 1; IL6, interleukin 6; LOX, lipoxygenase; LPS, lipopolysaccharide; LT, leukotriene; mPGES, mitochondrial prostaglandin E synthase; NFκB: nuclear factor-κB; NR: nuclear receptor; PG, prostaglandin; PGC1α, PPARγ coactivator-1α; PGDS, prostaglandin D synthase; PPAR, peroxisome proliferator-activated receptor; Resolv, resolvin; RXR, retinoic X receptor; SCD1, stearoyl-CoA desaturase 1; TNF, tumor necrosis factor; TX, thromboxane. (For color version of this figure, the reader is referred to the online version of this chapter.)

thereby affect the development of diseases such as cardiovascular disease (CVD) or cancer. Also, in order for these receptors to be involved in the beneficial effects of dietary fatty acids, they must be able to distinguish subtle differences in physical structure between the "good lipids" and "bad lipids" such as between ω3 and ω6, or between PGI_3 and PGI_2.

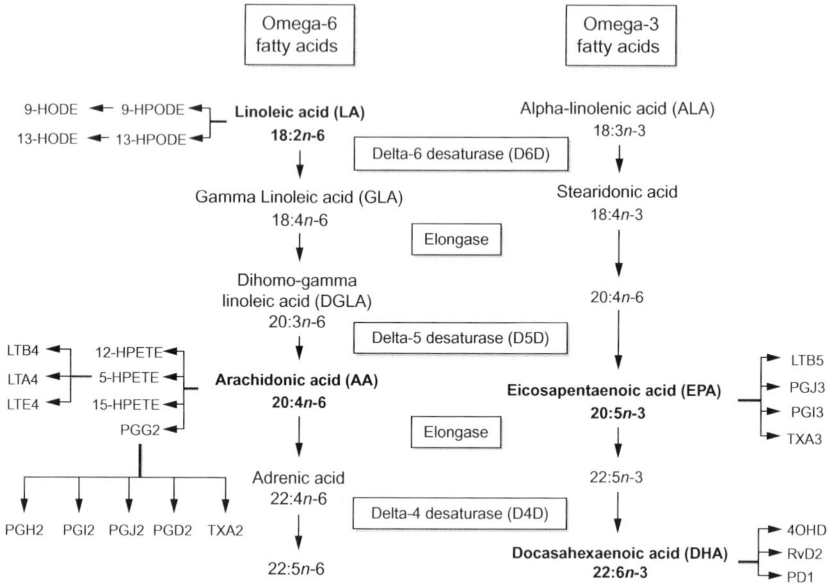

FIG. 2. Metabolism of ω6 and ω3 fatty acids. *Abbreviations*: 4OHD, 4-hydroxy DHA; LT, leukotriene; PG, prostaglandin; TX, thromboxane; HPODE, hydroxyperoxyoctadecadienoic acid; HODE, hydroxyoctadecadienoic acid; HPETE, hydroperoxyeicosatetraenoic acid; Rv, resolvin; PD, protectin.

A. Cardiovascular Disease

CVD is the leading cause of death in industrialized countries and is of rising concern worldwide. The relationship between CVD and diet has been studied for nearly 100 years, essentially since the first observation of high-fat and high-cholesterol diets producing atherosclerosis in rabbits.[4,5] Epidemiological studies have demonstrated that diets high in saturated fatty acids and/or cholesterol increase serum cholesterol and the risk of developing CVD. Correlations between diet and incidence of CVD across geographical boundaries and among emigrants have also been noted. These discoveries have led to the *diet–heart hypothesis*, which suggests that dietary saturated fat and cholesterol are the major causes of CVD and atherosclerosis in humans.[6] Although dietary fat has dominated the diet–heart hypothesis, there are many other foodstuffs and nutrients that may be involved in the etiology of CVD. Fiber, antioxidants, folic acid, calcium, and the carbohydrate content of food have an impact on heart disease and atherosclerosis as well.[4]

The type of fat in the diet, in particular the saturation of the fatty acid component, dramatically impacts the risk of developing several diseases in humans. For example, all three major classes of fatty acids (saturated, monounsaturated, and polyunsaturated) increase high-density lipoprotein (HDL) cholesterol in humans; however, saturated fatty acids increase and PUFAs decrease low-density lipoprotein (LDL) cholesterol.[7] The increased LDL-to-HDL ratio caused by saturated fat consumption is associated with an increased risk of developing CVD. Saturated fatty acids are generally considered to be atherogenic and to increase thrombosis.[8] *Trans*-fatty acids, found in vegetable shortening and deep-fried foods, raise LDL-to-HDL ratios to a much greater degree than saturated fat.[9] One potential mechanism by which *trans*-fats adversely affect insulin resistance, diabetes, and CVD risk is by inhibiting essential fatty acid metabolism.

PUFAs in general are important for maintaining membrane integrity and for providing precursors to bioactive prostaglandins that regulate inflammation, blood clotting, and lipid metabolism. Thus, it is necessary to have diets sufficient in PUFAs (ω3 and ω6) to maintain a variety of biological processes. Positive effects of diets high in ω3-PUFAs include decreased abdominal fat, cardiac arrhythmia, serum triacylglycerol (TAG), and thrombosis, as well as improved endothelial function. As noted by Hu and Willett,[10] several studies have shown an association between fish (high in EPA and DHA) and/or flaxseed oil (high in ALA) intake and decreased fatalities from CVD. Importantly, blood levels of EPA and DHA are strongly associated with decreased risk of death, myocardial infarction, and stroke.

B. Inflammatory Diseases

Chronic inflammation is a contributor to many human diseases including CVD. ω3-PUFAs play an important role in the regulation of inflammation by decreasing the production of inflammatory eicosanoids, cytokines, and reactive oxygen species and the expression of adhesion molecules (for review, see Refs. 11,12). Supplementation with EPA and DHA has proven effective in decreasing intestinal damage and improving gut histology in inflammatory bowel disease.[13,14] In other inflammatory conditions, fish oil supplementation decreases joint pain, number of tender and swollen joints, duration of morning stiffness, and, as a result, use of nonsteroidal anti-inflammatory drugs.[15]

C. Cancer

The association between ω3-PUFAs and cancer prevention has been an active area of research (e.g., Refs. 16–19), with breast cancer being one of the cancers most studied. For many years, high fat intake was considered the decisive explanation for the regional differences in breast cancer incidence. However, the results from many prospective cohort studies have not shown a general association between fat consumption *per se* and breast cancer risk[20,21] and have suggested that dietary sources and types of fat may be more important in cancer prevention. Correlational and emigrational epidemiology studies suggest a protective effect of dietary ω3-PUFAs and a promoting effect of ω6-PUFAs on breast cancer.[17,22] However, the results of such studies are mixed, in part because of the complex composition of dietary fat (source of ω3, ω3-to-ω6 ratio) and the presence of other bioactive molecules in the diet.[23,24]

Several review articles and meta-analyses have highlighted the benefits of marine-based ω3-PUFAs EPA and DHA in breast cancer prevention (e.g., Refs. 17,20,25,26). But much less is known about the plant-derived ALA (C18:3*n*-3), the most abundant of the ω3-PUFAs. The major dietary sources of this fatty acid include plant oils, seeds (flaxseed, canola, and perilla), dark leafy vegetables, and tree nuts, particularly walnuts.[17] As mentioned above, while mammals cannot synthesize ω3- or ω6-PUFAs *de novo*, mammalian cells can interconvert them by elongation, desaturation, and retroconversion. Support of retroconversion of ALA to other cancer-preventative ω3-PUFAs comes from our own work showing increases in membrane-associated ALA and EPA upon intervention with walnuts.[27–29] To explore the hypothesis that ALA inhibits breast cancer, two case-control studies used fatty acid levels in adipose breast tissue as a biomarker of past qualitative dietary intake of fatty acids.[30,31] Both studies showed that the relative risk of breast cancer for women in the highest quartile of ALA levels was significantly lower compared with those in the lowest quartile. This is highly suggestive of a protective effect of ALA on the risk of breast cancer in women.

D. Other Beneficial Effects

DHA plays a role in mediating the expression of at least 100 genes in the areas of neural development, function, and metabolism.[32,33] Between 1997 and 2006, 15 prospective cohort studies examined the association between ω3-PUFA intake and Alzheimer's disease; 13 showed a protective effect of increased DHA or fish intake on the risk of developing Alzheimer's disease.[34] DHA is an essential nutrient in the retina for visual acuity. Several studies have shown that preterm babies supplemented with DHA scored higher on visual acuity tests compared to those who did not receive DHA.[35] The impact of DHA on the visual acuity of nonpreterm infants is not clear because of inconsistent findings,[36] which may be due to methodological differences between studies (type, concentration, and duration of DHA supplementation; methods used for assessing outcomes).

II. Nutrigenomics

A. Transcriptional Response to ω3-PUFAs

Animal experiments and human studies have shown that ω3-PUFAs regulate genes in various tissues including adipose tissue and peripheral blood mononuclear cells (PBMCs). Several comprehensive analyses of transcription responses to ω3-PUFAs have been published, including in PBMCs following fish oil supplementation in humans,[37] in adipose tissue following a high-PUFA diet in humans[38] and mice,[39] in breast cancer cell lines treated with EPA and DHA,[40] and in colon cancer cells treated with DHA.[41] A comparison of the transcriptional responses of the human monocytic cell line THP-1 to equimolar ALA, DHA, and EPA (10 μM) is shown in Table I and Fig. 3. The results of these experiments demonstrate that the transcriptional responses to the three major ω3-PUFAs are very similar, although there are subtle qualitative and quantitative differences. Genes regulated by ALA, DHA, and EPA fall into three main ontological categories: inflammation; lipid and cholesterol metabolism; and cell differentiation and fate. The subsequent sections discuss the transcriptional responses seen in these three biological processes.

1. Inflammation

As stated earlier, clinical studies indicate that inflammation is at the root of many diseases including CVD, obesity, diabetes, and cancer. ω3-PUFAs, in particular EPA and DHA, have beneficial effects in these conditions, in part by suppressing the inflammatory response or augmenting the cellular defenses against oxidative damage. Some of the potential mechanisms by which ω3-PUFAs can induce these changes include activation of nuclear receptors (NRs) (discussed subsequently), reduction in NF-κB activity, and production of anti-inflammatory mediators such as resolvins and protectins. ALA decreases the production of inflammatory cytokines including interleukin 1 and 6 (IL1 and 6) and tumor necrosis factor-α (TNFα) in PBMCs following dietary intervention[94] and in lipopolysaccharide (LPS)-challenged THP-1 cells.[95] Similarly, DHA, EPA, or their combination in fish oil decreases these markers *in vivo*[37,96] and *in vitro*.[97] Other proinflammatory markers decreased by ω3-PUFA intervention include C-reactive protein (CRP),[98] intercellular adhesion molecule 1, and vascular cell adhesion molecule 1.[99] An increase has been seen in the expression of genes involved in cellular defense to oxidative stress, including heme oxygenase 1 (*HMOX1*)[100]; superoxide dismutase, extracellular (*SOD3*)[101]; and genes encoding glutathione transferases.[102]

2. Lipid and Cholesterol Metabolism

One of the most consistent effects of ω3-PUFA dietary supplementation is a lowering of circulating triglyceride levels.[103–106] In fact, this decrease is often seen in the absence of an anti-inflammatory response.[107] The predominant target tissue for the effects on lipid levels is the liver, although effects on the adipose tissue are also of importance.[108] In general, this benefit of ω3-PUFAs mainly results from a combination of decreased expression of lipogenesis-related genes and stimulation of fatty acid oxidation transcripts. Fish oil decreases the expression of the gene encoding sterol regulatory element-binding

TABLE I

Gene Regulated by ω3-PUFAs in THP-1 Cells

Gene symbol	Gene title	Fold change			Gene ontology biological process
		ALA	DHA	EPA	
TBC1D30	TBC1 domain family, member 30	3.6	1.5	1.3	Regulation of Rab GTPase activity
CALR	Calreticulin	1.7	2.4	2.9	Negative regulation of transcription from RNA polymerase II promoter
YWHAE	Tyrosine 3-monooxygenase/tryptophan 5-monooxygenase activation protein, epsilon polypeptide	1.4	2.1	2.7	Neuron migration
EIF2S3	Eukaryotic translation initiation factor 2, subunit 3 gamma, 52 kDa	1.2	1.4	2.6	Translation
EIF5A	Eukaryotic translation initiation factor 5A	1.4	3.1	3.4	mRNA export from nucleus
NUTF2	Nuclear transport factor 2	1.2	3.2	3.3	Protein import into nucleus
HNRNPL	Heterogeneous nuclear ribonucleoprotein L	1.2	3.1	2.9	Nuclear mRNA splicing processing
C9orf38	Chromosome 9 open reading frame 38	1.1	−2.9	−1.5	
ZNF117	Zinc finger protein 117	−1.1	−2.7	−1.5	Transcription
DHX9	DEAH (Asp-Glu-Ala-His) box polypeptide 9	−1.4	−3.1	−1.6	Nuclear mRNA splicing, via spliceosome
PLK4	Polo-like kinase 4	−1.4	−.2	−1.6	Protein amino acid phosphorylation
ABCB1	ATP-binding cassette, subfamily B (MDR/TAP), member 1	−1.6	−2.5	−1.8	Lipid metabolic process
ABCB4	ATP-binding cassette, subfamily B (MDR/TAP), member 4	−1.6	−2.8	−2.0	Lipid metabolic process
COPA	Coatomer protein complex, subunit alpha	−1.2	−3.5	−2.3	Protein folding
TPR	translocated promoter region (to activated MET oncogene)	−1.5	−3.6	−3.1	Translation
AXL	AXL receptor tyrosine kinase	−1.8	−3.1	−3.1	Protein amino acid phosphorylation
DKK2	Dickkopf homolog 2 (*Xenopus laevis*)	−1.8	−2.8	−2.6	Multicellular organismal development
GPR137B	G-protein-coupled receptor 137B	−1.2	−1.4	−2.8	
METTL7A	Methyltransferase like 7A	−1.2	−1.4	−2.9	Metabolic process
RASA4	RAS p21 protein activator 4	−1.2	−1.5	−2.7	Signal transduction
CCNG2	Cyclin G2	−1.5	−1.4	−2.7	Cell cycle checkpoint
ATP1B1	ATPase, Na+/K+ transporting, beta 1 polypeptide	−1.2	−1.9	−2.9	Response to hypoxia

SETD2	SET domain containing 2	−1.3	−2.1	−2.7	Transcription
GPNMB	Glycoprotein (transmembrane) nmb	−1.5	−1.8	−2.8	Cell adhesion
PNPLA3	Patatin-like phospholipase domain containing 3	−1.5	−1.7	−2.8	Lipid metabolic process
ACAT2	Acetyl-coenzyme A acetyltransferase 2	−1.7	−1.9	−2.7	Lipid metabolic process
N4BP2L2-IT1	N4BP2L2 intronic transcript 1 (nonprotein coding)	−1.7	−1.8	−2.5	
SAT1	Spermidine/spermine N1-acetyltransferase 1	−1.8	−1.6	−2.6	Polyamine metabolic process
LSS	Lanosterol synthase (2,3-oxidosqualene-lanosterol cyclase)	−1.8	−1.7	−2.7	Steroid biosynthetic process
PPARD	Peroxisome proliferator-activated receptor delta	−1.9	−1.9	−2.8	Fatty acid beta-oxidation
HMGCR	3-Hydroxy-3-methylglutaryl-coenzyme A reductase	−2.0	−1.7	−2.7	Steroid biosynthetic process
ACSL3	Acyl-coenzyme A synthetase long-chain family member 3	−2.0	−1.7	−2.7	Lipid metabolic process
IDI1	Isopentenyl-diphosphate delta isomerase 1	−2.1	−1.9	−2.5	Steroid biosynthetic process
ALOX5	Arachidonate 5-lipoxygenase	−2.2	−2.1	−3.1	Leukotriene metabolic process, inflammatory response
SQLE	Squalene epoxidase	−2.3	−2.3	−2.8	Sterol biosynthetic process
CCL3	Chemokine (C–C motif) ligand 3	−2.3	−2.3	−2.6	Cellular calcium ion homeostasis
ENPP2	Ectonucleotide pyrophosphatase/phosphodiesterase 2	−2.5	−2.4	−2.5	Phosphate metabolic process
BHLHE40	Basic helix-loop-helix family, member e40	−2.5	−2.3	−2.5	Transcription
GPR183	G-protein-coupled receptor 183	−2.4	−2.6	−2.6	Immune response
KLF2	Kruppel-like factor 2 (lung)	−2.7	−1.6	−1.5	Transcription
FADS2	Fatty acid desaturase 2	−3.0	−1.9	−2.3	Lipid metabolic process
DHCR7	7-Dehydrocholesterol reductase	−2.3	−2.3	−3.7	Steroid biosynthetic process
TUBA1A	Tubulin, alpha 1a	−2.5	−2.2	−3.7	Microtubule-based process
TMEM135	Transmembrane protein 135	−2.7	−2.1	−3.9	
CYP51A1	Cytochrome P450, family 51, subfamily A, polypeptide 1	−3.0	−2.1	−4.1	Steroid biosynthetic process
LDLR	Low-density lipoprotein receptor	−2.9	−1.7	−3.7	Lipid metabolic process
FADS1	Fatty acid desaturase 1	−2.6	−2.9	−4.6	Lipid metabolic process
HMGCS1	3-Hydroxy-3-methylglutaryl-coenzyme A synthase 1 (soluble)	−2.7	−2.4	−4.3	Lipid metabolic process
MSMO1	Methylsterol monooxygenase 1	−2.7	−1.9	−5.2	Fatty acid metabolic process
INSIG1	Insulin-induced gene 1	−3.8	−2.1	−4.7	Lipid metabolic process
TIMP3	TIMP metallopeptidase inhibitor 3	−2.7	−3.2	−1.8	Protein tyrosine kinase signaling pathway
PTX3	Pentraxin 3, long	−2.8	−2.9	−1.7	Inflammatory response

(*Continues*)

TABLE I (*Continued*)

Gene symbol	Gene title	Fold change			Gene ontology biological process
		ALA	DHA	EPA	
VEGFC	Vascular endothelial growth factor C	−2.9	−3.1	−2.6	Angiogenesis
AQP1	Aquaporin 1 (Colton blood group)	−3.6	−3.3	−3.3	Transport
TNF	Tumor necrosis factor	−3.9	−2.8	−2.3	Inflammatory response
C3AR1	Complement component 3a receptor 1	−4.5	−3.1	−2.1	Inflammatory response
SCD	Stearoyl-coenzyme A desaturase (delta-9 desaturase)	−4.2	−2.6	−5.9	Lipid metabolic process
ABCA1	ATP-binding cassette, subfamily A (ABC1), member 1	−4.3	−2.4	−7.6	Transport

John P. Vanden Heuvel (unpublished data). THP-1 cells were grown under standard conditions and treated with fatty acids (10 μM, as a BSA-conjugate) or control (BSA) for 24 h. RNA was extracted using Qiagen RNeasy and quality assessed by RNA Nano Chips on the Agilent Bioanalyzer. Each sample was labeled using the Affymetrix IVT Express Kit according to the manufacturer's protocol. The GeneChip Human Genome U133A 2.0 (Affymetrix), representing 14,500 well-characterized genes, was hybridized with the labeled RNA using GeneChip Hybridization Wash and Stain Kit (#702232) in the Affymetrix GeneChip Hybridization Oven 640, according to the manufacturer's instructions. Following hybridization, the arrays were washed and stained using the Affymetrix GeneChip Fluidics Station 450 according to the manufacturer's protocol and scanned using the GeneChip Scanner 3000 7G. The scanned image file (DAT) and the intensity data (CEL) was imported into GeneSpring 10.0 (Agilent Technologies). One-way ANOVA with asymptotic p value and Benjamini–Hochberg multiple corrections was performed. In a separate analysis, unpaired t-tests were used to find genes that were significantly regulated by the fatty acid relative to vehicle control (volcano plots, $p < 0.05$ with 2.5-fold differences).

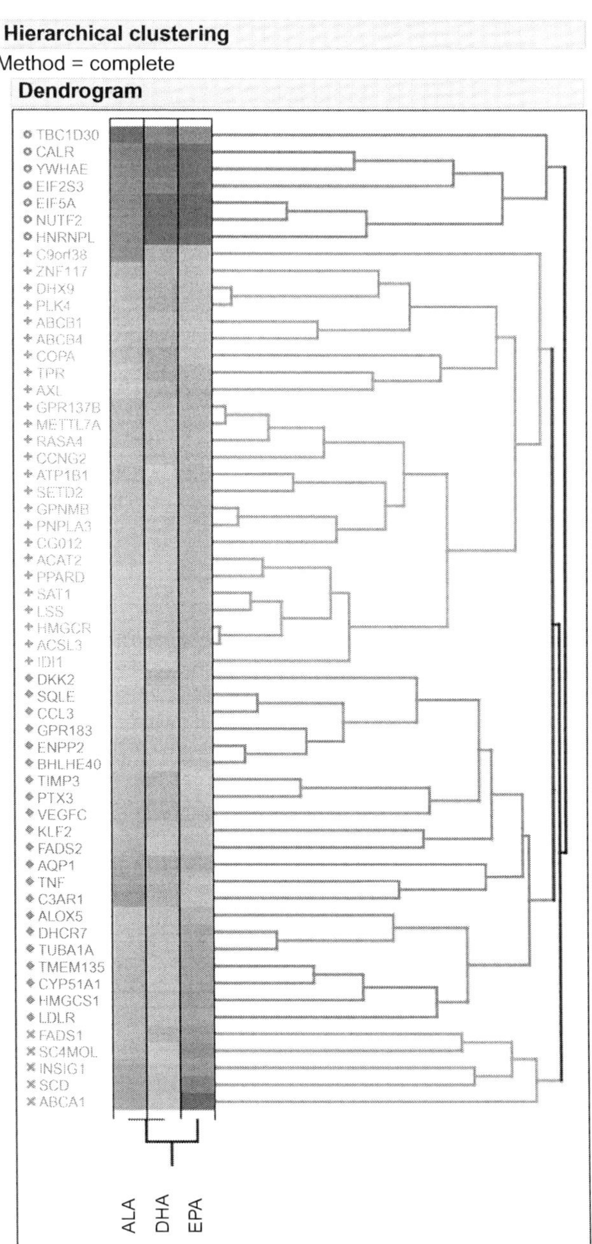

FIG. 3. Gene expression changes seen in THP-1 cells treated with ω3-PUFAs. See Table II for abbreviations. (See Color Insert.)

protein 1c (SREBP-1c),[109] a key enzyme in controlling lipogenesis. Other lipogenic genes decreased by ω3-PUFAs include fatty acid synthase (*FASN*), malic enzyme (*ME*), and glucose-6-phosphate dehydrogenase (*G6PD*).[110] In the adipose tissue of rats, ALA affects the expression of genes including sterol regulatory element-binding transcription factor 1 (*Srebf1*), lipoprotein lipase (*Lpl*), glycerol-3-phosphate dehydrogenase (*Gpd*), and adipokines leptin (*Lep*) and adiponectin, C1Q, and collagen domain containing (*Adipoq*).[111] In mice, stimulation of the expression of genes encoding regulatory factors for mitochondrial biogenesis and oxidative metabolism (peroxisome proliferator-activated receptor [PPAR], gamma, coactivator 1 alpha [*Pgc1α*, now known as *Ppargc1a*], and nuclear respiratory factor-1 [*Nrf1*], respectively) has been seen following EPA and DHA treatment.[39] In the liver of rodents, several genes involved in fatty acid metabolism are increased by PUFAs, including apolipoproteins A-I and A-II (*Apoa1*, *Apoa2*), acyl-coenzyme A (CoA) synthetase long-chain family member 1 (*Acsl1*), acyl-CoA oxidase (*ACO*, now known as *Acox*), liver fatty acid-binding protein (*Fabp1*), carnitine palmitoyltransferase 1 (*Cpt1*), and cytochrome P450, family 4, subfamily a, polypeptide 1 (*Cyp4a1*).[110]

Fatty acid elongation and desaturation are two key metabolic routes for the synthesis of saturated, monounsaturated, and polyunsaturated fatty acids (Fig. 2). Fatty acid desaturases (FADS) in particular have received considerable attention for their regulation by hormones and nutrients and their capacity to generate specific unsaturated fatty acids. One of these enzymes, stearoyl-CoA desaturase 1 (SCD1, also known as delta-9 desaturase [D9D]), has emerged as a key enzyme in the control of whole-body lipid composition.[112] Although oleate is found throughout the body, oleate derived endogenously from SCD is special in terms of its preferential trafficking through acyl-CoA: diacylglycerol *O*-acyltransferase 2 and its ability to drive triglyceride synthesis. ω3-PUFAs decrease SCD1 mRNA expression in liver, and this effect is correlated with decreased circulating triglycerides.[27] Decreased SCD1 expression in macrophages by ALA is associated with increased cholesterol efflux.[113] Delta-6 desaturase (D6D) and delta-5 desaturase (D5D) are the key enzymes for the synthesis of PUFAs such as AA and DHA. Saturated and monounsaturated fatty acids are elongated by fatty acid elongase 1 (ELOVL1, also known as Ssc1) and 6 (ELOVL6, also known as LCE, FACE, and rElo2). ω3-PUFAs decrease the expression of D6D and D5D (decreased levels of PUFAs increase their expression), as well as ELOVL6; however, the conversion of ALA to longer chain PUFAs is regulated by substrate levels to a greater extent than by expression of the synthetic enzymes.[114]

As can be ascertained from Table I, even in a tissue that is not considered a site of major lipid metabolism such as macrophages, the predominant genomic response to ω3-PUFAs is often altered fatty acid or sterol metabolism. In addition to *FADS1*, *FADS2*, and *SCD*, ATP-binding cassette, subfamily B, member 4

(*ABCB4*), acyl-CoA synthetase long-chain family member 3 (*ACSL3*), and 3-hydroxy-3-methylglutaryl-CoA synthase 1 (*HMGCS1*) are regulated by ALA, DHA, and EPA. As will be discussed in a later section, ω3-PUFAs may affect lipid metabolism at extrahepatic sites because of the expression and activation of specific transcription factors that control multiple pathways concurrently.

3. Cell Differentiation and Fate

Studies in human populations have linked high consumption of fish or fish oil to reduced risk of colon, prostate, and breast cancer[17] and increased effectiveness of chemotherapeutic agents.[115,116] This is consistent with the observations that ω3-PUFAs attenuate growth and induce apoptosis in a variety of human cancer cell lines derived from colonic, pancreatic, prostate, and breast cancer.[16] When four commercially available breast cancer epithelial cell lines were subjected to treatment with AA, LA, EPA, and DHA, cell growth-related genes were affected in a cell line- and fatty acid-dependent manner.[40] Several important genes were affected, including those responsible for regulating apoptosis (arachidonate 15-lipoxygenase [*ALOX15*], adenosine A1 receptor [*ADORA1*], Fas ligand [*FASLG*]); activating defense immunity (MCF.2 cell line-derived transforming sequence-like [*MCF2L*], bone marrow stromal cell antigen 1 [*BST1*], apolipoprotein B [*APOB*]); and controlling cell growth (*ALOX15*, protein tyrosine phosphatase [*PTP*], RAB4A, member RAS oncogene family [*RAB4A*]).[40] In cancer cell lines treated with ω3-PUFAs, pro-angiogenic vascular endothelial growth factor and levels of antiapoptotic Bcl-2 and Bcl-X(L) are decreased.[16]

In a convergence of altered lipid metabolism, inflammatory, and cellular growth pathways, ω3-PUFAs induce modifications in the activity of lipoxygenase (LOX) and cyclooxygenase (COX) enzymes that affect tumor growth. For example, a diet with a 1:1 ratio of ω6 to ω3 increased EPA and DHA levels but decreased the mRNA expression of FASN, COX-2, and 5-LOX in mammary tissues.[19] In a mouse mammary gland adenocarcinoma model, a diet rich in ALA simultaneously showed a decrease in 12-LOX, 15-LOX-2, 15-LOX-1, and prostaglandin E synthase activities that corresponded to higher apoptosis and lower mitosis in the tumor.[22] In colon cancer cells, DHA increases the expression of cyclin-dependent kinase inhibitors such as p21 (also known as waf1/cip1), p27, p57, p19, and growth arrest-specific proteins; this is consistent with the induction of apoptosis and inactivation of the B-cell CLL/lymphoma 2 (*BCL2*) family of genes by ω3-PUFAs.[41]

B. NRs as Sensors of Dietary Lipids

One of the key questions is how do ω3-PUFAs affect gene expression in a variety of tissues and modulate the activity of diverse biological programs? Using the gene expression patterns from microarray and other comprehensive

"-omic" studies, it is possible to mine the data for the involvement of known transcription factors or enzymes. Based on such an examination of the data from Table II (Fig. 4), members of the NR superfamily appear to be the predominant upstream regulators. The NRs identified as being transcriptional regulators of ω3-PUFA-responsive genes include liver X receptor (LXR) α and β (NR1H3 and 2); glucocorticoid receptor (NR3C1); thyroid hormone receptor α (TRα); peroxisome proliferator-activated receptor (PPAR) α, β/δ, and γ (NR1C1, 2, and 3); and estrogen-related receptor (ER) α (NR3B1). These ligand-activated transcription factors and several others identified or speculated to be fatty acid receptors were examined in whole-cell receptor assays for

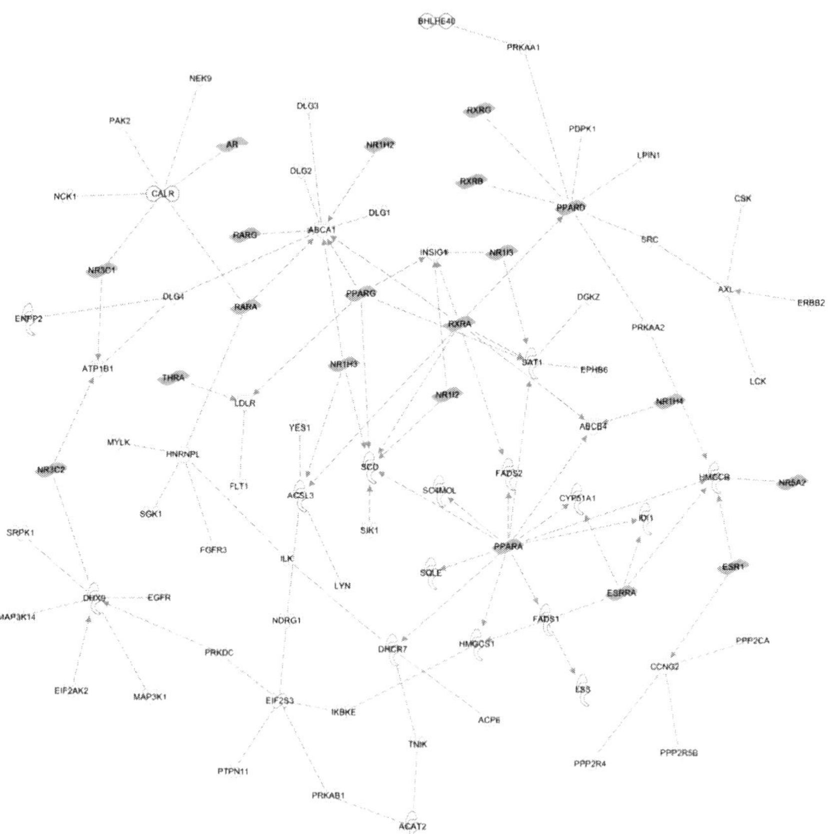

FIG. 4. Nuclear receptors as important upstream regulators of the genomic response to ω3-PUFAs. The genes depicted in Table II were examined for common transcriptional regulators using Ingenuity Pathway Analysis (IPA 9.0, Ingenuity Systems, Inc., Redwood City, CA).

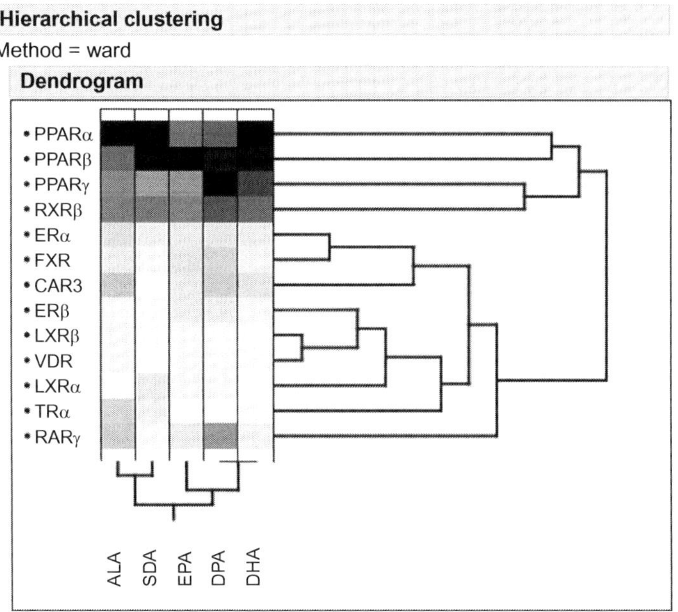

FIG. 5. Activation of nuclear receptors by ω3-PUFAs, determined by specific reporter cell lines (Indigo Biosciences, Inc., State College, PA). The expression observed following treatment for 24 h to 100 μM of the fatty acids is expressed relative to vehicle control and examined by hierarchical clustering using JMP (SAS Institute, Cary, NC). Shading is from white (relative activity, 0) to black (maximal activation 35-fold of PPARβ/δ by EPA).

activation by various ω3-PUFAs (Fig. 5). Significant activation by several fatty acids was seen for the PPARs (α, β/δ, and γ), retinoid X receptor α (RXRα), farnesoid X receptor (FXR), and retinoic acid receptor γ (RARγ). In contrast, ERα, ERβ, LXRα, LXRβ, TRα, vitamin D_3 receptor (VDR), and constitutive androstane receptor variant 3 (CAR3) activity was not affected. PPARα and PPARβ/δ were the most activated by ω3-PUFAs, followed by RXRα (sixfold), RARγ (threefold), and FXR (twofold).

NRs act as intracellular transcription factors that directly regulate gene expression in response to lipophilic molecules.[110,117–119] They affect a wide variety of cellular events, including fatty acid metabolism, inflammatory responses, cancer reproductive development, and detoxification of foreign substances. The fatty acid receptors PPAR, LXR, RXR, and FXR described above may be considered constituents of a large group of NRs, the "metabolic NRs," which act as overall sensors of metabolic intermediates, xenobiotics, and compounds in the diet and allow cells to respond to environmental changes by inducing the appropriate metabolic genes and pathways.[117,119] The subsequent sections describe the likely candidates for NRs that respond to dietary ω3-PUFAs and their metabolites and contribute to their biological responses, namely, PPAR, RXR, LXR, and FXR. The dietary and metabolic intermediates that activate these receptors, as well as the genes regulated by these NRs that contribute to prevention of disease, will be emphasized.

1. Peroxisome Proliferator-Activated Receptors

Of the several identified fatty acid receptors, perhaps the family that can best explain the effects of ω3-PUFAs is the PPAR family. The PPAR family of receptors was originally named on the basis of their ability to respond to xenobiotics (peroxisome proliferators); however, they were also the first to be examined as fatty acid receptors. It has now been well established that PPARs are ligand-activated transcription factors involved in gene expression in a tissue-, sex-, and species-dependent manner. The PPARs exist as three subtypes (α, β/δ, and γ) that vary in expression, ligand recognition, and biological function. PPARα was the first transcription factor identified as a prospective fatty acid receptor (for review, see Refs. 120–122). Based on numerous studies from the PPARα knockout mouse (PPAR$\alpha^{-/-}$), this receptor plays a role in the regulation of an extensive network of genes involved in glucose and lipid metabolism. In particular, PPARα regulates fatty acid transport; FABPs; fatty acyl-CoA synthesis; microsomal, peroxisomal, and mitochondrial fatty acid oxidation; ketogenesis; and fatty acid desaturation.

Several research groups have implicated saturated and unsaturated fatty acids as natural ligands for PPARα (for review, see Ref. 123). Natural PPARα ligands in human serum include palmitic acid, oleic acid, LA, and AA. Notably, PPARα is the only PPAR subtype that binds to a wide range of saturated fatty acids.[124] ω3-PUFAs such as ALA, EPA, and DHA activate PPARα with a similar potency and efficacy as LA.[124] Triglyceride-rich lipoproteins, including very low-density and low-density lipoproteins (VLDL and LDL), also contain PPARα ligands.[125] Activation of PPARα is seen when LPL is added to VLDL, showing that the endogenous ligands are probably fatty acids or their metabolites esterified into TAG. In rats, metabolism of AA by CYP4A results in a variety of PPARα ligands including 5,6-epoxyeicosatrienoic acid (5,6-EET); 8,9-EET; 11,12-EET; 14,14-EET; 20-hydroxyeicosatetraenoic acid (20-HETE); and 20-hydroxy-14,15-epoxyeicosatrienoic acid (20,14,15-HEET).[126] LTB_4 has also been reported to be a selective PPARα ligand.[127] Prostaglandins PGD2 and PGD1 activate PPARα in transient transfection reporter assay systems.[128] The LOX metabolite 8(S)-HETE is a high-affinity PPARα ligand, although it is not found at sufficient concentrations in the correct tissues to be characterized as a natural ligand. As no single high-affinity natural ligand has been identified, it is possible that the physiological role of PPARα may be to sense the total flux of fatty acids in metabolically active tissues.[129]

PPARγ is expressed in many tissues including adipose, muscle, vascular cells, macrophages, and epithelial cells of the mammary gland, prostate, and colon (for review, see Ref. 130). Activated PPARγ induces LPL and fatty acid transporters (CD36) and enhances adipocyte differentiation, in addition to inhibiting cytokine and *COX2* expression, perhaps by modulating NF-κB function. The PPARγ-null mouse is nonviable, implicating an important role for this protein in ontogeny[131] while also making the examination of this receptor's role in gene expression difficult.

Clinically relevant antidiabetic agents such as pioglitazone and rosiglitazone are potent PPARγ agonists (K_d in low nanomolar range). Several fatty acids and eicosanoid derivatives bind and activate PPARγ in the micromolar range.[127] Unlike the PPARα subtype, PPARγ has a clear preference for PUFAs.[124] The fatty acids LA, AA, and EPA bind PPARγ within the range of concentrations of free fatty acids found in human serum.[132] Although fatty acids are not particularly efficacious activators of PPARγ, intracellular conversion of fatty acids to eicosanoids, through enhanced expression of 15-LOX, greatly increase PPARγ-mediated transactivation.[132] Similar to PPARα, incubation of triglyceride-rich lipoproteins with LPL results in the production of PPARγ ligands.[125] In particular, oxidized LDL products, namely, 9-S-hydroxyoctadecadienoic acid (9-S-HODE) and 13-S-HODE, are good PPARγ activators. Potent PPARγ ligands also include phospholipids such as lysophosphatidic acid[133] and hexadecyl azelaoyl phosphatidylcholine.[134] The 5-LOX metabolite of DHA (4-hydroxy DHA)[135] and the COX-2 metabolites' electrophilic oxo-derivatives are[136] more potent PPARγ activators than the parent fatty acid and may be responsible for effects of this ω3-PUFA on angiogenesis and inflammation.

PPARβ/δ is ubiquitously expressed and is often found in higher abundance than PPARα or γ. Examination of the PPARβ/δ-null mouse has shown a role for this NR in development, myelination of the corpus callosum, lipid metabolism, and epidermal cell proliferation.[137] There has been some indication that PPARβ/δ is involved in adipogenesis,[137] although this has been refuted.[138] ω3-PUFAs are efficacious PPARβ/δ activators; in fact, this NR is the most highly regulated of those tested, albeit at higher doses of fatty acids[124] (see Fig. 5). Similar to PPARα and γ, incubation of triglyceride-rich lipoproteins with LPL results in the production of PPARβ/δ ligands.[125] Prostaglandins A1 (PGA1), PGD2, and PGD1 can activate PPARβ/δ in reporter assays.[128] Both the toxic lipid 4-hydroxynonenol and 15-HETE are also PPARβ/δ activators.[139]

The potential of highly potent PPAR activators in the treatment of atherosclerosis and dyslipidemia has been noted by other investigators.[110,140–143] The result of PPARα activation in rodent hepatocytes and certain other tissues is a dramatic increase in the peroxisomal enzymes, with a modest increase in mitochondrial oxidation of fatty acids. In addition, lipid transport proteins such as FABP and acyl-CoA-binding protein, as well as genes involved in fatty acid and cholesterol export, are under the control of PPARα. The targeted disruption of PPARα results in aberrant lipid metabolism, with fat droplets accumulating in liver cells. Not only is peroxisomal metabolism affected, but the constitutive levels of mitochondrial β-oxidation are lower in the PPARα-null mouse, showing the importance of this protein in overall fatty acid homeostasis.

The array of genes regulated by PPARγ in adipocytes is indicative of fatty acid accumulation. This regulation of gene expression is concomitant with increased differentiation of immature adipocytes into mature fat-storing cells.[144] These genes include *LPL*, adipocyte fatty acid-binding protein (aP2, now known as *FABP4*), and *CD36*.[145,146] Genes encoding adipocyte-secreted cytokines and hormones such as TNFα and leptin are also PPARγ targets.[147] The genes regulated by PPARγ in macrophages are similar to those in adipocytes and include *LPL* and *CD36*. Treatment of macrophages with PPARγ synthetic agonists inhibits the production of several cytokines such as IL1β and TNFα and may result in an anti-inflammatory response.[148,149] Another link between PPARγ and inflammation is the fact that the prostaglandin 15-deoxy-PGJ_2, a product of the COX pathway, and nonsteroidal anti-inflammatory drugs are potent activators of PPARγ.[150]

2. Retinoid X Receptors

RXRs are primarily involved in the transduction of the retinoid signaling pathway, although their role in the regulation of gene expression induced by ω3-PUFAs has garnered increasing attention. These receptors (α, β, or γ) can form homodimers, or they may serve as dimerization partners for other NRs including RARs, TR, VDR, and PPARs. As a heterodimerization partner, RXR is involved in the regulation of multiple cellular pathways. RXRα and β have ubiquitous distribution, whereas RXRγ is expressed in certain organs such as heart, skeletal muscle, and central nervous system structures.[151]

Although intensely studied for synthetic ligands, little is known about the natural activators of RXRs.[151] They are activated *in vitro* by the vitamin A metabolite 9-*cis* RA, but the levels of this molecule *in vivo* are extremely low. Several fatty acids including unsaturated, monounsaturated, and polyunsaturated fats such as AA and DHA have been identified as ligands of RXR, thus confirming the activation observed in reporter assays[124,152,153] (see Fig. 5). The effect of DHA was not observed in other NRs such as TR and VDR, although, as stated above, this fatty acid activates PPARα and β/δ. Docosatetraenoic acid, a compound structurally related to DHA, activates RXR with a much higher concentration.[153] Phytanic acid, a branched-chain fatty acid derived from chlorophyll, has also been reported to activate RXR albeit weakly.[154] Phytanic acid is capable of adipocyte differentiation and induces FABP4 mRNA in 3T3-L1 preadipocytes and may act as a natural rexinoid in 3T3-L1 cells.[154]

RXRα agonists are capable of reducing atherosclerosis in the apolipoprotein E (*Apoe*) knockout mouse, which is an established experimental model of atherosclerosis.[155] Retinoids are capable of increasing the expression of *ABCA1* a gene associated with reverse transport of cholesterol. Cholesterol efflux from peritoneal macrophages was significantly increased in an RXR-dependent fashion.[155] RXR-selective agonists counteract diabetes by decreasing hyperglycemia, hypertriglyceridemia, and hyperinsulinemia.[155] Null mutation of the *Rxra* gene resulted in developmental lethality in mice; they died *in utero* and demonstrated severe myocardial and ocular malformations.[156] The malformations resembled severe vitamin A syndrome, suggesting a physiological role of RXRα in retinol responses.[156]

3. Liver X Receptors

Liver X receptors (LXRα and LXRβ) are transcription factors commonly known as cholesterol sensors.[157,158] Although they are important regulators of the transport and metabolism of sterols and fatty acids, their potential role as direct sensors of ω3-PUFAs has been questioned.[124] Expression of LXRα is restricted, whereas LXRβ is ubiquitously present. LXRα is present in certain organs, namely, liver, kidney, intestine, adipose tissue, and adrenals. LXRα and β share a high degree of amino acid similarity (~80%) and are considered paralogs; as a result, there are very few subtype-specific agonists. Oxysterols, including 24(*S*), 25-epoxycholesterol, 22*R*-hydroxycholesterol, and 24(*S*)-hydroxycholesterol, are natural ligands of LXRs. Unsaturated fatty acids in addition to AA and other PUFAs competitively blocked activation of LXR by oxysterols.[159] This offers a potential mechanism for the ability of dietary PUFAs to decrease the synthesis and secretion of fatty acids and triglycerides in liver.[159] This suppressive effect can be eliminated by deletion and mutation of the LXR-responsive element located in the promoter region of *SREBF1*.[159] However, others have shown in rats that the unsaturated fatty acid suppression of SREBP-1 and its targeted lipogenic genes is independent of LXRα.[160] Perhaps the effects of fatty acids on LXR-mediated events are being affected by a direct interaction between PPARα and LXRα.[161] In fact, several xenobiotic PPARα ligands antagonize LXR's transcriptional activity.[162]

There is increasing interest in LXR agonists, whether dietary or pharmaceutical, in the prevention of inflammation and CVD.[163–165] The nonsteroidal LXR agonist GW3965 significantly reduced atherosclerosis in murine models of hyperlipidemia.[164] Several LXR-mediated genes include those associated with cholesterol and bile acid metabolism (e.g., *ABCA1*, *ABCG1*, *APOE*, *CYP7A*), as well as with fatty acid synthesis and regulation (*SREBF1*, *LPL*, *FASN*). Previous studies have shown that activation of PPARγ induced the expression of LXRα and ABCA1 and removed cholesterol from macrophages.[166] Hence, LXR was considered further downstream than PPARγ in reducing atherosclerosis.

Mice lacking the LXRα-encoding gene (*Nr1h3*) were unable to respond to dietary cholesterol and failed to induce cholesterol 7-hydroxylase (CYP7A), which is the rate-limiting enzyme for bile acid synthesis.[167] This resulted in excessive cholesterol accumulation in the liver, followed by impairment of functions. These knockout animals also have altered expression of genes associated with lipid metabolism. Interestingly, LXRβ (*Nr1h2*) knockout mice were unaffected when challenged with dietary cholesterol.[168] Selective bone marrow knockouts of macrophage LXRs increased atherosclerotic lesions in *Apoe*$^{-/-}$ and low-density lipoprotein receptor (*Ldlr*)$^{-/-}$ mice, suggesting LXR's role as an endogenous inhibitor of atherosclerosis.[164]

4. Farnesoid X Receptor

The farnesoid (or farnesol) X receptor is a member of the same subclass of NRs as LXRα and β, as well as the ecdysone receptor. It has been shown that FXR heterodimerizes to RXR and binds to DNA sequences consisting of an inverted repeat spaced by one nucleotide.[123] The FXR is mainly expressed in the liver, gut, kidney, and adrenal cortex. This receptor binds and is activated by several bile acids, including chenodeoxycholic acid, lithocholic acid, and deoxycholic acid. It received its name because it is activated by a large variety of endogenous isoprenoids, including farnesol. Other endogenous activators include all-*trans*-retinoic acid and, as mentioned above, fatty acids.[169] This NR binds to AA, ALA, and DHA (K_i in low micromolar range) but has little or no affinity for palmitic or stearic acid.[169]

Several studies have shown that FXR activation by ligands has an anti-atherogenic effect in animal models.[119,170,171] For example, in *Apoe*$^{-/-}$ mice, there was a decrease in aortic plaque formation when animals were fed the synthetic FXR ligand INT-74.[170] FXR activation by natural and synthetic ligands attenuates induction of the genes encoding IL1β, IL6, and TNFα in response to LPS, an effect reminiscent of what is commonly observed for PPAR activation. The receptor reduces cholesterol uptake on macrophages by regulation of *CD36* and *ABCA1* expression. A target gene for FXR is *NR0B2*, which encodes the small heterodimer partner (SHP), a regulator of cholesterol metabolism in its own right. Both FXR and SHP are expressed in the aorta and macrophages, and interest is growing in the potential for FXR ligands in the prevention and treatment of atherosclerotic lesions.[172] ALA and extracts of walnuts, rich in this ω3-PUFA, are activators of FXR and increase cholesterol efflux from macrophage-derived foam cells.[113] ALA-dependent gene expression via FXR and SHP includes altered expression of *Srebf1* and *Scd1*.[113]

III. Nutrigenetics

The Human Genome Project and the subsequent identification of single-nucleotide polymorphisms (SNPs) within populations have played a major role in predicting individual response to nutrients (nutrigenetics), leading to the concept of "personalized nutrition."[173] Phenotypic variability is based on interindividual genetic variation, which can be qualitative (base pair changes [SNPs], small deletions, duplications, or insertions) or quantitative (large duplications or deletions).[174] Qualitative variants can affect the regulatory region of a gene (i.e., the promoter region) or the coding/noncoding sequences; quantitative changes directly affect the level of expression. The inherited genotypic differences in DNA sequence contribute to phenotypic variation and to differences in disease risk in response to the environment, including the diet. Individual genotypic variations can alter nutrient metabolism, causing conditions that range from relatively mild but unpleasant ones such as lactase gene polymorphisms that result in lactose intolerance to potentially severe pathological conditions such as phenylketonuria.[174] There is ongoing intense research in the area of genetics and multifactorial diseases such as obesity[175] and CVD,[176] as well as response to diets.[48,177–180] This discussion of the nutrigenetic aspects of ω3-PUFAs focuses on genetic variation that affects their metabolism, molecular response, and clinical effects (Table II, also see reviews[42,45,48,178]).

TABLE II

GENETIC POLYMORPHISMS THAT AFFECT INTERINDIVIDUAL DIFFERENCES IN RESPONSE TO ω3-PUFAs

Protein	Gene	Variation	Potential effect[a]	References
Adiponectin	*ADIPOQ*	rs1501299	Effect	42,43
5-Lipoxygenase	*ALOX5*	Variable number of tandem Sp1 binding sites (3–8)	Metabolism	42,44
5-Lipoxygenase activating protein	*ALOX5P*	rs4076128	Effect	45–47
Apolipoprotein A-I	*APOA1*	−75G/A	Effect	48,49
Apolipoprotein A-IV	*APOA4*	Gln360His	Effect	48,49
		Thr347Ser	Effect/ intake	48,49
Apolipoprotein A-V	*APOA5*	rs3135506	Effect	42,50
		rs662799	Effect	42,50
Apolipoprotein C-III	*APOC3*	rs2854116	Effect	42,51
Apolipoprotein E	*APOE*	APOE2 Cys112 Cys158	Effect	42,52
		APOE4 Cysl30Arg	Effect	42,53
Complement component 3	*C3*	rs11569562	Effect	42,54
Fatty acid translocase	*CD36*	rs1049673	Uptake	42,55

(*Continues*)

TABLE II (*Continued*)

Protein	Gene	Variation	Potential effect[a]	References
		rs1527483	Uptake	42,56
		rs1761667	Uptake	42,57
		rs1984112	Uptake	42,57
CD44 antigen	*CD44*	Splice variant	Effect	42,58
Circadian locomoter output cycles protein kaput	*CLOCK*	rs1464490	Effect	59
		rs1801260	Effect	59
		rs3749474	Intake/ effect	59
		rs4580704	Intake/ effect	59
		rs4864548	Effect	59
Cytochrome P450 1A1	*CYP1A1*	rs1048943	Metabolism	42,60
		rs1799814	Metabolism	42,60
Fatty acid elongase 2	*ELOVL2*	rs953413	Metabolism	42,61
Coagulation factor VII	*F7*	exon Arg353Gln	Effect	42,62
Fatty acid-binding protein 2	*FABP2*	rs1799883	Uptake	42,63
Fatty acid coenzyme A ligase, long-chain 4	*ACSL4*	rs1324805	Metabolism	42,64
Delta-5 desaturase	*FADS1*	rs174537	Metabolism	42,61
		rs174544	Metabolism	42,45,65–67
		rs174545	Metabolism	42,66
		rs174553	Metabolism	42,45,67,68
		rs174556	Metabolism	42,69
		rs174561	Metabolism	42,45,67,69
Delta-6 desaturase	*FADS2*	rs174566	Metabolism	45,67
		rs174568	Metabolism	45,67
		rs174570	Metabolism	42,45,67,70
		rs174583	Metabolism	42,45,67,68
		rs174589	Metabolism	45,67
		rs2072114	Metabolism	45,67
		rs968567	Metabolism	45,67
		rs99780	Metabolism	42,45,67,69
Delta-5 and -6 desaturase, intragenic	*FADS3*	rs3834458	Metabolism	42,71
		rs1000778	Metabolism	42,72
Fibrinogen alpha chain	*FGA*	exon Thr312Ala	Effect	42,73
IL10	*IL10*	rs1800896	Effect	42
IL1β	*IL1B*	rs1143643	Effect	42
		rs16944	Effect	42
IL6	*IL6*	rs1800795	Effect	42
Potassium voltage-gated channel subfamily E member 1	*KCNE1*	rs1805127	Effect	42
Leptin receptor	*LEPR*	rs3790433	Effect	42
Hepatic lipase	*LIPC*	rs2070895	Effect	42
Lipoprotein A	*LPA*	C/T 93 5′ UTR Promoter	Effect	42

(*Continues*)

TABLE II (*Continued*)

Protein	Gene	Variation	Potential effect[a]	References
Lipoprotein lipase	*LPL*	S447X	Effect	48,49
Lymphotoxin-α	*LTA*	rs909253	Effect	42
Nuclear factor-κB	*NFKB1*	rs28720239	Effect	42
Endothelial nitric oxide synthase	*NOS3*	G > T 894 exon	Effect	42
		rs1799983	Effect	74
Farnesoid X receptor	*NR1H4*	rs10860603	Response	75,76
Prostaglandin E synthase	*PTGES*	rs7873087	Effect	42
Peroxisome proliferator-activated receptor α	*PPARA*	rs1800206	Response	42,77,78
		rs1800234	Response	42
		rs3892755	Response	42
		rs4253623	Response	79
		rs6008259	Response	42,80
Peroxisome proliferator-activated receptor β/δ	*PPARD*	c.−789C−>T	Response	75
		rs2016520	Response	81
		rs1053049	Response	82
		rs1883322	Response	83
		rs2016520	Response	84
		rs2076167	Response	84
		rs2267668	Response	82
		rs3734254	Response	83–85
		rs6902123	Response	85
		rs6922548	Response	83
		rs7769719	Response	83
Peroxisome proliferator-activated receptor γ	*PPARG*	rs1175543	Response	86
		rs1801282	Response/intake	48,87–89
		rs1805192	Response	42
		rs4684847	Response	86
		rs709158	Response	86
Peroxisome proliferator-activated receptor γ coactivator 1-α	*PPARGC1*	rs17574213	Response	84,90
		rs2970848	Response	84
		rs2970852	Response	84
		rs3774923	Response	84
		rs7665116	Response	84
		rs7672915	Response	84
		rs7695542	Response	84
		rs8192678	Response	84,91
Prostaglandin E2 receptor EP4 subtype	*PTGER4*	rs111866313	Effect	42
Cyclooxygenase 2	*PTGS2*	−1329A > G	Metabolism	75
		rs4648308	Metabolism	42
		rs4648310	Metabolism	
		rs4648310	Metabolism	42,45,92
		rs5275	Metabolism	42,45,93
		rs5277	Metabolism	42
		rs689466	Metabolism	42
		V102V	Metabolism	75

(*Continues*)

TABLE II (*Continued*)

Protein	Gene	Variation	Potential effect[a]	References
Plasminogen activator inhibitor 2	*SERPINB2*	rs1799889	Effect	42
Tumor necrosis factor α	*TNF*	rs1799724	Effect	42
		rs1800629	Effect	42
		rs361525	Effect	42

[a]The potential effect (metabolism, response, effect) is based on either the observed responses from clinical studies or by understanding the biological role of the polymorphic gene product. An effect on "metabolism" implies that a genetic variation will affect the biotransformation of an ω3-PUFA. A potential effect on "response" is typified by a gene product that recognizes or responds to levels or structure of the ω3-PUFA and is represented by receptors and transcriptional accessory proteins. Finally, polymorphisms that alter the "effect" of the ω3-PUFAs are generally products that characterize the biological response and include inflammatory markers and lipoproteins.

A. Metabolism of ω3-PUFAs

Genetic polymorphisms in the genes involved in ω3-PUFA metabolism can theoretically affect the conversion of ALA to EPA and DHA and the production of bioactive metabolites (see Fig. 1). It is estimated that genetic variation explains 40% or more of the interindividual variability in saturated, monounsaturated, and polyunsaturated fatty acid levels.[181] Additional evidence for the heritability of fatty acid composition comes from candidate gene studies. In particular, variation in the genes encoding D5D and D6D influences levels of several phospholipid fatty acids; *FADS1* and *FADS2* code for key enzymes in the conversion of the *n*-6 and *n*-3 C18 fatty acids to their respective C20 and C22 products (Fig. 2). The human *FADS1* and *FADS2* cDNAs have been cloned[182] and are located in a cluster on chromosome 11 (11q12–13.1) with a head-to-head orientation. Eighteen SNPs of the *FADS1-2* gene cluster have been identified in a German population.[67] Nine of these SNPs were associated with significant decreases in the percentage of EPA in serum fatty acids relative to wild type. In addition, a significant decrease in AA, EPA, and dihomo-γ-LA was associated with decreased production of strong inflammatory mediators. SNPs in *FADS1* and *FADS2* influence plasma phospholipid and erythrocyte ethanolamine phosphoglyceride ω6- and ω3-PUFAs in women during pregnancy and in breast milk during lactation.[68]

The enzymes 5-LOX and COX-2 are capable of metabolizing PUFAs to intermediates that influence inflammation, with both positive and negative effects. Dietary AA significantly enhances the apparent atherogenic effect of the *ALOX5* genotype, whereas increased dietary intake of ω3-PUFAs EPA and DHA blunts this effect (reviewed in Ref. 45). In addition, there is a significant interaction between the *ALOX5* polymorphism (rs4076128) and dietary LA intake as well a relationship with breast cancer risk.[47] Among women consuming a diet high in LA (top quartile of intake, >17.4 g/day), carrying the AA genotype was associated with higher breast cancer risk (age- and race-adjusted odds ratio, 1.8; 95% confidence interval, 1.2–2.9) compared with those carrying genotypes AG or GG. Studies have shown that genetic variants at the *COX2* gene (now known as prostaglandin-endoperoxide synthase-2 [*PTGS2*]) modify prostate inflammation and response to diet.[45] Increasing ω3 intake was associated with a decreased risk of aggressive prostate cancer, and this inverse association was even stronger among men with genetic variant rs4648310 (þ8897 A/G) flanking the 3′-region of *COX2*.[92] Individuals with the lowest intake of ω3-PUFAs and the genetic variant had the most aggressive tumor, whereas the ω3-PUFAs were protective; this effect was modified by the genetic variant.

B. Response to ω3-PUFAs

To date, several publications have examined the impact of the *PPARA* Leu162Val variant on the association between ω3-PUFA status and a range of outcomes related to lipid metabolism. In the first of these analyses, based on the Framingham cohort, the association between *PPARA* genotype and an array of plasma and lipoprotein, lipid, and apolipoprotein variables was examined. Overall, carriers of the Val allele had lower plasma TAG and apolipoprotein C3 (APOC3) concentrations relative to noncarriers when consuming a high-PUFA diet, with ω6-and ω3-PUFAs having a comparable effect. A subsequent analysis of the Atherosclerosis Risk in the Community (ARIC) cohort[80] failed to replicate the findings of the Framingham cohort and reported no impact of the *PPARA* Leu162Val variant on the association between PUFA intake and lipid levels.

In vitro studies have provided evidence that the Ala12 isoform of *PPARG* is associated with a reduced ability to induce transcription and adipogenesis.[183] In the Kuopio, Aarhus, Naples, Wollongong, Uppsala (KANWU) Study, researchers examined the impact of a diet rich in saturated fatty acid versus monounsaturated fatty acid, supplemented with or without 2.4 g EPA + DHA per day for 3 months, on a range of lipid and insulin sensitivity-related outcomes.[184] In individuals with a total fat and saturated fat intake of <37% and 10% of dietary energy, respectively, carriers of the Ala12 allele had significantly greater reductions in serum TAG levels in response to ω3-PUFA supplementation, suggesting that *PPARG* Pro12Ala genotype may contribute to the interindividual variability in the serum TAG response to EPA + DHA intervention. Thus, the *PPAR* genotype, in particular the *PPARA L162V* and *PPARG P12A* isoforms, may have an impact on the response to dietary intervention with ω3-PUFAs.

C. Effects of ω3-PUFAs

As outlined previously, the hypotriglyceride benefits of fish oils are well established, and the American Heart Association recommends 2–4 g per day as a potential alternative to fibrates. Apolipoprotein E (APOE) plays a major role in fatty acid and lipoprotein metabolism, and variations in this gene are associated with the risk of CVD and Alzheimer's disease.[42] For example, the *APOE4* ε4 (Arg 112, Arg 158) polymorphism is associated with a gene–diet interaction between ω3-PUFAs and serum TAG.[52]

In a cohort of hypertriglyceridemic individuals prospectively recruited according to *FABP2* Ala54Thr genotype, Thr carriers were found to be more responsive to supplementation with 1.8 g ω3-PUFAs per day, with 52% and 34% decreases in plasma TAG and APOC3 concentrations relative to 19% and 32% reduction in Ala homozygotes.[185]

In an observational study in 595 young healthy adults, Fontaine-Bisson and El-Sohemy[186] reported that the positive association between total PUFA and HDL cholesterol was evident only in noncarriers of the minor "A" allele for the *TNF* −238G>A and −308G>A loci. The effect was evident for both ω3- and ω6-PUFAs, but a significant genotype × diet effect was evident only for ω6-PUFAs. The same authors also reported significant PUFA × genotype × HDL interactions in a type 2 diabetic population.[187] Markovic *et al.*[188] investigated the impact of SNPs in several inflammatory genes on the response to intervention with 1.8 g EPA + DHA per day for 12 weeks. Although no significant impact of the above-mentioned *TNF* −308G>A was evident, a difficult-to-interpret interaction between BMI × *TNF-β* (now known as lymphotoxin alpha [*LTA*]) +252A>G × EPA was reported to influence the hypotriglyceridemic response. In the fight against CVD, ω3-PUFAs have beneficial effects in two main areas: decreasing circulating TAGs and ameliorating inflammation. An interaction between these ω3-PUFAs' beneficial effects and variation in several polymorphisms has been observed, notably in the *APOE*, *FABP2*, and *TNF* genes.

IV. Conclusions

Diets high in ω3-PUFAs have long been associated with decreased risk of CVD and prevention of certain types of cancer. ALA and its metabolites EPA and DHA are found in high concentrations in flaxseed and fish oils and are thought to improve heart health through decreasing thrombosis, inflammation, and plaque formation in arteries. The mechanism of these effects may be the result of regulation of gene expression via NRs, several of which are known to be "fatty acid receptors."

- PPARα and PPARβ/δ are receptors for unsaturated, monounsaturated, and polyunsaturated fatty acids, as well as several AA metabolites. Activation of PPARα is associated with increased fatty acid catabolism, decreased inflammation, and stimulation of the reverse cholesterol pathway. PPARγ has a clear preference for PUFAs and is also the target of AA metabolites. This receptor is involved in the storage of lipids in adipocytes, as well as decreasing inflammation and stimulating the reverse cholesterol pathway.
- RXRs are an important heterodimerization partner for NRs and hence can affect numerous metabolic pathways. DHA, as well as several other PUFAs, binds to and activates these central NRs.
- The role of LXRs as sensors of fatty acids is somewhat controversial, although they are clearly oxysterol receptors. Several studies have shown that fatty acids (unsaturated and saturated) antagonize LXR activity. These receptors are involved in fatty acid synthesis, bile acid synthesis, and reverse cholesterol transport, and synthetic agonists are being touted as anti-atherosclerosis agents.
- The FXR is the most recently identified member of the fatty acid receptor group and is activated by PUFAs. This NR is involved in hepatic bile acid clearance, and evidence is growing that it may be a potential target in other tissues, notably in the endothelial wall and in macrophages.

Taken together, these NRs represent potential targets for ω3-PUFAs that can help explain their mechanism of action in preventing CVD as well as certain cancers.

Our understanding of genetic differences in responsiveness to dietary intervention is continuing to grow. Polymorphisms in certain enzymes, transcription factors, inflammatory molecules, and lipoproteins have been associated with altered responsiveness to ω3-PUFAs:

- It is estimated that genetic variation explains a large portion of interindividual variability in ω3-PUFA levels. *FADS1* and *FADS2* code for key enzymes in the conversion of ω3-PUFAs to longer chain length products. Several SNPs in these genes are associated with significant decreases in the percentage of ω3-PUFA incorporated into serum lipids. Similarly, other enzymes involved in fatty acid metabolism such as 5-LOX and COX-2 are polymorphic in the human population, and their variation helps explain interindividual differences in levels and responsiveness to ω3-PUFAs.
- Several of the fatty acid receptors described above have prevalent SNPs that are associated with differential response to dietary ω3-PUFA intervention. For example, carriers of the 162Val variant of *PPARA* and the Ala12 isoform of *PPARG* generally respond to EPA and DHA supplementation with a greater reduction in serum triglycerides.

- Treatment with ω3-PUFAs is often associated with decreasing circulating triglycerides and inflammatory mediators. However, the molecules responsible for producing the beneficial responses vary within the population with polymorphic alleles in genes encoding lipoproteins such as APOE4 and TNFα, among others. In fact, several studies have shown an interaction between ω3-PUFAs' beneficial effects and polymorphisms in the *APOE*, *FABP2*, and *TNF* genes.

If we are to realize the dream of "personalized nutrition" in the context of dietary intervention, perhaps the ω3-PUFAs may serve as a case study to move the field forward. Much is known about how these fatty acids regulate gene expression, and we have identified several key mediators of their anti-inflammatory and cancer-preventive activities. In addition, the genetic variation in important pathways responsible for the metabolism and overall responsiveness to ω3-PUFAs is beginning to be realized. Perhaps it is now time to merge these areas of nutrigenomics and nutrigenics to provide guidance for the amount and type of ω3-PUFAs to be consumed in the diet or as a dietary supplement.

References

1. Gebauer SK, Heuvel JPV, Kris-Etherton PM, Gillies PJ. Integration of molecular biology and nutrition: the role of nutritional genomics in optimizing dietary guidance in lipids. *Future Lipidol* 2007;**2**:165–71.
2. Vanden Heuvel JP. Cardiovascular disease-related genes and regulation by diet. *Curr Atheroscler Rep* 2009;**11**:448–55.
3. Vanden Heuvel JP. Diet, fatty acids, and regulation of genes important for heart disease. *Curr Atheroscler Rep* 2004;**6**:432–40.
4. Renaud S, Lanzmann-Petithory D. Coronary heart disease: dietary links and pathogenesis. *Public Health Nutr* 2001;**4**:459–74.
5. Willett WC. The role of dietary n-6 fatty acids in the prevention of cardiovascular disease. *J Cardiovasc Med (Hagerstown)* 2007;**8**(Suppl. 1):S42–5.
6. Erkkilä A, de Mello VD, Risérus U, Laaksonen DE. Dietary fatty acids and cardiovascular disease: an epidemiological approach. *Prog Lipid Res* 2008;**47**:172–87.
7. Massaro M, Scoditti E, Carluccio MA, Montinari MR, De Caterina R. Omega-3 fatty acids, inflammation and angiogenesis: nutrigenomic effects as an explanation for anti-atherogenic and anti-inflammatory effects of fish and fish oils. *J Nutrigenet Nutrigenomics* 2008;**1**:4–23.
8. Lefevre M, Kris-Etherton PM, Zhao G, Tracy RP. Dietary fatty acids, hemostasis, and cardiovascular disease risk. *J Am Diet Assoc* 2004;**104**:410–9 quiz 492.
9. Gebauer SK, Psota TL, Kris-Etherton PM. The diversity of health effects of individual trans fatty acid isomers. *Lipids* 2007;**42**:787–99.
10. Hu FB, Willett WC. Optimal diets for prevention of coronary heart disease. *JAMA* 2002;**288**:2569–78.
11. Calder PC. N-3 polyunsaturated fatty acids and inflammation: from molecular biology to the clinic. *Lipids* 2003;**38**:343–52.

12. Simopoulos AP. Omega-3 fatty acids in inflammation and autoimmune diseases. *J Am Coll Nutr* 2002;**21**:495–505.
13. Calder PC. Polyunsaturated fatty acids, inflammatory processes and inflammatory bowel diseases. *Mol Nutr Food Res* 2008;**52**:885–97.
14. Wild GE, Drozdowski L, Tartaglia C, Clandinin MT, Thomson AB. Nutritional modulation of the inflammatory response in inflammatory bowel disease—from the molecular to the integrative to the clinical. *World J Gastroenterol* 2007;**13**:1–7.
15. James M, Proudman S, Cleland L. Fish oil and rheumatoid arthritis: past, present and future. *Proc Nutr Soc* 2010;**69**:316–23.
16. Wendel M, Heller AR. Anticancer actions of omega-3 fatty acids—current state and future perspectives. *Anticancer Agents Med Chem* 2009;**9**:457–70.
17. Berquin IM, Edwards IJ, Chen YQ. Multi-targeted therapy of cancer by omega-3 fatty acids. *Cancer Lett* 2008;**269**:363–77.
18. Fernández E, Gallus S, La Vecchia C. Nutrition and cancer risk: an overview. *J Br Menopause Soc* 2006;**12**:139–42.
19. Hardman WE. (n-3) fatty acids and cancer therapy. *J Nutr* 2004;**134**:3427S–30S.
20. Hanf V, Gonder U. Nutrition and primary prevention of breast cancer: foods, nutrients and breast cancer risk. *Eur J Obstet Gynecol Reprod Biol* 2005;**123**:139–49.
21. Pierce JP. Diet and breast cancer prognosis: making sense of the Women's Healthy Eating and Living and Women's Intervention Nutrition Study trials. *Curr Opin Obstet Gynecol* 2009;**21**:86–91.
22. Comba A, Maestri DM, Berra MA, Garcia CP, Das UN, Eynard AR, et al. Effect of ω-3 and ω-9 fatty acid rich oils on lipoxygenases and cyclooxygenases enzymes and on the growth of a mammary adenocarcinoma model. *Lipids Health Dis* 2010;**9**:112.
23. Thiébaut AC, Chajès V, Gerber M, Boutron-Ruault MC, Joulin V, Lenoir G, et al. Dietary intakes of omega-6 and omega-3 polyunsaturated fatty acids and the risk of breast cancer. *Int J Cancer* 2009;**124**:924–31.
24. Bougnoux P, Maillard V, Ferrari P, Jourdan ML, Chajès V. n-3 fatty acids and breast cancer. *IARC Sci Publ* 2002;**156**:337–41.
25. Binukumar B, Mathew A. Dietary fat and risk of breast cancer. *World J Surg Oncol* 2005;**3**:45.
26. Saadatian-Elahi M, Norat T, Goudable J, Riboli E. Biomarkers of dietary fatty acid intake and the risk of breast cancer: a meta-analysis. *Int J Cancer* 2004;**111**:584–91.
27. Velliquette RA, Gillies PJ, Kris-Etherton PM, Green JW, Zhao G, Vanden Heuvel JP. Regulation of human stearoyl-CoA desaturase by omega-3 and omega-6 fatty acids: implications for the dietary management of elevated serum triglycerides. *J Clin Lipidol* 2009;**3**:281–8.
28. Griel AE, Kris-Etherton PM. Beyond saturated fat: the importance of the dietary fatty acid profile on cardiovascular disease. *Nutr Rev* 2006;**64**:257–62.
29. Zhao G, Etherton TD, Martin KR, West SG, Gillies PJ, Kris-Etherton PM. Dietary alpha-linolenic acid reduces inflammatory and lipid cardiovascular risk factors in hypercholesterolemic men and women. *J Nutr* 2004;**134**:2991–7.
30. Klein V, Chajès V, Germain E, Schulgen G, Pinault M, Malvy D, et al. Low alpha-linolenic acid content of adipose breast tissue is associated with an increased risk of breast cancer. *Eur J Cancer* 2000;**36**:335–40.
31. Maillard V, Bougnoux P, Ferrari P, Jourdan ML, Pinault M, Lavillonnière F, et al. N-3 and N-6 fatty acids in breast adipose tissue and relative risk of breast cancer in a case-control study in Tours, France. *Int J Cancer* 2002;**98**:78–83.
32. Dyall SC, Michael-Titus AT. Neurological benefits of omega-3 fatty acids. *Neuromolecular Med* 2008;**10**:219–35.
33. Uauy R, Dangour AD. Nutrition in brain development and aging: role of essential fatty acids. *Nutr Rev* 2006;**64**:S24–33 discussion S72–91.

34. Pauwels EK, Volterrani D, Mariani G, Kairemo K. Fatty acid facts, Part IV: docosahexaenoic acid and Alzheimer's disease. A story of mice, men and fish. *Drug News Perspect* 2009;**22**:205–13.
35. Cakiner-Egilmez T. Omega 3 fatty acids and the eye. *Insight* 2008;**33**:20–5 quiz 26–7.
36. Neuringer M. Infant vision and retinal function in studies of dietary long-chain polyunsaturated fatty acids: methods, results, and implications. *Am J Clin Nutr* 2000;**71**:256S–67S.
37. Bouwens M, van de Rest O, Dellschaft N, Bromhaar MG, de Groot LC, Geleijnse JM, et al. Fish-oil supplementation induces antiinflammatory gene expression profiles in human blood mononuclear cells. *Am J Clin Nutr* 2009;**90**:415–24.
38. Radonjic M, van Erk MJ, Pasman WJ, Wortelboer HM, Hendriks HF, van Ommen B. Effect of body fat distribution on the transcription response to dietary fat interventions. *Genes Nutr* 2009;**4**:143–9.
39. Flachs P, Horakova O, Brauner P, Rossmeisl M, Pecina P, Franssen-van Hal N, et al. Polyunsaturated fatty acids of marine origin upregulate mitochondrial biogenesis and induce beta-oxidation in white fat. *Diabetologia* 2005;**48**:2365–75.
40. Hammamieh R, Chakraborty N, Miller SA, Waddy E, Barmada M, Das R, et al. Differential effects of omega-3 and omega-6 Fatty acids on gene expression in breast cancer cells. *Breast Cancer Res Treat* 2007;**101**:7–16.
41. Narayanan BA, Narayanan NK, Reddy BS. Docosahexaenoic acid regulated genes and transcription factors inducing apoptosis in human colon cancer cells. *Int J Oncol* 2001;**19**:1255–62.
42. Madden J, Williams CM, Calder PC, Lietz G, Miles EA, Cordell H, et al. The impact of common gene variants on the response of biomarkers of cardiovascular disease (CVD) risk to increased fish oil fatty acids intakes. *Annu Rev Nutr* 2011;**31**:203–34.
43. Alsaleh A, O'Dell SD, Frost GS, Griffin BA, Lovegrove JA, Jebb SA, et al. Single nucleotide polymorphisms at the ADIPOQ gene locus interact with age and dietary intake of fat to determine serum adiponectin in subjects at risk of the metabolic syndrome. *Am J Clin Nutr* 2011;**94**:262–9.
44. Stephensen CB, Armstrong P, Newman JW, Pedersen TL, Legault J, Schuster GU, et al. ALOX5 gene variants affect eicosanoid production and response to fish oil supplementation. *J Lipid Res* 2011;**52**:991–1003.
45. Simopoulos AP. Genetic variants in the metabolism of omega-6 and omega-3 fatty acids: their role in the determination of nutritional requirements and chronic disease risk. *Exp Biol Med (Maywood)* 2010;**235**:785–95.
46. Konstantinidou V, Khymenets O, Covas MI, de la Torre R, Muñoz-Aguayo D, Anglada R, et al. Time course of changes in the expression of insulin sensitivity-related genes after an acute load of virgin olive oil. *OMICS* 2009;**13**:431–8.
47. Wang J, John EM, Ingles SA. 5-Lipoxygenase and 5-lipoxygenase-activating protein gene polymorphisms, dietary linoleic acid, and risk for breast cancer. *Cancer Epidemiol Biomarkers Prev* 2008;**17**:2748–54.
48. Corella D, Ordovas JM. Single nucleotide polymorphisms that influence lipid metabolism: interaction with dietary factors. *Annu Rev Nutr* 2005;**25**:341–90.
49. Masson LF, McNeill G, Avenell A. Genetic variation and the lipid response to dietary intervention: a systematic review. *Am J Clin Nutr* 2003;**77**:1098–111.
50. Wang J, Ban MR, Zou GY, Cao H, Lin T, Kennedy BA, et al. Polygenic determinants of severe hypertriglyceridemia. *Hum Mol Genet* 2008;**17**:2894–9.
51. Petersen KF, Dufour S, Hariri A, Nelson-Williams C, Foo JN, Zhang XM, et al. Apolipoprotein C3 gene variants in nonalcoholic fatty liver disease. *N Engl J Med* 2010;**362**:1082–9.
52. Plourde M, Vohl MC, Vandal M, Couture P, Lemieux S, Cunnane SC. Plasma n-3 fatty acid response to an n-3 fatty acid supplement is modulated by apoE epsilon4 but not by the common PPAR-alpha L162V polymorphism in men. *Br J Nutr* 2009;**102**:1121–4.

53. Judson R, Brain C, Dain B, Windemuth A, Ruaño G, Reed C. New and confirmatory evidence of an association between APOE genotype and baseline C-reactive protein in dyslipidemic individuals. *Atherosclerosis* 2004;**177**:345–51.
54. Phillips CM, Goumidi L, Bertrais S, Ferguson JF, Field MR, Kelly ED, et al. Complement component 3 polymorphisms interact with polyunsaturated fatty acids to modulate risk of metabolic syndrome. *Am J Clin Nutr* 2009;**90**:1665–73.
55. Noel SE, Lai CQ, Mattei J, Parnell LD, Ordovas JM, Tucker KL. Variants of the CD36 gene and metabolic syndrome in Boston Puerto Rican adults. *Atherosclerosis* 2010; **211**:210–5.
56. Bokor S, Legry V, Meirhaeghe A, Ruiz JR, Mauro B, Widhalm K, et al. Single-nucleotide polymorphism of CD36 locus and obesity in European adolescents. *Obesity (Silver Spring)* 2010;**18**:1398–403.
57. Lecompte S, Szabo de Edelenyi F, Goumidi L, Maiani G, Moschonis G, Widhalm K, et al. Polymorphisms in the CD36/FAT gene are associated with plasma vitamin E concentrations in humans. *Am J Clin Nutr* 2011;**93**:644–51.
58. Madden J, Shearman CP, Dunn RL, Dastur ND, Tan RM, Nash GB, et al. Altered monocyte CD44 expression in peripheral arterial disease is corrected by fish oil supplementation. *Nutr Metab Cardiovasc Dis* 2009;**19**:247–52.
59. Garaulet M, Corbalán MD, Madrid JA, Morales E, Baraza JC, Lee YC, et al. CLOCK gene is implicated in weight reduction in obese patients participating in a dietary programme based on the Mediterranean diet. *Int J Obes (Lond)* 2010;**34**:516–23.
60. Schwarz D, Kisselev P, Chernogolov A, Schunck WH, Roots I. Human CYP1A1 variants lead to differential eicosapentaenoic acid metabolite patterns. *Biochem Biophys Res Commun* 2005;**336**:779–83.
61. Tanaka T, Shen J, Abecasis GR, Kisialiou A, Ordovas JM, Guralnik JM, et al. Genome-wide association study of plasma polyunsaturated fatty acids in the InCHIANTI Study. *PLoS Genet* 2009;**5**:e1000338.
62. Lindman AS, Pedersen JI, Hjerkinn EM, Arnesen H, Veierød MB, Ellingsen I, et al. The effects of long-term diet and omega-3 fatty acid supplementation on coagulation factor VII and serum phospholipids with special emphasis on the R353Q polymorphism of the FVII gene. *Thromb Haemost* 2004;**91**:1097–104.
63. Yamada Y, Kato K, Oguri M, Yoshida T, Yokoi K, Watanabe S, et al. Association of genetic variants with atherothrombotic cerebral infarction in Japanese individuals with metabolic syndrome. *Int J Mol Med* 2008;**21**:801–8.
64. Zeman M, Vecka M, Jáchymová M, Jirák R, Tvrzická E, Stanková B, et al. Fatty acid CoA ligase-4 gene polymorphism influences fatty acid metabolism in metabolic syndrome, but not in depression. *Tohoku J Exp Med* 2009;**217**:287–93.
65. Martinelli N, Girelli D, Malerba G, Guarini P, Illig T, Trabetti E, et al. FADS genotypes and desaturase activity estimated by the ratio of arachidonic acid to linoleic acid are associated with inflammation and coronary artery disease. *Am J Clin Nutr* 2008;**88**:941–9.
66. Malerba G, Schaeffer L, Xumerle L, Klopp N, Trabetti E, Biscuola M, et al. SNPs of the FADS gene cluster are associated with polyunsaturated fatty acids in a cohort of patients with cardiovascular disease. *Lipids* 2008;**43**:289–99.
67. Schaeffer L, Gohlke H, Muller M, Heid IM, Palmer LJ, Kompauer I, et al. Common genetic variants of the FADS1 FADS2 gene cluster and their reconstructed haplotypes are associated with the fatty acid composition in phospholipids. *Hum Mol Genet* 2006;**15**: 1745–56.
68. Xie L, Innis SM. Genetic variants of the FADS1 FADS2 gene cluster are associated with altered (n-6) and (n-3) essential fatty acids in plasma and erythrocyte phospholipids in women during pregnancy and in breast milk during lactation. *J Nutr* 2008;**138**:2222–8.

69. Mathias RA, Vergara C, Gao L, Rafaels N, Hand T, Campbell M, et al. FADS genetic variants and omega-6 polyunsaturated fatty acid metabolism in a homogeneous island population. *J Lipid Res* 2010;**51**:2766–74.
70. Lu Y, Feskens EJ, Dollé ME, Imholz S, Verschuren WM, Müller M, et al. Dietary n-3 and n-6 polyunsaturated fatty acid intake interacts with FADS1 genetic variation to affect total and HDL-cholesterol concentrations in the Doetinchem Cohort Study. *Am J Clin Nutr* 2010;**92**: 258–65.
71. Lattka E, Eggers S, Moeller G, Heim K, Weber M, Mehta D, et al. A common FADS2 promoter polymorphism increases promoter activity and facilitates binding of transcription factor ELK1. *J Lipid Res* 2010;**51**:182–91.
72. Kwak JH, Paik JK, Kim OY, Jang Y, Lee SH, Ordovas JM, et al. FADS gene polymorphisms in Koreans: association with ω6 polyunsaturated fatty acids in serum phospholipids, lipid peroxides, and coronary artery disease. *Atherosclerosis* 2011;**214**:94–100.
73. Vanschoonbeek K, Feijge MA, Paquay M, Rosing J, Saris W, Kluft C, et al. Variable hypocoagulant effect of fish oil intake in humans: modulation of fibrinogen level and thrombin generation. *Arterioscler Thromb Vasc Biol* 2004;**24**:1734–40.
74. Ferguson JF, Phillips CM, McMonagle J, Pérez-Martínez P, Shaw DI, Lovegrove JA, et al. NOS3 gene polymorphisms are associated with risk markers of cardiovascular disease, and interact with omega-3 polyunsaturated fatty acids. *Atherosclerosis* 2010;**211**:539–44.
75. Siezen CL, van Leeuwen AI, Kram NR, Luken ME, van Kranen HJ, Kampman E. Colorectal adenoma risk is modified by the interplay between polymorphisms in arachidonic acid pathway genes and fish consumption. *Carcinogenesis* 2005;**26**:449–57.
76. van den Berg SW, Dollé ME, Imholz S, van der A DL, van't Slot R, Wijmenga C, et al. Genetic variations in regulatory pathways of fatty acid and glucose metabolism are associated with obesity phenotypes: a population-based cohort study. *Int J Obes (Lond)* 2009;**33**: 1143–52.
77. Tai ES, Demissie S, Cupples LA, Corella D, Wilson PW, Schaefer EJ, et al. Association between the PPARA L162V polymorphism and plasma lipid levels: the Framingham Offspring Study. *Arterioscler Thromb Vasc Biol* 2002;**22**:805–10.
78. Gouni-Berthold I, Giannakidou E, Müller-Wieland D, Faust M, Kotzka J, Berthold HK, et al. Association between the PPARalpha L162V polymorphism, plasma lipoprotein levels, and atherosclerotic disease in patients with diabetes mellitus type 2 and in nondiabetic controls. *Am Heart J* 2004;**147**:1117–24.
79. Enquobahrie DA, Smith NL, Bis JC, Carty CL, Rice KM, Lumley T, et al. Cholesterol ester transfer protein, interleukin-8, peroxisome proliferator activator receptor alpha, and Toll-like receptor 4 genetic variations and risk of incident nonfatal myocardial infarction and ischemic stroke. *Am J Cardiol* 2008;**101**:1683–8.
80. Volcik KA, Nettleton JA, Ballantyne CM, Boerwinkle E. Peroxisome proliferator-activated receptor [alpha] genetic variation interacts with n-6 and long-chain n-3 fatty acid intake to affect total cholesterol and LDL-cholesterol concentrations in the Atherosclerosis Risk in Communities Study. *Am J Clin Nutr* 2008;**87**:1926–31.
81. Aberle J, Hopfer I, Beil FU, Seedorf U. Association of peroxisome proliferator-activated receptor delta +294T/C with body mass index and interaction with peroxisome proliferator-activated receptor alpha L162V. *Int J Obes (Lond)* 2006;**30**:1709–13.
82. Thamer C, Machann J, Stefan N, Schäfer SA, Machicao F, Staiger H, et al. Variations in PPARD determine the change in body composition during lifestyle intervention: a whole-body magnetic resonance study. *J Clin Endocrinol Metab* 2008;**93**:1497–500.
83. Deeken JF, Cormier T, Price DK, Sissung TM, Steinberg SM, Tran K, et al. A pharmacogenetic study of docetaxel and thalidomide in patients with castration-resistant prostate cancer using the DMET genotyping platform. *Pharmacogenomics J* 2010;**10**:191–9.

84. Helisalmi S, Vepsäläinen S, Hiltunen M, Koivisto AM, Salminen A, Laakso M, et al. Genetic study between SIRT1, PPARD, PGC-1alpha genes and Alzheimer's disease. *J Neurol* 2008; **255**:668–73.
85. Andrulionyte L, Peltola P, Chiasson JL, Laakso M. Single nucleotide polymorphisms of PPARD in combination with the Gly482Ser substitution of PGC-1A and the Pro12Ala substitution of PPARG2 predict the conversion from impaired glucose tolerance to type 2 diabetes: the STOP-NIDDM trial. *Diabetes* 2006;**55**:2148–52.
86. Gallicchio L, Kalesan B, Huang HY, Strickland P, Hoffman SC, Helzlsouer KJ. Genetic polymorphisms of peroxisome proliferator-activated receptors and the risk of cardiovascular morbidity and mortality in a community-based cohort in washington county, Maryland. *PPAR Res* 2008;**2008**:276581.
87. Memisoglu A, Hu FB, Hankinson SE, Manson JE, De Vivo I, Willett WC, et al. Interaction between a peroxisome proliferator-activated receptor gamma gene polymorphism and dietary fat intake in relation to body mass. *Hum Mol Genet* 2003;**12**:2923–9.
88. Vogels N, Mariman EC, Bouwman FG, Kester AD, Diepvens K, Westerterp-Plantenga MS. Relation of weight maintenance and dietary restraint to peroxisome proliferator-activated receptor gamma2, glucocorticoid receptor, and ciliary neurotrophic factor polymorphisms. *Am J Clin Nutr* 2005;**82**:740–6.
89. Rosado EL, Bressan J, Martins MF, Cecon PR, Martínez JA. Polymorphism in the PPAR-gamma2 and beta2-adrenergic genes and diet lipid effects on body composition, energy expenditure and eating behavior of obese women. *Appetite* 2007;**49**:635–43.
90. Stefan N, Thamer C, Staiger H, Machicao F, Machann J, Schick F, et al. Genetic variations in PPARD and PPARGC1A determine mitochondrial function and change in aerobic physical fitness and insulin sensitivity during lifestyle intervention. *J Clin Endocrinol Metab* 2007; **92**:1827–33.
91. Gallicchio L, Chang H, Christo DK, Thuita L, Huang HY, Strickland P, et al. Single nucleotide polymorphisms in inflammation-related genes and mortality in a community-based cohort in Washington county, Maryland. *Am J Epidemiol* 2008;**167**:807–13.
92. Fradet V, Cheng I, Casey G, Witte JS. Dietary omega-3 fatty acids, cyclooxygenase-2 genetic variation, and aggressive prostate cancer risk. *Clin Cancer Res* 2009;**15**:2559–66.
93. Hedelin M, Chang ET, Wiklund F, Bellocco R, Klint A, Adolfsson J, et al. Association of frequent consumption of fatty fish with prostate cancer risk is modified by COX-2 polymorphism. *Int J Cancer* 2007;**120**:398–405.
94. Zhao G, Etherton TD, Martin KR, Gillies PJ, West SG, Kris-Etherton PM. Dietary alpha-linolenic acid inhibits proinflammatory cytokine production by peripheral blood mononuclear cells in hypercholesterolemic subjects. *Am J Clin Nutr* 2007;**85**:385–91.
95. Zhao G, Etherton TD, Martin KR, Vanden Heuvel JP, Gillies PJ, West SG, et al. Anti-inflammatory effects of polyunsaturated fatty acids in THP-1 cells. *Biochem Biophys Res Commun* 2005;**336**:909–17.
96. Skulas-Ray AC, Kris-Etherton PM, Harris WS, Vanden Heuvel JP, Wagner PR, West SG. Dose-response effects of omega-3 fatty acids on triglycerides, inflammation, and endothelial function in healthy persons with moderate hypertriglyceridemia. *Am J Clin Nutr* 2010; **93**:243–52.
97. Li H, Ruan XZ, Powis SH, Fernando R, Mon WY, Wheeler DC, et al. EPA and DHA reduce LPS-induced inflammation responses in HK-2 cells: evidence for a PPAR-gamma-dependent mechanism. *Kidney Int* 2005;**67**:867–74.
98. Zhao A, Yu J, Lew JL, Huang L, Wright SD, Cui J. Polyunsaturated fatty acids are FXR ligands and differentially regulate expression of FXR targets. *DNA Cell Biol* 2004;**23**:519–26.
99. Goua M, Mulgrew S, Frank J, Rees D, Sneddon AA, Wahle KW. Regulation of adhesion molecule expression in human endothelial and smooth muscle cells by omega-3 fatty acids and

conjugated linoleic acids: involvement of the transcription factor NF-kappaB? *Prostaglandins Leukot Essent Fatty Acids* 2008;**78**:33–43.
100. Baker PR, Schopfer FJ, O'Donnell VB, Freeman BA. Convergence of nitric oxide and lipid signaling: anti-inflammatory nitro-fatty acids. *Free Radic Biol Med* 2009;**46**:989–1003.
101. Tanaka N, Zhang X, Sugiyama E, Kono H, Horiuchi A, Nakajima T, et al. Eicosapentaenoic acid improves hepatic steatosis independent of PPARα activation through inhibition of SREBP-1 maturation in mice. *Biochem Pharmacol* 2010;**80**:1601–12.
102. Takahashi M, Tsuboyama-Kasaoka N, Nakatani T, Ishii M, Tsutsumi S, Aburatani H, et al. Fish oil feeding alters liver gene expressions to defend against PPARalpha activation and ROS production. *Am J Physiol Gastrointest Liver Physiol* 2002;**282**:G338–48.
103. Duda MK, O'Shea KM, Stanley WC. Omega-3 polyunsaturated fatty acid supplementation for the treatment of heart failure: mechanisms and clinical potential. *Cardiovasc Res* 2009;**84**:33–41.
104. Tanaka N, Zhang X, Sugiyama E, Kono H, Horiuchi A, Nakajima T, et al. Eicosapentaenoic acid improves hepatic steatosis independent of PPARalpha activation through inhibition of SREBP-1 maturation in mice. *Biochem Pharmacol* 2010;**80**:1601–12.
105. Kjaer MA, Vegusdal A, Gjøen T, Rustan AC, Todorcević M, Ruyter B. Effect of rapeseed oil and dietary n-3 fatty acids on triacylglycerol synthesis and secretion in Atlantic salmon hepatocytes. *Biochim Biophys Acta* 2008;**1781**:112–22.
106. Skulas-Ray AC, West SG, Davidson MH, Kris-Etherton PM. Omega-3 fatty acid concentrates in the treatment of moderate hypertriglyceridemia. *Expert Opin Pharmacother* 2008;**9**:1237–48.
107. Skulas-Ray AC, Kris-Etherton PM, Harris WS, Vanden Heuvel JP, Wagner PR, West SG. Dose-response effects of omega-3 fatty acids on triglycerides, inflammation, and endothelial function in healthy persons with moderate hypertriglyceridemia. *Am J Clin Nutr* 2011;**93**:243–52.
108. Sun C, Wei ZW, Li Y. DHA regulates lipogenesis and lipolysis genes in mice adipose and liver. *Mol Biol Rep* 2010;**38**:731–7.
109. Ide T. Interaction of fish oil and conjugated linoleic acid in affecting hepatic activity of lipogenic enzymes and gene expression in liver and adipose tissue. *Diabetes* 2005;**54**:412–23.
110. Sampath H, Ntambi JM. Polyunsaturated fatty acid regulation of genes of lipid metabolism. *Annu Rev Nutr* 2005;**25**:317–40.
111. Muhlhausler BS, Cook-Johnson R, James M, Miljkovic D, Duthoit E, Gibson R. Opposing effects of omega-3 and omega-6 long chain polyunsaturated Fatty acids on the expression of lipogenic genes in omental and retroperitoneal adipose depots in the rat. *J Nutr Metab* 2010;.
112. Cohen P, Friedman JM. Leptin and the control of metabolism: role for stearoyl-CoA desaturase-1 (SCD-1). *J Nutr* 2004;**134**:2455S–63S.
113. Zhang J, Kris-Etherton PM, Thompson JT, Hannon DB, Gillies PJ, Vanden Heuvel JP. Alpha-linolenic acid increases cholesterol efflux in macrophage-derived foam cells by decreasing stearoyl CoA desaturase 1 expression: evidence for a farnesoid-X-receptor mechanism of action. *J Nutr Biochem* 2011;.
114. Tu WC, Cook-Johnson RJ, James MJ, Mühlhäusler BS, Gibson RA. Omega-3 long chain fatty acid synthesis is regulated more by substrate levels than gene expression. *Prostaglandins Leukot Essent Fatty Acids* 2010;**83**:61–8.
115. Manni A, Xu H, Washington S, Aliaga C, Cooper T, Richie JP, et al. The impact of fish oil on the chemopreventive efficacy of tamoxifen against development of N-Methyl-N-Nitrosourea-induced rat mammary carcinogenesis. *Cancer Prev Res (Phila)* 2010;**3**:322–30.
116. Shaikh IA, Brown I, Wahle KW, Heys SD. Enhancing cytotoxic therapies for breast and prostate cancers with polyunsaturated fatty acids. *Nutr Cancer* 2010;**62**:284–96.

117. Schulman IG. Nuclear receptors as drug targets for metabolic disease. *Adv Drug Deliv Rev* 2010;**62**:1307–15.
118. Beaven SW, Tontonoz P. Nuclear receptors in lipid metabolism: targeting the heart of dyslipidemia. *Annu Rev Med* 2006;**57**:313–29.
119. Francis GA, Fayard E, Picard F, Auwerx J. Nuclear receptors and the control of metabolism. *Annu Rev Physiol* 2003;**65**:261–311.
120. Hihi AK, Michalik L, Wahli W. PPARs: transcriptional effectors of fatty acids and their derivatives. *Cell Mol Life Sci* 2002;**59**:790–8.
121. Montagner A, Rando G, Degueurce G, Leuenberger N, Michalik L, Wahli W. New insights into the role of PPARs. *Prostaglandins Leukot Essent Fatty Acids* 2011;**85**:235–43.
122. Wahli W. Peroxisome proliferator-activated receptors (PPARs): from metabolic control to epidermal wound healing. *Swiss Med Wkly* 2002;**132**:83–91.
123. Khan SA, Vanden Heuvel JP. Role of nuclear receptors in the regulation of gene expression by dietary fatty acids (review). *J Nutr Biochem* 2003;**14**:554–67.
124. Latruffe N, Vamecq J. Peroxisome proliferators and peroxisome proliferator activated receptors (PPARs) as regulators of lipid metabolism. *Biochimie* 1997;**79**:81–94.
125. Ziouzenkova O, Perrey S, Asatryan L, Hwang J, MacNaul KL, Moller DE, et al. Lipolysis of triglyceride-rich lipoproteins generates PPAR ligands: evidence for an antiinflammatory role for lipoprotein lipase. *Proc Natl Acad Sci USA* 2003;**100**:2730–5.
126. Cowart LA, Wei S, Hsu MH, Johnson EF, Krishna MU, Falck JR, et al. The CYP4A isoforms hydroxylate epoxyeicosatrienoic acids to form high affinity peroxisome proliferator-activated receptor ligands. *J Biol Chem* 2002;**277**:35105–12.
127. Krey G, Braissant O, L'Horset F, Kalkhoven E, Perroud M, Parker MG, et al. Fatty acids, eicosanoids, and hypolipidemic agents identified as ligands of peroxisome proliferator-activated receptors by coactivator-dependent receptor ligand assay. *Mol Endocrinol* 1997;**11**:779–91.
128. Yu K, Bayona W, Kallen CB, Harding HP, Ravera CP, McMahon G, et al. Differential activation of peroxisome proliferator-activated receptors by eicosanoids. *J Biol Chem* 1995;**270**:23975–83.
129. Willson TM, Brown PJ, Sternbach DD, Henke BR. The PPARs: from orphan receptors to drug discovery. *J Med Chem* 2000;**43**:527–50.
130. Spiegelman BM, Hu E, Kim JB, Brun R. PPAR gamma and the control of adipogenesis. *Biochimie* 1997;**79**:111–2.
131. Barak Y, Nelson MC, Ong ES, Jones YZ, Ruiz-Lozano P, Chien KR, et al. PPAR gamma is required for placental, cardiac, and adipose tissue development. *Mol Cell* 1999;**4**: 585–95.
132. Willson TM, Wahli W. Peroxisome proliferator-activated receptor agonists. *Curr Opin Chem Biol* 1997;**1**:235–41.
133. McIntyre TM, Pontsler AV, Silva AR, St Hilaire A, Xu Y, Hinshaw JC, et al. Identification of an intracellular receptor for lysophosphatidic acid (LPA): LPA is a transcellular PPARgamma agonist. *Proc Natl Acad Sci USA* 2003;**100**:131–6.
134. Davies SS, Pontsler AV, Marathe GK, Harrison KA, Murphy RC, Hinshaw JC, et al. Oxidized alkyl phospholipids are specific, high affinity peroxisome proliferator-activated receptor gamma ligands and agonists. *J Biol Chem* 2001;**276**:16015–23.
135. Sapieha P, Stahl A, Chen J, Seaward MR, Willett KL, Krah NM, et al. 5-Lipoxygenase metabolite 4-HDHA is a mediator of the antiangiogenic effect of {omega}-3 polyunsaturated fatty acids. *Sci Transl Med* 2011;**3**:69ra12.
136. Groeger AL, Cipollina C, Cole MP, Woodcock SR, Bonacci G, Rudolph TK, et al. Cyclooxygenase-2 generates anti-inflammatory mediators from omega-3 fatty acids. *Nat Chem Biol* 2010;**6**:433–41.

137. Peters JM, Lee SS, Li W, Ward JM, Gavrilova O, Everett C, et al. Growth, adipose, brain, and skin alterations resulting from targeted disruption of the mouse peroxisome proliferator-activated receptor beta(delta). *Mol Cell Biol* 2000;**20**:5119–28.
138. Brun RP, Tontonoz P, Forman BM, Ellis R, Chen J, Evans RM, et al. Differential activation of adipogenesis by multiple PPAR isoforms. *Genes Dev* 1996;**10**:974–84.
139. Coleman JD, Prabhu KS, Thompson JT, Reddy PS, Peters JM, Peterson BR, et al. The oxidative stress mediator 4-hydroxynonenal is an intracellular agonist of the nuclear receptor peroxisome proliferator-activated receptor-beta/delta (PPARbeta/delta). *Free Radic Biol Med* 2007;**42**:1155–64.
140. Soskic SS, Dobutovic BD, Sudar EM, Obradovic MM, Nikolic DM, Zaric BL, et al. The peroxisome proliferator-activated receptors and atherosclerosis. *Angiology* 2011;**62**:523–34.
141. Zandbergen F, Plutzky J. PPARalpha in atherosclerosis and inflammation. *Biochim Biophys Acta* 2007;**1771**:972–82.
142. Seedorf U, Aberle J. Emerging roles of PPARdelta in metabolism. *Biochim Biophys Acta* 2007;**1771**:1125–31.
143. Shearer BG, Hoekstra WJ. Recent advances in peroxisome proliferator-activated receptor science. *Curr Med Chem* 2003;**10**:267–80.
144. Schoonjans K, Staels B, Auwerx J. The peroxisome proliferator activated receptors (PPARS) and their effects on lipid metabolism and adipocyte differentiation. *Biochim Biophys Acta* 1996;**1302**:93–109.
145. Heikkinen S, Auwerx J, Argmann CA. PPARgamma in human and mouse physiology. *Biochim Biophys Acta* 2007;**1771**:999–1013.
146. Széles L, Töröcsik D, Nagy L. PPARgamma in immunity and inflammation: cell types and diseases. *Biochim Biophys Acta* 2007;**1771**:1014–30.
147. Hammarstedt A, Andersson CX, Rotter Sopasakis V, Smith U. The effect of PPARgamma ligands on the adipose tissue in insulin resistance. *Prostaglandins Leukot Essent Fatty Acids* 2005;**73**:65–75.
148. Ricote M, Valledor AF, Glass CK. Decoding transcriptional programs regulated by PPARs and LXRs in the macrophage: effects on lipid homeostasis, inflammation, and atherosclerosis. *Arterioscler Thromb Vasc Biol* 2004;**24**:230–9.
149. Ricote M, Huang JT, Welch JS, Glass CK. The peroxisome proliferator-activated receptor (PPARgamma) as a regulator of monocyte/macrophage function. *J Leukoc Biol* 1999;**66**:733–9.
150. Lehmann JM, Lenhard JM, Oliver BB, Ringold GM, Kliewer SA. Peroxisome proliferator-activated receptors alpha and gamma are activated by indomethacin and other non-steroidal anti-inflammatory drugs. *J Biol Chem* 1997;**272**:3406–10.
151. Rastinejad F. Retinoid X receptor and its partners in the nuclear receptor family. *Curr Opin Struct Biol* 2001;**11**:33–8.
152. Lengqvist J, De Urquiza AM, Bergman AC, Wilson TM, Sjovall J, Perlmann T, et al. Polyunsaturated fatty acids including docosahexaenoic and arachidonic acid bind to the retinoid X receptor alpha ligand binding domain. *Mol Cell Proteomics* 2004;**3**:692–703.
153. de Urquiza AM, Liu S, Sjoberg M, Zetterstrom RH, Griffiths W, Sjovall J, et al. Docosahexaenoic acid, a ligand for the retinoid X receptor in mouse brain. *Science* 2000;**290**:2140–214.
154. Lemotte PK, Keidel S, Apfel CM. Phytanic acid is a retinoid X receptor ligand. *Eur J Biochem* 1996;**236**:328–33.
155. Claudel T, Leibowitz MD, Fievet C, Tailleux A, Wagner B, Repa JJ, et al. Reduction of atherosclerosis in apolipoprotein E knockout mice by activation of the retinoid X receptor. *Proc Natl Acad Sci USA* 2001;**98**:2610–5.
156. Kastner P, Grondona JM, Mark M, Gansmuller A, LeMeur M, Decimo D, et al. Genetic analysis of RXR alpha developmental function: convergence of RXR and RAR signaling pathways in heart and eye morphogenesis. *Cell* 1994;**78**:987–1003.

157. Lund EG, Menke JG, Sparrow CP. Liver X receptor agonists as potential therapeutic agents for dyslipidemia and atherosclerosis. *Arterioscler Thromb Vasc Biol* 2003;**23**:1169–77.
158. Millatt LJ, Bocher V, Fruchart JC, Staels B. Liver X receptors and the control of cholesterol homeostasis: potential therapeutic targets for the treatment of atherosclerosis. *Biochim Biophys Acta* 2003;**1631**:107–18.
159. Ou J, Tu H, Shan B, Luk A, DeBose-Boyd RA, Bashmakov Y, et al. Unsaturated fatty acids inhibit transcription of the sterol regulatory element-binding protein-1c (SREBP-1c) gene by antagonizing ligand-dependent activation of the LXR. *Proc Natl Acad Sci USA* 2001; **98**:6027–32.
160. Pawar A, Jump DB. Unsaturated fatty acid regulation of peroxisome proliferator-activated receptor alpha activity in rat primary hepatocytes. *J Biol Chem* 2003;**278**:35931–9.
161. Miyata KS, McCaw SE, Patel HV, Rachubinski RA, Capone JP. The orphan nuclear hormone receptor LXR alpha interacts with the peroxisome proliferator-activated receptor and inhibits peroxisome proliferator signaling. *J Biol Chem* 1996;**271**:9189–92.
162. Laffitte BA, Repa JJ, Joseph SB, Wilpitz DC, Kast HR, Mangelsdorf DJ, et al. LXRs control lipid-inducible expression of the apolipoprotein E gene in macrophages and adipocytes. *Proc Natl Acad Sci USA* 2001;**98**:507–12.
163. Joseph SB, Tontonoz P. LXRs: new therapeutic targets in atherosclerosis? *Curr Opin Pharmacol* 2003;**3**:192–7.
164. Joseph SB, McKilligin E, Pei L, Watson MA, Collins AR, Laffitte BA, et al. Synthetic LXR ligand inhibits the development of atherosclerosis in mice. *Proc Natl Acad Sci USA* 2002; **99**:7604–9.
165. Barish GD, Evans RM. PPARs and LXRs: atherosclerosis goes nuclear. *Trends Endocrinol Metab* 2004;**15**:158–65.
166. Chawla A, Barak Y, Nagy L, Liao D, Tontonoz P, Evans RM. PPAR-gamma dependent and independent effects on macrophage-gene expression in lipid metabolism and inflammation. *Nat Med* 2001;**7**:48–52.
167. Peet DJ, Turley SD, Ma W, Janowski BA, Lobaccaro JM, Hammer RE, et al. Cholesterol and bile acid metabolism are impaired in mice lacking the nuclear oxysterol receptor LXR alpha. *Cell* 1998;**93**:693–704.
168. Alberti S, Schuster G, Parini P, Feltkamp D, Diczfalusy U, Rudling M, et al. Hepatic cholesterol metabolism and resistance to dietary cholesterol in LXRbeta-deficient mice. *J Clin Invest* 2001;**107**:565–73.
169. McDonnell DP, Vegeto E, Gleeson MA. Nuclear hormone receptors as targets for new drug discovery. *Biotechnology (N Y)* 1993;**11**:1256–61.
170. Riccardi L, Mencarelli A, Renga B, Distrutti E, Fiorucci S. Anti-atherosclerotic effect of Farnesoid-X-Receptor in ApoE-/- mice. *Am J Physiol Heart Circ Physiol* 2008;.
171. Mencarelli A, Renga B, Distrutti E, Fiorucci S. Antiatherosclerotic effect of farnesoid X receptor. *Am J Physiol Heart Circ Physiol* 2009;**296**:H272–81.
172. Pelton PD, Patel M, Demarest KT. Nuclear receptors as potential targets for modulating reverse cholesterol transport. *Curr Top Med Chem* 2005;**5**:265–82.
173. Subbiah MT. Nutrigenetics and nutraceuticals: the next wave riding on personalized medicine. *Transl Res* 2007;**149**:55–61.
174. Corthésy-Theulaz I, den Dunnen JT, Ferré P, Geurts JM, Müller M, van Belzen N, et al. Nutrigenomics: the impact of biomics technology on nutrition research. *Ann Nutr Metab* 2005;**49**:355–65.
175. Ordovas JM, Shen J. Gene-environment interactions and susceptibility to metabolic syndrome and other chronic diseases. *J Periodontol* 2008;**79**:1508–13.
176. Smith CE, Ordovás JM. Fatty acid interactions with genetic polymorphisms for cardiovascular disease. *Curr Opin Clin Nutr Metab Care* 2010;**13**:139–44.

177. Ordovas JM. Nutrigenetics, plasma lipids, and cardiovascular risk. *J Am Diet Assoc* 2006;**106**:1074–81 quiz 1083.
178. Loktionov A. Common gene polymorphisms and nutrition: emerging links with pathogenesis of multifactorial chronic diseases (review). *J Nutr Biochem* 2003;**14**:426–51.
179. Vincent S, Planells R, Defoort C, Bernard MC, Gerber M, Prudhomme J, et al. Genetic polymorphisms and lipoprotein responses to diets. *Proc Nutr Soc* 2002;**61**:427–34.
180. Ordovas JM. The quest for cardiovascular health in the genomic era: nutrigenetics and plasma lipoproteins. *Proc Nutr Soc* 2004;**63**:145–52.
181. Lemaitre RN, Siscovick DS, Berry EM, Kark JD, Friedlander Y. Familial aggregation of red blood cell membrane fatty acid composition: the Kibbutzim Family Study. *Metabolism* 2008;**57**:662–8.
182. Marquardt A, Stohr H, White K, Weber BH. cDNA cloning, genomic structure, and chromosomal localization of three members of the human fatty acid desaturase family. *Genomics* 2000;**66**:175–83.
183. Masugi J, Tamori Y, Mori H, Koike T, Kasuga M. Inhibitory effect of a proline-to-alanine substitution at codon 12 of peroxisome proliferator-activated receptor-gamma 2 on thiazolidinedione-induced adipogenesis. *Biochem Biophys Res Commun* 2000;**268**:178–82.
184. Lindi V, Schwab U, Louheranta A, Laakso M, Vessby B, Hermansen K, et al. Impact of the Pro12Ala polymorphism of the PPAR-gamma2 gene on serum triacylglycerol response to n-3 fatty acid supplementation. *Mol Genet Metab* 2003;**79**:52–60.
185. Yanagisawa Y, Kawabata T, Tanaka O, Kawakami M, Hasegawa K, Kagawa Y. Improvement in blood lipid levels by dietary sn-1,3-diacylglycerol in young women with variants of lipid transporters 54T-FABP2 and -493g-MTP. *Biochem Biophys Res Commun* 2003;**302**:743–50.
186. Fontaine-Bisson B, Wolever TM, Chiasson JL, Rabasa-Lhoret R, Maheux P, Josse RG, et al. Genetic polymorphisms of tumor necrosis factor-alpha modify the association between dietary polyunsaturated fatty acids and fasting HDL-cholesterol and apo A-I concentrations. *Am J Clin Nutr* 2007;**86**:768–74.
187. Fontaine-Bisson B, El-Sohemy A. Genetic polymorphisms of tumor necrosis factor-alpha modify the association between dietary polyunsaturated fatty acids and plasma high-density lipoprotein-cholesterol concentrations in a population of young adults. *J Nutrigenet Nutrigenomics* 2008;**1**:215–23.
188. Markovic O, O'Reilly G, Fussell HM, Turner SJ, Calder PC, Howell WM, et al. Role of single nucleotide polymorphisms of pro-inflammatory cytokine genes in the relationship between serum lipids and inflammatory parameters, and the lipid-lowering effect of fish oil in healthy males. *Clin Nutr* 2004;**23**:1084–95.

Carbohydrate Intake

Armelle Leturque, Edith Brot-Laroche, and Maude Le Gall

Department of Physiology, Metabolism, Differentiation, Centre de Recherche des Cordeliers, Paris, France

Carbohydrates represent more than 50% of the energy sources present in most human diets. Sugar intake is regulated by metabolic, neuronal, and hedonic factors, and gene polymorphisms are involved in determining sugar preference. Nutrigenomic adaptations to carbohydrate availability have been evidenced in metabolic diseases, in the persistence of lactose digestion, and in amylase gene copy number. Furthermore, dietary oligosaccharides, fermentable by gut flora, can modulate the microbiotal diversity to the benefit of the host. Genetic diseases linked to mutations in the disaccharidase genes (sucrase-isomaltase, lactase) and in sugar transporter genes (sodium/glucose cotransporter 1, glucose transporters 1 and 2) severely impact carbohydrate intake. These diseases are revealed upon exposure to food containing the offending sugar, and withdrawal of this sugar from the diet prevents disease symptoms, failure to thrive, and premature death. Tailoring the sugar composition of diets to optimize wellness and to prevent the chronic occurrence of metabolic diseases is a future goal that may yet be realized through continued development of nutrigenetics and nutrigenomics approaches.

Progress in Molecular Biology
and Translational Science, Vol. 108
DOI: 10.1016/B978-0-12-398397-8.00005-8

I. Dietary Carbohydrates

Dietary carbohydrates, which make up more than half of the energy sources in most human diets, include monosaccharides (glucose, fructose, galactose), common disaccharides (sucrose, lactose, maltose), rare disaccharides (trehalose), polysaccharides, and oligosaccharides (starches). In addition, nondigestible polysaccharides classified as dietary fibers or fructans (pectins, cellulose) are present in food, and some can be fermented by gut microbiota.

According to the Food and Agriculture Organization of the United Nations, the major sources of carbohydrate consumed in both developed and developing countries are cereals (rice, wheat, maize), root crops, and, to a lesser extent, fruits, vegetables, and milk products. High-fructose corn syrup, a liquid alternative to sucrose that is composed of free glucose and fructose, was introduced in industrial food and beverages in the 1970s. The increased consumption of free monosaccharides is one of the environmental factors that have been linked to the increased prevalence of obesity.[1]

To be absorbed in the upper small intestine, polysaccharides and oligosaccharides must be hydrolyzed into their component monosaccharides by amylase, α-glucosidases, and disaccharidases (e.g., lactase for lactose, maltase-glucoamylase for maltose, sucrase-isomaltase for sucrose), which are all enzymes facing the intestinal lumen. In one report, the levels of intestinal oligo-/disaccharidases were reduced in fasted rats or in rats subjected to parenteral nutrition but, conversely, were increased when carbohydrates were in the lumen.[2] Transporters are necessary to promote glucose, galactose, and fructose entry into and exit from enterocytes. Between meals, that is, at low luminal concentrations, glucose and galactose are transported across the apical membrane into the enterocyte by the sodium/glucose cotransporter 1 (SGLT1), whereas fructose is taken up by the fructose transporter 5 (GLUT5). All three hexoses exit the enterocyte via glucose transporter 2 (GLUT2) in the basolateral membrane, a process that delivers these sugars to the blood. Following a sugar-rich meal, GLUT2 is translocated to the apical membrane of enterocytes, where its rate of sugar uptake can increase by threefold.[3] Thus, multiple steps and genes of intestinal carbohydrate intake are affected by nutrients and are at risk for pathological alterations.

II. Preference for Sweet Food

A. Sweet Taste

Food preference, food intake, and eating behavior are heavily influenced by taste. Density of taste buds on the tongue, genetic differences in taste receptors, and differences in taste receptor sensitivity all contribute to an individual's taste perception and to subsequent food preferences. There is an innate preference for sweeteners. The receptors for sweet taste, which are G-protein-coupled receptors encoded by two taste receptors, type 1 genes[4] are the heterodimeric TAS1R2:TAS1R3 taste receptors that respond specifically to sugars. Taste receptors are expressed not only in taste buds but also throughout the gastrointestinal tract.[3,5,6] Mutations affecting sweet taste detection may thus also affect endocrine and neuroendocrine responses to sugars and, consequently, food intake and metabolism.

Ligands of sweet taste receptors have been introduced in human food to reproduce sweet tastes. These noncarbohydrate, noncaloric sweeteners include the synthetic compounds saccharin, cyclamate, and acesulfame potassium and the naturally occurring compounds monellin, thaumatin, miraculin, and stevioside. These sweeteners can stimulate gene expression and transporter trafficking,[5,6] suggesting that the absence of calories might not be the sole factor to be considered when consuming these substances.

B. From Reward to Addiction?

Eating diets rich in carbohydrates reduces stress and produces feelings of gratification and pleasure. Carbohydrate consumption increases brain serotonin, inhibits corticotrophin-releasing factor, and affects opioid- and dopamine-mediated responses.[7] Sucrose in particular appears to be connected to the opioidergic system. Indeed, opioid receptor antagonists reduced sucrose intake by decreasing the pleasantness of sucrose consumption in humans[8] and in binge eaters.[9] Euphoric endorphins and dopamine are released within the nucleus accumbens, a hedonic region of the brain. If poorly regulated, this can lead to "sugar addiction."[10] In one report, 94% of rats preferred self-administration of saccharin or sucrose solutions rather than cocaine, suggesting that intense sweetness can surpass cocaine reward stimulation.[11] Data mining has identified several gene variants involved in carbohydrate craving behavior leading to obesity.[12] There is some hope for a treatment that modulates these affected pathways.

C. Polymorphisms Linked to Preference for Sugars

Linked to the selective intake of carbohydrates, several polymorphisms have been identified in genes involved in the control of food intake, rewards, sweet taste, sugar uptake, and obesity. Polymorphisms affecting an orexigenic neuropeptide encoded by the agouti-related protein gene (*AGRP*) (Ala67Thr) increased carbohydrate intake by 2.6% at the expense of fat intake.[13] The D2 dopamine receptor (DRD2) modulates the reward response to a high-sugar diet, and the C957T single-nucleotide polymorphism (SNP) in the *DRD2* gene is associated with decreased sucrose consumption. Indeed, men with C/C, C/T, and T/T *DRD2* genotypes consumed 60, 48, and 39 g of sugar per day, respectively, without varying their total energy intake.[14] More expectedly, a polymorphism in the sweet taste receptor gene *TAS1R2* (Ile191Val) was associated with an 18% reduction in sugar consumption.[15] In addition, noncoding polymorphisms in a regulatory region of the *TAS1R3* promoter reduced gene expression and could explain sucrose taste sensitivity for 16% of the population.[16] The sugar transporter and detector GLUT2 contributes to the regulation of food intake.[17,18] Subjects bearing the Thr110Ile allele in the GLUT2 gene (now known as solute carrier family 2, member 2 [*SLC2A2*]) eat 30% more sugar than those bearing the common allele.[19] The mechanism responsible for this phenotype probably involves a deficiency in glucose sensing rather than transport[20] because the Thr110Ileu mutation did not affect the rate of glucose transport.[21] The functional consequences of these variations in sugar preference are not yet known.

In contrast, the consequences of mutations in the tubby homolog (*TUB*) and potassium channel tetramerization domain containing 15 (*KCTD15*) genes have been characterized. At least one *TUB* polymorphism is associated with an increase in sugar intake.[22] The mutation occurs in the splice donor site and results in 44 carboxy-terminal residues being substituted by 24 intron-encoded amino acids. Interestingly, this mutation has been reported to cause obesity in tubby mice.[23] In addition, in a genome-wide association study of genetic susceptibility loci for obesity, carriers of a risk allele in an SNP in or near the *KCTD15* gene were found to eat more total carbohydrate (per allele: 2.50 g/day) and more mono- and disaccharides (per allele: 2.62 g/day)[24] than non-carriers. This suggests that some obesity loci are potentially associated with macronutrient intake preferences.

III. Carbohydrate Regulation of Gene Expression

A. Nutrients Affecting Gene Expression

By studying the regulation of the lactose operon, Jacob and Monod discovered in 1961 that *E. coli* produce the proteins required for lactose catabolism (beta-galactoside permease [LacY], beta-galactosidase [LacZ], and beta-galactoside transacetylase [LacA]) only when lactose is available in the culture medium. The lactose operon is controlled by the action of a lactose metabolite on gene targets.[25] Since this key finding, carbohydrate availability has been shown to be involved in the regulation of many genes in humans (Fig. 1).

Glucose is a potent regulator of gene expression. High glucose concentrations stimulate gene expression through the binding of the carbohydrate response element binding protein (ChREBP)/max-like protein X (Mlx) complex to a responsive sequence element in target promoters,[31] whereas low glucose concentrations induce the expression of a different set of genes.

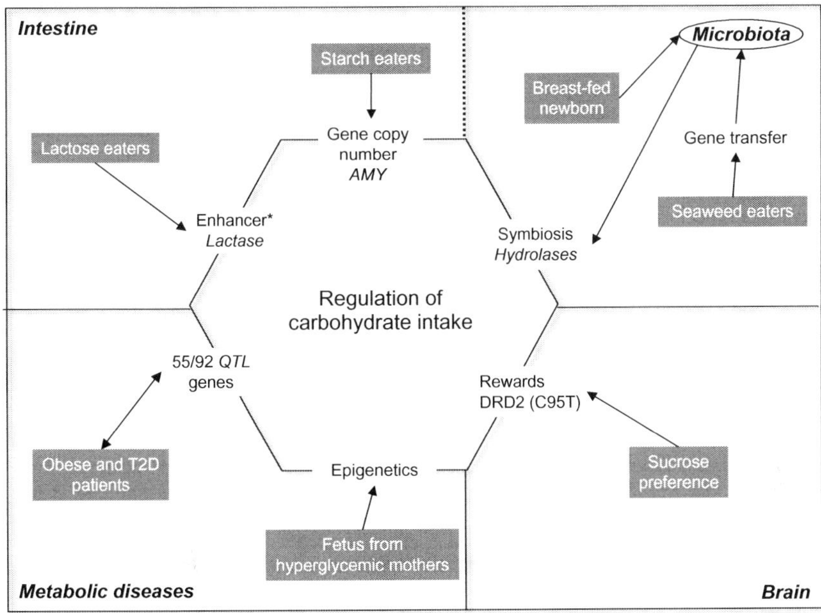

FIG. 1. Sugar-induced regulation of carbohydrate intake. Lactose eaters retain a mutation* in the lactase enhancer which allows lactase persistency and lactose digestion in adulthood.[26] Starch eaters bear high amylase (*AMY1*) copy numbers.[27] Through adaptations in the gut microbiota of breast-fed newborns[28] or seaweed eaters,[29] complex polysaccharides can be hydrolyzed and digested. An SNP in the dopamine receptor (*DRD2*) amplifies reward in the brain of carbohydrate eaters.[14] Metabolic diseases promote epigenetic regulation. QTLs (quantitative trait loci) in carbohydrate metabolism- and transport-related genes are associated with type 2 diabetes (T2D).[30]

However, the signaling pathways that mediate glucose detection to the nucleus are not fully understood. A glucose metabolic pathway is triggered by a metabolite whose identity is still under debate. Other pathways are generated at the plasma membrane through transporter–detectors such as GLUT2 or detectors such as TAS1R2:TAS1R3.[5]

To establish links between carbohydrate intake and health, long-term nutritional intervention trials are needed to determine the effects of the total amount and frequency of carbohydrate consumption, as well as the nature of the carbohydrates consumed (free or complex), and to generate complete biomarker profiles. Furthermore, the role of carbohydrates as potential regulators of energy balance, appetite, energy expenditure, and other processes needs to be defined before carbohydrate intake recommendations can be proposed.

B. Dietary Sugar and Metabolic Diseases

High consumption of refined carbohydrates is often accompanied by the development of metabolic diseases. Varma *et al.* used a data-mining approach to evaluate the role of carbohydrate metabolic pathway genes in the development of obesity and type 2 diabetes (T2D). Of the 92 genes known to be involved in carbohydrate metabolism and transport, 67 were associated with one or more quantitative trait loci (QTL) for the two diseases, 55 were associated with QTL for body weight, and many were associated with QTL for T2D.[30]

There is strong evidence for the differential impact of excessive fructose consumption as compared to glucose consumption.[32] Although weight gains were similar for glucose and fructose excesses, fructose induces specific metabolic alterations, dyslipidemia, and insulin resistance, resulting in elevated levels of fasting triglycerides and visceral fat deposits. This highlights the pathological risks induced by high fructose consumption in diabetic or obese subjects and supports the reduction of fructose from industrial foods or beverages.

A SNP in the gene-encoding GLUT2 has been reported; the mutation that results in Thr110Ile has been associated with a significant increased risk for T2D[33,34] and hypercholesterolemia.[35] The polymorphism did not affect sugar transport activity when expressed in *Xenopus* oocytes,[21] but as mentioned earlier, this *SLC2A2* variant is associated with high daily consumption of sugars, suggesting a defect in glucose sensing by the brain which contributes to a defective regulation of food intake.[19]

C. Evolutionary/Environmental Adaptation to Dietary Sugars

The ability to digest the milk disaccharide lactose declines after weaning in most mammals. In certain human populations that do not consume dairy products, intestinal lactase (lactase-phlorizin hydrolase) activity decreases during the first 4 years of life to 10% of its original level.[36] Furthermore, as revealed by epidemiological studies, the remaining levels of disaccharidase activity in lactose-intolerant subjects vary among ethnic groups.[37] Lactose intolerance is a common condition for more than half of adult humans, but it differs from congenital lactase deficiency (MIM #223000), which is a rare and severe gastrointestinal disorder.[37]

However, in human populations that consume fresh milk and dairy products, lactase persistence allows adult subjects to digest lactose. The shift from lactose intolerance to lactase persistence constitutes an elegant example of evolutionary adaptation discovered by combining epidemiological and archaeological approaches.

Lactose intolerance is a recessive trait, whereas lactase persistence is dominant. The lactase persistence locus is between intron 13 and exon 17 of the minichromosome maintenance complex component 6 gene (*MCM6*) located upstream of the lactase gene (*LCT*). The T to C variant at −3712, together with the European C to T variant at −13910, shows greater transcription factor (Oct-1) binding than the ancestral variants. When both SNPs are present, the activity of an upstream enhancer of the *LCT* promoter is elevated and enables persistent lactase production into adulthood.[26,38] An evolutionary advantage of lactase persistency could be enhanced calcium absorption, which can prevent osteoporosis.

Symptoms of lactose intolerance are diarrhea, gas bloat, and abdominal pain caused by the fermentation of undigested lactose in the distal intestine, but these symptoms do not appear in all lactose malabsorbers. The diagnosis of lactase intolerance is confirmed by a hydrogen breath test after lactose absorption or by genetic analysis. Eliminating all lactose sources from the diet is not recommended; rather, the quantity of lactose in milk, ice cream, cheese, yogurt, etc., that can be supported by the subject must be determined by personal experience.

D. Epigenetic Adaptation to Dietary Sugars

The adaptation to carbohydrate intake might be heritable because of DNA methylation or histone acetylation causing suppression of gene expression without changes in DNA sequence. Epigenetic mechanisms have primarily been studied in fat intake and have been less frequently documented in carbohydrate intake.

Fructose is an important source of calories, and its consumption is increasing in the form of corn syrup. Exploiting the fact that fructose transporter GLUT5 expression is controlled by dietary fructose and not by glucose,[39] researchers have studied epigenetic regulation of GLUT5 expression *in vivo* in mice. In correlation with increased GLUT5 expression, histone H3 acetylation increased with fructose consumption more so than with glucose feeding at two regions of the GLUT5 gene (now known as *SLC2A5*) promoter (−1600 to −1400 and −1200 to −1000).[40] This was not the case for H3 acetylation at the promoters of two other sugar-sensitive genes.[40] Similarly, induction of mouse sucrase-isomaltase gene expression by a diet rich in carbohydrate is associated with acetylation of histones H3 and H4.[41] The epigenetic changes in carbohydrate-related gene expression induced by sugar consumption are just beginning to be defined.

Epigenetic regulation has been investigated in hyperglycemic conditions. In human vascular cells, transient hyperglycemia can induce persistent epigenetic changes that activate gene expression despite restoration of normoglycemia.[42] In addition, genome-wide studies have demonstrated that cell-type-specific changes in histone methylation patterns can occur under diabetic conditions.[43]

Genetics studies in people conceived during famine have revealed that prenatal malnutrition can impact life as an adult. Increased incidences of glucose intolerance, diabetes, and obesity in 40-year-old adults who were exposed to famine *in utero* were observed following the Nigerian civil war[44] and a Chinese famine.[45] Furthermore, as a result of the Dutch famine (1944–1945), neonatal adiposity and poor health in later life were reported to be increased in the offspring of subjects exposed to famine *in utero*.[46] These findings support the need to study how an abundance of glucose (e.g., prenatal exposure to hyperglycemic diabetic mothers) affects the metabolic memory of adult offspring.

E. Selection of Gene Copy Number Induced by Dietary Sugars

Variable copy numbers of the salivary amylase gene (*AMY1*) have been reported among individuals.[47] *AMY1* gene repeats increase enzyme production and thus carbohydrate processing. Evolutionary biologists looking for selection of genetic mutations have shown that humans eating high-starch diets have additional copies of *AMY1* compared to those eating traditionally low-starch diets.[27] This suggests the existence of diet-related positive selective pressure on the copy number variants of the *AMY1* locus.

F. Symbiotic Adaptation to Dietary Sugars

Gut microbiota can supply the host with energy from dietary carbohydrates that the host cannot metabolize alone.[48,49] Carbohydrates are major nutrients for hosts and for commensal microbiota. Only simple sugars are absorbed in the proximal jejunum. Disaccharides (sucrose, lactose, maltose) and starch must be hydrolyzed into their constituent monosaccharides prior to jejunal absorption. Polysaccharides from plants are poorly hydrolyzed and reach the distal gut undigested. Nevertheless, gut microbes are equipped with enzymes that break down the wide variety of these polysaccharide linkages. Gut commensal bacteria can grow on this abundant carbon source and can provide some energy to the host by hydrolysis of plant polysaccharides. The energy is provided to the host in the form of short-chain fatty acids (propionate, butyrate, acetate) in the lumen of the distal intestine. The link between the microbiotal composition and energy-harvesting capacity has been intensively studied in obesity.[48,49] Fermentable polysaccharides constitute prebiotics and might be critical carbohydrates for obese subjects.

This symbiotic adaptation has been recently confirmed and extended by a study showing that Japanese individuals can easily digest seaweed carbohydrates. This is due to a lateral gene transfer from algae bacteria (*Zobellia galactanivorans*) to gut bacteria (*Bacteroides plebeius*). Indeed, through the transfer of a marine glycosyl hydrolase, gut bacteria acquire the capacity to produce porphyranase and agarase enzymes and to digest sulfated polysaccharides from nori. This adaptation is not observed in North Americans, whose seaweed consumption is recent and rare rather than traditional.[29]

While human milk contains 7% lactose, which is a glucose–galactose disaccharide, it also contains 1% oligosaccharides that are nondigestible because of their complex structures. Mass spectrometry has identified more than 200 different oligosaccharides in human milk.[28,50] Each mother can produce about 100 different oligosaccharides, the identities of which differ among mothers and vary during the course of lactation. Though these carbohydrates cannot be digested by infants, they constitute prebiotics that help certain bacteria to colonize the infant gut.[51] Indeed, *Bifidobacterium longum* subsp. *infantis* (*B. infantis*) can utilize oligosaccharides from human milk. Thus, oligosaccharides from human milk influence in part the composition of infant microbiota, which benefits infant health.[51]

The variety of carbohydrates in food contributes to the diversity of gut microbiota, and their combined genetic diversity deeply affects the way carbohydrates are metabolized in humans.

IV. Genetic Diseases Affecting Carbohydrate Intake

Genetic disorders that affect carbohydrate intake are detected at birth or at weaning when newborns or infants are exposed to food containing the offending carbohydrates. These are generally rare genetic disorders and affect several crucial hydrolases or glucose transporters (Fig. 2).

A. Hydrolase Deficiency

Several hydrolase deficiencies prevent the breakdown of complex carbohydrates to monosaccharides able to be readily transported through the intestine. Congenital lactase deficiency (MIM #223000) is a severe gastrointestinal disorder characterized by watery diarrhea in infants fed breast milk or lactose-containing formulas.[57] Monosaccharide formula is given to these infants, and administration of the lacking enzyme, lactase, is a possible treatment. Sucrase-isomaltase deficiency (MIM #222900) is an autosomal recessive disorder[58] detected at weaning when infants are first exposed to sucrose. The inability to hydrolyze sucrose into fructose and glucose results in

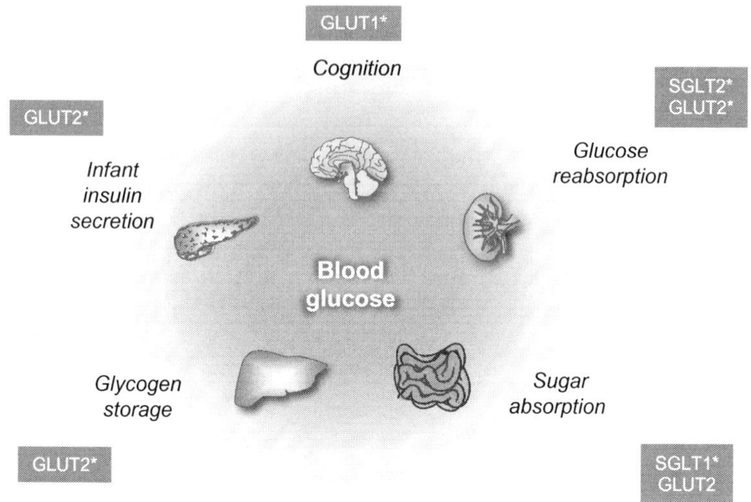

FIG. 2. Mutations in glucose transporter genes affecting carbohydrate intake. Mutations* in the *SLC2A1* gene coding for GLUT1 affect brain energy metabolism and deteriorate cognition.[52] Mutations° in the GLUT2 gene, *SLC2A2*, affect renal glucose reabsorption and glycogen storage, intestinal glucose absorption, hepatic glucose production and storage,[53] and pancreatic insulin secretion by neonates.[54] A mutation° in SGLT2[55] affects renal glucose reabsorption, and mutations in *SLC5A1* coding for SGLT1 affect intestinal glucose–galactose absorption.[56]

diarrhea and failure to thrive. It can be caused by an isolated sucrase deficiency, with normal levels of isomaltase activity, or by a combined sucrase and isomaltase deficiency.[59]

B. Glucose–Galactose Malabsorption

Glucose–galactose malabsorption (MIM #606824) is usually identified in infancy and is a very rare disease with an autosomal recessive mode of inheritance. The first mutation identified in the *SLC5A1* gene prevents SGLT1 processing and insertion of the protein into the brush border membrane of enterocytes, which leads to its accumulation in Golgi compartments.[60] This prevents intestinal milk sugars from being absorbed because the SGLT1's function is to actively transport free glucose and galactose through the enterocyte apical membrane into the cytoplasm.[56] Hundreds of different mutations that impair the transport functions of SGLT1 have been identified.

The diagnosis of glucose–galactose malabsorption is considered when a sugar-containing diarrhea presents during the neonatal period. The disorder may be fatal unless the sources of glucose and galactose (i.e., milk lactose) are omitted from the diet. However, patients are able to digest fructose-base formulas. Despite permanent SGLT1 inactivity, a symptomatic remission can occur with age, although the rescue mechanism has yet to be identified. Additionally, the long-term consequences of a high-fructose diet on the liver functions of these patients have not been reported.

C. GLUT1 Deficiency Syndrome

GLUT1 deficiency syndrome (GLUT1-DS) (MIM #606777) is characterized by severe neurologic alterations, seizures, microcephaly, delays in mental and motor development, ataxia, and dystonia. It is due to mutations in the *SLC2A1* gene coding for GLUT1 protein[61] and displays autosomal dominant inheritance. GLUT1-DS manifestations are explained by deficient glucose transport in the brain, where glucose is the major energy source. Incidence and prevalence cannot be estimated since only about 100 cases have been described worldwide. The severity of GLUT1-DS symptoms is related to the level of residual transport capacity of GLUT1. Total inactivity is an embryonic-lethal condition.[52] Thus, most mutant proteins identified in GLUT1-DS subjects have shown up to 50% decrease of erythrocyte 3-OMG uptakes. Mutation hotspots have been identified in the *SLC2A1* gene in three different families.[62–64] Other mutations are private and distributed throughout the protein.

GLUT1-DS is diagnosed when glucose concentrations in the cerebrospinal fluid are lower than 40 mg/dl. The cerebrospinal fluid glucose/blood glucose concentration ratios in GLUT1-DS patients are about 0.45 (normal ratio: 0.65 ± 0.01). Clinical improvement is obtained by feeding patients a ketogenic diet. Indeed, a very low-carbohydrate, high-fat, and normoproteic diet promotes the formation of ketone bodies, which are an alternative energy source for the brain and are usually produced during starvation. The diet must be personalized to produce enough ketone bodies for the correct fueling of the individual patient's brain. For some patients, cognitive functions remain altered despite a ketogenic diet.[65]

D. Fanconi–Bickel Syndrome

Several mutations have been reported for the gene-encoding GLUT2 (MIM #138160) and are found in patients with Fanconi–Bickel syndrome (MIM #227810).[53] These patients suffer from glycogen accumulation (hepatomegaly, nephromegaly), glucose–galactose malabsorption, gross urinary loss of glucose, failure to thrive, and, in some patients, diabetes.[53,54] *SLC2A2* mutations likely inactivate GLUT2 functions. Indeed, the first patient described by Fanconi and Bickel presented a homozygous mutation, giving rise to a half-truncated protein that lacked the entire sugar channel. Since then, 33 mutations have been reported in families spread around the world, but no clear hotspots in the *SLC2A2* gene have been identified. Only double heterozygous or homozygous mutations lead to Fanconi–Bickel syndrome, suggesting that a single unaltered allele is sufficient to fulfill vital protein functions. This is consistent with an autosomal recessive pattern of inheritance. The diet of patients must be adapted to compensate for calorie loss in urine and is administrated by frequent boluses of slowly absorbed carbohydrates. Uncooked cornstarch has been reported to restore patient growth.[66]

The functional impacts of these *SLC2A2* point mutations have not been determined, except for Val197Ile, which abolishes transport activity and was discovered as a single-allele mutation in a T2D patient.[21]

V. Concluding Remarks

Interactions between carbohydrate nutrients and genes deeply impact glucose homeostasis and health and can modulate the risks of developing metabolic diseases. Whereas personalized nutrition can already be achieved in patients suffering from monogenic diseases of carbohydrate intake, this has yet to become a reality for patients suffering from many metabolic diseases and awaits further nutrigenetics and nutrigenomics research. Tailoring diets to optimize wellness or to prevent chronic metabolic diseases are future goals that may yet be realized through further nutritional and genetics research.

References

1. Bray GA, Nielsen SJ, Popkin BM. Consumption of high-fructose corn syrup in beverages may play a role in the epidemic of obesity. *Am J Clin Nutr* 2004;**79**:537–43.
2. Nelson DW, Murali SG, Liu X, Koopmann MC, Holst JJ, Ney DM. Insulin-like growth factor I and glucagon-like peptide-2 responses to fasting followed by controlled or ad libitum refeeding in rats. *Am J Physiol Regul Integr Comp Physiol* 2008;**294**:R1175–84.
3. Kellett GL, Brot-Laroche E, Mace OJ, Leturque A. Sugar absorption in the intestine: the role of GLUT2. *Annu Rev Nutr* 2008;**28**:35–54.
4. Nelson G, Hoon MA, Chandrashekar J, Zhang Y, Ryba NJ, Zuker CS. Mammalian sweet taste receptors. *Cell* 2001;**106**:381–90.
5. Le Gall M, Tobin V, Stolarczyk E, Dalet V, Leturque A, Brot-Laroche E. Sugar sensing by enterocytes combines polarity, membrane bound detectors and sugar metabolism. *J Cell Physiol* 2007;**213**:834–43.
6. Mace OJ, Affleck J, Patel N, Kellett GL. Sweet taste receptors in rat small intestine stimulate glucose absorption through apical GLUT2. *J Physiol* 2007;**582**:379–92.
7. Levine AS, Kotz CM, Gosnell BA. Sugars: hedonic aspects, neuroregulation, and energy balance. *Am J Clin Nutr* 2003;**78**:834S–842S.
8. Fantino M, Hosotte J, Apfelbaum M. An opioid antagonist, naltrexone, reduces preference for sucrose in humans. *Am J Physiol* 1986;**251**:R91–6.
9. Drewnowski A, Krahn DD, Demitrack MA, Nairn K, Gosnell BA. Taste responses and preferences for sweet high-fat foods: evidence for opioid involvement. *Physiol Behav* 1992;**51**:371–9.

10. Avena NM, Rada P, Hoebel BG. Evidence for sugar addiction: behavioral and neurochemical effects of intermittent, excessive sugar intake. *Neurosci Biobehav Rev* 2008;**32**:20–39.
11. Lenoir M, Serre F, Cantin L, Ahmed SH. Intense sweetness surpasses cocaine reward. *PLoS One* 2007;**2**:e698.
12. Downs BW, Chen AL, Chen TJ, Waite RL, Braverman ER, Kerner M, et al. Nutrigenomic targeting of carbohydrate craving behavior: can we manage obesity and aberrant craving behaviors with neurochemical pathway manipulation by Immunological Compatible Substances (nutrients) using a Genetic Positioning System (GPS) Map? *Med Hypotheses* 2009;**73**:427–34.
13. Loos RJ, Rankinen T, Rice T, Rao DC, Leon AS, Skinner JS, et al. Two ethnic-specific polymorphisms in the human Agouti-related protein gene are associated with macronutrient intake. *Am J Clin Nutr* 2005;**82**:1097–101.
14. Eny KM, Corey PN, El-Sohemy A. Dopamine D2 receptor genotype (C957T) and habitual consumption of sugars in a free-living population of men and women. *J Nutrigenet Nutrigenomics* 2009;**2**:235–42.
15. Eny KM, Wolever TM, Corey PN, El-Sohemy A. Genetic variation in TAS1R2 (Ile191Val) is associated with consumption of sugars in overweight and obese individuals in 2 distinct populations. *Am J Clin Nutr* 2010;**92**:1501–10.
16. Fushan AA, Simons CT, Slack JP, Manichaikul A, Drayna D. Allelic polymorphism within the TAS1R3 promoter is associated with human taste sensitivity to sucrose. *Curr Biol* 2009; **19**:1288–93.
17. Bady I, Marty N, Dallaporta M, Emery M, Gyger J, Tarussio D, et al. Evidence from glut2-null mice that glucose is a critical physiological regulator of feeding. *Diabetes* 2006;**55**:988–95.
18. Stolarczyk E, Guissard C, Michau A, Even PC, Grosfeld A, Serradas P, et al. Detection of extracellular glucose by GLUT2 contributes to hypothalamic control of food intake. *Am J Physiol Endocrinol Metab* 2010;**298**:E1078–87.
19. Eny KM, Wolever TM, Fontaine-Bisson B, El-Sohemy A. Genetic variant in the glucose transporter type 2 is associated with higher intakes of sugars in two distinct populations. *Physiol Genomics* 2008;**33**:355–60.
20. Leturque A, Brot-Laroche E, Le Gall M. GLUT2 mutations, translocation, and receptor function in diet sugar managing. *Am J Physiol Endocrinol Metab* 2009;**296**:E985–92.
21. Mueckler M, Kruse M, Strube M, Riggs AC, Chiu KC, Permutt MA. A mutation in the Glut2 glucose transporter gene of a diabetic patient abolishes transport activity. *J Biol Chem* 1994; **269**:17765–7.
22. van Vliet-Ostaptchouk JV, Onland-Moret NC, Shiri-Sverdlov R, van Gorp PJ, Custers A, Peeters PH, et al. Polymorphisms of the TUB gene are associated with body composition and eating behavior in middle-aged women. *PLoS One* 2008;**3**:e1405.
23. Kleyn PW, Fan W, Kovats SG, Lee JJ, Pulido JC, Wu Y, et al. Identification and characterization of the mouse obesity gene tubby: a member of a novel gene family. *Cell* 1996;**85**:281–90.
24. Bauer F, Elbers CC, Adan RA, Loos RJ, Onland-Moret NC, Grobbee DE, et al. Obesity genes identified in genome-wide association studies are associated with adiposity measures and potentially with nutrient-specific food preference. *Am J Clin Nutr* 2009;**90**:951–9.
25. Jacob F, Monod J. Genetic regulatory mechanisms in the synthesis of proteins. *J Mol Biol* 1961;**3**:318–56.
26. Enattah NS, Jensen TG, Nielsen M, Lewinski R, Kuokkanen M, Rasinpera H, et al. Independent introduction of two lactase-persistence alleles into human populations reflects different history of adaptation to milk culture. *Am J Hum Genet* 2008;**82**:57–72.
27. Perry GH, Dominy NJ, Claw KG, Lee AS, Fiegler H, Redon R, et al. Diet and the evolution of human amylase gene copy number variation. *Nat Genet* 2007;**39**:1256–60.
28. Sela DA, Li Y, Lerno L, Wu S, Marcobal AM, German JB, et al. An infant-associated bacterial commensal utilizes breast milk sialyloligosaccharides. *J Biol Chem* 2011;**286**:11909–18.

29. Hehemann JH, Correc G, Barbeyron T, Helbert W, Czjzek M, Michel G. Transfer of carbohydrate-active enzymes from marine bacteria to Japanese gut microbiota. *Nature* 2010; **464**:908–12.
30. Varma V, Wise C, Kaput J. Carbohydrate metabolic pathway genes associated with quantitative trait loci (QTL) for obesity and type 2 diabetes: identification by data mining. *Biotechnol J* 2010;**5**:942–9.
31. Uyeda K, Repa JJ. Carbohydrate response element binding protein, ChREBP, a transcription factor coupling hepatic glucose utilization and lipid synthesis. *Cell Metab* 2006;**4**:107–10.
32. Stanhope KL, Schwarz JM, Keim NL, Griffen SC, Bremer AA, Graham JL, et al. Consuming fructose-sweetened, not glucose-sweetened, beverages increases visceral adiposity and lipids and decreases insulin sensitivity in overweight/obese humans. *J Clin Invest* 2009;**119**:1322–34.
33. Gaulton KJ, Willer CJ, Li Y, Scott LJ, Conneely KN, Jackson AU, et al. Comprehensive association study of type 2 diabetes and related quantitative traits with 222 candidate genes. *Diabetes* 2008;**57**:3136–44.
34. Willer CJ, Bonnycastle LL, Conneely KN, Duren WL, Jackson AU, Scott LJ, et al. Screening of 134 single nucleotide polymorphisms (SNPs) previously associated with type 2 diabetes replicates association with 12 SNPs in nine genes. *Diabetes* 2007;**56**:256–64.
35. Igl W, Johansson A, Wilson JF, Wild SH, Polasek O, Hayward C, et al. Modeling of environmental effects in genome-wide association studies identifies SLC2A2 and HP as novel loci influencing serum cholesterol levels. *PLoS Genet* 2010;**6**:e1000798.
36. Simoons FJ. Primary adult lactose intolerance and the milking habit: a problem in biologic and cultural interrelations. II. A culture historical hypothesis. *Am J Dig Dis* 1970;**15**:695–710.
37. Jarvela I, Torniainen S, Kolho KL. Molecular genetics of human lactase deficiencies. *Ann Med* 2009;**41**:568–75.
38. Jensen TG, Liebert A, Lewinsky R, Swallow DM, Olsen J, Troelsen JT. The -14010°C variant associated with lactase persistence is located between an Oct-1 and HNF1alpha binding site and increases lactase promoter activity. *Hum Genet* 2011;**130**:483–93.
39. Douard V, Ferraris RP. Regulation of the fructose transporter GLUT5 in health and disease. *Am J Physiol Endocrinol Metab* 2008;**295**:E227–37.
40. Suzuki T, Douard V, Mochizuki K, Goda T, Ferraris RP. Diet-induced epigenetic regulation in vivo of the intestinal fructose transporter Glut5 during development of rat small intestine. *Biochem J* 2011;**435**:43–53.
41. Honma K, Mochizuki K, Goda T. Carbohydrate/fat ratio in the diet alters histone acetylation on the sucrase-isomaltase gene and its expression in mouse small intestine. *Biochem Biophys Res Commun* 2007;**357**:1124–9.
42. El-Osta A, Brasacchio D, Yao D, Pocai A, Jones PL, Roeder RG, et al. Transient high glucose causes persistent epigenetic changes and altered gene expression during subsequent normoglycemia. *J Exp Med* 2008;**205**:2409–17.
43. Villeneuve LM, Reddy MA, Natarajan R. Epigenetics: deciphering its role in diabetes and its chronic complications. *Clin Exp Pharmacol Physiol* 2011;**38**:401–9.
44. Hult M, Tornhammar P, Ueda P, Chima C, Bonamy AK, Ozumba B, et al. Hypertension, diabetes and overweight: looming legacies of the Biafran famine. *PLoS One* 2011;**5**:e13582.
45. Li Y, He Y, Qi L, Jaddoe VW, Feskens EJ, Yang X, et al. Exposure to the Chinese famine in early life and the risk of hyperglycemia and type 2 diabetes in adulthood. *Diabetes* 2010; **59**:2400–6.
46. Painter RC, Osmond C, Gluckman P, Hanson M, Phillips DI, Roseboom TJ. Transgenerational effects of prenatal exposure to the Dutch famine on neonatal adiposity and health in later life. *BJOG* 2008;**115**:1243–9.
47. Iafrate AJ, Feuk L, Rivera MN, Listewnik ML, Donahoe PK, Qi Y, et al. Detection of large-scale variation in the human genome. *Nat Genet* 2004;**36**:949–51.

48. Turnbaugh PJ, Ley RE, Mahowald MA, Magrini V, Mardis ER, Gordon JI. An obesity-associated gut microbiome with increased capacity for energy harvest. *Nature* 2006;**444**:1027–31.
49. Musso G, Gambino R, Cassader M. Interactions between gut microbiota and host metabolism predisposing to obesity and diabetes. *Annu Rev Med* 2011;**62**:361–80.
50. Wu S, Grimm R, German JB, Lebrilla CB. Annotation and structural analysis of sialylated human milk oligosaccharides. *J Proteome Res* 2011;**10**:856–68.
51. Sela DA, Mills DA. Nursing our microbiota: molecular linkages between bifidobacteria and milk oligosaccharides. *Trends Microbiol* 2010;**18**:298–307.
52. De Vivo DC, Wang D. Glut1 deficiency: CSF glucose. How low is too low? *Rev Neurol (Paris)* 2008;**164**:877–80.
53. Santer R, Steinmann B, Schaub J. Fanconi-Bickel syndrome—a congenital defect of facilitative glucose transport. *Curr Mol Med* 2002;**2**:213–27.
54. Taha D, Al-Harbi N, Al-Sabban E. Hyperglycemia and hypoinsulinemia in patients with Fanconi-Bickel syndrome. *J Pediatr Endocrinol Metab* 2008;**21**:581–6.
55. van den Heuvel LP, Assink K, Willemsen M, Monnens L. Autosomal recessive renal glucosuria attributable to a mutation in the sodium glucose cotransporter (SGLT2). *Hum Genet* 2002;**111**:544–7.
56. Wright EM, Turk E, Martin MG. Molecular basis for glucose-galactose malabsorption. *Cell Biochem Biophys* 2002;**36**:115–21.
57. Levin B, Abraham JM, Burgess EA, Wallis PG. Congenital lactose malabsorption. *Arch Dis Child* 1970;**45**:173–7.
58. Hauri HP, Roth J, Sterchi EE, Lentze MJ. Transport to cell surface of intestinal sucrase-isomaltase is blocked in the Golgi apparatus in a patient with congenital sucrase-isomaltase deficiency. *Proc Natl Acad Sci USA* 1985;**82**:4423–7.
59. Sander P, Alfalah M, Keiser M, Korponay-Szabo I, Kovacs JB, Leeb T, et al. Novel mutations in the human sucrase-isomaltase gene (SI) that cause congenital carbohydrate malabsorption. *Hum Mutat* 2006;**27**:119.
60. Turk E, Zabel B, Mundlos S, Dyer J, Wright EM. Glucose/galactose malabsorption caused by a defect in the Na+/glucose cotransporter. *Nature* 1991;**350**:354–6.
61. Wang D, Kranz-Eble P, De Vivo DC. Mutational analysis of GLUT1 (SLC2A1) in Glut-1 deficiency syndrome. *Hum Mutat* 2000;**16**:224–31.
62. Brockmann K. The expanding phenotype of GLUT1-deficiency syndrome. *Brain Dev* 2009;**31**:545–52.
63. Ho YY, Yang H, Klepper J, Fischbarg J, Wang D, De Vivo DC. Glucose transporter type 1 deficiency syndrome (Glut1DS): methylxanthines potentiate GLUT1 haploinsufficiency in vitro. *Pediatr Res* 2001;**50**:254–60.
64. Pascual JM, Wang D, Yang R, Shi L, Yang H, De Vivo DC. Structural signatures and membrane helix 4 in GLUT1: inferences from human blood-brain glucose transport mutants. *J Biol Chem* 2008;**283**:16732–42.
65. Slaughter L, Vartzelis G, Arthur T. New GLUT-1 mutation in a child with treatment-resistant epilepsy. *Epilepsy Res* 2009;**84**:254–6.
66. Lee PJ, Van't Hoff WG, Leonard JV. Catch-up growth in Fanconi-Bickel syndrome with uncooked cornstarch. *J Inherit Metab Dis* 1995;**18**:153–6.

Genetic and Epigenomic Footprints of Folate

J. Michael Salbaum* and Claudia Kappen†

*Regulation of Gene Expression Laboratory, Pennington Biomedical Research Center, Baton Rouge, Louisiana, USA

†Developmental Biology Laboratory, Pennington Biomedical Research Center, Baton Rouge, Louisiana, USA

Dietary micronutrient composition has long been recognized as a determining factor for human health. Historically, biochemical research has successfully unraveled how vitamins serve as essential cofactors for enzymatic reactions in the biochemical machinery of the cell. Folate, also known as vitamin B9, follows this paradigm as well. Folate deficiency is linked to adverse health conditions, and dietary supplementation with folate has proven

Progress in Molecular Biology
and Translational Science, Vol. 108
DOI: 10.1016/B978-0-12-398397-8.00006-X

highly beneficial in the prevention of neural tube defects. With its function in single-carbon metabolism, folate levels affect nucleotide synthesis, with implications for cell proliferation, DNA repair, and genomic stability. Furthermore, by providing the single-carbon moiety in the synthesis pathway for *S*-adenosylmethionine, the main methyl donor in the cell, folate also impacts methylation reactions. It is this capacity that extends the reach of folate functions into the realm of epigenetics and gene regulation. Methylation reactions play a major role for several modalities of the epigenome. The specific methylation status of histones, noncoding RNAs, transcription factors, or DNA represents a significant determinant for the transcriptional output of a cell. Proper folate status is therefore necessary for a broad range of biological functions that go beyond the biochemistry of folate. In this review, we examine evolutionary, genetic, and epigenomic footprints of folate and the implications for human health.

I. Folate and Single-Carbon Metabolism

Folate is an essential micronutrient with a central function in single-carbon transfer reactions, and folate status and metabolism are of significant interest to public health. Folate has proven highly successful in the prevention of neural tube defects,[1–3] so much so that the United States prescribed a mandatory fortification of grain products with folate that began in 1998. This program led to a significantly reduced prevalence of birth defects such as spina bifida, which is a severe and disabling birth defect that comes with significant psychological and financial hardships for afflicted families. Fortification with folate has been a clear public health success story[4] and a victory for preventive medicine.

While folate is thought to be highly beneficial for the prevention of birth defects, the relationship between folate and cancer is more complex. Folate plays an essential role in genome stability[5]; high folate status may therefore be beneficial in preventing genome instability, a key event of neoplastic transformation,[6] while folate insufficiency may actually support the early steps of carcinogenesis. Furthermore, high folate levels may, in turn, promote the cancer progression of existing neoplasms,[7] where it can act as a mitogen. In fact, antifolates are used in cancer chemotherapy,[8] while folate is utilized as a targeting moiety to deliver cytotoxic drugs to tumors that overexpress a folate receptor (*FOLR*) gene.[9] Yet, because of the dual relationship in preventing and promoting neoplastic disease, the benefits of folate fortification, supplementation, and high dietary intake are not unequivocal.

Folate provides a methyl group for two streams of methyl trafficking in the cell: nucleotide synthesis to maintain the nucleotide pool for DNA replication or repair, which ultimately affects genomic stability and formation of methionine, the precursor for *S*-adenosylmethionine, which serves as primary methyl group donor for the majority of methylation reactions in the cell. Methylation of nucleic acids, proteins, and lipids is therefore impacted by the folate level. Higher organisms have lost the ability to synthesize folate and must acquire it via diet; typically, leafy green vegetables represent a suitable source. Dietary folate therefore provides a link to maintenance and stability of the genome via the nucleotide synthesis pathway, as well as to functional aspects such as the regulation of gene expression via the methylation of DNA and histones. In this review, we will discuss the evolutionary aspects of folate transport and metabolism, examine the consequences of genetic defects in folate pathway genes in mice and humans, and explore implications of the nutritional methyl-donor supply for epigenomics of health and disease.

II. Genes of the Folate Cycle: Biochemical and Evolutionary Aspects

A simplified schematic overview of the folate cycle, as shown in Fig. 1, consists of mechanisms of uptake and transport into the cell, enzymatic trapping and processing of folate for discharge of the methyl group, and recharging of folate with a methyl group from intracellular sources. It should be noted that, because of biochemical complexity, we will use "folate" as an umbrella term rather than refer to each of the specific biochemical derivatives. Several mechanisms of folate uptake from dietary sources exist. Transporter-based means of folate import into the cell are either through the reduced folate carrier (RFC1; encoded by solute carrier family 19, member 1 [*SLC19A1*]) or through the proton-coupled folate transporter (PCFT; encoded by solute carrier family 46, member 1 [*SLC46A1*]). Transport of folate into mitochondria is achieved via the mitochondrial folate transporter/carrier (encoded by *SLC25A32*). The genes for these transporter molecules show deep evolutionary conservation within the animal kingdom: they can be detected in the genomes of *Pseudocoelomata* such as the nematode *Caenorhabditis elegans*, and within *Coelomata* they are present in both *Protostomia* and *Deuterostomia*. In the branch of *Protostomia*, homologous genes are present in the genome of the fruit fly *Drosophila melanogaster*. In *Deuterostomia*, conserved genes for folate transporters can be detected in *Tunicata* such as the sea squirt *Ciona intestinalis*, as well as in all higher branches of *Craniata*, as evidenced by the folate transporter genes of *Mus*

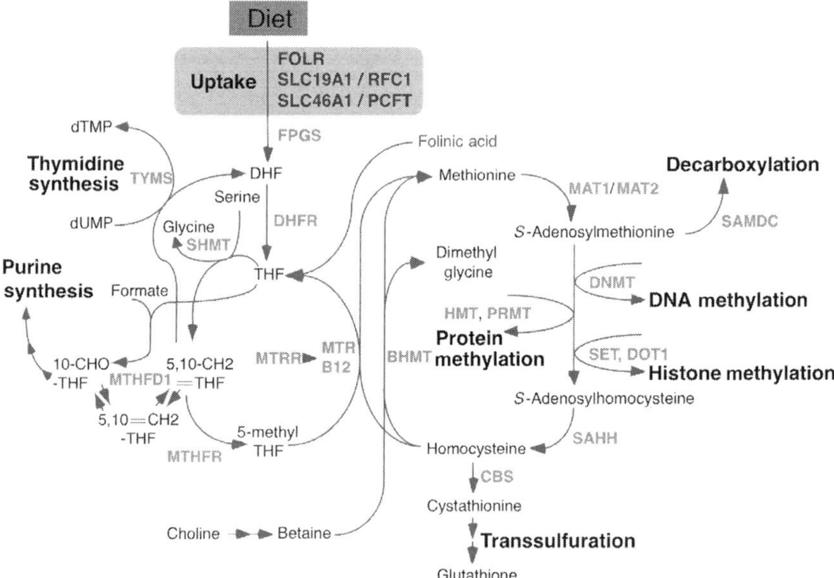

Fig. 1. Schematic representation of core aspects of the folate cycle. *Abbreviations*: 5,10-CH2=THF, 5,10-methylenetetrahydrofolate; 5,10=CH2-THF, 5,10-methenyltetrahydrofolate; 10-CHO-THF, 10-formyl tetrahydrofolate; B12, vitamin B12; BHMT, betaine homocysteine methyltransferase; CBS, choline betaine synthetase; DHF, dihydrofolate; DHFR, dihydrofolate reductase; DNMT, DNA methyltransferase; dTMP, deoxythymidine monophosphate; dUMP, deoxyuridine monophosphate; FPGS, folylpolyglutamate synthase; FOLR, folate receptor; HMT, histone methyltransferase; MAT1, methionine adenosyltransferase II; MAT2, methionine adenosyltransferase II; MTHFD1, methylenetetrahydrofolate dehydrogenase (NADP+ dependent) 1, methenyltetrahydrofolate cyclohydrolase, formyltetrahydrofolate synthetase; MTHFR, methylenetetrahydrofolate reductase; 5-methylTHF, 5-methyltetrahydrofolate; MTR, methionine synthase; MTRR, methionine synthase reductase; PCFT, proton-coupled folate transporter (SLC46A1); PRMT, protein arginine methyltransferase; RFC1, reduced folate carrier (SLC19A1); SAHH, *S*-adenosylhomocysteine hydrolase; SAMDC, *S*-adenosylmethionine decarboxylase; SET, DOT1, histone H3 methyltransferase; SHMT, serine hydroxymethyltransferase; THF, tetrahydrofolate; TYMS, thymidylate synthase. (For color version of this figure, the reader is referred to the online version of this chapter.)

musculus or *Homo sapiens*. The high conservation of folate-transport gene sequences between animal species indicates that the diet dependence on folate is likely of evolutionary significance.

Following transport into the cell, folate is polyglutamylated by folylpolyglutamate synthase (FPGS) and converted by dihydrofolate reductase (DHFR) to tetrahydrofolate (THF). When required, loading THF with the crucial single-carbon group is achieved by serine hydroxymethyltransferase (SHMT) in a reaction that converts the amino acid serine (from nutritional or other proteolytic sources) to glycine and results in the formation of 5,10-methylenetetrahydrofolate. In fact, serine serves as the major source of single-carbon groups in the cell, whereas folate constitutes the essential shuttle vehicle for that methyl group. 5,10-Methylenetetrahydrofolate represents a central node from which several directions for single-carbon metabolism are possible:

(i) Purine synthesis via MTHFD1 (trifunctional methylenetetrahydrofolate dehydrogenase (NADP+ dependent) 1, methenyltetrahydrofolate cyclohydrolase, formyltetrahydrofolate synthetase) which converts 5,10-methylenetetrahydrofolate through a series of reversible reactions into 10-formyl THF, the methyl-donor substrate for *de novo* purine synthesis.
(ii) Thymidine synthesis through the action of thymidylate synthase, which utilizes 5,10-methylenetetrahydrofolate for the conversion of deoxyuridine monophosphate to deoxythymidine monophosphate.
(iii) Methionine synthesis in the irreversible reaction catalyzed by methylenetetrahydrofolate reductase (MTHFR) to yield 5-methyltetrahydrofolate, the methyl-donor substrate in the vitamin B12-dependent conversion of homocysteine to methionine via 5-methyltetrahydrofolate–homocysteine methyltransferase reductase (MTRR, methionine synthase reductase) and 5-methyltetrahydrofolate-homocysteine methyltransferase (MTR, methionine synthase). Through the action of methionine adenosyltransferases (MAT1A, MAT2A, MAT2B), methionine is converted to S-adenosylmethionine, the main methyl-donor substrate channeled toward the major enzymatic methylation reactions in the cell. Synthesis of methionine leaves THF, as the methyl group is discharged.
(iv) Finally, single-carbon recharging of THF is again achieved by SHMT as mentioned above. This reaction completes the cycle for replenishment of the methyl-donor substrate pool. Methionine is converted to S-adenosylmethionine, and execution of a methylation reaction leaves S-adenosylhomocysteine, which is converted to homocysteine, the methyl acceptor not only from the folate cycle but also from the choline/betaine stream of the methyl-donor supply.

All genes that encode enzymes of the folate cycle show deep evolutionary conservation; they are not just present in genomes of the animal kingdom but also can be detected in genomes of primitive *Eukaryota* such as the yeast *Saccharomyces cerevisiae*. This is not surprising given the fundamental role of the folate cycle in the essential single-carbon biochemistry of the cell.

The picture on folate uptake is, however, not complete without giving consideration to folate receptors. In contrast to folate transporters that are transmembrane proteins, folate receptors are glycolipid-anchored cell surface proteins that bind folates with very high affinity. Transport into the cell occurs by endocytosis, fusion of the resulting vesicle with lysosomes, pH-mediated release of folates from the receptor, and folate entry into the cytoplasm through the membrane-situated folate transporter molecules discussed earlier. Folate receptors are recycled to the cell surface and essentially serve to critically enrich folates for transporter-based cytoplasmic import.[10] Four genes encoding folate receptors can be distinguished in the human genome, while three are currently recognized in the mouse genome. In contrast to all the other folate-related genes, folate receptor (FOLR) genes do not show deep conservation across evolutionary phyla. Rather, it appears that they are an invention of *Chordata*, as they seem to be missing from genomes of *Protostomia* (e.g., *D. melanogaster*) or *Pseudocoelomata* (e.g., *C. elegans*). Genes encoding folate receptors can be found as low on the *Chordata* branch as *Tunicata*: species of *Ciona* have *FOLR* genes in their genomes, so do all higher branches of *Chordata*. One may speculate that the presence of genes encoding these high-affinity receptors for folates permits a highly efficient extraction of folate from nutrition sources. Given the fundamental importance of folate for cell biology, such receptors may have yielded specific evolutionary advantages, and one can imagine that such advantages could have been in the realm of overall genome stability or in the possible elaboration and expansion of methylation-based genomic regulatory mechanisms.

While the cellular uptake of folate has received much attention, far less is known about the export of folate from one cell to the next or its transport to serum or lymph. Uptake of folate is certainly of high importance for the organism, yet not every cell in the mammalian body that needs folate is located at an interface where folate is readily available. Interfaces where folate is available include the intestinal epithelium, where folate is imported from the diet; the boundary to blood and lymph, where folate is in circulation; and the contact surface to cerebrospinal fluid. Cells that are not in direct contact with such interfaces would have difficulty acquiring folates, and it is reasonable to assume that folate transport mechanisms between cells must exist, and that cells at folate uptake interfaces would have export mechanisms that make folates available to other cells in the body. Recent evidence has accumulated indicating that folate export from cells may occur via members of the superfamily of ATP-binding cassette (ABC) proteins. These transmembrane proteins mediate ATP-dependent transport of various molecules across extra- and intracellular membranes. Members of this family implicated in folate export are P-glycoprotein 1 (ABCB1)[11,12] and multidrug resistance protein 3 (ABCC3).[13] It is also thought that folate can leave the cell via RFC1/SLC19A1, a protein initially thought to be involved only in folate uptake. How folate is transported once it leaves the cell is less clear; the major form of folate circulating in blood is 5-methyltetrahydrofolate in the monoglutamylated form. Further research is needed to clarify how folate can travel across tissues; this would be of particular interest for parenchyma of the early developing embryo before establishment of a functional vascular capillary system, and for poorly vascularized tumors.

III. Genetic Footprints of Folate Pathway Genes

With the success of folate in the prevention of birth defects, folate has taken on a significant role in public health. Similarly, the role of folate in cancer, either as risk factor or as therapeutic target, has brought much attention to this micronutrient. For both birth defects and cancer, folate is thought to represent a direct interface between nutrition and pathology, with the inferred prospect that therapeutic interventions may be achieved simply via altered nutrition or dietary supplementation. The molecular correlate of this interface is the folate pathway and its genes, which were thought to be primary candidates in increased disease susceptibilities through disturbance of the folate pathway. They have therefore received strong interest in the biomedical research community. Considerable effort has been directed at determining the biological function of these genes for normal development, for birth defect phenotypes, and for contributions to cancer in animal models. Concomitantly, variations in human folate pathway genes have been investigated for possible associations with birth defect prevalence, cancer incidence, and aging; the variant *MTHFR* C677T has received the most attention in this context. In this section, we briefly review the genetic footprint of these genes with respect to mouse model systems as well as human genetics. To date, over 5000 sequence variants in folate pathway genes have been described (available in detail at www.ensembl.org) and are summarized in Table I.

A. *FOLR1* (Folate Receptor 1)

Despite being a rather recent evolutionary occurrence, *Folr1* (originally termed folate receptor alpha in humans and folate binding protein 1 in mice) constitutes an essential gene for mice. Targeted mutation of this gene results in embryonic lethality shortly after gastrulation, and embryos present with severe morphogenetic abnormalities.[14] Expression of *Folr1* in the mouse embryo has been reported in the vicinity of neural tube closure sites[15] and, most notably, in the visceral endoderm, a tissue that provides critical nutritional support to the embryo at a time when the placenta is yet to be established.[16] Two well-characterized promoters contribute to the expression of *FOLR1*,[17] and to date, this is the only gene in the folate pathway where a transcriptional enhancer element with *in vivo* activity has been reported.[16] Interestingly, the developmental lethality of the null mutation can be rescued by supplementation with folinic acid,[18] indicating a redundancy of folate transport mechanisms. Genetic

TABLE I

Mutations in Genes of the Folate Cycle

Variation	FOLR1	FOLR2	FOLR3	FOLR4	SLC19A1	SLC46A1	SLC25A32	FPGS	DHFR	MTHFD1	MTHFD2	MTHFR	MTRR	MTR	SHMT1	SHMT2
Essential splice site	0	0	7	0	0	2	0	0	0	5	0	0	0	4	0	28
Stop gained	0	0	0	0	0	0	0	0	0	0	0	4	6	0	0	21
Stop lost	0	0	0	0	0	0	0	0	0	0	0	0	0	0	0	0
Complex in/del	0	0	5	0	0	0	0	0	0	0	0	0	0	0	0	0
Frameshift coding	0	0	5	0	0	6	0	0	0	2	0	0	0	8	0	8
Nonsynonymous coding	16	46	8	4	25	26	8	23	3	56	10	82	61	64	27	119
Splice site	3	0	8	0	13	2	6	1	0	18	7	10	7	17	16	41
Partial codon	0	0	0	0	0	0	0	0	0	0	0	0	0	0	0	0
Synonymous coding	8	3	27	0	40	5	0	25	3	3	19	77	62	38	8	52
Coding unknown	0	0	3	0	0	0	0	0	0	0	0	0	0	0	0	0
Within mature miRNA	0	0	0	0	0	0	0	0	0	0	0	0	0	0	0	0
Intronic	50	66	43	2	366	20	111	403	57	275	126	295	554	320	162	375
NMD transcript	0	0	39	0	0	0	89	0	0	0	44	0	401	0	0	287
Within noncoding gene	0	0	10	0	99	0	5	296	29	142	30	0	190	100	0	160
Upstream	2	1	16	2	11	0	6	6	16	7	14	7	27	6	3	6
Downstream	0	10	7	0	14	0	0	19	12	5	9	5	26	16	3	38
5′UTR	10	9	6	1	16	0	8	0	29	4	5	76	22	12	10	13
3′UTR	0	0	14	0	39	0	54	37	56	0	26	130	155	90	31	124
ALL	86	135	133	7	564	58	192	586	188	410	227	676	981	619	244	876

NMD, nonsense-mediated decay; UTR, untranslated region.

variations in the human gene are not strongly associated with morbidities such as neural tube defects.[19,20] However, loss-of-function mutations in *FOLR1* have been detected in patients with cerebral folate transport deficiency[21]; the lack of functional *FOLR1* expression in the choroid plexus presents a plausible mechanism for the folate deficit in cerebrospinal fluid in these patients.

B. *FOLR2* (Folate Receptor 2)

Mice with a targeted mutation of the *Folr2* gene are viable but respond to exposure to valproic acid or arsenic[22] with a higher rate of developmental defects. Expression in the developing embryo occurs in the cartilaginous anlagen of the skeleton.[23] Human genetic variation in this gene has received little attention; this is also the case for *FOLR3* and *FOLR4*.

C. *SLC19A1* (Reduced Folate Carrier, RFC1)

The human sequence variant G80A of the *SLC19A1* gene conveys increased risk for acute lymphoblastic leukemia[24] and for Alzheimer's disease.[25] In mice, a targeted mutation of the *Slc19A1* gene leads to early developmental lethality[26,27]; death of homozygote mutant embryos can be postponed by supplementation with folinic acid in a dose-dependent manner, but perinatal lethality of the mutation cannot be overcome by supplementation.[27] *SLC19A1* is a widely expressed gene.[28] Besides regulating intracellular concentrations of folates, this carrier also can transport methotrexate,[29] a chemotherapeutic drug that acts as antifolate by blocking DHFR and several other enzymes in the folate pathway.

D. *SLC46A1* (Proton-Coupled Folate Transporter, PCFT)

Loss of function in the human *SLC46A1* gene due to a mutation that results in skipping exon 3 during mRNA splicing leads to hereditary familial folate malabsorption,[30] that is, a lack of adequate folate uptake from the gastrointestinal tract. Mice homozygous for a targeted mutation of this gene are viable but display elevated levels of homocysteine and exhibit severe hematopoietic deficits.[31,32] Consistent with a role in *intestinal* absorption of folate, supplementation via intraperitoneal injection is indeed successful, whereas *oral* supplementation fails to rescue the hematopoietic phenotype.[32] Parenteral supplementation with folinic acid[33,34] of human infants afflicted by *SLC46A1* mutations[35] is an effective treatment that permits normal development of these children.

E. *SLC25A32* (Mitochondrial Folate Transporter/ Carrier) and *FPGS* (Folylpolyglutamate Synthase)

Currently, mouse strains bearing mutations in these two genes are not available, and the consequences of human genetic variation have not been explored.

F. *DHFR* (Dihydrofolate Reductase)

Similar to *Slc25a32* and *Fpgs*, mouse mutations for this gene have not been reported. However, mutations in the human *DHFR* gene, in particular A458T[36] and C238T,[37] have recently been found as the cause for severe DHFR deficiency; these mutations were associated with megaloblastic anemia and cerebral folate deficiency. Therapeutic intervention is achieved by treatment with folinic acid,[37] a 5-formyl derivative of tetrahydrofolic acid that can be converted to other reduced folates while bypassing DHFR. This gene is a target for antifolate cancer chemotherapy with compounds such as methotrexate,[38] and folinic acid is used as part of a methotrexate chemotherapy regimen in order to save normal, nontransformed cells from the effects of methotrexate. In that context, the sequence variant C829T is of note, revealing the regulation of *DHFR* by microRNAs. It renders a binding site for microRNA-24 ineffective,[39] leading to increased expression of *DHFR* and subsequent methotrexate resistance. The gene is also a target in the fight against the malaria parasite *Plasmodium falciparum*; several mutations in *dhfr* have been detected that render *P. falciparum* resistant to antimalaria drugs.[40]

G. *MTHFD1* and *MTHFD2*

A single-nucleotide polymorphism in the human *MTHFD1* gene has shown genetic association with several folate-dependent pathologies. Homozygosity for the rare allele at G1958A appears to constitute a maternal risk for neural tube defects[41] and heart defects in the infant, severe placental abruption, and late pregnancy loss.[42] Loss of the *Mthfd1* gene via a gene-trap mutation in mice also has severe consequences leading to embryonic lethality.[43] A targeted mutation in the *Mthfd2* gene, which encodes the mitochondrial version of the enzyme, also leads to embryonic death at mid-gestation,[44] thereby revealing an essential developmental function for folate metabolism in mitochondria.

H. *MTHFR* (Methylene Tetrahydrofolate Reductase)

A mutation in the human *MTHFR* gene (C677T) represents the most common genetic cause for elevated homocysteine levels. This sequence variant yields a hypomorphic allele of the gene,[45] with the T configuration showing significantly reduced, but not absent, enzyme activity. The result is hyperhomocysteinemia, a condition thought to lead to vascular pathology.[46] The *MTHFR*

C677T mutation has received attention in many lines of investigation. It is associated with increased risk for neural tube defects[47,48] and further pregnancy complications,[49] increased risk for gastric cancer,[50] and decreased risk for childhood acute lymphoblastic leukemia[51] and colon cancer[52]; no association to prostate cancer[53] or lung cancer[54] was found. However, recent meta-analyses have called some of these risk associations into question.[55–57] Furthermore, the relationship of this mutation to congenital heart defects[58] and coronary heart disease[59] has been questioned.[60] The *MTHFR* C677T variant may affect the risk for migraine,[61,62] and a link to autism-spectrum disorder has been postulated.[63,64] Mice lacking the *Mthfr* gene[65] display delayed development, impaired growth, and increased morbidity and mortality in the early postnatal period. Interestingly, supplementation of these mice with betaine, a methyl donor derived from choline that is capable of fueling the methylation reaction from homocysteine to methionine, can at least partially rescue the mortality phenotype and can ameliorate neuronal proliferation and differentiation deficits associated with the lack of MTHFR.[66] Furthermore, MTHFR deficiency can be protective against the adverse developmental effects brought about by excessively high folate intake.[67]

MTHFR catalyzes an irreversible reaction, directing the single-carbon stream toward generation of *S*-adenosylmethionine and general methylation reactions; lack of MTHFR activity is therefore more likely to compromise that aspect rather than compromising purine or pyrimidine synthesis. Hence, one may speculate that phenotypes resulting from MTHFR deficiency are likely to involve a methylation deficit, potentially in the epigenome.

I. *MTR* (Methionine Synthase)

Human patients with a deficiency in the *MTR* gene show altered levels of methionine and homocysteine[68] and are affected by megaloblastic anemia that is sometimes associated with neural dysfunction and mental retardation.[69] The mutation *MTR* A2756G, which may constitute a gain-of-function allele, exhibits an association with elevated risk for prostate cancer.[53] Mice lacking the *Mtr* gene die *in utero*[70]; homozygous embryos survive past the implantation stage but succumb a short time after that. This suggests that the human mutation may not be as severe as the null mutation of the mouse model, and that residual MTR activity may be retained in humans carrying a genetic burden at the *MTR* locus.

J. *MTRR* (Methionine Synthase Reductase)

Mutations of the human *MTRR* gene, encoding an enzyme necessary for the activation of MTR, are associated with a higher risk for birth defects,[71] including neural tube defects.[72] The mutation A66G shows association to decreased risk for childhood acute lymphoblastic leukemia.[73] Mice carrying a hypomorphic allele at the *mtrr* locus exhibit hyperhomocysteinemia.[74] Such mice are viable but are burdened by adversely affected cardiac development and reduced overall reproductive success.

K. *SHMT* (Serine Hydroxymethyltransferase)

Heterozygosity at *SHMT1* C1420T in humans appears to confer a lower risk for childhood acute lymphoblastic leukemia,[53,75] and a weak genetic association was detected between the *SHMT1* C1420T variant and increased risk for prostate cancer.[53] *SHMT2* is a target for the microRNA miR-193b.[76] Mice lacking the *Shmt1* gene encoding a cytoplasmic version of the SHMT enzyme[77] appear healthy but suffer from abnormalities in hepatic levels of *S*-adenosylmethionine and in uracil incorporation into genomic DNA. Furthermore, at low maternal folate and choline status, these mice show neural tube defects due to impaired *de novo* thymidylate synthesis.[78] A second gene termed *Shmt2* encodes two transcripts, with one encoding a version of the enzyme that is exclusively localized in mitochondria and the other giving rise to a cytoplasmic version of the enzyme that is functionally redundant for the product of *Shmt1*.[79] For *Shmt2*, the Knockout Mouse Project (KOMP) Repository (www.komp.org) currently reports targeted embryonic stem cells, but no mutant mice.

Deficiencies in many of the genes of the folate cycle are associated with embryonic lethality. Although specific phenotypes differ and manifest at various stages of embryogenesis, it is noteworthy that such phenotypes can be found in almost all aspects of folate metabolism. Hematopoietic phenotypes, where they have been characterized, appear to depend on the folate moiety itself, as no such phenotypes can be observed downstream of MTR where the methyl group is carried by molecules other than folate. Together, the existing mouse mutations in these genes underscore the essential position of the folate pathway for functional cell biology. The picture available from human mutations, however, is more complex, as several instances exist where a single mutation can confer not only increased risk for one type of morbidity but also decreased risk for another type of morbidity. The molecular mechanisms that mediate those risks are thought to involve two general realms: (i) DNA synthesis and repair for cell proliferation and (ii) regulation of gene expression via the epigenome.

IV. Folate and the Epigenome

Because of folate's role in methylation reactions, the relationship between folate and the epigenome has received increased attention in recent years. The epigenome—the combination of DNA methylation, histone modification, transcription factor function, and noncoding RNA expression—constitutes domains in the genome that permit gene transcription. The term is derived from the classical definition of epigenetics—the generation of different and stable phenotypes without changes in the underlying DNA sequence[80]; the epigenome represents the molecular instantiation of this concept. Interest in the biological role of epigenetic modifications has intensified with the development of technologies that allow assessment on a genome-wide scale[81,82] rather than at single genes (hence the term epigenomics). The connection to folate comes from the fact that the various molecular modalities of the epigenome rely on methylation reactions to define the status of the respective epigenomic marks, or to regulate activities of epigenomic factors. While these methylation reactions all use *S*-adenosylmethionine as the direct methyl donor, the supply of *S*-adenosylmethionine, and thereby the efficacy of the pertinent methylation reaction, is dependent on the rate of replenishment of methionine from homocysteine; this reaction, in turn, depends mainly on the availability of 5-methyltetrahydrofolate, or to some extent on the choline/betaine pathway. This dependence establishes a direct link between nutrient availability and epigenetic modifications and hence between regulation of gene expression and phenotypes of health and disease.

A. DNA Methylation

DNA methylation has long been thought of as the main conduit to establish, maintain, and transmit epigenetic information. Methylation of cytosine residues in the CpG dinucleotide sequence forms a covalent, stable alteration of DNA, which is generally associated with silencing of gene transcription.[83,84] This methylation mark is established by *de novo* DNA methyltransferases (DNMTs) during development,[85] remains on the DNA through mitosis, and serves as a template for the maintenance DNA methyltransferase DNMT1 to establish the correct cell-type-specific pattern of methylation marks on the as-of-yet-unmethylated daughter strand after passage of the replication fork. In this fashion, epigenetic marks can be transmitted through mitosis and maintained over many cell generations.

DNA methylation was long believed to be permanent.[86] However, recent findings support the notion that DNA methylation is in fact reversible,[87,88] and active DNA demethylation may occur through the base-excision repair mechanism.[89] Folate deficiency and the resulting lack of methyl donors for DNA methylation may lead to passive loss of the DNA methylation mark based on a "failure-to-maintain" mechanism, and thereby to a loss or a change of epigenetic information. This is of particular interest in situations where rapid cell proliferation occurs, requiring high activity from DNMT1 after each mitotic division, with a concomitantly high demand for methyl-donor substrates. Examples of such situations are (i) embryonic development, with its rapid cell cycles during growth of the developing embryo; (ii) hematopoiesis; (iii) intestinal cell regeneration; (iv) the swift cell proliferation necessary to mount a successful immune response (e.g., to a pathogen challenge); or (v) the proliferation of transformed cells in the progression of cancer. Failure of DNMT1 to keep up with demand due to low levels of *S*-adenosylmethionine under methyl-donor deficiency would result in hemimethylated DNA sites following the first mitotic cycle, and in the absence of template information for the next cycle of maintenance DNA methylation, with under- or unmethylated DNA as consequence. In this fashion, epigenetic information in the form of DNA methylation patterns can quickly be diluted, or lost altogether. It is generally thought that DNA hypomethylation may lead to a loss of gene silencing and to ectopic gene expression.

B. Histone Methylation

Histone modifications are another realm of the epigenome where methylation reactions take center stage. The N-termini of histones H3 and H4 form the so-called histone tails that are amenable to covalent modifications. Methylation of lysine residues is one prominent modification; however, acetylation, phosphorylation, ubiquitination, and ATP-ribosylation occur as well. Collectively, these modifications are referred to as the "histone code"[90]; in adhering to the context of folate and methylation reactions, we limit our discussion to histone methylation. In contrast to DNA methylation, where the presence of the methylation mark is normally associated with transcriptional silencing, the situation for histone methylation is more complex. Depending on the specific nature of the methylation mark, histone methylation can be associated either with active transcription or with gene silencing; in fact, many proteins with histone-modifying activities were originally described as transcriptional coactivators or corepressors.[91–93] In this context, it is important to know not only which lysine residue on the histone tail carries the mark, but also the extent of methylation: mono-, di-, and trimethylation of the terminal amino group of the respective lysine are possible.

While discussion of each histone methylation mark is beyond the scope of this review, a few paradigmatic examples serve to illustrate the complexity.

- Trimethylation at lysine 4 of histone H3 (H3K4me3) occurs primarily at promoters of actively transcribed genes,[94,95] whereas monomethylation at the same residue is more likely to be found at DNA sequences with transcriptional enhancer function.[96]
- Trimethylation at lysine 36 of histone H3 is also associated with active transcription[97]; this mark is typically found within the transcribed body of a gene, and it appears to be enriched on exons.
- In contrast, trimethylation at lysine 27 of histone H3 is associated with repression of transcription[98,99]; in a similar fashion, trimethylation at lysine 9 of histone H3 is concomitant with silencing[100] and subsequent formation of heterochromatin. These two trimethylation marks are of particular interest

because acetylation at these specific lysine residues is typically correlated with transcriptional activation. Histone H3 lysine 27 acetylation is a mark for transcriptional enhancer activity,[100] and changing the nature of the modification there can in fact serve as a regulatory switch from activation to silencing. Therefore, alteration of these histone marks can have profound effects on transcriptional output and, consequently, on the phenotype of a given cell.

Transcriptional effects of histone modifications are the result of a balance between transcription-activating methylation marks, transcription-silencing methylation marks, and transcription-activating acetylation marks. Histone acetylation status can be affected by the supply of acetyl-CoA dependent on metabolism and nutrition.[101] Likewise, histone methylation status may be influenced by the dietary methyl-donor supply.[102,103] These histone modifications are therefore a direct target of nutrition, with epigenetic consequences for the organism.

The mechanism for the transmission of histone methylation marks through mitosis is not as well understood as the transmission of DNA methylation information. It is thought to employ an analogous "read and write" mechanism, because the enzymes that generate the respective histone methylation marks tend to remain associated with the DNA replication fork to regenerate the histone mark after the replication fork has passed and nucleosomes are reassembled.[104] Similar to the passive loss of DNA methylation, if the supply of methyl donors is limited, histone methylation marks are susceptible to mitotic dilution or complete loss due to a failure-to-maintain mechanism. In addition, histone methylation marks are subject to active enzymatic removal by histone demethylases.[105]

The role of folate in histone methylation extends beyond that of being the methyl carrier toward methylation reactions. A recent finding revealed a new relationship between folate and histone methylation, as folate was found to be an enzymatic cofactor for the lysine-specific histone demethylase 1A (KDM1A).[106] Specifically, THF can serve as an acceptor for formaldehyde which is generated during the oxidative demethylation of histone tails. Such a reaction would serve not only to trap and convert formaldehyde, which is a toxic compound, but also to recharge THF with a single-carbon moiety that can then be used for future methylation reactions. Mouse embryos lacking KDM1A fail to gastrulate, resulting in embryonic lethality at 7.5 days of gestation.[107] Loss of the *kdm1* homologue in *C. elegans* can lead to transgenerational effects,[108] underscoring the contribution of histone methylation marks to epigenetic information.

C. Transcription Factors

Regulation of epigenomic activities by methylation is not restricted to DNA and histone methylation. Recent evidence demonstrates that transcription factors can also be regulated in their activity by direct methylation. Transcription factors are not always discussed in the context of the epigenome, but rather in the framework of direct transcriptional regulation.[104] However, they also fit the definition of epigenetics that we use here (generation of differential and stable phenotypes without changes in the genomic DNA sequence). Transcription factors establish stable phenotypic differences between cell types,[109] and activation of gene loci can be propagated through mitosis.

The interplay between methylation and transcription factors was originally established by the finding that transcriptional coactivators or corepressors (proteins that associate with transcription factors to form larger complexes and thereby modulate transcriptional output) have histone-modifying activities[93] and establish or modify histone acetylation or methylation in the immediate vicinity of the genomic binding site for the transcription factor complex. By now, it has also been established that transcription factor proteins can be directly methylated,[110,111] which affects the transcriptional output driven by these factors. What is not clear at this time, however, is whether this methylation-dependent regulatory mechanism has general applicability, or whether it is restricted to specific transcription factors. It is also difficult to estimate how a methyl-donor deficiency may influence this regulatory mechanism, and how broad the effect on overall transcriptional output may be.

D. Noncoding RNA

Finally, noncoding RNAs are subjected to methylation reactions as well. RNA methylation has long been described, with tRNA as the prime example. Recent results suggest that methylation of tRNA[112] (interestingly, by DNMT2, originally considered a DNA methyltransferase but recently renamed TRDMT1, tRNA aspartic acid methyltransferase 1) may afford protection from ribonuclease degradation, thereby extending the half-life and functionality of each molecule. Less is known about how methylation reactions may affect the epigenetic function of noncoding RNAs, such as microRNAs or chromatin-associated long noncoding RNAs, both associated with reduced transcriptional output.[113] However, generation of microRNAs and small interfering RNAs in plants, production of small interfering RNAs in *D. melanogaster*, and biosynthesis of RNAs of the Piwi-interacting family of small noncoding RNAs (piRNAs)[114] in mammalian germ cells include a methylation reaction at the 3′-end.[115] How these methylation reactions impact the epigenetic function of small RNAs represents an emerging field of biological inquiry. Yet, piRNAs also serve to illustrate the intricate relationship between the different molecular modalities of the epigenome in general and the various methylation reactions in particular, as methylated piRNAs direct DNA methylation in the process of transcriptional silencing of retrotransposons in developing mammalian germ cells.[116]

Methylation reactions therefore play a fundamental role in the different molecular modalities of the epigenome and contribute significantly to the way the epigenome shapes the output of genomic information in the form of gene expression. The majority of methylation reactions in the epigenome are associated with reduced transcriptional output and even transcriptional silencing of large regions of the genome; however, the correlation between particular histone methylation marks and transcriptionally active promoters or enhancers argues against such a categorical view.

All the respective enzymes that catalyze methylation reactions in the epigenome critically depend on a sufficient supply of their methyl-donor substrate, typically *S*-adenosylmethionine. Folate feeds into this methyl-donor supply, and therefore, folate levels are thought to have a profound effect on the efficacy of methylation reactions. Yet, it is important to note that while the folate pathway is the major source for replenishing *S*-adenosylmethionine levels in the cell for future methylation reactions, it is not the sole source. The choline/betaine route of the methyl-donor supply also feeds into the cellular pool of *S*-adenosylmethionine[117] and may interact with folate deficiency.[118,119] When considering epigenomic effects with respect to nutritional folate deficiency or to genetic deficits affecting enzymes of the folate cycle, such effects must be interpreted in the context of the entire methyl-donor stream, and not based on isolated pathways.

V. Epigenomic Footprint of Folate

Historically, the biological relevance of folate for epigenetic effects has been derived from the views that folate is critical for the supply of methyl groups in the cell and that epigenetic effects are mediated principally by DNA methylation. Therefore, the simplest assumption was that folate levels would essentially constitute the rate-limiting factor for DNA methylation by dictating the status of the methyl-donor pool. Based on such a feed-forward model, a direct and positive correlation between folate levels and DNA methylation status could be postulated: folate deficiency results in reduced DNA methylation; conversely, high folate status leads to increased DNA methylation, and hence, folate status determines epigenetic events. Because folate status is dependent on nutrition, this would provide a direct conduit between nutrition, epigenetics, and the resulting phenotypes. In addition, the prevailing view for a long time was that DNA methylation was essentially a stable modification, and therefore, epigenetic events, once established, were nearly permanent or at least very long-lasting biological phenomena. Such epigenetic events could be changed only because of insufficient means to maintain DNA methylation patterns, such as in the case of folate deficiency. As a consequence, the loss of DNA methylation was thought to lead to increased levels of gene expression, which in the case of oncogenes may contribute to the development of cancer. However, this picture has dramatically changed in recent years.

A. Epigenetics and Plasticity

The first aspect of this change is the recognition that epigenetic events are not permanent, but epigenetic marks are actually subject to plasticity beyond the passive failure-to-maintain mechanisms through mitotic dilution. Active DNA demethylation can occur via the base-excision repair pathway,[89] and histone demethylation may ensue through specific enzymes.[105] Less is known about noncoding RNA or transcription factor methylation, but cellular turnover may erase these marks. Very strong corroboration for epigenomic plasticity has arisen from nutritional studies. Mice exposed to methyl-donor diets show diet-dependent DNA methylation patterns at specific gene loci, together with differences in gene expression.[120–122] Choline in the diet can affect DNA methylation and histone methylation, as well as the expression of histone methyltransferases (HMTs) such as G9a[102] (see also chapter "The Nutrigenetics and Nutrigenomics of the Dietary Requirement for Choline"). High levels of methyl donors in the diet affect both DNA and HMTs, with concomitantly higher levels of DNA and histone methylation.[122] Dietary methyl deficiencies can alter histone methylation as well.[123] Besides nutritional studies, it is thought that the capacity for epigenomic plasticity is the basis for induction of pluripotency via a specific set of transcription factors.[124] These examples demonstrate that epigenetic structures can be actively changed, not just by degrading the epigenetic marks through deficient maintenance mechanisms.

B. Relationship Between Epigenetic Marks and Transcription

The second aspect of change is in regard to the nature and diversity of epigenetic modifications. As discussed previously, epigenetics is no longer viewed as being synonymous with DNA methylation, but the concept of the epigenome has expanded to several molecular dimensions. While methylation reactions play a central role in setting the various epigenetic marks in each of the realms, the categorical idea that methylation status is negatively correlated with transcription does not extend through all epigenomic modalities: in the world of histones, methylation status can be associated with active transcription or with silencing of gene expression, depending on the specific methylation mark (see Section IV.B). While there may be a positive correlation between methyl-donor supply and methylation levels, subsequent conclusions about the transcriptional status—and thereby about the potential phenotype—of a cell are no longer straightforward.

C. Folate Status and DNA Methylation

To complicate matters further, experimental evidence on the relationship between folate status and DNA methylation has suggested higher levels of complexity than would be predicted on the basis of a simple feed-forward model. The consequences of folate deficiency can range from global DNA hypomethylation,[125] to global DNA hypomethylation coupled with DNA hypermethylation at specific genes or promoters,[126,127] to global DNA hypermethylation.[128,129] Hypermethylation, in particular, is in stark contrast to predictions for folate deficiency if one follows the simple supply model; one potential explanation for this conundrum is the induction of DNMTs in response to low methyl-donor levels.[130] This may reveal why there are increased levels of methylation, but it does not explain why promoters or CpG islands (typically regions of low or no DNA methylation) are no longer protected, and are subjected to *de novo* methylation. Additional complications come from findings that folate effects can be tissue specific,[131] and that not just the extent but also the duration of folate deficiency can have consequences on DNA methylation[132]; evidence has also been obtained that DNA methylation may in fact not change at all in response to folate status[133–136] despite folate-responsive changes in gene expression. It appears that a linear feed-forward model from folate level, to methyl-donor status, to DNA methylation, and finally to gene expression does not suffice to explain the various experimental observations, and that more complex regulatory relationships must be considered.

D. Folate and the Regulation of Transcription

In contrast to the historical view that folate affects gene expression solely via DNA methylation, it would appear that changes in gene transcription may in fact precede changes at the level of DNA methylation. A recent study on folate supplementation, deficiency, and repletion in human subjects at risk for colon cancer reported significant effects of folate on the regulation of gene expression, but neither genomic nor promoter methylation was affected under conditions of high or low folate.[135] Interestingly, the experiments provided support for the notion that high folate status may constitute a risk factor for colorectal carcinogenesis by affecting proinflammatory pathways. It should be noted that the altered transcriptome can exert immediate effect on cellular phenotype and function.

We suggest that, in this scenario, it is conceivable that changes in DNA methylation, rather than being the primary means of translating folate effects, are the final outcome of a multilayered regulatory circuitry that may engage other molecular realms of the epigenome, such as histone methylation[103] or microRNAs,[137,138] in order to manifest the effects of folate. Evidence for this concept has emerged recently in a report on the effects of folate on histone methylation at the hairy and enhancer of split 1 (*Hes1*) and neurogenin 2 (*Neurog2*) gene loci.[139] In mouse embryos with paired box protein Pax-3 deficiency, histone H3K27 dimethylation, a repressive epigenomic mark, was increased at the promoters for both genes, whereas treatment with folate reduced the respective methylation marks to normal levels. This result demonstrated that folate is capable of modulating epigenetic chromatin marks. However, the fact that an *increase* in folate leads to a *decrease* of the methylation status underscores the notion that the feed-forward model is insufficient to explain such a change, and that the regulatory mechanism underlying this methylation change is more complex. It has been proposed that in this scenario, folate acts through microRNAs to increase expression of the demethylase KDM6B, which then mediates the reduction of histone H3K27 dimethylation levels. As more such studies are conducted, one would expect a clearer picture of how folate affects the epigenome and transcriptome and ultimately cellular phenotypes.

We therefore suggest that the relationship between folate and the epigenome is far more complex than previously anticipated, yet it has significant ramifications for the biology of health and disease. It is therefore necessary to explore that relationship in more detail, taking advantage of genome-wide tools that have become available in recent years.

VI. A Roadmap for Folate and the Epigenome

One particular caveat to the interpretations of the relationship between folate and the epigenome is that many results were obtained before the advent of genome-wide technologies. Such studies initially measured global DNA methylation levels[134,140] (e.g., as overall content of 5-methylcytosine but without positional information in the genome) or resorted to candidate gene studies of gene- or promoter-specific DNA methylation.[141] Technological limitations dictated this approach, as methods that would have allowed gene-specific attributions in a whole-genome context were not available. While candidate gene approaches were successful in unraveling the role of folate and DNA methylation in specific paradigms, the lack of a whole-genome context made generalizable biological interpretations difficult.

One such example is the methyl-donor-dependent transcription change at the A^{vY} allele: methylation patterns change at an IAP element (intracisternal A-particle) near the gene. As a remnant of a viral sequence with a long-terminal repeat element that can activate transcription, the IAP can override the normal regulatory machinery at the gene.[121] Similarities have been detected at the $Axin^{Fu}$ allele.[122] With approximately 8000 IAP elements with intact long-terminal repeats in the mouse genome,[142] IAP methylation-based mechanisms may affect many more genes, and therefore may account for a significant fraction of methyl-donor-conferred phenotypic changes. Although the studies on A^{vY} and Ax^{Fu} were paradigmatic for this mechanism, the general conclusion has been questioned recently.[143] A whole-genome context will be required to determine the entire epigenomic footprint of methyl-donor diets on IAPs so that a correlation with the respective phenotypes can be made. Furthermore, IAPs are thought to be a feature of rodent genomes, and mechanistic parallels to the human genome need to be established. We therefore propose that it is necessary to make the whole-genome context a primary focus for new investigations of molecular consequences of folate status.

Genome-wide technologies to determine changes in gene expression patterns in response to folate status are well established. Microarray hybridization has been used extensively to characterize transcriptomic responses in various tissues and cell types[126,144–146]; folate-responsive gene expression incorporates common pathways such as proliferation, inflammation, and apoptosis in folate-responsive transcriptomic changes. The focus of such experiments was primarily on known protein-coding genes. Genome-wide technologies to assess the epigenomic footprints of folate, mostly based on next-generation sequencing, have become available not just for DNA methylation,[81,147] but also for histone methylation,[148–150] as well as other epigenomics modalities. In addition, new sequencing-based methods[151] now make it possible to overcome the prior restriction of expression analyses to known protein-coding genes and permit genome-wide assessments of the entire transcriptome, including antisense and noncoding RNAs,[152] in relation to folate status.

To get a better general understanding of the effects of folate on epigenome, transcriptome, and ultimately cellular phenotypes, it will be necessary to conduct epigenomics studies in conjunction with transcriptomics assessments and to apply several of these new genome-wide technologies *within the same paradigm*. Only the combination of such approaches will allow the concomitant definition of folate-sensitive target genes and reveal how the epigenome landscape of those target genes is affected by folate status, and how the epigenome through such targets manifests phenotype. While this appears to be an ambitious goal that will require substantial efforts and resources, we believe that only such integrative studies will reveal the true epigenomic footprint of folate and establish the biological context that is necessary to interpret folate effects upstream of gene expression with reference to DNA methylation, histone modifications, noncoding RNAs, and transcription factors. Furthermore, such analyses will permit "downstream" correlations of this epigenomic footprint to functional outcomes beyond transcription at various "–omics" levels, all the way to assessments for health and disease. We suggest that such an experimental approach will be highly productive, bear many surprises, and lead to novel insights into the physiology and pathophysiology of folate.

References

1. Smithells RW, Ankers C, Carver ME, Lennon D, Schorah CJ, Sheppard S. Maternal nutrition in early pregnancy. *Br J Nutr* 1977;**38**:497–506.
2. Smithells RW, Sheppard S, Schorah CJ. Vitamin deficiencies and neural tube defects. *Arch Dis Child* 1976;**51**:944–50.
3. Smithells RW, Sheppard S, Schorah CJ, Seller MJ, Nevin NC, Harris R, et al. Apparent prevention of neural tube defects by periconceptional vitamin supplementation. *Arch Dis Child* 1981;**56**:911–8.
4. Obican SG, Finnell RH, Mills JL, Shaw GM, Scialli AR. Folic acid in early pregnancy: a public health success story. *FASEB J* 2010;**24**:4167–74.
5. Fenech M. The role of folic acid and Vitamin B12 in genomic stability of human cells. *Mutat Res* 2001;**475**:57–67.
6. Sieber O, Heinimann K, Tomlinson I. Genomic stability and tumorigenesis. *Semin Cancer Biol* 2005;**15**:61–6.
7. Kim YI. Role of folate in colon cancer development and progression. *J Nutr* 2003;**133**: 3731S–3739S.
8. Goldman ID, Chattopadhyay S, Zhao R, Moran R. The antifolates: evolution, new agents in the clinic, and how targeting delivery via specific membrane transporters is driving the development of a next generation of folate analogs. *Curr Opin Investig Drugs* 2010;**11**: 1409–23.
9. Low PS, Kularatne SA. Folate-targeted therapeutic and imaging agents for cancer. *Curr Opin Chem Biol* 2009;**13**:256–62.
10. Kamen BA, Smith AK. A review of folate receptor alpha cycling and 5-methyltetrahydrofolate accumulation with an emphasis on cell models in vitro. *Adv Drug Deliv Rev* 2004;**56**:1085–97.
11. Hooijberg JH, Broxterman HJ, Kool M, Assaraf YG, Peters GJ, Noordhuis P, et al. Antifolate resistance mediated by the multidrug resistance proteins MRP1 and MRP2. *Cancer Res* 1999;**59**:2532–5.
12. Hooijberg JH, Jansen G, Assaraf YG, Kathmann I, Pieters R, Laan AC, et al. Folate concentration dependent transport activity of the Multidrug Resistance Protein 1 (ABCC1). *Biochem Pharmacol* 2004;**67**:1541–8.
13. Kitamura Y, Hirouchi M, Kusuhara H, Schuetz JD, Sugiyama Y. Increasing systemic exposure of methotrexate by active efflux mediated by multidrug resistance-associated protein 3 (mrp3/abcc3). *J Pharmacol Exp Ther* 2008;**327**:465–73.
14. Piedrahita JA, Oetama B, Bennett GD, van Waes J, Kamen BA, Richardson J, et al. Mice lacking the folic acid-binding protein Folbp1 are defective in early embryonic development. *Nat Genet* 1999;**23**:228–32.
15. Saitsu H, Ishibashi M, Nakano H, Shiota K. Spatial and temporal expression of folate-binding protein 1 (Fbp1) is closely associated with anterior neural tube closure in mice. *Dev Dyn* 2003;**226**:112–7.
16. Salbaum J, Finnell R, Kappen C. Regulation of Folate receptor 1 gene expression in the visceral endoderm. *Birth Defects Res A Clin Mol Teratol* 2009;**85**:303–13.
17. Elwood PC, Nachmanoff K, Saikawa Y, Page ST, Pacheco P, Roberts S, et al. The divergent 5′ termini of the alpha human folate receptor (hFR) mRNAs originate from two tissue-specific promoters and alternative splicing: characterization of the alpha hFR gene structure. *Biochemistry* 1997;**36**:1467–78.
18. Finnell RH, Spiegelstein O, Wlodarczyk B, Triplett A, Pogribny IP, Melnyk S, et al. DNA methylation in Folbp1 knockout mice supplemented with folic acid during gestation. *J Nutr* 2002;**132**:2457S–2461S.

19. Barber R, Shalat S, Hendricks K, Joggerst B, Larsen R, Suarez L, et al. Investigation of folate pathway gene polymorphisms and the incidence of neural tube defects in a Texas hispanic population. *Mol Genet Metab* 2000;**70**:45–52.
20. Barber RC, Shaw GM, Lammer EJ, Greer KA, Biela TA, Lacey SW, et al. Lack of association between mutations in the folate receptor-alpha gene and spina bifida. *Am J Med Genet* 1998;**76**:310–7.
21. Steinfeld R, Grapp M, Kraetzner R, Dreha-Kulaczewski S, Helms G, Dechent P, et al. Folate receptor alpha defect causes cerebral folate transport deficiency: a treatable neurodegenerative disorder associated with disturbed myelin metabolism. *Am J Hum Genet* 2009;**85**:354–63.
22. Spiegelstein O, Gould A, Wlodarczyk B, Tsie M, Lu X, Le C, et al. Developmental consequences of in utero sodium arsenate exposure in mice with folate transport deficiencies. *Toxicol Appl Pharmacol* 2005;**203**:18–26.
23. Kappen C, Mello MA, Finnell RH, Salbaum JM. Folate modulates Hox gene-controlled skeletal phenotypes. *Genesis* 2004;**39**:155–66.
24. Laverdiere C, Chiasson S, Costea I, Moghrabi A, Krajinovic M. Polymorphism G80A in the reduced folate carrier gene and its relationship to methotrexate plasma levels and outcome of childhood acute lymphoblastic leukemia. *Blood* 2002;**100**:3832–4.
25. Bi XH, Zhao HL, Zhang ZX, Zhang JW. Association of RFC1 A80G and MTHFR C677T polymorphisms with Alzheimer's disease. *Neurobiol Aging* 2009;**30**:1601–7.
26. Ma DW, Finnell RH, Davidson LA, Callaway ES, Spiegelstein O, Piedrahita JA, et al. Folate transport gene inactivation in mice increases sensitivity to colon carcinogenesis. *Cancer Res* 2005;**65**:887–97.
27. Zhao R, Russell RG, Wang Y, Liu L, Gao F, Kneitz B, et al. Rescue of embryonic lethality in reduced folate carrier-deficient mice by maternal folic acid supplementation reveals early neonatal failure of hematopoietic organs. *J Biol Chem* 2001;**276**:10224–8.
28. Maddox DM, Manlapat A, Roon P, Prasad P, Ganapathy V, Smith SB. Reduced-folate carrier (RFC) is expressed in placenta and yolk sac, as well as in cells of the developing forebrain, hindbrain, neural tube, craniofacial region, eye, limb buds and heart. *BMC Dev Biol* 2003;**3**:6.
29. Dixon KH, Lanpher BC, Chiu J, Kelley K, Cowan KH. A novel cDNA restores reduced folate carrier activity and methotrexate sensitivity to transport deficient cells. *J Biol Chem* 1994;**269**: 17–20.
30. Qiu A, Jansen M, Sakaris A, Min SH, Chattopadhyay S, Tsai E, et al. Identification of an intestinal folate transporter and the molecular basis for hereditary folate malabsorption. *Cell* 2006;**127**:917–28.
31. Jakubowski H, Perla-Kajan J, Finnell RH, Cabrera RM, Wang H, Gupta S, et al. Genetic or nutritional disorders in homocysteine or folate metabolism increase protein N-homocysteinylation in mice. *FASEB J* 2009;**23**:1721–7.
32. Salojin KV, Cabrera RM, Sun W, Chang WC, Lin C, Duncan L, et al. A mouse model of hereditary folate malabsorption: deletion of the PCFT gene leads to systemic folate deficiency. *Blood* 2011;**117**:4895–904.
33. Atabay B, Turker M, Ozer EA, Mahadeo K, Diop-Bove N, Goldman ID. Mutation of the proton-coupled folate transporter gene (PCFT-SLC46A1) in Turkish siblings with hereditary folate malabsorption. *Pediatr Hematol Oncol* 2010;**27**:614–9.
34. Borzutzky A, Crompton B, Bergmann AK, Giliani S, Baxi S, Martin M, et al. Reversible severe combined immunodeficiency phenotype secondary to a mutation of the proton-coupled folate transporter. *Clin Immunol* 2009;**133**:287–94.
35. Zhao R, Min SH, Qiu A, Sakaris A, Goldberg GL, Sandoval C, et al. The spectrum of mutations in the PCFT gene, coding for an intestinal folate transporter, that are the basis for hereditary folate malabsorption. *Blood* 2007;**110**:1147–52.

36. Cario H, Smith DE, Blom H, Blau N, Bode H, Holzmann K, et al. Dihydrofolate reductase deficiency due to a homozygous DHFR mutation causes megaloblastic anemia and cerebral folate deficiency leading to severe neurologic disease. *Am J Hum Genet* 2011;**88**:226–31.
37. Banka S, Blom HJ, Walter J, Aziz M, Urquhart J, Clouthier CM, et al. Identification and characterization of an inborn error of metabolism caused by dihydrofolate reductase deficiency. *Am J Hum Genet* 2011;**88**:216–25.
38. Rajagopalan PT, Zhang Z, McCourt L, Dwyer M, Benkovic SJ, Hammes GG. Interaction of dihydrofolate reductase with methotrexate: ensemble and single-molecule kinetics. *Proc Natl Acad Sci USA* 2002;**99**:13481–6.
39. Mishra PJ, Humeniuk R, Longo-Sorbello GS, Banerjee D, Bertino JR. A miR-24 microRNA binding-site polymorphism in dihydrofolate reductase gene leads to methotrexate resistance. *Proc Natl Acad Sci USA* 2007;**104**:13513–8.
40. Sridaran S, McClintock SK, Syphard LM, Herman KM, Barnwell JW, Udhayakumar V. Antifolate drug resistance in Africa: meta-analysis of reported dihydrofolate reductase (dhfr) and dihydropteroate synthase (dhps) mutant genotype frequencies in African Plasmodium falciparum parasite populations. *Malar J* 2010;**9**:247.
41. Parle-McDermott A, Kirke PN, Mills JL, Molloy AM, Cox C, O'Leary VB, et al. Confirmation of the R653Q polymorphism of the trifunctional C1-synthase enzyme as a maternal risk for neural tube defects in the Irish population. *Eur J Hum Genet* 2006;**14**:768–72.
42. Parle-McDermott A, Mills JL, Kirke PN, Cox C, Signore CC, Kirke S, et al. MTHFD1 R653Q polymorphism is a maternal genetic risk factor for severe abruptio placentae. *Am J Med Genet A* 2005;**132**:365–8.
43. MacFarlane AJ, Perry CA, Girnary HH, Gao D, Allen RH, Stabler SP, et al. Mthfd1 is an essential gene in mice and alters biomarkers of impaired one-carbon metabolism. *J Biol Chem* 2009;**284**:1533–9.
44. Di Pietro E, Sirois J, Tremblay ML, MacKenzie RE. Mitochondrial NAD-dependent methylenetetrahydrofolate dehydrogenase-methenyltetrahydrofolate cyclohydrolase is essential for embryonic development. *Mol Cell Biol* 2002;**22**:4158–66.
45. Frosst P, Blom HJ, Milos R, Goyette P, Sheppard CA, Matthews RG, et al. A candidate genetic risk factor for vascular disease: a common mutation in methylenetetrahydrofolate reductase. *Nat Genet* 1995;**10**:111–3.
46. Austin RC, Lentz SR, Werstuck GH. Role of hyperhomocysteinemia in endothelial dysfunction and atherothrombotic disease. *Cell Death Differ* 2004;**11**(Suppl. 1):S56–64.
47. van der Put NM, Steegers-Theunissen RP, Frosst P, Trijbels FJ, Eskes TK, van den Heuvel LP, et al. Mutated methylenetetrahydrofolate reductase as a risk factor for spina bifida. *Lancet* 1995;**346**:1070–1.
48. Munoz JB, Lacasana M, Cavazos RG, Borja-Aburto VH, Galaviz-Hernandez C, Garduno CA. Methylenetetrahydrofolate reductase gene polymorphisms and the risk of anencephaly in Mexico. *Mol Hum Reprod* 2007;**13**:419–24.
49. Nurk E, Tell GS, Refsum H, Ueland PM, Vollset SE. Associations between maternal methylenetetrahydrofolate reductase polymorphisms and adverse outcomes of pregnancy: the Hordaland Homocysteine Study. *Am J Med* 2004;**117**:26–31.
50. Boccia S, Hung R, Ricciardi G, Gianfagna F, Ebert MP, Fang JY, et al. Meta- and pooled analyses of the methylenetetrahydrofolate reductase C677T and A1298C polymorphisms and gastric cancer risk: a huge-GSEC review. *Am J Epidemiol* 2008;**167**:505–16.
51. Yan J, Yin M, Dreyer ZE, Scheurer ME, Kamdar K, Wei Q, et al. A meta-analysis of MTHFR C677T and A1298C polymorphisms and risk of acute lymphoblastic leukemia in children. *Pediatr Blood Cancer* 2011;**58**:513–8.

52. Le Marchand L, Wilkens LR, Kolonel LN, Henderson BE. The MTHFR C677T polymorphism and colorectal cancer: the multiethnic cohort study. *Cancer Epidemiol Biomarkers Prev* 2005;**14**:1198–203.
53. Collin SM, Metcalfe C, Zuccolo L, Lewis SJ, Chen L, Cox A, et al. Association of folate-pathway gene polymorphisms with the risk of prostate cancer: a population-based nested case-control study, systematic review, and meta-analysis. *Cancer Epidemiol Biomarkers Prev* 2009;**18**:2528–39.
54. Mao R, Fan Y, Jin Y, Bai J, Fu S. Methylenetetrahydrofolate reductase gene polymorphisms and lung cancer: a meta-analysis. *J Hum Genet* 2008;**53**:340–8.
55. Wang J, Zhan P, Chen B, Zhou R, Yang Y, Ouyang J. MTHFR C677T polymorphisms and childhood acute lymphoblastic leukemia: a meta-analysis. *Leuk Res* 2010;**34**:1596–600.
56. Vollset SE, Igland J, Jenab M, Fredriksen A, Meyer K, Eussen S, et al. The association of gastric cancer risk with plasma folate, cobalamin, and methylenetetrahydrofolate reductase polymorphisms in the European Prospective Investigation into Cancer and Nutrition. *Cancer Epidemiol Biomarkers Prev* 2007;**16**:2416–24.
57. Boyles AL, Billups AV, Deak KL, Siegel DG, Mehltretter L, Slifer SH, et al. Neural tube defects and folate pathway genes: family-based association tests of gene-gene and gene-environment interactions. *Environ Health Perspect* 2006;**114**:1547–52.
58. van Beynum IM, den Heijer M, Blom HJ, Kapusta L. The MTHFR 677C->T polymorphism and the risk of congenital heart defects: a literature review and meta-analysis. *QJM* 2007;**100**:743–53.
59. Klerk M, Verhoef P, Clarke R, Blom HJ, Kok FJ, Schouten EG. MTHFR 677C–>T polymorphism and risk of coronary heart disease: a meta-analysis. *JAMA* 2002;**288**:2023–31.
60. Lewis SJ, Ebrahim S, Davey Smith G. Meta-analysis of MTHFR 677C->T polymorphism and coronary heart disease: does totality of evidence support causal role for homocysteine and preventive potential of folate? *BMJ* 2005;**331**:1053.
61. Liu A, Menon S, Colson NJ, Quinlan S, Cox H, Peterson M, et al. Analysis of the MTHFR C677T variant with migraine phenotypes. *BMC Res Notes* 2010;**3**:213.
62. Schurks M, Zee RY, Buring JE, Kurth T. Interrelationships among the MTHFR 677C>T polymorphism, migraine, and cardiovascular disease. *Neurology* 2008;**71**:505–13.
63. Pasca SP, Dronca E, Kaucsar T, Craciun EC, Endreffy E, Ferencz BK, et al. One carbon metabolism disturbances and the C677T MTHFR gene polymorphism in children with autism spectrum disorders. *J Cell Mol Med* 2009;**13**:4229–38.
64. Goin-Kochel RP, Porter AE, Peters SU, Shinawi M, Sahoo T, Beaudet AL. The MTHFR 677C–>T polymorphism and behaviors in children with autism: exploratory genotype-phenotype correlations. *Autism Res* 2009;**2**:98–108.
65. Chen Z, Karaplis AC, Ackerman SL, Pogribny IP, Melnyk S, Lussier-Cacan S, et al. Mice deficient in methylenetetrahydrofolate reductase exhibit hyperhomocysteinemia and decreased methylation capacity, with neuropathology and aortic lipid deposition. *Hum Mol Genet* 2001;**10**:433–43.
66. Schwahn BC, Laryea MD, Chen Z, Melnyk S, Pogribny I, Garrow T, et al. Betaine rescue of an animal model with methylenetetrahydrofolate reductase deficiency. *Biochem J* 2004;**382**:831–40.
67. Pickell L, Brown K, Li D, Wang XL, Deng L, Wu Q, et al. High intake of folic acid disrupts embryonic development in mice. *Birth Defects Res A Clin Mol Teratol* 2011;**91**:8–19.
68. Watkins D, Ru M, Hwang HY, Kim CD, Murray A, Philip NS, et al. Hyperhomocysteinemia due to methionine synthase deficiency, cblG: structure of the MTR gene, genotype diversity, and recognition of a common mutation, P1173L. *Am J Hum Genet* 2002;**71**:143–53.

69. Zavadakova P, Fowler B, Zeman J, Suormala T, Pristoupilova K, Kozich V, et al. CblE type of homocystinuria due to methionine synthase reductase deficiency: clinical and molecular studies and prenatal diagnosis in two families. *J Inherit Metab Dis* 2002;**25**:461–76.
70. Swanson DA, Liu ML, Baker PJ, Garrett L, Stitzel M, Wu J, et al. Targeted disruption of the methionine synthase gene in mice. *Mol Cell Biol* 2001;**21**:1058–65.
71. Zhu H, Wicker NJ, Shaw GM, Lammer EJ, Hendricks K, Suarez L, et al. Homocysteine remethylation enzyme polymorphisms and increased risks for neural tube defects. *Mol Genet Metab* 2003;**78**:216–21.
72. van der Linden IJ, den Heijer M, Afman LA, Gellekink H, Vermeulen SH, Kluijtmans LA, et al. The methionine synthase reductase 66A>G polymorphism is a maternal risk factor for spina bifida. *J Mol Med* 2006;**84**:1047–54.
73. Gast A, Bermejo JL, Flohr T, Stanulla M, Burwinkel B, Schrappe M, et al. Folate metabolic gene polymorphisms and childhood acute lymphoblastic leukemia: a case-control study. *Leukemia* 2007;**21**:320–5.
74. Elmore CL, Wu X, Leclerc D, Watson ED, Bottiglieri T, Krupenko NI, et al. Metabolic derangement of methionine and folate metabolism in mice deficient in methionine synthase reductase. *Mol Genet Metab* 2007;**91**:85–97.
75. Vijayakrishnan J, Houlston RS. Candidate gene association studies and risk of childhood acute lymphoblastic leukemia: a systematic review and meta-analysis. *Haematologica* 2010;**95**:1405–14.
76. Leivonen SK, Rokka A, Ostling P, Kohonen P, Corthals GL, Kallioniemi O, et al. Identification of miR-193b targets in breast cancer cells and systems biological analysis of their functional impact. *Mol Cell Proteomics* 2011;**10**: M110 005322.
77. MacFarlane AJ, Liu X, Perry CA, Flodby P, Allen RH, Stabler SP, et al. Cytoplasmic serine hydroxymethyltransferase regulates the metabolic partitioning of methylenetetrahydrofolate but is not essential in mice. *J Biol Chem* 2008;**283**:25846–53.
78. Beaudin AE, Abarinov EV, Noden DM, Perry CA, Chu S, Stabler SP, et al. Shmt1 and de novo thymidylate biosynthesis underlie folate-responsive neural tube defects in mice. *Am J Clin Nutr* 2011;**93**:789–98.
79. Anderson DD, Stover PJ. SHMT1 and SHMT2 are functionally redundant in nuclear de novo thymidylate biosynthesis. *PLoS One* 2009;**4**:e5839.
80. Van Speybroeck L. From epigenesis to epigenetics: the case of C. H. Waddington. *Ann N Y Acad Sci* 2002;**981**:61–81.
81. Bock C, Tomazou EM, Brinkman AB, Muller F, Simmer F, Gu H, et al. Quantitative comparison of genome-wide DNA methylation mapping technologies. *Nat Biotechnol* 2010;**28**:1106–14.
82. van Steensel B, Henikoff S. Epigenomic profiling using microarrays. *Biotechniques* 2003;**35**: 346–50 352–344, 356–347.
83. Comb M, Goodman HM. CpG methylation inhibits proenkephalin gene expression and binding of the transcription factor AP-2. *Nucleic Acids Res* 1990;**18**:3975–82.
84. Nan X, Campoy FJ, Bird A. MeCP2 is a transcriptional repressor with abundant binding sites in genomic chromatin. *Cell* 1997;**88**:471–81.
85. Goll MG, Bestor TH. Eukaryotic cytosine methyltransferases. *Annu Rev Biochem* 2005;**74**:481–514.
86. Razin A, Riggs AD. DNA methylation and gene function. *Science* 1980;**210**:604–10.
87. Ito S, D'Alessio AC, Taranova OV, Hong K, Sowers LC, Zhang Y. Role of Tet proteins in 5mC to 5hmC conversion, ES-cell self-renewal and inner cell mass specification. *Nature* 2010;**466**:1129–33.
88. Ma DK, Jang MH, Guo JU, Kitabatake Y, Chang ML, Pow-Anpongkul N, et al. Neuronal activity-induced Gadd45b promotes epigenetic DNA demethylation and adult neurogenesis. *Science* 2009;**323**:1074–7.

89. Hajkova P, Jeffries SJ, Lee C, Miller N, Jackson SP, Surani MA. Genome-wide reprogramming in the mouse germ line entails the base excision repair pathway. *Science* 2010;**329**:78–82.
90. Jenuwein T, Allis CD. Translating the histone code. *Science* 2001;**293**:1074–80.
91. Lee DY, Northrop JP, Kuo MH, Stallcup MR. Histone H3 lysine 9 methyltransferase G9a is a transcriptional coactivator for nuclear receptors. *J Biol Chem* 2006;**281**:8476–85.
92. Spencer TE, Jenster G, Burcin MM, Allis CD, Zhou J, Mizzen CA, et al. Steroid receptor coactivator-1 is a histone acetyltransferase. *Nature* 1997;**389**:194–8.
93. Xu L, Glass CK, Rosenfeld MG. Coactivator and corepressor complexes in nuclear receptor function. *Curr Opin Genet Dev* 1999;**9**:140–7.
94. Eissenberg JC, Shilatifard A. Histone H3 lysine 4 (H3K4) methylation in development and differentiation. *Dev Biol* 2010;**339**:240–9.
95. Shilatifard A. Molecular implementation and physiological roles for histone H3 lysine 4 (H3K4) methylation. *Curr Opin Cell Biol* 2008;**20**:341–8.
96. Heintzman ND, Hon GC, Hawkins RD, Kheradpour P, Stark A, Harp LF, et al. Histone modifications at human enhancers reflect global cell-type-specific gene expression. *Nature* 2009;**459**:108–12.
97. Krogan NJ, Kim M, Tong A, Golshani A, Cagney G, Canadien V, et al. Methylation of histone H3 by Set2 in Saccharomyces cerevisiae is linked to transcriptional elongation by RNA polymerase II. *Mol Cell Biol* 2003;**23**:4207–18.
98. Peters AH, Kubicek S, Mechtler K, O'Sullivan RJ, Derijck AA, Perez-Burgos L, et al. Partitioning and plasticity of repressive histone methylation states in mammalian chromatin. *Mol Cell* 2003;**12**:1577–89.
99. Plath K, Fang J, Mlynarczyk-Evans SK, Cao R, Worringer KA, Wang H, et al. Role of histone H3 lysine 27 methylation in X inactivation. *Science* 2003;**300**:131–5.
100. Lachner M, O'Carroll D, Rea S, Mechtler K, Jenuwein T. Methylation of histone H3 lysine 9 creates a binding site for HP1 proteins. *Nature* 2001;**410**:116–20.
101. Wellen KE, Hatzivassiliou G, Sachdeva UM, Bui TV, Cross JR, Thompson CB. ATP-citrate lyase links cellular metabolism to histone acetylation. *Science* 2009;**324**:1076–80.
102. Davison JM, Mellott TJ, Kovacheva VP, Blusztajn JK. Gestational choline supply regulates methylation of histone H3, expression of histone methyltransferases G9a (Kmt1c) and Suv39h1 (Kmt1a), and DNA methylation of their genes in rat fetal liver and brain. *J Biol Chem* 2009;**284**:1982–9.
103. Piyathilake CJ, Macaluso M, Celedonio JE, Badiga S, Bell WC, Grizzle WE. Mandatory fortification with folic acid in the United States appears to have adverse effects on histone methylation in women with pre-cancer but not in women free of pre-cancer. *Int J Womens Health* 2010;**1**:131–7.
104. Salbaum JM, Kappen C. Diabetic embryopathy: a role for the epigenome? *Birth Defects Res A Clin Mol Teratol* 2011;**91**:770–80.
105. Agger K, Christensen J, Cloos PA, Helin K. The emerging functions of histone demethylases. *Curr Opin Genet Dev* 2008;**18**:159–68.
106. Luka Z, Moss F, Loukachevitch LV, Bornhop DJ, Wagner C. Histone demethylase LSD1 is a folate-binding protein. *Biochemistry* 2011;**50**:4750–6.
107. Wang J, Hevi S, Kurash JK, Lei H, Gay F, Bajko J, et al. The lysine demethylase LSD1 (KDM1) is required for maintenance of global DNA methylation. *Nat Genet* 2009; **41**:125–9.
108. Katz DJ, Edwards TM, Reinke V, Kelly WG. A C. elegans LSD1 demethylase contributes to germline immortality by reprogramming epigenetic memory. *Cell* 2009;**137**:308–20.
109. Tapscott SJ. The circuitry of a master switch: Myod and the regulation of skeletal muscle gene transcription. *Development* 2005;**132**:2685–95.

110. Yamagata K, Daitoku H, Takahashi Y, Namiki K, Hisatake K, Kako K, et al. Arginine methylation of FOXO transcription factors inhibits their phosphorylation by Akt. *Mol Cell* 2008;**32**:221–31.
111. Stark GR, Wang Y, Lu T. Lysine methylation of promoter-bound transcription factors and relevance to cancer. *Cell Res* 2011;**21**:375–80.
112. Goll MG, Kirpekar F, Maggert KA, Yoder JA, Hsieh CL, Zhang X, et al. Methylation of tRNAAsp by the DNA methyltransferase homolog Dnmt2. *Science* 2006;**311**:395–8.
113. Malecova B, Morris KV. Transcriptional gene silencing through epigenetic changes mediated by non-coding RNAs. *Curr Opin Mol Ther* 2010;**12**:214–22.
114. Siomi MC, Sato K, Pezic D, Aravin AA. PIWI-interacting small RNAs: the vanguard of genome defence. *Nat Rev Mol Cell Biol* 2011;**12**:246–58.
115. Huang Y, Ji L, Huang Q, Vassylyev DG, Chen X, Ma JB. Structural insights into mechanisms of the small RNA methyltransferase HEN1. *Nature* 2009;**461**:823–7.
116. Kuramochi-Miyagawa S, Watanabe T, Gotoh K, Totoki Y, Toyoda A, Ikawa M, et al. DNA methylation of retrotransposon genes is regulated by Piwi family members MILI and MIWI2 in murine fetal testes. *Genes Dev* 2008;**22**:908–17.
117. Zeisel SH. Nutritional genomics: defining the dietary requirement and effects of choline. *J Nutr* 2011;**141**:531–4.
118. Kim YI, Miller JW, da Costa KA, Nadeau M, Smith D, Selhub J, et al. Severe folate deficiency causes secondary depletion of choline and phosphocholine in rat liver. *J Nutr* 1994; **124**:2197–203.
119. Maloney CA, Hay SM, Rees WD. Folate deficiency during pregnancy impacts on methyl metabolism without affecting global DNA methylation in the rat fetus. *Br J Nutr* 2007;**97**: 1090–8.
120. Wolff GL, Kodell RL, Moore SR, Cooney CA. Maternal epigenetics and methyl supplements affect agouti gene expression in Avy/a mice. *FASEB J* 1998;**12**:949–57.
121. Waterland RA, Jirtle RL. Transposable elements: targets for early nutritional effects on epigenetic gene regulation. *Mol Cell Biol* 2003;**23**:5293–300.
122. Waterland RA, Dolinoy DC, Lin JR, Smith CA, Shi X, Tahiliani KG. Maternal methyl supplements increase offspring DNA methylation at Axin Fused. *Genesis* 2006;**44**:401–6.
123. Dobosy JR, Fu VX, Desotelle JA, Srinivasan R, Kenowski ML, Almassi N, et al. A methyl-deficient diet modifies histone methylation and alters Igf2 and H19 repression in the prostate. *Prostate* 2008;**68**:1187–95.
124. Takahashi K, Yamanaka S. Induction of pluripotent stem cells from mouse embryonic and adult fibroblast cultures by defined factors. *Cell* 2006;**126**:663–76.
125. Balaghi M, Wagner C. DNA methylation in folate deficiency: use of CpG methylase. *Biochem Biophys Res Commun* 1993;**193**:1184–90.
126. Jhaveri MS, Wagner C, Trepel JB. Impact of extracellular folate levels on global gene expression. *Mol Pharmacol* 2001;**60**:1288–95.
127. Pogribny IP, James SJ. De novo methylation of the p16INK4A gene in early preneoplastic liver and tumors induced by folate/methyl deficiency in rats. *Cancer Lett* 2002;**187**:69–75.
128. Sohn KJ, Stempak JM, Reid S, Shirwadkar S, Mason JB, Kim YI. The effect of dietary folate on genomic and p53-specific DNA methylation in rat colon. *Carcinogenesis* 2003; **24**:81–90.
129. Song J, Sohn KJ, Medline A, Ash C, Gallinger S, Kim YI. Chemopreventive effects of dietary folate on intestinal polyps in Apc+/-Msh2-/- mice. *Cancer Res* 2000;**60**:3191–9.
130. Ghoshal K, Li X, Datta J, Bai S, Pogribny I, Pogribny M, et al. A folate- and methyl-deficient diet alters the expression of DNA methyltransferases and methyl CpG binding proteins involved in epigenetic gene silencing in livers of F344 rats. *J Nutr* 2006;**136**:1522–7.

131. Linhart HG, Troen A, Bell GW, Cantu E, Chao WH, Moran E, et al. Folate deficiency induces genomic uracil misincorporation and hypomethylation but does not increase DNA point mutations. *Gastroenterology* 2009;**136**:227–35 e223.
132. Kim YI. Nutritional epigenetics: impact of folate deficiency on DNA methylation and colon cancer susceptibility. *J Nutr* 2005;**135**:2703–9.
133. Duthie SJ, Narayanan S, Brand GM, Grant G. DNA stability and genomic methylation status in colonocytes isolated from methyl-donor-deficient rats. *Eur J Nutr* 2000;**39**:106–11.
134. Kim YI, Christman JK, Fleet JC, Cravo ML, Salomon RN, Smith D, et al. Moderate folate deficiency does not cause global hypomethylation of hepatic and colonic DNA or c-myc-specific hypomethylation of colonic DNA in rats. *Am J Clin Nutr* 1995;**61**:1083–90.
135. Protiva P, Mason JB, Liu Z, Hopkins ME, Nelson C, Marshall JR, et al. Altered folate availability modifies the molecular environment of the human colorectum: implications for colorectal carcinogenesis. *Cancer Prev Res (Phila)* 2011;**4**:530–43.
136. Crott JW, Liu Z, Keyes MK, Choi SW, Jang H, Moyer MP, et al. Moderate folate depletion modulates the expression of selected genes involved in cell cycle, intracellular signaling and folate uptake in human colonic epithelial cell lines. *J Nutr Biochem* 2008;**19**:328–35.
137. Davis CD, Ross SA. Evidence for dietary regulation of microRNA expression in cancer cells. *Nutr Rev* 2008;**66**:477–82.
138. Shookhoff JM, Gallicano GI. A new perspective on neural tube defects: folic acid and microRNA misexpression. *Genesis* 2010;**48**:282–94.
139. Ichi S, Costa FF, Bischof JM, Nakazaki H, Shen YW, Boshnjaku V, et al. Folic acid remodels chromatin on Hes1 and Neurog2 promoters during caudal neural tube development. *J Biol Chem* 2010;**285**:36922–32.
140. Piyathilake CJ, Johanning GL, Macaluso M, Whiteside M, Oelschlager DK, Heimburger DC, et al. Localized folate and vitamin B-12 deficiency in squamous cell lung cancer is associated with global DNA hypomethylation. *Nutr Cancer* 2000;**37**:99–107.
141. Pogribny IP, Poirier LA, James SJ. Differential sensitivity to loss of cytosine methyl groups within the hepatic p53 gene of folate/methyl deficient rats. *Carcinogenesis* 1995;**16**:2863–7.
142. Qin C, Wang Z, Shang J, Bekkari K, Liu R, Pacchione S, et al. Intracisternal A particle genes: distribution in the mouse genome, active subtypes, and potential roles as species-specific mediators of susceptibility to cancer. *Mol Carcinog* 2010;**49**:54–67.
143. Cropley JE, Suter CM, Beckman KB, Martin DI. CpG methylation of a silent controlling element in the murine Avy allele is incomplete and unresponsive to methyl donor supplementation. *PLoS One* 2010;**5**:e9055.
144. Garcia-Crespo D, Knock E, Jabado N, Rozen R. Intestinal neoplasia induced by low dietary folate is associated with altered tumor expression profiles and decreased apoptosis in mouse normal intestine. *J Nutr* 2009;**139**:488–94.
145. Gelineau-van Waes J, Maddox JR, Smith LM, van Waes M, Wilberding J, Eudy JD, et al. Microarray analysis of E9.5 reduced folate carrier (RFC1; Slc19a1) knockout embryos reveals altered expression of genes in the cubilin-megalin multiligand endocytic receptor complex. *BMC Genomics* 2008;**9**:156.
146. Katula KS, Heinloth AN, Paules RS. Folate deficiency in normal human fibroblasts leads to altered expression of genes primarily linked to cell signaling, the cytoskeleton and extracellular matrix. *J Nutr Biochem* 2007;**18**:541–52.
147. Bormann Chung CA, Boyd VL, McKernan KJ, Fu Y, Monighetti C, Peckham HE, et al. Whole methylome analysis by ultra-deep sequencing using two-base encoding. *PLoS One* 2010;**5**:e9320.
148. Barski A, Cuddapah S, Cui K, Roh TY, Schones DE, Wang Z, et al. High-resolution profiling of histone methylations in the human genome. *Cell* 2007;**129**:823–37.

149. Barski A, Zhao K. Genomic location analysis by ChIP-Seq. *J Cell Biochem* 2009;**107**:11–8.
150. Mikkelsen TS, Ku M, Jaffe DB, Issac B, Lieberman E, Giannoukos G, et al. Genome-wide maps of chromatin state in pluripotent and lineage-committed cells. *Nature* 2007;**448**: 553–60.
151. Mortazavi A, Williams BA, McCue K, Schaeffer L, Wold B. Mapping and quantifying mammalian transcriptomes by RNA-Seq. *Nat Methods* 2008;**5**:621–8.
152. Kuchen S, Resch W, Yamane A, Kuo N, Li Z, Chakraborty T, et al. Regulation of microRNA expression and abundance during lymphopoiesis. *Immunity* 2010;**32**:828–39.

The Nutrigenetics and Nutrigenomics of the Dietary Requirement for Choline

KAREN D. CORBIN AND
STEVEN H. ZEISEL

University of North Carolina at Chapel Hill, Nutrition Research Institute, Kannapolis, North Carolina, USA

Advances in nutrigenetics and nutrigenomics have been instrumental in demonstrating that nutrient requirements vary among individuals. This is exemplified by studies of the nutrient choline, in which gender, single-nucleotide polymorphisms, estrogen status, and gut microbiome composition have been shown to influence its optimal intake level. Choline is an essential nutrient with a wide range of biological functions, and current studies are aimed at refining our understanding of its requirements and, importantly, on defining the molecular mechanisms that mediate its effects in instances of suboptimal dietary intake. This chapter introduces the reader to challenges in developing individual nutrition recommendations, the biological function of choline, current and future research paradigms to fully understand the consequences of inadequate choline nutrition, and some forward thinking about the potential for individualized nutrition recommendations to become a tangible application for improved health.

Progress in Molecular Biology
and Translational Science, Vol. 108
DOI: 10.1016/B978-0-12-398397-8.00007-1

I. Introduction

A. Opportunities and Challenges in Nutrigenetics and Nutrigenomics

In the post-genomics era, there is the potential to utilize biological signatures to determine the underlying mechanisms governing metabolic variation and optimal nutrient requirements for individuals. This gives us an unprecedented opportunity to develop tailored interventions that meet an individual's nutritional needs, with the goal of preventing or treating disease. In order to turn this potential into reality, we must first decipher how our genetic code and other biological parameters influence metabolism and thereby nutrient requirements. In addition, we will have to understand how nutrients impact gene expression, metabolite profiles, epigenetic patterns, and protein function; this will necessitate development of computational tools that can be integrated across all these effects of nutrients. Before these questions can be addressed, we must surmount several barriers that now hinder the ability to utilize cutting-edge nutrition science in developing individualized nutrition recommendations.

One challenge is the common misconception that only very large studies can detect the effects of common genetic variants at single bases (single-nucleotide polymorphisms, SNPs) on nutrient demands. This mistaken idea arises because of the belief that the effect size of SNPs on nutrient requirements is very small and because it is common practice to measure every SNP that is easy to measure rather than to select a mechanistically targeted, smaller set of SNPs for analysis. Genome-wide association studies (GWAS) comprehensively address the role of millions of SNPs on phenotypes of interest, but they make many comparisons and have little power unless tens of thousands of people are genotyped. This is because, to avoid false discoveries, such studies must use very stringent probability values that are impossible to attain in studies of smaller size. The assumption that only large studies can detect metabolically functional SNPs is not true: when scientists limit the number of comparisons made by prospectively focusing on a targeted set of SNPs of interest and when the effect size of the functional SNP is relatively large, clinically sized studies are sufficient to detect SNPs that alter nutrient requirements.[1]

Another challenge lies in the increasing complexity of the factors that modulate nutrient–gene interactions. Reductionist approaches to science simplify analyses but often lead to linear thinking (A causes B which causes C). Full understanding of the factors that mediate variation in metabolism and nutrition requirements will necessitate a more integrated approach that recognizes the multifaceted networks of interactions involved. Optimal nutrient intakes are dictated by an individual's capacity to metabolize and utilize ingested nutrients and, in the case of nutrients with endogenous synthesis pathways, nutrition requirements are also highly dependent on the proper function of those pathways. Part of an individual's metabolic capacity is mediated by the inherited genetic code, and many nutrient–gene interactions have been uncovered. Examples range from classic monogenic interactions, such as deficiency in phenylalanine hydroxylase leading to phenylketonuria to the more subtle regulation of gene expression by polyunsaturated fatty acids.[2]

Although gene–nutrient interactions are highly important for interindividual variability in the requirements and effects of nutrients, further complexity is introduced by the recognition that epigenetic and posttranslational mechanisms are concurrently involved.[1] Because exposure to nutrients during critical developmental windows of time can permanently alter gene expression, nutrition science is presented with a tremendous hurdle, particularly in the context of the multiple nutrients that humans are exposed to differentially throughout a lifetime. From the womb through adulthood, environmental exposures can alter the context within which genetic architecture exerts its effects on metabolism.

B. Clinical-Scale Studies in Nutrigenetics

Some insights into human disease have been gleaned from GWAS, such as the identification of SNPs in the patatin-like phospholipase domain containing protein 3 (*PNPLA3*) gene that are highly associated with fatty liver, liver damage, and liver disease progression and the discovery of several key loci, including the fat mass and obesity associated (*FTO*) gene, that are associated with obesity and diabetes.[3–6] However, most scientists agree that GWAS have not resulted in as many important gene–nutrition–disease linkages as had been expected when the human genome was first sequenced. Because of the nature of some nutrition research that involves controlled diets and complex phenotyping, it would be very useful if smaller clinical-sized studies could be used to generate data on metabolic individuality.

The size of the study needed to detect nutrient–gene interactions is defined by the design of the study; clinical-sized studies ($\sim$100 subjects) are viable if careful attention is given to several factors. First, limit the number of statistical comparisons that will be made, thereby increasing the power of the study (fewer corrections need to be made to avoid false discoveries). Rather than measuring millions of SNPs that are on the commercially available chips, select SNPs on the basis of knowledge of the underlying metabolic process. For instance, for the nutrients of interest, select genes in pathways responsible for endogenous production, metabolism, and elimination of the nutrients, as well as genes in intersecting pathways. Because there are likely to be many SNPs in each of these genes, further focus can be attained by selecting SNPs that lead to defective protein products, such as SNPs that alter amino acid sequence in regions likely to impact function (e.g., catalytic or targeting domains), or by selecting SNPs in regions of genes that control expression (e.g., near transcription factor binding sites). These are the SNPs that are most likely to have functional effects. In addition, SNPs that are more commonly present in the population can be given priority over rarely occurring SNPs.[1]

Precise assessment of gene sequence and expression is not enough; the variance in measurement of phenotype and/or diet exposure must also be minimized. The outcome measure (phenotype) selected should be one that can be accurately measured. Unfortunately, the assessment of diet exposure is often the weakest link, with large potential for errors in diet assessment being the accepted norm.[7] For this reason, many investigators carefully control diets using monitored-meal paradigms in hospitalized or outpatient volunteers.[8] It is important to consider diet when conducting studies searching for functional SNPs that influence nutrient requirements and metabolism. It is likely that many such SNPs cause metabolic inefficiencies that can be overcome if dietary intake of the nutrient is high. For example, the common SNP in the methylenetetrahydrofolate reductase (*MTHFR*; rs1801133) gene, which encodes an enzyme important for the remethylation of homocysteine to methionine, results in approximately a 50% reduction in enzyme activity in people homozygous for the variant allele.[9] This SNP would be important in individuals eating diets low in folate but would be no problem in people with high folate intake. An analysis that did not consider diet intake might miss the effect in the low-folate group when it was averaged across groups eating more folate. With careful consideration of rationale, hypotheses, and study design, it is possible for a clinical-sized study to unravel some of the intricacies that regulate individual nutrition requirements.[1]

C. Choline Research: A Platform for Exploring Metabolic Variation

The case study of the effects of genetic variation on dietary choline requirements provides an excellent example of how clinical nutrigenetics/nutrigenomics can be used to understand individual nutrition needs. A comprehensive translational research platform for choline has been instrumental in moving our understanding to a point where it is increasingly evident that optimal choline intake is predictable but varies greatly between individuals. The study of choline gives us a glimpse of the way one can utilize nutrigenetics and nutrigenomics to find individualized, applied solutions to nutrition-related health problems.

II. Choline Biology

A. Choline Functions

Choline is an essential water-soluble nutrient needed for the normal function of all cells. It is metabolized into several compounds that exert a wide range of biological functions including cell signaling, cholinergic neurotransmission, cell membrane structure, mitochondrial function, cholesterol/lipid transport and metabolism, and methylation reactions.[10] Betaine, phosphatidylcholine, sphingomyelin, and acetylcholine are examples of choline-containing compounds of physiologic relevance (Fig. 1).

- The conversion of choline to betaine is irreversible and occurs in the mitochondria.[11] Betaine is a methyl group donor that influences gene expression via epigenetic mechanisms[12] and is an osmolyte used in the glomerulus of the kidney to help reabsorb water.[13]
- Phosphatidylcholine (also called lecithin) is the predominant phospholipid (>50%) in most mammalian membranes and is also important for hepatic lipid packaging and export.[14] Sphingomyelin is a phospholipid needed for membrane and myelin formation. Both phosphatidylcholine and sphingomyelin are sources of second messengers (including diacylglycerol, arachidonic acid, and ceramide) that alter cell function by potentiating signaling cascades.[15,16]
- Acetylcholine is a neurotransmitter important in brain functions such as memory and mood. It is the neurotransmitter most often used by neurons interfacing between brain and the periphery; the nerves controlling skeletal muscles, heart rate, breathing, sweating, and salivation all use acetylcholine.

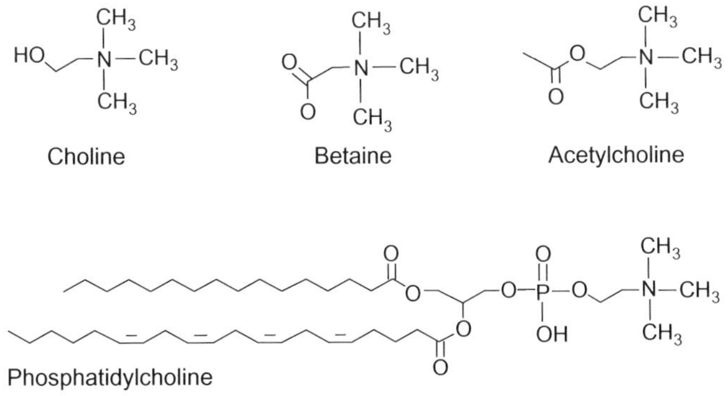

FIG. 1. Structures of several important choline-containing molecules.

For additional details on choline metabolism, the reader is referred to a recent comprehensive review.[10]

B. Dietary Choline Requirements

Most of the foods we eat contain varying amounts of choline, choline esters, and betaine. The foods with the greatest abundance of choline are of animal origin, especially egg yolk and liver. Other sources of choline are wheat germ, soy, nuts, and other legumes. Many foods also have lecithin as an additive, usually used for its emulsifying properties, and this provides a significant amount of choline. Details about the choline content of foods are in the most current U.S. Department of Agriculture database (http://www.ars.usda.gov/SP2UserFiles/Place/12354500/Data/Choline/Choln02.pdf). Human breast milk is also a good source of free choline and choline esters, and the manufacturers of infant formulas have recently modified the content of choline compounds to levels similar to those in breast milk.

Aside from dietary sources, humans possess a pathway to make choline moiety *de novo,* as part of a phosphatidylcholine molecule, via the phosphatidylethanolamine *N*-methyltransferase (PEMT) pathway.[17] Based on the presence of this endogenous production pathway, for many years choline was classified as a nonessential nutrient. This was the case until 1998 when the U.S. Institute of Medicine (Food and Nutrition Board) established for the first time adequate intake (AI) and tolerable upper intake limit values for choline, based on limited human studies in which dietary choline deficiency was associated with liver disease in humans.[18] Table I lists the AI levels for choline.

The dietary intake of choline has declined in recent years as people have been avoiding foods high in cholesterol (like eggs), which are the richest sources of choline. It is not surprising that the 2005 National Health and Nutrition Examination Survey (NHANES) data show that only a small percentage of people in the United States, approximately 10%, achieve the recommended dietary intakes for choline.[19] Several studies report that 20–25% of Americans eat one-third to half of the recommended AI for choline (<203 mg/day in the Framingham Heart Study[20], <217 mg/day in the Atherosclerosis Risk in Communities Study[21], <293 mg/day in the Nurses' Health Study[22]).

III. Utilizing Nutrigenetics and Nutrigenomics Approaches to Understand Choline Requirements

A. Nutrigenetics of Choline Requirements

One could surmise, based on the availability of an AI level, that the effects of insufficient or excess choline are well understood and that the current AI level is adequate to meet the needs of a broad range of individuals.

TABLE I
ADEQUATE INTAKE (AI) VALUES FOR CHOLINE (MILLIGRAMS PER DAY)

Life stage	Age	Males	Females
Infants	0–6 months	125	125
	7–12 months	150	150
Children	1–3 years	200	200
	4–8 years	250	250
	9–13 years	375	375
Adolescents	14–18 years	550	400
Adults	19 years and older	550	425
Pregnancy	All ages	–	450
Breast feeding	All ages	–	550

Modified from Ref. 18.

The research platform for choline has advanced significantly since the establishment of the AI in 1998, and it is now clear that choline requirements are influenced by several factors. This knowledge has been gained through a multifaceted clinical, epidemiological, and basic science approach that has taken advantage of several nutrigenetics and nutrigenomics platforms including genomics, epigenetics, and metabolomics.

In a clinical-sized nutrigenetics study, adult men and women (pre- and postmenopausal) aged 18–70 years were hospitalized and fed a standard diet containing the AI for choline (550 mg/70 kg/day; baseline phase) for 10 days. On day 11, the subjects were placed on a diet containing <50 mg/day of choline for up to 42 days (depletion phase). Blood and urine were collected throughout the study to measure various parameters of dietary choline status, markers of organ dysfunction, and liver fat. If, during the study, functional markers indicated an adverse effect associated with choline deficiency, subjects were switched to a diet containing increasing amounts of choline until repleted (repletion phase). This study demonstrated that 77% of men, 80% of postmenopausal women, and 44% of premenopausal women developed elevated liver fat, increased liver enzymes, or increased muscle enzymes (creatine phosphokinase) when placed on a choline-deficient diet. In addition, 10% of subjects needed 850 mg/day of choline to avoid these same signs of deficiency. Importantly, the symptoms associated with choline deficiency were fully reversed when choline was reintroduced into the diet.[8,23,24] These were clear indications that choline is indeed a required nutrient and that the requirement is not the same in all people.

Dietary choline requirements are governed by an individual's ability to make choline *de novo* and by the rates of choline utilization. Genes in choline, methionine, and folate metabolism are intertwined in the pathways for choline production and utilization (Fig. 2), and normal function of these genes is required for optimal choline status.

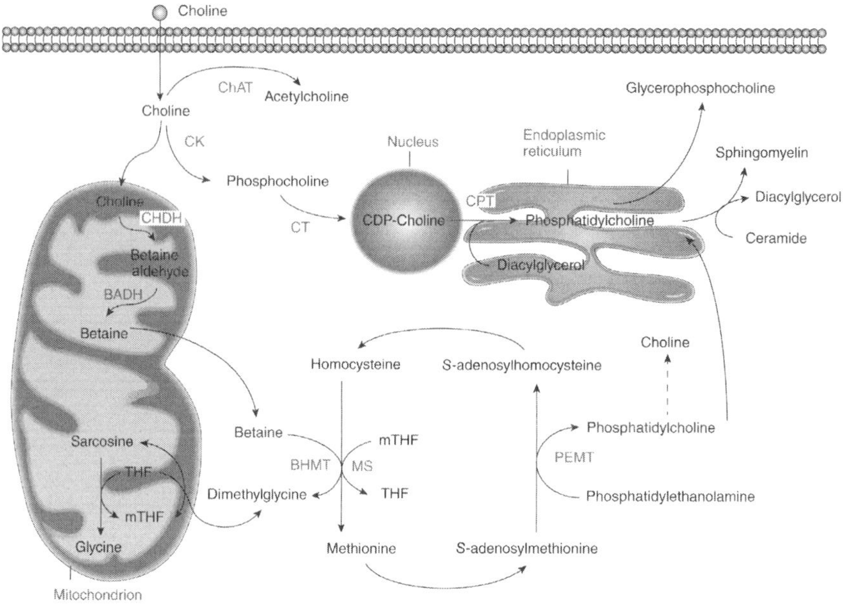

FIG. 2. Choline, folate, and homocysteine metabolism are closely interrelated. The pathways for the metabolism of these three nutrients intersect at the formation of methionine from homocysteine. BADH, betaine-aldehyde dehydrogenase; BHMT, betaine–homocysteine methyltransferase; CDP, cytidine diphosphate; ChAT, choline acetyltransferase; CHDH, choline dehydrogenase; CK, choline kinase; CPT, choline phosphotransferase; CT, cytidine triphosphate (CTP):phosphocholine cytidylyltransferase; MS, methionine synthase; mTHF, methyltetrahydrofolate; PEMT, phosphatidylethanolamine *N*-methyltransferase; THF, tetrahydrofolate. From Ref. 25, with permission.

Humans have variation in the sequences for these genes (SNPs) and in the number of copies of these genes (copy number variation). While copy number variation in these genes has not yet been proven to be important, SNPs in these genes strongly influence dietary choline demands.[8,23,26] One of the SNPs identified as having a role in choline-deficiency-related organ dysfunction is in the promoter region of the *PEMT* gene (rs12325817) and is very common (the North Carolina, United States, population is 18% homozygous variant [VV], 26% homozygous ancestral alleles [WW][8,24]). In women with ancestral alleles, estrogen induces the *PEMT* gene, thereby providing an endogenous source of choline; thus, premenopausal women are relatively resistant to developing organ dysfunction when fed a low-choline diet.[24,26,27] However, the *PEMT* rs12325817 SNP marks a haplotype with decreased estrogen-responsive induction of *PEMT*.[27] Women carriers of the variant allele developed organ dysfunction when fed a low-choline diet (odds ratio = 25). Another SNP in the *PEMT* gene (rs7946) is present more often in people with fatty liver.

This SNP was functional (altered dietary choline requirements) in individuals also eating a diet low in folate.[28] Premenopausal women who are carriers of an SNP in the gene-encoding 5,10-methylenetetrahydrofolate dehydrogenase (*MTHFD1*; rs2236225) are more than 85 times as likely as noncarriers to develop choline-deficiency-induced organ dysfunction. This also is a very common polymorphism (62% of the North Carolina population have one allele, and 11% are homozygous for the variant allele).[23] Another important enzyme in the choline metabolism pathway is choline dehydrogenase (CHDH). We observed that an SNP in the *CHDH* gene (rs12676; 37–42% of the population have the variant allele and ~9% are homozygous for the variant[29]) increased by 20-fold the susceptibility to develop organ dysfunction when premenopausal women were fed low-choline diets.[8] As studies progress, additional SNPs are being identified that mediate individual choline requirements. The relevance of SNPs on optimal nutrient intakes is different depending on gender and ethnicity and probably on other factors that we do not yet understand.

B. Mouse Models for Studying Choline Requirements

The studies in humans described above identified multiple functional SNPs in choline metabolism. The nature of human studies makes it difficult to assess all of the potential effects of these SNPs. Genetic manipulation of mice is relatively easy, and it is possible to delete genes of interest in mice to determine their functional significance. It is important to realize that the SNPs of interest mark a haplotype, often with multiple SNPs in linkage disequilibrium, and that the actual SNP causing the functional effect may not be the one identified. In addition, patterns of SNPs in multiple functionally related genes likely converge to lead to observed phenotypes. Gene deletions result in a total loss of activity, while gene polymorphisms at worst result in 100% loss of gene function but generally result in intermediate degrees of loss (sometimes they result in a gain of function). With these caveats, genetically manipulated mouse models can be a very useful translation from bedside to bench research. They enhance our ability to design future nutrigenetics studies by allowing us to carefully define relevant mechanisms that can be studied in humans.

The *Pemt* knockout mouse (*Pemt*$^{-/-}$) seizes and dies when deprived of choline,[30] it hypermethylates proteins and DNA,[31] it has greatly diminished omega-3 fatty acids in membrane phosphatidylcholine,[32,33] and it has abnormal brain development.[31,33] Importantly, it also has fatty liver because of the inability to maintain normal concentrations of all choline metabolites,[34] making it a useful model for studying mechanisms that might be important for choline-deficiency-mediated fatty liver in humans. The *Chdh* knockout mouse (*Chdh*$^{-/-}$) is remarkable. In the tissues that express this gene, its deletion resulted in abnormal mitochondrial morphology and function.[11] In addition, there was infertility due to diminished sperm motility in the *Chdh*$^{-/-}$ males secondary to reduced ATP production by mitochondria. Each of the phenotypes discovered in mouse models is now being investigated in humans with the relevant SNPs.

C. Choline, Epigenetics, and Brain Development

A human being has a fixed genetic code but attains some measure of genetic flexibility through epigenetic modifications (usually methylation of DNA or methylation/acetylation of histones associated with DNA), which integrate environmental cues (including diet) with the genome.[35] The fetal environment transfers signals to the fetus that reflect the expected environment after birth. These signals lead to a shift in gene expression and modification of biological pathways in a direction that should confer a survival benefit in the expected postnatal environment.[36]

The DNA methyltransferases, which catalyze the transfer of a methyl group to DNA, all use *S*-adenosylmethionine as the methyl donor. Choline influences DNA methylation because it is a major dietary methyl donor. As shown in Fig. 2, it is at the methylation of homocysteine that the choline, betaine, folate, and methionine metabolic pathways intersect. Homocysteine methylation occurs by two parallel pathways; in the first, vitamins B_{12} and folic acid are involved in a reaction catalyzed by methionine synthase.[37] The alternative pathway for the methylation of homocysteine to form methionine is catalyzed by betaine–homocysteine methyltransferase (BHMT).[38] These pathways may be fungible, in that perturbations in one are compensated for by adjustments in the parallel pathway. Rats treated with the antifolate methotrexate had diminished pools of choline metabolites in liver.[39,40] Rats ingesting a choline-deficient diet had diminished tissue concentrations of methionine and *S*-adenosylmethionine[41] and doubled plasma homocysteine concentrations.[42] Humans depleted of choline had diminished capacity to methylate homocysteine and developed elevated homocysteine concentrations in plasma after a methionine loading test.[43]

DNA methylation occurs at cytosine bases that are followed by a guanosine (CpG-sites).[44] In mammals, most CpG-sites in DNA are methylated (90–98%[45]), but there are specific CpG-rich areas of DNA where most CpGs are not methylated; these are called CpG islands.[46] It was initially believed that CpG islands that span the 5′-end of the regulatory regions of genes were the only places where DNA methylation occurred, and that when these CpGs were methylated, gene expression was usually suppressed or silenced.[46,47] We are learning that epigenetic marks are not limited to CpG islands. They are also found in shores of these islands, exons, and intergenic DNA. Methylation may be the default state for genes, and though the purpose for methylation of intragenic DNA is unclear at this time, it may protect against expression of unwanted genes.[45]

Pregnancy and the early postnatal period are important times when choline, via epigenetic mechanisms, has been shown to play a significant role in the development and health outcomes later in life. During pregnancy, there is both a high demand for and an increased production of choline via the PEMT pathway, as the gene is induced by estrogen.[27,48] In general, this additional production is insufficient to meet choline requirements, and in rodent models choline pools are depleted during pregnancy.[49] Large amounts of choline are transferred to the fetus across the placenta, and fetal choline concentrations are many fold higher than adult concentrations.[50] During critical periods of gestation, low choline availability can lead to poor brain development and long-term cognitive and behavioral impairments in rodents.[51–53] In addition, several human studies have shown increased incidence of neural tube defects and orofacial cleft defects in infants when pregnant mothers consumed a diet deficient in choline.[54–56] Interestingly, choline supplementation can improve the behavioral and developmental consequences of maternal alcohol intake on fetal development in rodent models.[57–60] Choline also protects against seizure-induced memory impairment in rodent models.[61] These data suggest that choline has a protective effect during a suboptimal fetal environment.

One plausible mechanism mediating the role of choline (a methyl donor) in development is via the modification of epigenetic marks that regulate gene expression. In brain and other tissues, a choline/methionine-deficient diet directly altered methylation in CpG islands within several genes.[62] After feeding pregnant rat dams a choline/methionine-deficient diet, neural progenitor cells of the fetal hippocampus proliferated half as fast compared to fetal brain from mothers fed a control diet. The choline-deficient neural progenitor cells had decreased gene-specific DNA methylation in CpG islands of *Cdkn3*. This gene encodes kinase-associated phosphatase, and when hypomethylated, the gene is overexpressed, resulting in increased phosphatase levels and subsequent activation of the retinoblastoma protein pathway that inhibits cyclin-dependent kinase.[63,64]

Epigenetic marks on histones (the proteins around which DNA is wound) also regulate gene expression. In mice, maternal choline deficiency during pregnancy altered methylation of lysine residues on histone H3 in fetal neural progenitor cells from gestational day 17 in the areas of the hippocampus where neurogenesis was occurring.[65] Monomethyl and dimethyl lysine 9 on histone H3 residues are usually associated with silencing of genes, whereas dimethyl and trimethyl lysine 4 on histone H3 are enriched in areas with transcriptionally active chromatin. Thus, choline availability modulates histone methylation and thereby gene expression and proliferation in fetal neural progenitor cells.

Similarly, maternal dietary choline intake modulates angiogenesis, the formation of new blood vessels, in fetal brain.[66] In mice, maternal diets low in choline were associated with diminished proliferation of endothelial progenitor cells in fetal hippocampus, and the mechanism involved decreased DNA methylation in fetal brain within the promoter of two genes that are important regulators of angiogenesis (vascular endothelial growth factor C, *Vegfc* and angiopoietin 2, *Angpt2*).[66] This shows that choline's effects on brain development are not limited to neuronal cells, but extend to endothelial cells and blood vessel formation in the fetal hippocampus.

As noted earlier, maternal dietary choline deficiency increases neural tube closure defects in rodent and human fetuses.[56,67–69] Later in gestation, maternal choline deficiency alters the development of fetal hippocampus by decreasing neural progenitor cell proliferation and by increasing apoptosis and expression of markers of differentiation.[70–72] The offspring of choline-deficient mothers exhibited insensitivity to long-term potentiation when they were adult animals[73] and decremented visuospatial and auditory memory.[74] More choline (about four times higher than normal) during days 11–17 of gestation in rodents increased hippocampal progenitor cell proliferation,[75,76] decreased apoptosis in these cells,[75,76] enhanced long-term potentiation in the offspring when they were adult animals,[73,77,78] and enhanced visuospatial and auditory memory by as much as 30% throughout their lifetimes.[51,53,74,79–82] Indeed, adult rodents decline in memory as they age, and offspring exposed to extra choline *in utero* do not show this "senility."[80,81] Thus, choline supplementation during a critical period in pregnancy causes lifelong changes in brain structure and function, probably by changing epigenetic marking.

There are other good examples of how powerful diet can be in changing epigenetic marks. Feeding pregnant pseudoagouti Avy/a mouse dams a choline-, methionine-, and folate-supplemented diet altered epigenetic regulation of agouti expression in their offspring, as indicated by increased methylation of the involved gene and by agouti/black mottling of their coats.[12,83] In another example,[84] there was increased DNA methylation of the fetal gene axin 1 fused ($Axin1^{Fu}$) after methyl donor supplementation of female mice before and during pregnancy, which reduced by 50% the incidence of tail kinking in $Axin^{Fu/+}$ offspring. It is clear that the dietary manipulation of methyl donors (either deficiency or supplementation) can have a profound impact on gene expression via changes in epigenetic marks.

D. Metabolomics, Gut Microbiome, and Choline Requirements

Metabolomics, the characterization of small-molecule metabolites in a cell or tissue, can give us an overall understanding of the function of biological pathways. This is another scientific discipline that is an important part of nutrigenomics because it allows for the characterization of health status and the monitoring of the efficacy of nutrition interventions. In an effort to define metabolites that could be used as biomarkers to predict which people are at risk for developing choline-deficiency symptoms, a study was developed using targeted and untargeted metabolomics. This study demonstrated that a set of molecules, including choline and folate metabolites, amino acids, and acylcarnitines, could accurately predict which humans would develop fatty liver when placed on a choline-deficient diet.[85] This study highlighted the value of using metabolomics to define signatures relevant to nutrition requirements.

It is now well accepted that human biology is intricately intertwined with the function of the trillions of bacteria (gut flora or gut microbiome) that inhabit the intestinal tract. The bacterial genome and its active products are participants in host health. In particular, the gut flora play an important role in the breakdown and absorption of many nutrients. This is an important factor to consider in the design and interpretation of nutrigenetics and nutrigenomics research studies because it influences nutritional individuality. In a metabolomics study in mice, it was shown that certain gut bacteria break down choline before it is absorbed, essentially mimicking a choline-deficient scenario. Mice with a gut microbiota that can hypercatabolize choline developed fatty liver and impaired glucose metabolism.[86] In one of only a limited number of studies related to the gut microbiome in humans, a choline deficient diet led to alterations in gut microbes that correlated with the development of fatty liver. This effect was strengthened when combined with knowledge of genotypes in choline-related genes.[87]

A recent paper characterized metabolomic profiles that could predict risk for cardiovascular disease in mice and humans. In that study, elevated levels of choline, betaine, and trimethylamine-*N*-oxide (a metabolite of choline) were associated with increased macrophage cholesterol and foam cell formation, which are hallmarks of the atherosclerotic process.[88] Although it is difficult to discern whether alterations in choline metabolites are a cause or consequence of cardiovascular disease, it is certainly reasonable to postulate an important role for choline in cardiovascular health.

This exciting new field holds much promise for giving us a comprehensive understanding of nutrition systems biology, and extensive research is needed to fully understand the role of the gut microbiome in human health. These examples of metabolomics and gut microbiome studies highlight important nutrigenomics tools that can be applied and integrated to develop a more comprehensive picture of the causes and consequences of nutrient inadequacies.

IV. Summary and Future Directions

A. Lessons from Choline

The research platform built to understand the individual nutrition requirements for choline has brought to light several important concepts that should prompt a shift in standard clinical nutrition paradigms. Although nutrition recommendations developed for populations are valuable for helping the majority of people avoid the consequences of nutritional imbalances, there will undoubtedly be a subset of individuals who are unable to reach optimal health through these requirements and, more importantly, who could suffer harm from deficiency or toxicity. Of great importance is the fact that, under certain metabolic backgrounds and depending on exposure, choline could have both positive and negative effects on health outcomes. This paradox is actually quite common in nutrition, where nutrients are found to be beneficial for some and not others.

Rather than looking at outliers or paradoxes as a source of "noise" in our datasets, we should instead see them as the source of better mechanistic insight into the function of a nutrient. By studying the reason for hypo- or hypersensitivity to a nutrient, we can gain valuable information about key functional mechanisms that mediate the nutrient's biological effects. It is also evident from choline research that better health outcomes are the result of a diet that is well matched both to the environment during fetal development and to an individual's genotype. This strongly supports the notion that dietary recommendations need to be tailored to developmental needs and biological signatures.

Additional valuable insights from choline research are that a well-designed clinical study can lead to a great deal of information about individual nutrition requirements and that SNPs can have a very large impact on phenotype. Choline research also clearly demonstrates how nutrients directly impact biological pathways on many levels that go beyond changes in transcription of genes, such as epigenetics, organelle function, and metabolic flux.

B. From Research Platforms to Applied Solutions

Despite the great progress in nutrition science since the sequencing of the human genome, we are still not at a point where we can utilize advanced diagnostic tools to prescribe individual diets based on unique biological signatures. Several important solutions are needed to overcome the hurdles facing nutrigenetics and nutrigenomics. Technology is a key solution in the pursuit of defining nutritional individuality. The development and widespread availability of cutting-edge methodologies to interrogate biological systems is a central application of technology in nutrigenomics. In recent years, many such methods have become accessible to study gene expression, epigenetic marks, genetic variation, metabolite profiles, and protein expression. Sophisticated new bioinformatics tools have been instrumental in managing large nutrition data sets. These advances have been central to the progress in nutrigenomics thus far, and many more advances will be needed to continue this forward motion. Perhaps the most important way to move nutrigenetics and nutrigenomics from an exciting set of ideas into applied reality is the integration of scientific findings into clinical settings so that people can utilize the most current biomedical knowledge to guide their lifestyle choices. In order to implement biomedical science findings into clinical practice, it is clear that nutrition science requires a translational platform where one can seamlessly shift knowledge back and forth from bench to bedside and that utilizes multiple cutting-edge and classical methodologies. It is with this concept of individualized nutrition recommendations, learned in part from our understanding of choline requirements, that we can move toward optimally effective nutrition interventions.

References

1. Zeisel SH. Choline: clinical nutrigenetic/nutrigenomic approaches for identification of functions and dietary requirements. *World Rev Nutr Diet* 2010;**101**:73–83.
2. Ordovas JM, Mooser V. Nutrigenomics and nutrigenetics. *Curr Opin Lipidol* 2004;**15**:101–8.
3. Romeo S, Kozlitina J, Xing C, Pertsemlidis A, Cox D, Pennacchio LA, et al. Genetic variation in PNPLA3 confers susceptibility to nonalcoholic fatty liver disease. *Nat Genet* 2008;**40**:1461–5.
4. Wagenknecht LE, Palmer ND, Bowden DW, Rotter JI, Norris JM, Ziegler J, et al. Association of PNPLA3 with non-alcoholic fatty liver disease in a minority cohort: the Insulin Resistance Atherosclerosis Family Study. *Liver Int* 2011;**31**:412–6.
5. Liu G, Zhu H, Lagou V, Gutin B, Stallmann-Jorgensen IS, Treiber FA, et al. FTO variant rs9939609 is associated with body mass index and waist circumference, but not with energy intake or physical activity in European- and African-American youth. *BMC Med Genet* 2010;**11**:57.
6. Frayling TM, Timpson NJ, Weedon MN, Zeggini E, Freathy RM, Lindgren CM, et al. A common variant in the FTO gene is associated with body mass index and predisposes to childhood and adult obesity. *Science* 2007;**316**:889–94.
7. Thompson FE, Byers T. Dietary assessment resource manual. *J Nutr* 1994;**124**:2245S–317S.
8. da Costa KA, Kozyreva OG, Song J, Galanko JA, Fischer LM, Zeisel SH. Common genetic polymorphisms affect the human requirement for the nutrient choline. *FASEB J* 2006;**20**:1336–44.
9. Frosst P, Blom HJ, Milos R, Goyette P, Sheppard CA, Matthews RG, et al. A candidate genetic risk factor for vascular disease: a common mutation in methylenetetrahydrofolate reductase. *Nat Genet* 1995;**10**:111–3.
10. Zeisel SH. Choline: critical role during fetal development and dietary requirements in adults. *Annu Rev Nutr* 2006;**26**:229–50.
11. Johnson AR, Craciunescu CN, Guo Z, Teng YW, Thresher RJ, Blusztajn JK, et al. Deletion of murine choline dehydrogenase results in diminished sperm motility. *FASEB J* 2010;**24**:2752–61.
12. Wolff GL, Kodell RL, Moore SR, Cooney CA. Maternal epigenetics and methyl supplements affect agouti gene expression in Avy/a mice. *FASEB J* 1998;**12**:949–57.
13. Guder WG, Beck FX, Schmolke M. Regulation and localization of organic osmolytes in mammalian kidney. *Klin Wochenschr* 1990;**68**:1091–5.
14. Yao ZM, Vance DE. Reduction in VLDL, but not HDL, in plasma of rats deficient in choline. *Biochem Cell Biol* 1990;**68**:552–8.
15. Exton JH. Phosphatidylcholine breakdown and signal transduction. *Biochim Biophys Acta* 1994;**1212**:26–42.
16. Spiegel S, Merrill AJ. Sphingolipid metabolism and cell growth regulation. *FASEB J* 1996;**10**:1388–97.

17. Ridgway ND, Yao Z, Vance DE. Phosphatidylethanolamine levels and regulation of phosphatidylethanolamine N-methyltransferase. *J Biol Chem* 1989;**264**:1203–7.
18. Institute of Medicine & National Academy of Sciences USA . *Choline. Dietary reference intakes for folate, thiamin, riboflavin, niacin, vitamin B12, pantothenic acid, biotin, and choline*. vol. 1. Washington, DC: National Academy Press; 1998; pp. 390–422.
19. Jensen HH, Batres-Marquez SP, Carriquiry A, Schalinske KL. Choline in the diets of the U.S. population: NHANES, 2003-2004. *FASEB J* 2007;**21**:lb219.
20. Cho E, Zeisel SH, Jacques P, Selhub J, Dougherty L, Colditz GA, et al. Dietary choline and betaine assessed by food-frequency questionnaire in relation to plasma total homocysteine concentration in the Framingham Offspring Study. *Am J Clin Nutr* 2006;**83**:905–11.
21. Bidulescu A, Chambless LE, Siega-Riz AM, Zeisel SH, Heiss G. Usual choline and betaine dietary intake and incident coronary heart disease: the Atherosclerosis Risk in Communities (ARIC) study. *BMC Cardiovasc Disord* 2007;**7**:20.
22. Cho E, Willett WC, Colditz GA, Fuchs CS, Wu K, Chan AT, et al. Dietary choline and betaine and the risk of distal colorectal adenoma in women. *J Natl Cancer Inst* 2007;**99**:1224–31.
23. Kohlmeier M, da Costa KA, Fischer LM, Zeisel SH. Genetic variation of folate-mediated one-carbon transfer pathway predicts susceptibility to choline deficiency in humans. *Proc Natl Acad Sci USA* 2005;**102**:16025–30.
24. Fischer LM, daCosta K, Kwock L, Stewart P, Lu T-S, Stabler S, et al. Sex and menopausal status influence human dietary requirements for the nutrient choline. *Am J Clin Nutr* 2007;**85**:1275–85.
25. Zeisel SH and Corbin KD. Choline. In: Present Knowledge in Nutrition, 10th ed. Erdman JWK, MacDonald I, and Zeisel SH, editors. 2012.
26. Fischer LM, da Costa KA, Galanko J, Sha W, Stephenson B, Vick J, et al. Choline intake and genetic polymorphisms influence choline metabolite concentrations in human breast milk and plasma. *Am J Clin Nutr* 2010;**92**:336–46.
27. Resseguie ME, da Costa KA, Galanko JA, Patel M, Davis IJ, Zeisel SH. Aberrant estrogen regulation of PEMT results in choline deficiency-associated liver dysfunction. *J Biol Chem* 2011;**286**:1649–58.
28. Ivanov A, Nash-Barboza S, Hinkis S, Caudill MA. Genetic variants in phosphatidylethanolamine N-methyltransferase and methylenetetrahydrofolate dehydrogenase influence biomarkers of choline metabolism when folate intake is restricted. *J Am Diet Assoc* 2009;**109**:313–8.
29. Xu X, Gammon MD, Zeisel SH, Lee YL, Wetmur JG, Teitelbaum SL, et al. Choline metabolism and risk of breast cancer in a population-based study. *FASEB J* 2008;**22**:2045–52.
30. Walkey CJ, Donohue LR, Bronson R, Agellon LB, Vance DE. Disruption of the murine gene encoding phosphatidylethanolamine N-methyltransferase. *Proc Natl Acad Sci USA* 1997;**94**:12880–5.
31. Zhu X, Mar MH, Song J, Zeisel SH. Deletion of the Pemt gene increases progenitor cell mitosis, DNA and protein methylation and decreases calretinin expression in embryonic day 17 mouse hippocampus. *Brain Res Dev Brain Res* 2004;**149**:121–9.
32. Watkins SM, Zhu X, Zeisel SH. Phosphatidylethanolamine-N-methyltransferase activity and dietary choline regulate liver-plasma lipid flux and essential fatty acid metabolism in mice. *J Nutr* 2003;**133**:3386–91.
33. da Costa KA, Rai KS, Craciunescu CN, Parikh K, Mehedint MG, Sanders LM, et al. Dietary docosahexaenoic acid supplementation modulates hippocampal development in the Pemt−/− mouse. *J Biol Chem* 2010;**285**:1008–15.
34. Zhu X, Song J, Mar MH, Edwards LJ, Zeisel SH. Phosphatidylethanolamine N-methyltransferase (PEMT) knockout mice have hepatic steatosis and abnormal hepatic

choline metabolite concentrations despite ingesting a recommended dietary intake of choline. *Biochem J* 2003;**370**:987–93.
35. Jirtle RL, Skinner MK. Environmental epigenomics and disease susceptibility. *Nat Rev Genet* 2007;**8**:253–62.
36. Godfrey KM, Lillycrop KA, Burdge GC, Gluckman PD, Hanson MA. Epigenetic mechanisms and the mismatch concept of the developmental origins of health and disease. *Pediatr Res* 2007;**61**:5R–10R.
37. Weisberg IS, Jacques PF, Selhub J, Bostom AG, Chen Z, Curtis Ellison R, et al. The 1298A–>C polymorphism in methylenetetrahydrofolate reductase (MTHFR): in vitro expression and association with homocysteine. *Atherosclerosis* 2001;**156**:409–15.
38. Sunden S, Renduchintala M, Park E, Miklasz S, Garrow T. Betaine-Homocysteine methyltransferase expression in porcine and human tissues and chromosomal localization of the human gene. *Arch Biochem Biophys* 1997;**345**:171–4.
39. Pomfret EA, da Costa K, Zeisel SH. Effects of choline deficiency and methotrexate treatment upon rat liver. *J Nutr Biochem* 1990;**1**:533–41.
40. Selhub J, Seyoum E, Pomfret EA, Zeisel SH. Effects of choline deficiency and methotrexate treatment upon liver folate content and distribution. *Cancer Res* 1991;**51**:16–21.
41. Zeisel SH, Zola T, daCosta K, Pomfret EA. Effect of choline deficiency on S-adenosylmethionine and methionine concentrations in rat liver. *Biochem J* 1989;**259**:725–9.
42. Varela-Moreiras G, Ragel C, Perez de Miguelsanz J. Choline deficiency and methotrexate treatment induces marked but reversible changes in hepatic folate concentrations, serum homocysteine and DNA methylation rates in rats. *J Am Coll Nutr* 1995;**14**:480–5.
43. da Costa KA, Gaffney CE, Fischer LM, Zeisel SH. Choline deficiency in mice and humans is associated with increased plasma homocysteine concentration after a methionine load. *Am J Clin Nutr* 2005;**81**:440–4.
44. Holliday R, Grigg GW. DNA methylation and mutation. *Mutat Res* 1993;**285**:61–7.
45. Suzuki MM, Bird A. DNA methylation landscapes: provocative insights from epigenomics. *Nat Rev Genet* 2008;**9**:465–76.
46. Jeltsch A. Beyond Watson and Crick: DNA methylation and molecular enzymology of DNA methyltransferases. *Chembiochem* 2002;**3**:382.
47. Bird AP. CpG-rich islands and the function of DNA methylation. *Nature* 1986;**321**:209–13.
48. Resseguie M, Song J, Niculescu MD, da Costa KA, Randall TA, Zeisel SH. Phosphatidylethanolamine N-methyltransferase (PEMT) gene expression is induced by estrogen in human and mouse primary hepatocytes. *FASEB J* 2007;**21**:2622–32.
49. Zeisel SH, Mar M-H, Zhou Z-W, da Costa K-A. Pregnancy and lactation are associated with diminished concentrations of choline and its metabolites in rat liver. *J Nutr* 1995;**125**:3049–54.
50. Zeisel SH. Choline: critical role during fetal development and dietary requirements in adults. *Annu Rev Nutr* 2006;**26**:229–50.
51. Meck W, Williams C. Perinatal choline supplementation increases the threshold for chunking in spatial memory. *Neuroreport* 1997;**8**:3053–9.
52. Meck WH, Smith RA, Williams CL. Organizational changes in cholinergic activity and enhanced visuospatial memory as a function of choline administered prenatally or postnatally or both. *Behav Neurosci* 1989;**103**:1234–41.
53. Meck WH, Smith RA, Williams CL. Pre- and postnatal choline supplementation produces long-term facilitation of spatial memory. *Dev Psychobiol* 1988;**21**:339–53.
54. Shaw GM, Finnell RH, Blom HJ, Carmichael SL, Vollset SE, Yang W, et al. Choline and risk of neural tube defects in a folate-fortified population. *Epidemiology* 2009;**20**:714–9. doi:10.1097/EDE.0b013e3181ac9fe7.
55. Shaw GM, Carmichael SL, Laurent C, Rasmussen SA. Maternal nutrient intakes and risk of orofacial clefts. *Epidemiology* 2006;**17**:285–91.

56. Shaw GM, Carmichael SL, Yang W, Selvin S, Schaffer DM. Periconceptional dietary intake of choline and betaine and neural tube defects in offspring. *Am J Epidemiol* 2004;**160**:102–9.
57. Thomas JD, Idrus NM, Monk BR, Dominguez HD. Prenatal choline supplementation mitigates behavioral alterations associated with prenatal alcohol exposure in rats. *Birth Defects Res A Clin Mol Teratol* 2010;**88**:827–37.
58. Thomas JD, Biane JS, O'Bryan KA, O'Neill TM, Dominguez HD. Choline supplementation following third-trimester-equivalent alcohol exposure attenuates behavioral alterations in rats. *Behav Neurosci* 2007;**121**:120–30.
59. Thomas JD, O'Neill TM, Dominguez HD. Perinatal choline supplementation does not mitigate motor coordination deficits associated with neonatal alcohol exposure in rats. *Neurotoxicol Teratol* 2004;**26**:223–9.
60. Thomas JD, Garrison M, O'Neill TM. Perinatal choline supplementation attenuates behavioral alterations associated with neonatal alcohol exposure in rats. *Neurotoxicol Teratol* 2004;**26**: 35–45.
61. Holmes GL, Yang Y, Liu Z, Cermak JM, Sarkisian MR, Stafstrom CE, et al. Seizure-induced memory impairment is reduced by choline supplementation before or after status epilepticus. *Epilepsy Res* 2002;**48**:3–13.
62. Alonso-Aperte E, Varela-Moreiras G. Brain folates and DNA methylation in rats fed a choline deficient diet or treated with low doses of methotrexate. *Int J Vitam Nutr Res* 1996;**66**:232–6.
63. Niculescu MD, Craciunescu CN, Zeisel SH. Dietary choline deficiency alters global and gene-specific DNA methylation in the developing hippocampus of mouse fetal brains. *FASEB J* 2006;**20**:43–9.
64. Niculescu MD, Yamamuro Y, Zeisel SH. Choline availability modulates human neuroblastoma cell proliferation and alters the methylation of the promoter region of the cyclin-dependent kinase inhibitor 3 gene. *J Neurochem* 2004;**89**:1252–9.
65. Mehedint MG, Niculescu MD, Craciunescu CN, Zeisel SH. Choline deficiency alters global histone methylation and epigenetic marking at the Re1 site of the calbindin 1 gene. *FASEB J* 2010;**24**:184–95.
66. Mehedint MG, Craciunescu CN, Zeisel SH. Maternal dietary choline deficiency alters angiogenesis in fetal mouse hippocampus. *Proc Natl Acad Sci USA* 2010;**107**:12834–9.
67. Fisher MC, Zeisel SH, Mar MH, Sadler TW. Inhibitors of choline uptake and metabolism cause developmental abnormalities in neurulating mouse embryos. *Teratology* 2001;**64**: 114–22.
68. Fisher MC, Zeisel SH, Mar MH, Sadler TW. Perturbations in choline metabolism cause neural tube defects in mouse embryos in vitro. *FASEB J* 2002;**16**:619–21.
69. Blom HJ, Shaw GM, den Heijer M, Finnell RH. Neural tube defects and folate: case far from closed. *Nat Rev Neurosci* 2006;**7**:724–31.
70. Craciunescu CN, Albright CD, Mar MH, Song J, Zeisel SH. Choline availability during embryonic development alters progenitor cell mitosis in developing mouse hippocampus. *J Nutr* 2003;**133**:3614–8.
71. Albright CD, Siwek DF, Craciunescu CN, Mar MH, Kowall NW, Williams CL, et al. Choline availability during embryonic development alters the localization of calretinin in developing and aging mouse hippocampus. *Nutr Neurosci* 2003;**6**:129–34.
72. Niculescu MD, Craciunescu CN, Zeisel SH. Gene expression profiling of choline-deprived neural precursor cells isolated from mouse brain. *Brain Res* 2005;**134**:309–22.
73. Jones JP, Meck W, Williams CL, Wilson WA, Swartzwelder HS. Choline availability to the developing rat fetus alters adult hippocampal long-term potentiation. *Brain Res* 1999;**118**: 159–67.
74. Meck WH, Williams CL. Choline supplementation during prenatal development reduces proactive interference in spatial memory. *Brain Res* 1999;**118**:51–9.

75. Albright CD, Friedrich CB, Brown EC, Mar MH, Zeisel SH. Maternal dietary choline availability alters mitosis, apoptosis and the localization of TOAD-64 protein in the developing fetal rat septum. *Brain Res* 1999;**115**:123–9.
76. Albright CD, Tsai AY, Friedrich CB, Mar MH, Zeisel SH. Choline availability alters embryonic development of the hippocampus and septum in the rat. *Brain Res* 1999;**113**:13–20.
77. Pyapali G, Turner D, Williams C, Meck W, Swartzwelder HS. Prenatal choline supplementation decreases the threshold for induction of long-term potentiation in young adult rats. *J Neurophysiol* 1998;**79**:1790–6.
78. Montoya DA, White AM, Williams CL, Blusztajn JK, Meck WH, Swartzwelder HS. Prenatal choline exposure alters hippocampal responsiveness to cholinergic stimulation in adulthood. *Brain Res Dev Brain Res* 2000;**123**:25–32.
79. Meck W, Williams C. Characterization of the facilitative effects of perinatal choline supplementation on timing and temporal memory. *Neuroreport* 1997;**8**:2831–5.
80. Meck W, Williams C. Simultaneous temporal processing is sensitive to prenatal choline availability in mature and aged rats. *Neuroreport* 1997;**8**:3045–51.
81. Meck WH, Williams CL. Metabolic imprinting of choline by its availability during gestation: implications for memory and attentional processing across the lifespan. *Neurosci Biobehav Rev* 2003;**27**:385–99.
82. Williams CL, Meck WH, Heyer DD, Loy R. Hypertrophy of basal forebrain neurons and enhanced visuospatial memory in perinatally choline-supplemented rats. *Brain Res* 1998;**794**:225–38.
83. Cooney CA, Dave AA, Wolff GL. Maternal methyl supplements in mice affect epigenetic variation and DNA methylation of offspring. *J Nutr* 2002;**132**:2393S–400S.
84. Waterland RA, Dolinoy DC, Lin JR, Smith CA, Shi X, Tahiliani KG. Maternal methyl supplements increase offspring DNA methylation at Axin Fused. *Genesis* 2006;**44**:401–6.
85. Sha W, da Costa KA, Fischer LM, Milburn MV, Lawton KA, Berger A, et al. Metabolomic profiling can predict which humans will develop liver dysfunction when deprived of dietary choline. *FASEB J* 2010;**24**:2962–75.
86. Dumas ME, Barton RH, Toye A, Cloarec O, Blancher C, Rothwell A, et al. Metabolic profiling reveals a contribution of gut microbiota to fatty liver phenotype in insulin-resistant mice. *Proc Natl Acad Sci USA* 2006;**103**:12511–6.
87. Spencer MD, Hamp TJ, Reid RW, Fischer LM, Zeisel SH, Fodor AA. Association between composition of the human gastrointestinal microbiome and development of fatty liver with choline deficiency. *Gastroenterology* 2011;**140**:976–86.
88. Wang Z, Klipfell E, Bennett BJ, Koeth R, Levison BS, Dugar B, et al. Gut flora metabolism of phosphatidylcholine promotes cardiovascular disease. *Nature* 2011;**472**:57–63.

Genetic Determinants of Dietary Antioxidant Status

Laura A. Da Costa,* Bibiana García-Bailo,*,† Alaa Badawi,† and Ahmed El-Sohemy*

*Department of Nutritional Sciences, Faculty of Medicine, University of Toronto, Toronto, Ontario, Canada

†Office of Biotechnology, Genomics and Population Health, Public Health Agency of Canada, Toronto, Ontario, Canada

Oxidative stress refers to a physiological state in which an imbalance between pro-oxidants and antioxidants results in oxidative damage. Oxidative stress has been associated with the development of numerous chronic diseases such as type 2 diabetes, cardiovascular disease (CVD), osteoporosis, and cancer. Endogenous production of free radicals occurs during normal physiological processes, such as aerobic metabolism, oxidation of biological molecules, and enzymatic activity. Environmental factors such as ultraviolet radiation, air pollution, and cigarette smoking can also contribute to the accumulation of free radicals in the body. Excess free radicals can damage tissues and promote the upregulation of disease-related pathways such as inflammation. Modulating oxidative stress by dietary supplementation with antioxidant micronutrients such as vitamins C and E or phytochemicals such as different carotenoids may help prevent or delay the development of certain diseases. However, research on antioxidant supplementation and disease has yielded inconsistent findings, which may be due, in part, to interindividual genetic variation. Polymorphisms in genes coding for endogenous antioxidant enzymes or proteins responsible for the absorption, transport, distribution, or metabolism of dietary antioxidants have been shown to affect antioxidant status and response to supplementation. These genetic variants may also interact with environmental factors, such as diet, to determine an individual's overall antioxidant status.

Progress in Molecular Biology
and Translational Science, Vol. 108
DOI: 10.1016/B978-0-12-398397-8.00008-3

1877-1173/12 $35.00

This chapter examines current knowledge of the relationship between genetic variation and dietary antioxidant status.

I. Radical Production, Antioxidants, and Oxidative Stress

Oxidative stress is defined as the imbalance between the production of pro-oxidant free radicals and the body's ability to defend against them. Oxidative stress has been implicated in several chronic diseases including cancer, arthritis, osteoporosis, type 2 diabetes, neurodegenerative diseases (including Alzheimer's and Parkinson's), and cardiovascular disease (CVD).[1–4] Free radicals, characterized by the presence of unpaired electrons, are highly reactive molecular species that react with other compounds, initiating cytotoxic oxidative chain reactions. The superoxide anion (O_2^-), the most common free radical in the body, is constitutively produced during the process of aerobic respiration in the mitochondria as a result of electron leakage along the electron transport chain.[5] Other unstable and reactive compounds that do not contain unpaired electrons can also trigger the production of free radicals by acting as oxidizing agents. These include the reactive oxygen species (ROS), such as hydrogen peroxide (H_2O_2) or singlet oxygen ($^1O_2$), and reactive nitrogen species (RNS), which include nitric oxide (NO) and nitrogen dioxide (NO_2).

ROS and RNS can substantially disrupt physiological processes through participation in reduction–oxidation (redox) reactions involving the loss (oxidation) or addition of electrons (reduction). Polyunsaturated fatty acids (PUFAs), which are characterized by conjugated double carbon bonds that can act as a source of electrons, are particularly susceptible to oxidative damage. Cell membranes are rich in PUFAs, and free radical damage to PUFAs can reduce the integrity of the membrane.[3] Oxidation of lipids, including low-density lipoprotein (LDL), may also play an important role in the development of CVD. Indeed, uptake of oxidized LDL into macrophages at sites of vascular inflammation can lead to the formation of foam cells and the development of atherosclerotic plaques.[6,7] ROS damage to proteins, such as various receptors, enzymes, and transporters, can result in their inactivation and disrupt a variety of important physiological functions.[3] Free radicals can also damage DNA by oxidizing bases, which may result in the introduction of mutations and contribute to aging and development of age-related diseases such as cancer.[4,5] These mutations may be particularly carcinogenic when occurring in tumor suppressor genes or proto-oncogenes.[8] Beyond direct macromolecule damage, ROS production associated with cardiometabolic risk factors such as hyperglycemia, increases in plasma levels of free fatty acids, and hyperinsulinemia may also result in aberrant signaling and gene expression. ROS and RNS activate nuclear factor-κB (NF-κB), a proinflammatory transcription factor that triggers a signaling cascade leading to the chronic subclinical inflammation associated with conditions such as obesity and the metabolic syndrome.[9]

Despite their potential for damage, ROS and RNS play an important role in normal physiological processes, including the immune response, apoptosis, cellular signaling, and gene transcription.[10,11] The generation of ROS within macrophages and neutrophils, for example, is critical for initiating phagocytosis of invading pathogens and providing the "oxidative burst" used for their destruction.[12,13] The RNS nitric oxide (NO) is also an important regulator of vascular homeostasis.[14] By binding to and activating the guanylate cyclase enzyme, NO induces vasorelaxation in smooth muscle cells and inhibits platelet aggregation.[10] Because of the critical functions of ROS at lower levels and their potential for damage at higher levels, physiological regulation of ROS production and effects is tightly controlled through a complex network of antioxidant defense systems.

Physiological ROS regulation is mediated by endogenous and exogenous antioxidants. Antioxidants react with ROS to inhibit oxidation chain reactions, reducing their accumulation and ultimately limiting damage to other compounds, cells, and tissues. The body produces numerous antioxidants endogenously, but endogenous production is often insufficient to prevent oxidative stress. Exogenous antioxidants, including nutrients and phytochemicals found in the diet, can supplement the endogenous system in free radical defense. However, under conditions of excessive free radical production, this too may be insufficient to prevent damage. Each individual has a different capacity to manage ROS and oxidative stress, which is a function of endogenous and exogenous antioxidant systems and exposure to internal and environmental stimuli such as inflammation, air pollution (ozone [O_3], NO_2), and cigarette smoking. In addition, genetic variation may affect an individual's ability to defend against oxidative stress (Fig. 1).[15]

II. Endogenous Antioxidants

The body's natural defense system against oxidative stress includes enzymes such as glutathione peroxidase (GPX), thioredoxin reductase, superoxide dismutase (SOD), and catalase, as well as nonenzymatic compounds such as glutathione and thioredoxin (Table I). However, the antioxidant activity of some of these compounds is mediated by minerals derived from the diet, such as selenium, copper, manganese, and zinc, which act as critical cofactors.[16] Deficiency in one or more of these minerals may result in a decreased capacity to manage oxidative stress. Transfer proteins, such as transferrin, lactoferrin,

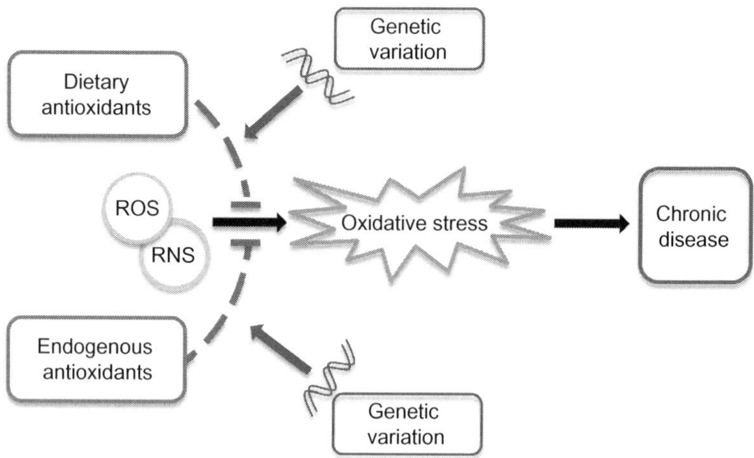

FIG. 1. The relationship between reactive oxygen species (ROS) and reactive nitrogen species (RNS), dietary and endogenous antioxidants, oxidative stress, and chronic disease. ROS and RNS are produced as a part of normal physiological processes (e.g., aerobic respiration, immune function) and are acquired from the external environment (e.g., air pollution, ultraviolet radiation). The production of ROS/RNS is regulated by endogenous antioxidants (e.g., superoxide dismutase, glutathione) and dietary antioxidants (e.g., vitamin C, vitamin E, carotenoids, polyphenols). Genetic variation modifies the ability of the endogenous and dietary antioxidant systems to manage ROS and RNS. When ROS and RNS production exceeds the ability of the antioxidant networks to manage them, oxidative stress results. Macromolecule damage and abnormal cellular signaling and gene expression associated with oxidative stress may then lead to chronic disease development (e.g., type 2 diabetes, cardiovascular disease, and cancer). (For color version of this figure, the reader is referred to the online version of this chapter.)

ceruloplasmin, and albumin, also act as antioxidants by sequestering pro-oxidant metals including copper and iron, making them unavailable to participate in redox reactions.[17]

Endogenous antioxidant enzymes work together to neutralize ROS, reducing them to forms that are stable and can be excreted or recycled. For example, manganese-containing superoxide dismutase (Mn-SOD), found in mitochondria, is an endogenous antioxidant enzyme that scavenges and converts superoxide radicals to hydrogen peroxide.[18] Hydrogen peroxide then is further reduced to water through the action of enzymes such as the selenium-containing GPX1.[18] Secondary to the antioxidant enzymes are a variety of detoxification enzymes that catabolize toxic metabolites to facilitate their excretion. These are termed phase II enzymes, which are involved in xenobiotic and drug metabolism, and they include the glutathione *S*-transferase (GST) family of enzymes, γ-glutamylcysteine synthetase, and NAD(P)H: quinone reductase.[19]

TABLE I

EXAMPLES OF ENDOGENOUS AND DIETARY ANTIOXIDANTS

Endogenous antioxidants	Dietary antioxidants
Antioxidant enzymes	Vitamin
Superoxide dismutase (SOD)	C (ascorbic acid/ascorbate)
Cytosolic copper/zinc SOD	E (tocopherols, tocotrienols)
Extracellular copper/zinc SOD	Carotenoids
Mitochondrial manganese	α- and β-carotene
Glutathione peroxidase (GPX)	Lutein
Cytosolic GPX1	Zeaxanthin
Cytosolic GPX2	Lycopene
Plasma-based GPX3	β-Cryptoxanthin
Phospholipid hydroperoxidase (GPX 4)	Trace elements
Catalase	Selenium
Glutathione reductase	Zinc
Thioredoxin reductase	
Glutathione *S*-transferase family	Polyphenols
GSTM1	Flavonols
GSTP1	Quercetin
GSTT1	Myricetin
γ-Glutamyl cysteine synthetase	Kaempferol
NAD(P)H:quinone reductase	Flavanols
Glucose-6-phosphate dehydrogenase	Epigallocatechin
Aldo-keto reductase	Epicatechin
Peroxiredoxin	Catechin
Paraoxanase	Anthocyanidins
	Delphinidin
Nonenzymatic antioxidants[a]	Cyanidin
Glutathione	Apigenidin
Thioredoxin	Isoflavones
Lipoic acid	Genistein
Uric acid	Daidzein
NADPH	Glycitein
Bilirubin	Flavanones
Carnosine	Taxifolin
Melatonin	Naringenin
Ubiquinol	Eriodictyol
	Flavones
Transition metal-binding proteins	Luteolin
Ceruloplasmin	Apignen
Ferritin	Phenolic acids
Lactoferrin	Chlorogenic acid
Transferrin	Gallic acid
Metallothionein	Stilbenoids
Albumin	Resveratrol

[a]Several endogenous antioxidants including ubiquinol, carnosine, and lipoic acid can also be obtained from supplements and food sources.

regardless of genotype. This interaction was not replicated in populations with higher serum ascorbic acid levels.[41] Nonetheless, these results suggest that the GST enzymes may protect against serum ascorbic acid deficiency when the diet provides insufficient vitamin C.

HP is an acute-phase protein that binds free hemoglobin in the circulation to prevent free radical generation and oxidative damage produced by the iron-containing heme of the hemoglobin molecule.[54] A common polymorphism in the *HP* gene results in two different alleles, *HP1* and *HP2*, which give rise to three different genotypes, *HP1-1*, *HP1-2*, and *HP2-2*.[55,56] The *HP1-1* genotype produces the most functional version of the protein, which is more effective at binding free hemoglobin than the other forms. The *HP1-2* genotype produces a protein of intermediate functionality, while the *HP2-2* genotype produces the least functional version of the HP protein.[57] Increased utilization of ascorbic acid may compensate for the reduced antioxidant activity of the HP protein among those with the *HP2-2* genotype.[58] Indeed, three separate studies have shown that serum ascorbic acid is significantly lower among subjects with the *HP2-2* genotype.[43,45,59] Interestingly, a gene–diet interaction was also observed for the *HP* polymorphism. Subjects who met the RDA for vitamin C or who consumed the recommended amounts of fruits and vegetables were not at increased risk of serum ascorbic acid deficiency, regardless of *HP* genotype.[46] However, individuals with the *HP2-2* genotype were at greater risk of deficiency with insufficient intake. Ascorbic acid deficiency ($<11\ \mu M$)[35,60] is quite common among young adults in North America[61,62] and the United Kingdom.[63,64] Understanding the complex interactions between dietary vitamin C, genetic variation in ascorbic acid metabolism, and serum ascorbic acid may help identify individuals and subpopulations at risk of deficiency.

B. Vitamin E

Vitamin E encompasses a group of eight compounds, including α, β, γ, and δ tocopherols and α, β, γ, and δ tocotrienols, with differing biological activities. Each compound contains a hydroxyl-containing chromanol ring with a varying number and position of methyl groups between the α, β, γ, and δ forms.[65] While both the tocopherols and tocotrienols have hydrophobic side chains, these are unsaturated in the case of tocotrienols and saturated among the tocopherols. Vitamin E is an essential micronutrient regularly consumed in the diet from vegetable oils and products derived from vegetables oils, including margarine, salad dressing, and mayonnaise. While some oils, such as soybean oil, contain a mix of tocopherols, others, such as sunflower oil, contain almost exclusively α-tocopherol, and palm oil is one of the few vegetable oils containing tocotrienols.[66] Additional sources of vitamin E in the diet include animal fats and meats, whole grains, nuts, and seeds.

Vitamin E compounds in the diet are hydrolyzed in the duodenum by pancreatic lipases, incorporated into micelles, and absorbed either by passive diffusion or actively via the cholesterol transporter scavenger receptor class B type 1 (SR-B1).[67] All forms of vitamin E are equally absorbed and incorporated into chylomicrons by microsomal triglyceride transfer protein, released into the circulation, and eventually taken up by the liver hepatocytes as chylomicron remnants.[68,69] The selective α-tocopherol transfer protein (α-TTP) incorporates α-tocopherol into very LDLs, which are released into the circulation for uptake and use by peripheral tissues.[70] Other forms of vitamin E are mainly metabolized or excreted unaltered in the bile and urine.[71,72] The α-TTP, therefore, is responsible for maintaining much higher circulating levels of α-tocopherol in comparison to the other forms of vitamin E, despite γ-tocopherol being the most abundant form in the diet.[73] Intracellular transport of α-tocopherol by cells does not appear to involve the α-TTP, which has been attributed to other proteins, including the tocopherol-associated proteins (TAPs).[74,75]

1. Physiological Functions

Vitamin E has several functions in the body, including membrane stabilization and regulation of gene expression and cellular signaling, in addition to its well-recognized antioxidant properties.[76,77] Vitamin E is a chain-breaking antioxidant capable of interrupting radical chain reactions produced by the lipid peroxy radicals (LOO•).[78] In addition, it directly scavenges superoxide radicals and singlet oxygen, both of which can initiate oxidation reactions. Its antioxidant activity is attributed to its chromanol rings, which contain hydroxyl groups capable of donating hydrogen to ROS.[65] The α-tocopheroxyl radical (α-TO•) formed is relatively stable, since the unpaired electron can be delocalized across the molecule.[79,80] Regeneration of α-tocopherol can occur via redox reactions with vitamin C, reduced glutathione, or coenzyme Q.[32,81–83] Since vitamin E is a lipid-soluble molecule, its antioxidant functions are important in the protection of membrane lipids against peroxidation.

2. Genetic Determinants of Vitamin E Status

Circulating levels of vitamin E, and specifically α-tocopherol, are used as short-term biomarkers of vitamin E status.[84] There are large interindividual differences that have been observed in circulating levels of vitamin E, and the correlations between intake and circulating concentrations are relatively weak.[85–87] Interindividual genetic variation in absorption and metabolism of vitamin E may explain some of this variability. It has been estimated that approximately 22% of the total variance in circulating α-tocopherol concentrations is attributable to variation in key genes.[88] Biologically plausible candidates are those involved in vitamin E binding, including *TTPA*, which encodes α-TTP, and SEC14-like 2 (*SEC14L2*), which encodes TAP. Indeed, a rare mutation in

TTPA has been associated with ataxia with vitamin E deficiency, which is characterized by extremely low concentrations of plasma α-tocopherol.[89–92] Common variants in both the *TTPA* and *SEC14L2* genes have also been associated with serum α-tocopherol levels.[93]

In a recent genome-wide association study (GWAS), three loci were significantly associated with circulating concentrations of α-tocopherol. These findings were replicated in two additional populations, with meta-analysis of the data confirming the associations that were observed.[94] The analyses were conducted using 4014 male smokers from Finland who participated in the Alpha-Tocopherol, Beta-Carotene (ATBC) Cancer Prevention Study; 992 caucasian male smokers from the multicenter U.S. Prostate, Lung, Colorectal, and Ovarian Cancer Screening Trial; and 2775 women from the U.S. Nurses' Health Study. The strongest association identified was for a polymorphism near BUD13 homolog (*BUD13*), zinc finger protein 259 (*ZNF259*), and the apolipoprotein A-I/C-III/A-IV/A-V (*APOA1/APOC3/APOA4/APOA5*) gene cluster. Interestingly, the only other GWAS to examine serum α-tocopherol reported a significant association with a polymorphism near the *APOA5* gene.[95] The importance of variants affecting lipid and lipoprotein metabolism on circulating vitamin E levels is supported by several candidate gene studies that have identified variants in *APOC3*, *APOA4*, *APOA5*, *APOE*, *SCARB1* (encoding SR-B1), and cholesteryl ester transfer protein (*CETP*), which were associated with α-tocopherol levels,[96–100] and variants in *APOA4*, *APOE*, *SCARB1*, and hepatic lipase (*LIPC*), which were associated with γ-tocopherol levels.[96–98] Interestingly, many of these associations were sex dependent, which may reflect estrogenic effects on lipoprotein metabolism.[96,98,101]

The SR-B1 also may have direct effects on vitamin E levels, as it is likely involved in intestinal uptake of vitamin E and carotenoids.[67,102–106] Involvement of the SR-B1 protein was confirmed in the most recent GWAS analysis, which also identified the *SCARB1* and the cytochrome P450, family 4, subfamily F, polypeptide 2 (*CYP4F2*) genes as loci associated with circulating α-tocopherol concentration,[94] the latter of which is involved in vitamin E metabolism.[107,108] These studies collectively indicate that genetic variants affecting lipid transport and metabolism, as well as vitamin E uptake and metabolism, are important determinants of circulating vitamin E levels.

Genetic variants could also be used to identify individuals in the population who may benefit from vitamin E supplementation. Several double-blind placebo-controlled trials assessing the effects of α-tocopherol supplementation on cardiovascular risk factors and outcomes among diabetic patients reported beneficial effects of supplementation, which were limited to those with the *HP2-2* genotype.[109–111] Two variants in the *SEC14L2* gene have also been shown to modify the association between α-tocopherol supplementation and incident prostate cancer.[93] In an examination of 982 incident cases of prostate cancer and 851 controls from the ATBC Cancer Prevention Study, there was a reduced risk with supplementation among those homozygous for the common allele of each *SEC14L2* variant and a nonsignificant increased risk among those carrying the variant alleles.[93] However, a limitation of the study is that cases and controls were matched for supplement use, which effectively removes the previously observed protective effect of supplementation. In a recent study examining results from the Carotene and Retinol Efficacy Trial (CARET), high serum α-tocopherol was associated with a reduced risk of aggressive prostate cancer among smokers, but this association was modified by a polymorphism in the gene encoding the ROS-producing myloperoxidase enzyme (*MPO*).[112] These studies suggest that, based on genotype, certain population subgroups may benefit from vitamin E supplementation, yet the role of adequate dietary intake remains to be investigated.

C. Carotenoids

Carotenoids are a large group of lipophilic, light-absorbing plant pigments. Over 600 natural carotenoids have been identified. However, only approximately 60 of these are normally found in the diet, and only a small number of those have been isolated from human blood and tissues.[113,114] The most abundant forms in human plasma include α-carotene, β-carotene, lycopene, lutein, zeaxanthin, and β-cryptoxanthin. Carotenoids are related in structure by a 40-carbon skeleton which includes cyclical structures at one or both ends and a polyene chain of conjugated double bonds.[115] The xanthophylls (zeaxanthin, β-cryptoxanthin, and lutein) are differentiated by the presence of hydroxyl groups, increasing their polarity in comparison to the other carotenoids.

Carotenoids exist as pigments in plants and algae. In addition, they are synthesized by photosynthetic bacteria. Carotenoids provide their characteristic color to many fruits and vegetables, such as the orange in carrots and the red color of tomatoes. Carotenoids can also be found in dark green vegetables; however, the abundance of chlorophyll masks their colors.[116] Dietary sources of carotenoids are mainly fruits and vegetables, with tomatoes and tomato products being the primary source of lycopene. Lutein and zeaxanthin can also be obtained from egg yolk, which is a highly bioavailable source of these xanthophylls.[117,118] Bioavailability of some carotenoids is also significantly improved by cooking, chopping, and the presence of dietary fat.[114,119–122]

Ingested carotenoids enter the small intestine, where they are solubilized with lipids and bile components, incorporated into micelles, and taken up by the mucosal cells.[123] Intestinal absorption of carotenoids was traditionally considered to be a passive process, but accumulating evidence points to an active process involving SR-B1.[102–106] Absorbed carotenoids are packaged into chylomicrons for release into the lymphatic system and eventual transport into the bloodstream.[114,123] Chylomicrons taken up by the liver are repackaged into lipoproteins, which are the main source of circulating carotenoids. In the circulation, the carotenes and lycopene are found within LDL particles. Meanwhile, the xanthophylls are found in both LDL and high-density lipoprotein (HDL), where they are distributed for use by the peripheral tissues.[123]

1. Physiological Functions

The most well studied function of carotenoids is their role as precursors for vitamin A. This, however, applies only to the carotenes and cryptoxanthins, as lutein, zeaxanthin, and lycopene cannot be converted to retinol. Lutein and zeaxanthin are the only carotenoids found in the macula of the retina,[124,125] where they absorb harmful blue light entering the eye and may play an important role in preventing the development of age-related macular degeneration and cataracts.[126,127] Many roles have been proposed for carotenoids, including inhibition of tumor growth, protection against genotoxicity, and modulation of the immune system.[128]

Carotenoids are effective free radical scavengers, particularly of singlet oxygen ($^1O_2$).[129] Their antioxidant abilities have been attributed to their electron-rich polyene chain with extensive conjugated double bonds. As carotenoids are lipid soluble, their antioxidant properties may be particularly important for protection of lipids and may work synergistically with vitamin E to prevent lipid peroxidation.[130,131] Carotenoids also may play an important role in the protection of DNA damage from ROS attack while additionally modulating DNA repair mechanisms.[132,133]

2. Genetic Determinants of Carotenoid Status

Considerable interindividual variability in absorption, circulating levels of carotenoids, and response to carotenoid supplementation has been reported and may be related to individual genetic differences affecting absorption, uptake, transport, and metabolism.[134–136] Because the carotenoids are hydrophobic antioxidants transported in the circulation associated to lipoproteins, many of the early attempts to identify genetic determinants of circulating carotenoids using a candidate gene approach focused on genes involved in lipoprotein metabolism. These initial studies identified variants in several genes that encode proteins related to lipid and lipoprotein homeostasis, including APOE, APOB, APOA4,[96,97] SR-B1,[96] intestinal fatty acid binding protein, and hepatic lipase.[98]

Two variants in the *BCMO1* gene, encoding the β-carotene 15,15'-monooxygenase 1, have been shown to be associated with fasting β-carotene concentrations and the ability of this enzyme to convert β-carotene to vitamin A in a β-carotene supplementation study.[137] This was supported by a GWAS, which identified variants in the *BCMO1* gene as being significantly associated with circulating levels of carotenoids.[95] In a meta-analysis of GWAS data from three large cohorts, a variant in *BCMO1* was significantly associated with circulating levels of α- and β-carotene, lycopene, lutein, and zeaxanthin.[95] Interestingly, only the provitamin A carotenoids α- and β-carotene were associated with the *BCMO1* polymorphism, while the provitamin carotenoid β-cryptoxanthin and circulating levels of retinol were not. The association to the non-provitamin A carotenoids (lycopene, lutein, and zeaxanthin) may reflect their ability to enhance or inhibit *BCMO1* activity or interact with the absorption or transport of the provitamin A carotenoids.[95,138–140]

In a recent comparative intervention study, 29 subjects were given a supplement containing lutein or placebo for 6 months after a 3-week lutein-poor diet.[141] Measurements of plasma lutein and macular pigment optical density, which is an indicator of lutein content of the macula of the eye, were taken before and after the intervention. Variants in the *BCMO1* and fatty acid translocase (*CD36*) genes were associated with circulating levels of lutein, and they affected the plasma and macular pigment optical density response to lutein supplementation. In both genes, variants associated with lower baseline lutein levels were also associated with the greatest response to supplementation.[141] These results suggest that variation in the *BCMO1* gene may affect both circulating levels of carotenoids and the response to supplementation.

Variation in genes that encode endogenous antioxidant enzymes may also help identify subgroups that could benefit from carotenoid supplementation. Serum paraoxonase/arylesterase 1 (PON1) is a hydrolyzing enzyme found in the circulation associated with HDL and plays an important role in protecting LDL from oxidation.[142–144] Two functional polymorphisms in the *PON1* gene have been associated with reduced activity of the enzyme.[145–147] Individuals with these variants may be more susceptible to oxidative stress. However, serum lycopene has been shown to modify the relationships between these genetic variants and measures of oxidative stress and bone turnover, suggesting that lycopene found in LDL may replace the lost PON1 activity.[148] Findings of studies such as these could help identify subgroups that benefit from carotenoid supplementation, such as individuals at risk for eye disease or osteoporosis.

IV. Conclusion

Oxidative stress has been implicated in the development of chronic disease, although the findings from various lines of research remain inconclusive. Antioxidant-rich diets high in fruits and vegetables, which may help reduce oxidative stress, tend to be inversely associated with the development and progression of cancer[149,150] and CVD.[151,152] However, results from clinical trials involving supplementation of dietary antioxidants have been mostly null,[153–158] and in some cases, adverse effects have been observed.[159–161] Common genetic variants affecting uptake, distribution, transport, or metabolism of dietary antioxidants have been linked to variation in antioxidant serum levels and response to supplementation[40,46,94,95,141] and may help explain some of the inconsistencies between observational and intervention studies. Furthermore, genetic variants may modulate the relationship between the endogenous antioxidant defense system and external environmental factors, such as dietary antioxidant intake, and therefore may affect susceptibility to oxidative stress and disease development. Elucidating the relationship between common genetic variants and antioxidant status may have important public health implications through the identification of individuals and subgroups that benefit most from dietary intervention or supplementation with antioxidants.

References

1. Gutteridge JMC. Invited review. Free radicals in disease processes: a compilation of cause and consequence. *Free Radic Res Commun* 1993;**19**:141–58.
2. Kehrer JP. Free radicals as mediators of tissue injury and disease. *Crit Rev Toxicol* 1993; **23**:21–48.
3. Halliwell B. Oxidative stress and neurodegeneration: where are we now? *J Neurochem* 2006;**97**:1634–58.
4. Halliwell B. Oxidative stress and cancer: have we moved forward? *Biochem J* 2007;**401**:1–11.
5. Ames BN, Shigenaga MK, Hagen TM. Oxidants, antioxidants, and the degenerative diseases of aging. *Proc Natl Acad Sci USA* 1993;**90**:7915–22.
6. Goldstein JL, Ho YK, Basu SK, Brown MS. Binding site on macrophages that mediates uptake and degradation of acetylated low density lipoprotein, producing massive cholesterol deposition. *Proc Natl Acad Sci USA* 1979;**76**:333–7.
7. Lusis AJ. Atherosclerosis. *Nature* 2000;**407**:233–41.
8. Marnett LJ. Oxyradicals and DNA damage. *Carcinogenesis* 2000;**21**:361–70.
9. Garcia-Bailo B, El-Sohemy A, Haddad P, Arora P, BenZaied F, Karmali M, et al. Vitamins D, C, and E in the prevention of type 2 diabetes mellitus: modulation of inflammation and oxidative stress. *Biologics* 2011;**5**:1–13.
10. Dröge W. Free radicals in the physiological control of cell function. *Physiol Rev* 2002; **82**:47–95.
11. Seifried HE, Anderson DE, Fisher EI, Milner JA. A review of the interaction among dietary antioxidants and reactive oxygen species. *J Nutr Biochem* 2007;**18**:567–79.
12. Babior BM, Kipnes RS, Curnutte JT. Biological defense mechanisms. The production by leukocytes of superoxide, a potential bactericidal agent. *J Clin Invest* 1973;**52**:741–4.
13. Krinsky NI. Singlet excited oxygen as a mediator of the antibacterial action of leukocytes. *Science* 1974;**186**:363–5.
14. Carr A, Frei B. The role of natural antioxidants in preserving the biological activity of endothelium-derived nitric oxide. *Free Radic Biol Med* 2000;**28**:1806–14.
15. Wang XL, Rainwater DL, VandeBerg JF, Mitchell BD, Mahaney MC. Genetic contributions to plasma total antioxidant activity. *Arterioscler Thromb Vasc Biol* 2001;**21**:1190–5.
16. Klotz LO, Kröncke KD, Buchczyk DP, Sies H. Role of copper, zinc, selenium and tellurium in the cellular defense against oxidative and nitrosative stress. *J Nutr* 2003; **133**:1448S–1451S.
17. Young IS, Woodside JV. Antioxidants in health and disease. *J Clin Pathol* 2001;**54**:176–86.

18. Evans JL, Goldfine ID, Maddux BA, Grodsky GM. Oxidative stress and stress-activated signaling pathways: a unifying hypothesis of type 2 diabetes. *Endocr Rev* 2002;**23**:599–622.
19. Blomhoff R. Dietary antioxidants and cardiovascular disease. *Curr Opin Lipidol* 2005;**16**:47–54.
20. Carlsen H, Myhrstad MCW, Thoresen M, Moskaug JÃ, Blomhoff R. Berry intake increases the activity of the y-glutamylcysteine synthetase promoter in transgenic reporter mice. *J Nutr* 2003;**133**:2137–40.
21. Breinholt V, Lauridsen ST, Daneshvar B, Jakobsen J. Dose-response effects of lycopene on selected drug-metabolizing and antioxidant enzymes in the rat. *Cancer Lett* 2000; **154**:201–10.
22. Moskaug JÃ, Carlsen H, Myhrstad M, Blomhoff R. Molecular imaging of the biological effects of quercetin and quercetin-rich foods. *Mech Ageing Dev* 2004;**125**:315–24.
23. Nishikimi M, Fukuyama R, Minoshima S, Shimizu N, Yagi K. Cloning and chromosomal mapping of the human nonfunctional gene for L- gulono-γ-lactone oxidase, the enzyme for L-ascorbic acid biosynthesis missing in man. *J Biol Chem* 1994;**269**:13685–8.
24. Tsukaguchi H, Tokui T, Mackenzle B, Berger UV, Chen XZ, Wang Y, et al. A family of mammalian Na+-dependent L-ascorbic acid transporters. *Nature* 1999;**399**:70–5.
25. Padayatty SJ, Katz A, Wang Y, Eck P, Kwon O, Lee JH, et al. Vitamin C as an antioxidant: evaluation of its role in disease prevention. *J Am Coll Nutr* 2003;**22**:18–35.
26. Wannamethee SG, Lowe GDO, Rumley A, Bruckdorfer KR, Whincup PH. Associations of vitamin C status, fruit and vegetable intakes, and markers of inflammation and hemostasis. *Am J Clin Nutr* 2006;**83**:567–74.
27. Griffiths HR, Lunec J. Ascorbic acid in the 21st century—more than a simple antioxidant. *Environ Toxicol Pharmacol* 2001;**10**:173–82.
28. Fraga CG, Oteiza PI. Iron toxicity and antioxidant nutrients. *Toxicology* 2002;**180**:23–32.
29. Rietjens IMCM, Boersma MG, Haan LD, Spenkelink B, Awad HM, Cnubben NHP, et al. The pro-oxidant chemistry of the natural antioxidants vitamin C, vitamin E, carotenoids and flavonoids. *Environ Toxicol Pharmacol* 2002;**11**:321–33.
30. Winkler BS, Orselli SM, Rex TS. The redox couple between glutathione and ascorbic acid: a chemical and physiological perspective. *Free Radic Biol Med* 1994;**17**:333–49.
31. May JM, Mendiratta S, Hill KE, Burk RF. Reduction of dehydroascorbate to ascotbate by the selenoenzyme thioredoxin reductase. *J Biol Chem* 1997;**272**:22607–10.
32. Packer JE, Slater TF, Willson RL. Direct observation of a free radical interaction between vitamin E and vitamin C. *Nature* 1979;**278**:737–8.
33. May JM, Qu ZC, Mendiratta S. Protection and recycling of α-tocopherol in human erythrocytes by intracellular ascorbic acid. *Arch Biochem Biophys* 1998;**349**:281–9.
34. Levine M, Conry-Cantilena C, Wang Y, Welch RW, Washko PW, Dhariwal KR, et al. Vitamin C pharmacokinetics in healthy volunteers: evidence for a recommended dietary allowance. *Proc Natl Acad Sci USA* 1996;**93**:3704–9.
35. Loria CM, Whelton PK, Caulfield LE, Szklo M, Klag MJ. Agreement among indicators of vitamin C status. *Am J Epidemiol* 1998;**147**:587–96.
36. Block G, Mangels AR, Patterson BH, Levander OA, Norkus EP, Taylor PR. Body weight and prior depletion affect plasma ascorbate levels attained on identical vitamin C intake: a controlled-diet study. *J Am Coll Nutr* 1999;**18**:628–37.
37. Cahill LE, El-Sohemy A. Vitamin C transporter gene polymorphisms, dietary vitamin C and serum ascorbic acid. *J Nutrigenet Nutrigenomics* 2009;**2**:292–301.
38. Timpson NJ, Forouhi NG, Brion MJ, Harbord RM, Cook DG, Johnson P, et al. Genetic variation at the SLC23A1 locus is associated with circulating concentrations of L-ascorbic acid (vitamin C): evidence from 5 independent studies with >15,000 participants. *Am J Clin Nutr* 2010;**92**:375–82.

39. Dusinská M, Ficek A, Horská A, Raslová K, Petrovská H, Vallová B, et al. Glutathione S-transferase polymorphisms influence the level of oxidative DNA damage and antioxidant protection in humans. *Mutat Res* 2001;**482**:47–55.
40. Cahill LE, Fontaine-Bisson B, El-Sohemy A. Functional genetic variants of glutathione S-transferase protect against serum ascorbic acid deficiency. *Am J Clin Nutr* 2009;**90**:1411–7.
41. Block G, Shaikh N, Jensen CD, Volberg V, Holland N. Serum vitamin C and other biomarkers differ by genotype of phase 2 enzyme genes *GSTM1* and *GSTT1*. *Am J Clin Nutr* 2011; **94**:929–37.
42. Horska A, Mislanova C, Bonassi S, Ceppi M, Volkovova K, Dusinska M. Vitamin C levels in blood are influenced by polymorphisms in glutathione S-transferases. *Eur J Clin Nutr* 2011;**50**:437–46.
43. Langlois MR, Delanghe JR, De Buyzere ML, Bernard DR, Ouyang J. Effect of haptoglobin on the metabolism of vitamin C. *Am J Clin Nutr* 1997;**66**:606–10.
44. Lee DH, Folsom AR, Harnack L, Halliwell B, Jacobs Jr. DR. Does supplemental vitamin C increase cardiovascular disease risk in women with diabetes? *Am J Clin Nutr* 2004; **80**:1194–200.
45. Na N, Delanghe JR, Taes YEC, Torck M, Baeyens WRG, Ouyang J. Serum vitamin C concentration is influenced by haptoglobin polymorphism and iron status in Chinese. *Clin Chim Acta* 2006;**365**:319–24.
46. Cahill LE, El-Sohemy A. Haptoglobin genotype modifies the association between dietary vitamin C and serum ascorbic acid deficiency. *Am J Clin Nutr* 2010;**92**:1494–500.
47. Strange RC, Spiteri MA, Ramachandran S, Fryer AA. Glutathione-S-transferase family of enzymes. *Mutat Res* 2001;**482**:21–6.
48. Montecinos V, Guzman P, Barra V, Villagran M, Munoz-Montesino C, Sotomayor K, et al. Vitamin C is an essential antioxidant that enhances survival of oxidatively stressed human vascular endothelial cells in the presence of a vast molar excess of glutathione. *J Biol Chem* 2007;**282**:15506–15.
49. Xu SJ, Wang YP, Roe B, Pearson WR. Characterization of the human class Mu glutathione S-transferase gene cluster and the GSTM1 deletion. *J Biol Chem* 1998;**273**:3517–27.
50. Bruhn C, Brockmöller J, Kerb R, Roots I, Borchert HH. Concordance between enzyme activity and genotype of glutathione S-transferase theta (GSTT1). *Biochem Pharmacol* 1998; **56**:1189–93.
51. Pemble S, Schroeder KR, Spencer SR, Meyer DJ, Hallier E, Bolt HM, et al. Human glutathione S-transferase Theta (GSTT1): cDNA cloning and the characterization of a genetic polymorphism. *Biochem J* 1994;**300**:271–6.
52. Zimniak P, Nanduri B, Pikula S, Bandorowicz-Pikula J, Singhal S, Srivastava SK, et al. Naturally occurring human glutathione S-transferase GSTP1-1 isoforms with isoleucine and valine in position 104 differ in enzymic properties. *Eur J Biochem* 1994;**224**:893–9.
53. Ali-Osman F, Akande O, Antoun G, Mao JX, Buolamwini J. Molecular cloning, characterization, and expression in Escherichia coli of full-length cDNAs of three human glutathione S-transferase Pi gene variants: evidence for differential catalytic activity of the encoded proteins. *J Biol Chem* 1997;**272**:10004–12.
54. Gutteridge JMC. The antioxidant activity of haptoglobin towards haemoglobin-stimulated lipid peroxidation. *Biochim Biophys Acta* 1987;**917**:219–23.
55. Asleh R, Levy AP. In vivo and in vitro studies establishing haptoglobin as a major susceptibility gene for diabetic vascular disease. *Vasc Health Risk Manag* 2005;**1**:19–28.
56. Carter K, Worwood M. Haptoglobin: a review of the major allele frequencies worldwide and their association with diseases. *Int J Lab Hematol* 2007;**29**:92–110.
57. Langlois MR, Delanghe JR. Biological and clinical significance of haptoglobin polymorphism in humans. *Clin Chem* 1996;**42**:1589–600.

58. Delanghe JR, Langlois MR, De Buyzere ML, Torck MA. Vitamin C deficiency and scurvy are not only a dietary problem but are codetermined by the haptoglobin polymorphism. *Clin Chem* 2007;**53**:1397–400.
59. Lee YW, Min WK, Chun S, Lee W, Park H, Lee YK, et al. Lack of association between oxidized LDL-cholesterol concentrations and haptoglobin phenotypes in healthy subjects. *Ann Clin Biochem* 2004;**41**:485–7.
60. Jacob RA. Assessment of human vitamin C status. *J Nutr* 1990;**120**:1480–5.
61. Cahill L, Corey PN, El-Sohemy A. Vitamin C deficiency in a population of young canadian adults. *Am J Epidemiol* 2009;**170**:464–71.
62. Schleicher RL, Carroll MD, Ford ES, Lacher DA. Serum vitamin C and the prevalence of vitamin C deficiency in the United States: 2003–2004 National Health and Nutrition Examination Survey (NHANES). *Am J Clin Nutr* 2009;**90**:1252–63.
63. Wrieden WL, Hannah MK, Bolton-Smith C, Tavendale R, Morrison C, Tunstall-Pedoe H. Plasma vitamin C and food choice in the third Glasgow MONICA population survey. *J Epidemiol Community Health* 2000;**54**:355–60.
64. Mosdøl A, Erens B, Brunner EJ. Estimated prevalence and predictors of vitamin C deficiency within UK's low-income population. *J Public Health (Oxf)* 2008;**30**:456–60.
65. Schneider C. Chemistry and biology of vitamin E. *Mol Nutr Food Res* 2005;**49**:7–30.
66. Singh U, Devaraj S, Jialal I. Vitamin E, oxidative stress, and inflammation. *Annu Rev Nutr* 2005;**25**:151–74.
67. Reboul E, Klein A, Bietrix F, Gleize B, Malezet-Desmoulins C, Schneider M, et al. Scavenger receptor class B type I (SR-BI) is involved in vitamin E transport across the enterocyte. *J Biol Chem* 2006;**281**:4739–45.
68. Anwar K, Kayden HJ, Hussain MM. Transport of vitamin E by differentiated Caco-2 cells. *J Lipid Res* 2006;**47**:1261–73.
69. Anwar K, Iqbal J, Hussain MM. Mechanisms involved in vitamin E transport by primary enterocytes and in vivo absorption. *J Lipid Res* 2007;**48**:2028–38.
70. Hosomi A, Arita M, Sato Y, Kiyose C, Ueda T, Igarashi O, et al. Affinity for α-tocopherol transfer protein as a determinant of the biological activities of vitamin E analogs. *FEBS Lett* 1997;**409**:105–8.
71. Brigelius-Flohé R. Vitamin E and drug metabolism. *Biochem Biophys Res Commun* 2003;**305**:737–40.
72. Eggermont E. Recent advances in vitamin E metabolism and deficiency. *Eur J Pediatr* 2006;**165**:429–34.
73. Brigelius-Flohé R. Bioactivity of vitamin E. *Nutr Res Rev* 2006;**19**:174–86.
74. Zimmer S, Stocker A, Sarbolouki MN, Spycher SE, Sassoon J, Azzi A. A novel human tocopherol-associated protein: cloning, in vitro expression, and characterization. *J Biol Chem* 2000;**275**:25672–80.
75. Zingg JM, Kempna P, Paris M, Reiter E, Villacorta L, Cipollone R, et al. Characterization of three human sec14p-like proteins: alpha-tocopherol transport activity and expression pattern in tissues. *Biochimie* 2008;**90**:1703–15.
76. Wang X, Quinn PJ. The location and function of vitamin E in membranes (review). *Mol Membr Biol* 2000;**17**:143–56.
77. Traber MG. Vitamin E regulation. *Curr Opin Gastroenterol* 2005;**21**:223–7.
78. Thomas SR, Stocker R. Molecular action of vitamin E in lipoprotein oxidation: implications for atherosclerosis. *Free Radic Biol Med* 2000;**28**:1795–805.
79. Burton GW, Traber MG. Vitamin E: antioxidant activity, biokinetics, and bioavailability. *Annu Rev Nutr* 1990;**10**:357–82.
80. Kamal-Eldin A, Appelqvist LÅ. The chemistry and antioxidant properties of tocopherols and tocotrienols. *Lipids* 1996;**31**:671–701.

81. Vatassery GT, Smith WE, Quach HT. Ascorbic acid, glutathione and synthetic antioxidants prevent the oxidation of vitamin E in platelets. *Lipids* 1989;**24**:1043–7.
82. Kagan VE, Serbinova EA, Forte T, Scita G, Packer L. Recycling of vitamin E in human low density lipoproteins. *J Lipid Res* 1992;**33**:385–97.
83. Shi H, Noguchi N, Niki E. Comparative study on dynamics of antioxidative action of alpha-tocopheryl hydroquinone, ubiquinol, and alpha-tocopherol against lipid peroxidation. *Free Radic Biol Med* 1999;**27**:334–46.
84. Morrissey PA, Sheehy PJA. Optimal nutrition: vitamin E. *Proc Nutr Soc* 1999;**58**:459–68.
85. Tangney CC, Shekelle RB, Raynor W, Gale M, Betz EP. Intra- and interindividual variation in measurements of beta-carotene, retinol, and tocopherols in diet and plasma. *Am J Clin Nutr* 1987;**45**:764–9.
86. Roxborough HE, Burton GW, Kelly FJ. Inter- and intra-individual variation in plasma and red blood cell vitamin E after supplementation. *Free Radic Res* 2000;**33**:437–45.
87. El-Sohemy A, Baylin A, Ascherio A, Kabagambe E, Spiegelman D, Campos H. Population-based study of α- and γ-tocopherol in plasma and adipose tissue as biomarkers of intake in Costa Rican adults. *Am J Clin Nutr* 2001;**74**:356–63.
88. Gueguen S, Leroy P, Gueguen R, Siest G, Visvikis S, Herbeth B. Genetic and environmental contributions to serum retinol and α-tocopherol concentrations: the Stanislas Family Study. *Am J Clin Nutr* 2005;**81**:1034–44.
89. Ouahchi K, Arita M, Kayden H, Hentati F, Hamida MB, Sokol R, et al. Ataxia with isolated vitamin E deficiency is caused by mutations in the α-tocopherol transfer protein. *Nat Genet* 1995;**9**:141–5.
90. Gotoda T, Arita M, Arai H, Inoue K, Yokota T, Fukuo Y, et al. Adult-onset spinocerebellar dysfunction caused by a mutation in the gene for the α-tocopherol-transfer protein. *N Engl J Med* 1995;**333**:1313–8.
91. Hentati A, Deng HX, Hung WY, Nayer M, Said Ahmed M, He X, et al. Human α-tocopherol transfer protein: gene structure and mutations in familial vitamin E deficiency. *Ann Neurol* 1996;**39**:295–300.
92. Cavalier L, Ouahchi K, Kayden HJ, Di Donato S, Reutenauer L, Mandel JL, et al. Ataxia with isolated vitamin E deficiency: heterogeneity of mutations and phenotypic variability in a large number of families. *Am J Hum Genet* 1998;**62**:301–10.
93. Wright ME, Peters U, Gunter MJ, Moore SC, Lawson KA, Yeager M, et al. Association of variants in two vitamin e transport genes with circulating vitamin e concentrations and prostate cancer risk. *Cancer Res* 2009;**69**:1429–38.
94. Major JM, Yu K, Wheeler W, Zhang H, Cornelis MC, Wright ME, et al. Genome-wide association study identifies common variants associated with circulating vitamin E levels. *Hum Mol Genet* 2011;**20**:3876–83.
95. Ferrucci L, Perry JRB, Matteini A, Perola M, Tanaka T, Silander K, et al. Common variation in the β-carotene 15,15′-monooxygenase 1 gene affects circulating levels of carotenoids: a genome-wide association study. *Am J Hum Genet* 2008;**84**:123–33.
96. Borel P, Moussa M, Reboul E, Lyan B, Defoort C, Vincent-Baudry S, et al. Human plasma levels of vitamin E and carotenoids are associated with genetic polymorphisms in genes involved in lipid metabolism. *J Nutr* 2007;**137**:2653–9.
97. Ortega H, Castilla P, Gómez-Coronado D, Garcés C, Benavente M, Rodríguez-Artalejo F, et al. Influence of apolipoprotein E genotype on fat-soluble plasma antioxidants in Spanish children. *Am J Clin Nutr* 2005;**81**:624–32.
98. Borel P, Moussa M, Reboul E, Lyan B, Defoort C, Vincent-Baudry S, et al. Human fasting plasma concentrations of vitamin E and carotenoids, and their association with genetic variants in apo C-III, cholesteryl ester transfer protein, hepatic lipase, intestinal fatty acid binding protein and microsomal triacylglycerol transfer protein. *Br J Nutr* 2009;**101**:680–7.

99. Girona J, Guardiola M, Cabre A, Manzanares JM, Heras M, Ribalta J, et al. The apolipoprotein A5 gene -1131T->C polymorphism affects vitamin E plasma concentrations in type 2 diabetic patients. *Clin Chem Lab Med* 2008;**46**:453–7.
100. Sundl I, Guardiola M, Khoschsorur G, Sola R, Vallve JC, Godas G, et al. Increased concentrations of circulating vitamin E in carriers of the apolipoprotein A5 gene 1131T>C variant and associations with plasma lipids and lipid peroxidation. *J Lipid Res* 2007;**48**:2506–13.
101. Henriksson P, Stamberger M, Eriksson M, Rudling M, Diczfalusy U, Berglund L, et al. Oestrogen-induced changes in lipoprotein metabolism: role in prevention of atherosclerosis in the cholesterol-fed rabbit. *Eur J Clin Invest* 1989;**19**:395–403.
102. Reboul E, Abou L, Mikail C, Ghiringhelli O, Andre M, Portugal H, et al. Lutein transport by Caco-2 TC-7 cells occurs partly by a facilitated process involving the scavenger receptor class B type I (SR-BI). *Biochem J* 2005;**387**:455–61.
103. Van Bennekum A, Werder M, Thuahnai ST, Han CH, Duong P, Williams DL, et al. Class B scavenger receptor-mediated intestinal absorption of dietary β-carotene and cholesterol. *Biochemistry* 2005;**44**:4517–25.
104. During A, Dawson HD, Harrison EH. Carotenoid transport is decreased and expression of the lipid transporters SR-BI, NPC1L1, and ABCA1 is downregulated in caco-2 cells treated with ezetimibe. *J Nutr* 2005;**135**:2305–12.
105. During A, Harrison EH. Mechanisms of provitamin A (carotenoid) and vitamin A (retinol) transport into and out of intestinal Caco-2 cells. *J Lipid Res* 2007;**48**:2283–94.
106. Moussa M, Landrier JF, Reboul E, Ghiringhelli O, Comera C, Collet X, et al. Lycopene absorption in human intestinal cells and in mice involves scavenger receptor class B type I but not Niemann-Pick C1-like. *J Nutr* 2008;**138**:1432–6.
107. Sontag TJ, Parker RS. Cytochrome P450 ω-hydroxylase pathway of tocopherol catabolism: novel mechanism of regulation of vitamin E status. *J Biol Chem* 2002;**277**:25290–6.
108. Hardwick JP. Cytochrome P450 omega hydroxylase (CYP4) function in fatty acid metabolism and metabolic diseases. *Biochem Pharmacol* 2008;**75**:2263–75.
109. Milman U, Blum S, Shapira C, Aronson D, Miller-Lotan R, Anbinder Y, et al. Vitamin E supplementation reduces cardiovascular events in a subgroup of middle-aged individuals with both type 2 diabetes mellitus and the haptoglobin 2-2 genotype: a prospective double-blinded clinical trial. *Arterioscler Thromb Vasc Biol* 2008;**28**:341–7.
110. Asleh R, Levy AP. Divergent effects of α-tocopherol and vitamin c on the generation of dysfunctional HDL associated with diabetes and the Hp 2-2 genotype. *Antioxid Redox Signal* 2010;**12**:209–18.
111. Farbstein D, Blum S, Pollak M, Asaf R, Viener HL, Lache O. Vitamin E therapy results in a reduction in HDL function in individuals with diabetes and the haptoglobin 2-1 genotype. *Atherosclerosis* 2011;**219**:240–4.
112. Cheng TY, Barnett MJ, Kristal AR, Ambrosone CB, King IB, Thornquist MD, et al. Genetic variation in myeloperoxidase modifies the association of serum alpha-tocopherol with aggressive prostate cancer among current smokers. *J Nutr* 2011;**141**:1731–7.
113. Olson JA, Krinsky NI. Introduction: the colorful, fascinating world of the carotenoids: important physiologic modulators. *FASEB J* 1995;**9**:1547–50.
114. Yonekura L, Nagao A. Intestinal absorption of dietary carotenoids. *Mol Nutr Food Res* 2007; **51**:107–15.
115. Britton G. Structure and properties of carotenoids in relation to function. *FASEB J* 1995; **9**:1551–8.
116. Johnson LJ, Meacham SL, Kruskall LJ. The antioxidants—vitamin C, vitamin E, selenium, and carotenoids. *J Agromedicine* 2003;**9**:65–82.

117. Handelman GJ, Nightingale ZD, Lichtenstein AH, Schaefer EJ, Blumberg JB. Lutein and zeaxanthin concentrations in plasma after dietary supplementation with egg yolk. *Am J Clin Nutr* 1999;**70**:247–51.
118. Chung HY, Rasmussen HM, Johnson EJ. Lutein bioavailability is higher from lutein-enriched eggs than from supplements and spinach in men. *J Nutr* 2004;**134**:1887–93.
119. Stahl W, Sies H. Uptake of lycopene and its geometrical isomers is greater from heat-processed than from unprocessed tomato juice in humans. *J Nutr* 1992;**122**:2161–6.
120. Gartner C, Stahl W, Sies H. Lycopene is more bioavailable from tomato paste than from fresh tomatoes. *Am J Clin Nutr* 1997;**66**:116–22.
121. Rock CL, Lovalvo JL, Emenhiser C, Ruffin MT, Flatt SW, Schwartz SJ. Bioavailability of β-carotene is lower in raw than in processed carrots and spinach in women. *J Nutr* 1998; **128**:913–6.
122. Böhm V, Bitsch R. Intestinal absorption of lycopene from different matrices and interactions to other carotenoids, the lipid status, and the antioxidant capacity of human plasma. *Eur J Nutr* 1999;**38**:118–25.
123. Parker RS. Absorption, metabolism, and transport of carotenoids. *FASEB J* 1996; **10**:542–51.
124. Bone RA, Landrum JT, Tarsis SL. Preliminary identification of the human macular pigment. *Vision Res* 1985;**25**:1531–5.
125. Handelman GJ, Dratz EA, Reay CC, Van Kuijk FJGM. Carotenoids in the human macula and whole retina. *Invest Ophthalmol Vis Sci* 1988;**29**:850–5.
126. Beatty S, Koh HH, Phil M, Henson D, Boulton M. The role of oxidative stress in the pathogenesis of age-related macular degeneration. *Surv Ophthalmol* 2000;**45**:115–34.
127. Ma L, Lin XM. Effects of lutein and zeaxanthin on aspects of eye health. *J Sci Food Agric* 2010;**90**:2–12.
128. Krinsky NI, Johnson EJ. Carotenoid actions and their relation to health and disease. *Mol Aspects Med* 2005;**26**:459–516.
129. Young AJ, Lowe GM. Antioxidant and prooxidant properties of carotenoids. *Arch Biochem Biophys* 2001;**385**:20–7.
130. Palozza P, Krinsky NI. β-Carotene and α-tocopherol are synergistic antioxidants. *Arch Biochem Biophys* 1992;**297**:184–7.
131. Niki E, Noguchi N, Tsuchihashi H, Gotoh N. Interaction among vitamin C, vitamin E, and β-carotene. *Am J Clin Nutr* 1995;**62**:1322S–1326S.
132. Astley SB, Elliott RM, Archer DB, Southon S. Evidence that dietary supplementation with carotenoids and carotenoid-rich foods modulates the DNA damage: repair balance in human lymphocytes. *Br J Nutr* 2004;**91**:63–72.
133. Zhao X, Aldini G, Johnson EJ, Rasmussen H, Kraemer K, Woolf H, et al. Modification of lymphocyte DNA damage by carotenoid supplementation in postmenopausal women. *Am J Clin Nutr* 2006;**83**:163–9.
134. Stahl W, Schwarz W, Sundquist AR, Sies H. cis-trans isomers of lycopene and β-carotene in human serum and tissues. *Arch Biochem Biophys* 1992;**294**:173–7.
135. Carughi A, Hooper FG. Plasma carotenoid concentrations before and after supplementation with a carotenoid mixture. *Am J Clin Nutr* 1994;**59**:896–9.
136. Borel P, Grolier P, Mekki N, Boirie Y, Rochette Y, Le Roy B, et al. Low and high responders to pharmacological doses of β-carotene: proportion in the population, mechanisms involved and consequences on β-carotene metabolism. *J Lipid Res* 1998;**39**:2250–60.
137. Leung WC, Hessel S, Méplan C, Flint J, Oberhauser V, Tourniaire F, et al. Two common single nucleotide polymorphisms in the gene encoding β-carotene 15,15′-monoxygenase alter β-carotene metabolism in female volunteers. *FASEB J* 2009;**23**:1041–53.

138. Wahlqvist ML, Wattanapenpaiboon N, Macrae FA, Lambert JR, MacLennan R, Hsu-Hag BHH, et al. Changes in serum carotenoids in subjects with colorectal adenomas after 24 mo of β-carotene supplementation. *Am J Clin Nutr* 1994;**60**:936–43.
139. Kostic D, White WS, Olson JA. Intestinal absorption, serum clearance, and interactions between lutein and β-carotene when administered to human adults in separate or combined oral doses. *Am J Clin Nutr* 1995;**62**:604–10.
140. Van den Berg H, Van Vliet T. Effect of simultaneous, single oral doses of β-carotene with lutein or lycopene on the β-carotene and retinyl ester responses in the triacylglycerol-rich lipoprotein fraction of men. *Am J Clin Nutr* 1998;**68**:82–9.
141. Borel P, De Edelenyi FS, Vincent-Baudry S, Malezet-Desmoulin C, Margotat A, Lyan B, et al. Genetic variants in BCMO1 and CD36 are associated with plasma lutein concentrations and macular pigment optical density in humans. *Ann Med* 2011;**43**:47–59.
142. Mackness MI, Arrol S, Abbott C, Durrington PN. Protection of low-density lipoprotein against oxidative modification by high-density lipoprotein associated paraoxonase. *Atherosclerosis* 1993;**104**:129–35.
143. Mackness MI, Arrol S, Mackness B, Durrington PN. Alloenzymes of paraoxonase and effectiveness of high-density lipoproteins in protecting low-density lipoprotein against lipid peroxidation. *Lancet* 1997;**349**:851–2.
144. Jaouad L, Milochevitch C, Khalil A. PON1 paraoxonase activity is reduced during HDL oxidation and is an indicator of HDL antioxidant capacity. *Free Radic Res* 2003;**37**:77–83.
145. MacKness B, MacKness MI, Arrol S, Turkie W, Julier K, Abuasha B, et al. Serum paraoxonase (PON1) 55 and 192 polymorphism and paraoxonase activity and concentration in non-insulin dependent diabetes mellitus. *Atherosclerosis* 1998;**139**:341–9.
146. Mackness B, Mackness MI, Durrington PN, Arrol S, Evans AE, McMaster D, et al. Paraoxonase activity in two healthy populations with differing rates of coronary heart disease. *Eur J Clin Invest* 2000;**30**:4–10.
147. Rantala M, Silaste ML, Tuominen A, Kaikkonen J, Salonen JT, Alfthan G, et al. Dietary modifications and gene polymorphisms alter serum paraoxonase activity in healthy women. *J Nutr* 2002;**132**:3012–7.
148. MacKinnon ES, El-Sohemy A, Rao AV, Rao LG. Paraoxonase 1 polymorphisms 172T>A and 584A>G modify the association between serum concentrations of the antioxidant lycopene and bone turnover markers and oxidative stress parameters in women 25–70 years of age. *J Nutrigenet Nutrigenomics* 2010;**3**:1–8.
149. Gandini S, Merzenich H, Robertson C, Boyle P. Meta-analysis of studies on breast cancer risk and dietthe role of fruit and vegetable consumption and the intake of associated micronutrients. *Eur J Cancer* 2000;**36**:636–46.
150. Aune D, Lau R, Chan DSM, Vieira R, Greenwood DC, Kampman E, et al. Nonlinear reduction in risk for colorectal cancer by fruit and vegetable intake based on meta-analysis of prospective studies. *Gastroenterology* 2011;**141**:106–18.
151. He FJ, Nowson CA, MacGregor GA. Fruit and vegetable consumption and stroke: meta-analysis of cohort studies. *Lancet* 2006;**367**:320–6.
152. Dauchet L, Amouyel P, Hercberg S, Dallongeville J. Fruit and vegetable consumption and risk of coronary heart disease: a meta-analysis of cohort studies. *J Nutr* 2006;**136**:2588–93.
153. Vivekananthan DP, Penn MS, Sapp SK, Hsu A, Topol EJ. Use of antioxidant vitamins for the prevention of cardiovascular disease: meta-analysis of randomised trials. *Lancet* 2003;**361**:2017–23.
154. Bleys J, Miller Iii ER, Pastor-Barriuso R, Appel LJ, Guallar E. Vitamin-mineral supplementation and the progression of atherosclerosis: a meta-analysis of randomized controlled trials. *Am J Clin Nutr* 2006;**84**:880–7.

155. Huang HY, Caballero B, Chang S, Alberg AJ, Semba RD, Schneyer CR, et al. The efficacy and safety of multivitamin and mineral supplement use to prevent cancer and chronic disease in adults: a systematic review for a National Institutes of Health state-of-the-science conference. *Ann Intern Med* 2006;**145**:372–85.
156. Chong EWT, Wong TY, Kreis AJ, Simpson JA, Guymer RH. Dietary antioxidants and primary prevention of age related macular degeneration: systematic review and meta-analysis. *Br Med J* 2007;**335**:755–9.
157. Gallicchio L, Boyd K, Matanoski G, Tao X, Chen L, Lam TK, et al. Carotenoids and the risk of developing lung cancer: a systematic review. *Am J Clin Nutr* 2008;**88**:372–83.
158. Jiang L, Yang KH, Tian JH, Guan QL, Yao N, Cao N, et al. Efficacy of antioxidant vitamins and selenium supplement in prostate cancer prevention: a meta-analysis of randomized controlled trials. *Nutr Cancer* 2010;**62**:719–27.
159. Bjelakovic G, Nikolova D, Simonetti RG, Gluud C. Antioxidant supplements for prevention of gastrointestinal cancers: a systematic review and meta-analysis. *Lancet* 2004;**364**:1219–28.
160. Miller 3rd ER, Pastor-Barriuso R, Dalal D, Riemersma RA, Appel LJ, Guallar E. Meta-analysis: high-dosage vitamin E supplementation may increase all-cause mortality. *Ann Intern Med* 2005;**142**:37–46 + I-40.
161. Bjelakovic G, Nikolova D, Gluud LL, Simonetti RG, Gluud C. Mortality in randomized trials of antioxidant supplements for primary and secondary prevention: systematic review and meta-analysis. *J Am Med Assoc* 2007;**297**:842–57.

Mineral Intake

Maria G. Stathopoulou,[°] Stavroula Kanoni,[°,†] George Papanikolaou,[°] Smaragdi Antonopoulou,[°] Tzortzis Nomikos,[°] and George Dedoussis[°]

[°]*Department of Dietetics and Nutritional Science, Harokopio University of Athens, Athens, Greece*

[†]*Genetics of complex traits in humans (Team 147), Wellcome Trust Sanger Institute, Hinxton, Cambridge, UK*

Progress in Molecular Biology
and Translational Science, Vol. 108
DOI: 10.1016/B978-0-12-398397-8.00009-5

1877-1173/12 $35.00

Minerals play a key role in the regulation of metabolic and physiological pathways. Adequate intake is required to maintain homeostasis, cell protection, functionality, and health, while deficiencies are associated with specific illnesses. Among the minerals, calcium, copper, iron, selenium, and zinc are considered especially important because of their physiological roles and their participation in a variety of biological processes. Also, these elements are associated with genetic diseases and are known to interact with genetic variants in a wide range of diseases.

I. Introduction

Minerals in nutrition are inorganic elements essential for human health. Their biological roles are wide and include serving as enzyme cofactors, stabilizers of organic molecules, structural components of body tissues such as bones, second messengers, regulators of the acid–base balance, participants in redox reactions, and participants in the preservation of cellular pH and electrical gradients. Therefore, they play a key role in the regulation of metabolic and physiological pathways. Adequate intake is required to maintain homeostasis, cell protection, functionality, and health, while deficiencies are associated with specific illnesses.

In recent decades, minerals have assumed great public health importance. Consequently, considerable research has been carried out concerning their metabolism, physiological role, requirements, and health consequences from deficiency or toxicity. The fields of biochemistry, cell biology, and molecular genetics are being widely used for the elucidation of several aspects of mineral metabolism. In particular, the genetics approach is expected to increase our understanding of mineral biology. Natural variation in the genome could influence nutritionally relevant phenotypes, such as tissue mineral levels or mineral metabolism. This concept has led to the investigation of interactions between genetic variants and mineral intake and metabolism. This research aims to define and establish the principles of personalized nutrition, based on individuals' genetic variations.

The scope of this chapter is to review the most recent knowledge concerning minerals' biological roles, metabolism, and genetics. The progress made in the field of gene–minerals interactions is extensively discussed. The review focuses on calcium, copper, iron, selenium, and zinc, which are considered especially important because of their physiological roles and their participation in a variety of biological processes. Also, these elements are associated with genetic diseases and are known to interact with genetic variants in a wide range of diseases.

II. Calcium (Ca)

A. Dietary Sources

The basic dietary sources for calcium are milk and other dairy products (yogurt and cheese). Additional food sources include certain fish (salmon and small fish consumed with bones such as sardines), seafood (clams and oysters), vegetables (turnip and mustard greens, broccoli, cauliflower, and kale), legumes and legume products (tofu), and dried fruits. Also, a variety of enriched foods (fruit juices and bread) can provide significant amounts of calcium.[1] Dietary reference intakes for calcium recommend the following: children 1–3 years, 500 mg/day; children 4–8 years, 800 mg/day; adolescents, 1100 mg/day; adults 19–50 years, 800 mg/day; men 51–70, 800 mg/day; and women 51 to >70 and men >70 years, 1000 mg/day.[2] However, given the significant role of calcium in the maintenance of bone health, as well as in the prevention of chronic diseases such as cardiovascular disease (CVD) and some types of cancer,[3–5] recent data suggest that intakes should be higher (1200–1500 mg/day) for men >51 years and postmenopausal women.[1]

B. Biological Roles

The mineralization of bone is calcium's most important action. Apart from the formation of skeleton, calcium is essential for several significant physiological processes, including nerve impulse transmission, muscle contraction, regulation of blood clotting and pressure, enzyme regulation, membrane permeability, and cellular metabolism.[1,4,5]

C. Metabolism

Calcium is the most abundant divalent cation of the body. It is stored almost exclusively in bone and teeth as hydroxyapatite (99%) and the rest is distributed in intra- and extracellular fluids.[1,4] Vitamin D, along with the combined actions of parathormone and calcitonin, plays a determinant role in calcium absorption and homeostasis. This vitamin acts through its receptor (vitamin D receptor, VDR) in the intestines, bone cells, parathyroid glands, and kidneys to regulate blood calcium levels.[6–8]

D. Calcium, Genetics, and Human Diseases

1. Calcium and Bone Phenotypes

Numerous studies have shown interactions of calcium intake with single-nucleotide polymorphisms (SNPs) of the *VDR* gene. This gene encodes a nuclear receptor of the steroid hormone receptor family.[9] It is the gene that actually initiated the study of the genetics of osteoporosis.[10] The first report on this gene was published in 1995, in a study of calcium supplementation

(500 mg/day) in Caucasian postmenopausal women.[11] This study demonstrated that in the placebo group, the BB genotype (minor allele) of the BsmI polymorphism was associated with increased bone loss, while in the intervention group, all genotypes had decreased rates of bone loss, similar for all genotypes.[11] Although the sample size of the particular work was limited, the results showed that homozygous subjects for the minor allele of the BsmI SNP had better clinical response with higher calcium intake levels.

In another study of calcium supplementation (800 mg/day for 18 months) in a small sample of elderly Caucasians, only the Bb genotype of the BsmI SNP was associated with differences in bone mineral density (BMD) changes, while for the other genotypes, calcium intake did not seem to play a significant role.[12] Salamone and colleagues showed that femoral neck BMD in premenopausal Caucasian women was significantly higher in carriers of the B allele of the BsmI SNP with higher calcium intake (>1036 mg/day), while the BMD of the bb genotype was not modified according to calcium intake.[13] In a large population of elderly individuals, the bb genotype of the same SNP was associated with higher trochanter BMD compared to the BB genotype only in those with calcium intake >800 mg/day, while this association was reversed in intake <500 mg/day.[14] In a 1-year intervention study of foods enriched with calcium in prepubertal girls, carriers of the B allele of the BsmI SNP had initially lower BMD compared to the bb genotype; however, they responded more favorably in the intervention. In contrast, the BMD of bb subjects was not modified by the calcium-enriched foods.[15]

Apart from the BsmI SNP, other *VDR* polymorphisms seem to interact with calcium intake in bone phenotypes. In a longitudinal study (6 years) of postmenopausal Caucasian women, the rate of bone loss in the hip was higher for the carriers of the t allele (minor allele) of the TaqI SNP compared to TT genotype individuals only in those with low calcium intake. No difference was observed in the BMD between genotypes with higher calcium intake levels.[16] The most recent results concern the Cdx-2 SNP of the *VDR* gene. In particular, Stathopoulou and colleagues[17] demonstrated that the minor allele A was associated with lower lumbar spine BMD in Caucasian postmenopausal women only in the lower calcium intake group (<680 mg/day). Furthermore, the B allele of the BsmI SNP and the t allele of the TaqI SNP were associated with osteoporosis in the same group. In contrast, in the higher calcium intake group, none of the SNPs was associated with osteoporosis or BMD.[17] In another study, Fang and colleagues[18] showed a suggestive protective effect of the A allele of the Cdx-2 SNP over the BMD of a large number of adults (>55 years) in those with low calcium intake only (<600 mg/day), but the difference did not reach statistical significance. Also, another study showed a trend of calcium intake*FokI SNP, such that ff individuals (girls and premenopausal women) with higher calcium intake had lower lumbar spine and femoral neck BMD (marginally significant) and FF genotype girls had better response to calcium supplementation concerning BMD increase.[19]

The rs4516035 SNP in the promoter of the *VDR* gene has been shown to affect serum calcium levels only in the (pediatric) population with lower calcium intake (493 mg/day).[20] However, there have been some studies that did not identify interactions between calcium intake and *VDR* SNPs.[21,22] The discrepancy between studies may be due to the different levels of calcium intake being examined in each population[14]; therefore, the assessment of these interactions should preferably be performed in each population based on defined intake levels.

Apart from *VDR*, other genes have been shown to interact with calcium intake in the determination of bone phenotypes. Recently, the rs4988321 SNP of the low-density lipoprotein receptor-related protein 5 (*LRP5*) gene was shown to affect BMD on the basis of calcium intake levels in postmenopausal women. In particular, those with the A allele demonstrated significantly lower lumbar spine BMD only with lower calcium intake (<680 mg/day), whereas with higher intake, there were no differences in BMD between genotypes.[23] In a study by Fang and colleagues[24] in elderly Caucasians, the homozygous individuals for haplotype 1 of the D site of albumin promoter binding protein (*DBP*) gene had a 47% increased risk for clinical fractures compared to non-carriers only in the low calcium intake (< 1090 mg/day) group, while no statistical interaction was found for BMD and serum vitamin D levels.[24] The interleukin 6 (*IL6*) gene has also been implicated in interactions with calcium intake. In a pre-menarche Chinese population, the G allele of the -634C/G SNP was associated with lower total-body bone mineral content compared to CC individuals only in the group with lower calcium intake (<460 mg/day), but not in those with higher intake.[25] However, the study sample was relatively small. Also, the *IL6* -174G/C SNP was associated with lower hip Ward's area BMD in GG individuals in the group with lower calcium intake (<941 mg/day).[26]

2. Calcium and Cancer

In a case–control study concerning renal cell carcinoma, two SNPs (rs3118538 and rs10776909) of the retinoid X receptor (*RXR*) gene were assessed for possible interaction with consumption of calcium-rich foods. For both SNPs, the study demonstrated a positive association between increased renal cell carcinoma risk and increased frequency of yogurt and calcium-rich food consumption in the wild-type genotypes.[27] Interactions with calcium intake and the *VDR* gene have been demonstrated in a matched case–control study for breast cancer. The bb and TT genotypes for the BsmI and TaqI SNPs, respectively, had a decreased risk for breast cancer in those with higher calcium intake (>902 mg/day) compared to carriers of the minor alleles. A similar association was observed for the poly(A)LL tail of the poly(A) variation.[28] Finally, numerous studies have assessed possible interactions of calcium with genetic variants and the risk for colorectal cancer. In a Chinese case–control study, the f allele of the FokI SNP of *VDR* was weakly associated with increased risk in the group with calcium intake of <388 mg/day (borderline significance).[29] In another case–control study, the V variant of the D1822V SNP of the adenomatous polyposis coli (*APC*) gene was associated with decreased risk in cases with higher calcium intake.[30] Furthermore, the risk of colorectal adenoma was greater in carriers of the A allele of the Thr1482Ile SNP of the transient receptor potential cation channel, subfamily M, member 7 (*TRPM7*) gene in diets with higher calcium/magnesium intake. Similar results were observed for the risk of hyperplastic polyps.[31] However, there have been studies that showed no interactions between *VDR*, calcium-sensing receptor (*CASR*), or *DBP* genes and calcium intake in colorectal cancer.[32–35]

3. Calcium and Other Phenotypes

The sirtuin 1 (*SIRT1*) gene is associated with body mass index (BMI) and obesity. Recently, two SNPs (rs7895833 and rs1467568) of this gene have been shown to interact with calcium intake on BMI in a large population of elderly individuals. However, these results were not significant after Bonferroni correction for multiple testing.[36]

4. Animal Models and *In Vitro* Studies

In a diet-induced obesity mice model, a high-calcium diet for 21 weeks altered the expression of 129 genes in the adipose tissue, particularly those participating in the biological pathways of insulin and adipocytokine signaling and fatty acid metabolism.[37] A high-calcium diet for 14 days also altered the expression of 10 genes in the colon of rats, especially the mucosal pentraxin (*Mptx*) gene, which is colon specific and is associated with colon cell turnover and disease.[38] A comparative microarray approach in three different mouse models of colon cancer fed a Western diet showed that increasing dietary calcium and vitamin D can be effective in inhibiting tumor formation.[39] Also, low calcium was associated with alteration in gene expression in mice and human adenocarcinoma-derived Caco-2 cells on markers of inflammation, detoxification, and the vitamin D system (protection against tumorigenesis).[40]

III. Copper (Cu)

A. Dietary Sources

The primary source of copper in mammals is the diet, and ingested copper is absorbed from the intestinal lumen to the circulatory system through specific transporters. It is estimated that the efficiency of copper absorption in humans ranges from 12% to 60%[41] depending on copper intake, presence of dietary factors that may promote or inhibit its absorption, and the copper status of the individual. The consumption of foods rich in copper such as whole cereals, liver, oysters, cocoa products, nuts (particularly cashews), dried fruits, and legumes[42] is associated with retention of copper levels within the physiological range in healthy subjects. On the other hand, formulating nutritional recommendations for specific subgroups at risk of experiencing adverse effects from moderate copper deficiency or excess is a challenge that requires better knowledge of early changes associated with high and low copper intakes. Changes in food group intake may influence copper levels; for example, higher milk intake has been associated with increased copper levels in children,[43] and plasma levels of copper have been positively correlated with the percentage of energy provided by fat or carbohydrates and the intake of monounsaturated and polyunsaturated fats.[44] Factors such as season (copper concentration is higher in greener seasons), soil quality, geography, water source, and use of fertilizers influence the final copper content in food.[45]

Several studies have assessed the actual consumption of copper; most of them have been conducted in the United States.[46,47] The Third National Health and Nutrition Examination Survey (NHANES III) demonstrated that the mean intake of copper from food alone for men and women 19–70 years of age was 1.54–1.70 and 1.13–1.18 mg/day, respectively.[48] Comparable copper intakes have been reported for Europe, where different countries have reported mean dietary copper intakes ranging from 1.0 to 2.3 mg/day in adult men and 0.9–1.8 mg/day in women.[49] The EURRECA (EURopean micronutrient RECommendations Aligned) Network of Excellence (www.eurreca.org) is working toward the development of aligned recommendations across different countries. A protocol was required to assign resources to those micronutrients for which recommendations are most in need of alignment. In the context of EURRECA, various status markers used to assess intake, exposure, and body levels of each micronutrient were evaluated.[50]

B. Biological Roles

Copper is an essential trace element for all living organisms, as it functions as a cofactor for several enzymes. Copper's ability to exist in two oxidation states enables it to be an important cofactor for various enzymes that require redox chemistry for their function. These enzymes include Cu/Zn superoxide dismutase, which protects the cell against free radicals, and the cytochrome *C* oxidase complex, which is an essential component of the mitochondrial respiratory chain. However, copper is toxic when its concentration is too high because it generates excess amounts of reactive oxygen species (ROS), which damage proteins, lipids, and nucleic acids.[51]

C. Metabolism

Circulating copper is mainly transported to the liver, but to a lesser extent, it is transported to other organs and tissues as well. At the cellular level, a precise system of copper transport and delivery to intracellular proteins is necessary for normal physiology. Copper uptake across the cellular membrane is thought to most commonly depend on copper transporter 1 found in the membrane.[52,53] In the cytoplasm, free copper ions are toxic; thus copper is bound to metallothioneins (MTs) or copper-specific chaperone proteins. Up to 12 copper atoms can be bound by one molecule of MTs in a stable complex that appears to interchange with copper bound to glutathione.[54]

While copper's ability to shift between oxidation states contributes to its necessity, it also renders copper highly toxic when in excess because of the potential to generate ROS. Therefore, a tight regulation of copper homeostasis is required. Under normal homeostatic conditions in mammals, the uptake of copper through the intestines ensures that essential needs are met, whereas excretion through the bile prevents toxicity.[55]

In mammals, multiple proteins associated with copper metabolism are regulated by posttranslational mechanisms.[56] However, regulation also occurs at the transcriptional level. Expression of genes involved in the protection against copper overload involves specific responses, such as the induction of copper-binding MTs, and general stress responses, such as the oxidative stress response and proteasomal degradation.[57] ATPases ATP7A and ATP7B are copper transporters that sequester copper from the cytosol into the trans-Golgi network for loading onto copper-requiring enzymes.[58] ATP7A is expressed in the majority of tissues except of the liver, while ATP7B expression is found mainly in the liver but also in the kidney and placenta.[58,59] Under elevated copper levels in polarized cells, ATP7A relocates toward the basolateral plasma membranes, while ATP7B travels to the apical side of the membrane to export the metal from the cell. Recent studies have shown that COMMD1 (copper metabolism [Murr1] domain-containing protein 1), whose deficiency causes canine copper toxicosis,[60] is able to bind to the N-terminal copper-binding region of ATP7B but not that of ATP7A.[61] These data suggest that COMMD1 and ATP7B cooperate in the excretion of copper from the hepatocyte and that abolishment of the interaction between these proteins underlies the pathophysiology of canine copper toxicosis. Transient knockdown of *COMMD1* in HEK293 cells leads to increased cellular copper levels,[62] supporting the role of COMMD1 as a regulator of copper homeostasis.

Copper uptake in fish gills has been shown to be inhibited by elevated sodium levels[63] and vice versa, linking sodium transport to copper transport. The exact link between copper and sodium transport needs further characterization but has been postulated to involve the amiloride-sensitive epithelial sodium channel.[64] This channel constitutes the rate-limiting step for sodium reabsorption in epithelial cells that line the distal part of the renal tubule, the distal colon, the duct of several exocrine glands, and the lung.[65] Considering the critical role of COMMD1 in copper metabolism, we could speculate that regulation of amiloride-sensitive sodium channel activity by COMMD1 might provide the link between sodium and copper transport.

Physiological copper levels are regulated by balancing the rates of biliary copper excretion and dietary copper absorption. Dietary copper absorption requires active transport across the basolateral membrane.[66] In contrast, the biliary secretion of copper requires an active transport step across the apical membrane of hepatocytes.[67] One hypothesis for the differential expression of ATP7A and ATP7B in intestinal epithelial cells and hepatocytes, respectively, is to permit the trafficking of the copper pump to the appropriate membrane.

D. Copper, Genetics, and Human Diseases

1. Copper and Genetic Diseases

Various genetic diseases of copper metabolism are characterized by either depletion or accumulation of copper, underlining the importance of a tight regulation. Menkes disease[68] is characterized by a general copper deficiency due to malabsorption of dietary copper and results in early growth retardation, peculiar ("kinky") hair, and focal cerebral and cerebellar degeneration. The disease is caused by mutations in the *ATP7A* gene,[59,69,70] which encodes the copper-translocating P-type ATPase. Wilson disease is a copper-overload disorder[71] caused by mutations in the *ATP7B* gene.[72–74] In Wilson disease, copper accumulates in the liver and brain, causing extensive hepatic and neurological abnormalities. The highest prevalence of Wilson disease has been reported in a mountainous area of Crete,[75] underlying the necessity of a prevention program in certain areas. Other non-Wilsonian forms of hepatic copper-overload syndromes have been described,[76,77] including Indian childhood cirrhosis,[78] endemic Tyrolean infantile cirrhosis,[77] and sporadic cases occurring worldwide grouped together as idiopathic copper toxicosis.[66] These non-Wilsonian copper-overload disorders are fatal at an early age because of liver failure as a consequence of chronic liver cirrhosis. Many of these disorders can be classified as ecogenetic disorders, in the sense that both an excessive copper intake and a genetic defect underlie their pathology.[77,78]

Copper affects the fertilizing capacity of human spermatozoa by interfering with sperm migration, viability, and acrosomal reaction *in vitro*.[79] Homeobox (*HOX*) genes encode evolutionarily conserved transcription factors, which are essential to embryonic development, endometrial development, and endometrial receptivity.[80] Decreased endometrial HOXA10 expression is associated with use of the copper intrauterine device.[81]

2. Copper and Other Phenotypes

Recent results support the notion that metal-related abnormalities accompany cognitive deterioration, as suggested by the correlation between an increase in serum copper not bound to ceruloplasmin ("free" copper) and worsening on the Mini–Mental Status Examination.[82]

Copper, manganese, and selenium are required in small amounts as components of antioxidant enzymes; they are actively involved in protecting the body against oxidative stress.[83] Therefore, determining their plasma levels may contribute to improved assessment of the health and nutritional status of certain populations.

The mean plasma levels of copper, manganese, and selenium among the healthy adult population living in Andalusia were found to be similar to those measured among comparable populations in earlier studies and within the range of normality.[44] With greater age, the mean plasma values of copper tended to rise. In general, the anthropometric parameters measured did not significantly modify the plasma levels of the elements studied, except that there was a slight tendency for copper levels to decrease with rising BMI. Thus, the sedentary population presented plasma levels of copper that were slightly higher than those of the more active population.

Various factors have been proposed as possibly affecting plasma levels of copper, including age. It has been shown that there is a negative correlation between age and copper levels in both men and women.[43] Although copper deficiency is not common among humans, in a recent study, 4.4% (15 persons) of the total population (340 persons) had hypocupremia (plasma concentrations < 75 μg/dL).[44] This might be explained by the low copper content in the geographic area of Andalusia.[84] Among the study population, there was a tendency for obese persons to present lower plasma levels of copper. This could be related to the fact that, in this particularly obese population, there was a lower consumption of polyunsaturated fats.[85] Gender also influences plasma copper. Women presented significantly higher mean plasma copper levels compared to men in a study by Johnson.[86] This finding could be due to the fact that adult women (aged 20–59 years) demonstrate higher levels of absorption and thus a more rapid turnover of copper.[86] Furthermore, it has been reported that smoking increases plasma levels of copper and that these levels are positively correlated with CuZn–SOD activity. Both plasma copper and CuZn–SOD activities have been shown to increase in response to the chronic inflammation of the respiratory tract found in smokers.[83]

It was recently demonstrated that the zinc-finger domain of transcription factor Sp1 functions as a sensor of copper which regulates copper transporter 1 up and down in response to copper concentration variations.[87]

IV. Iron (Fe)

A. Dietary Sources

An average Western diet contains approximately 10–20 mg of iron per day; 90% of dietary iron is in the form of nonheme iron, which despite its high abundance is less bioavailable than heme iron, which represents approximately 40% of the total absorbed iron in populations with high meat intake. Absorption of nonheme iron is affected by iron inhibitors in the diet such as phytates, calcium, and tannins, while ascorbic acid, alcohol, and meat products increase the absorption. Gastric acidity enhances the chelation of iron to soluble, low-molecular-weight compounds (certain amines, amino acids, and sugars), preventing the formation of insoluble iron complexes. Coadministration of medicines (tetracyclines, proton-pump inhibitors, and antacids) decreases iron absorption, and prolonged achlorhydria leads to iron deficiency.

Daily iron requirements for adult men and nonmenstruating women are normally equal to obligatory daily iron losses, estimated at approximately 1 mg/day. Iron losses occur through sloughed enterocytes, occult blood, and shedding of skin cells. Menstrual iron losses vary among women, and the distribution of iron excretion was found to highly skewed for menstruating women, with a median of 1.58 mg/day.[88] In periods of rapid growth (e.g., during infancy, childhood, adolescence), iron requirements are especially high. Increased iron absorption compensates for the increase in erythrocyte mass in puberty and through the second and third trimester of pregnancy, when increased iron is required for both fetal growth and maternal red cells.

B. Biological Roles

Because of its chemical properties, iron is indispensable for living organisms. Iron can readily accept and donate electrons, functioning as an oxidant or reductant. Iron associates with proteins, binds to oxygen, and mediates catalytic reactions, acting as a cofactor in vital biochemical activities such as oxygen transport, energy production, and cellular proliferation. The most important chemical moieties containing iron are iron–sulfur clusters (ISC) and heme. Through the Haber–Weiss–Fenton reaction, iron catalyzes the production of ROS which can damage cellular macromolecules such as lipid membranes, nucleic acids, and proteins.

The potential toxicity of excess iron denotes the biological importance of accurate homeostatic mechanisms regulating iron acquisition, distribution, and storage at the cellular and systemic level. The iron-responsive element/iron-regulating protein (IRE/IRP) system is the key regulator of cellular iron handling, while the recently discovered iron-regulating peptide hepcidin and its receptor, the cellular iron exporter ferroportin, maintain systemic iron balance in humans.

C. Metabolism

Oxidized (Fe^{+3}) iron is practically insoluble, and its transport across the apical membrane of duodenal cells requires its reduction to ferrous iron (Fe^{+2}). Iron is reduced with the assistance of protein reductases located in the brush border, such as duodenal cytochrome *B*. Ferrous iron crosses the apical membrane of the enterocytes through the ferrous proton-coupled iron transporter divalent metal ion transporter 1 (DMT1), which also transports other divalent metals such as zinc, copper, and cobalt. Animals with genetic defects in DMT1 (mk/mk) mice and Belgrade rats) have impaired iron absorption and recycling and microcytic anemia.

Absorption of heme iron remains less well defined. Current evidence suggests that heme iron absorption is facilitated by a heme receptor. A candidate heme transporter (heme carrier protein 1, also referred to as proton-coupled folate transporter) was recently described, but it was subsequently demonstrated that it transports folate more efficiently than heme. Therefore, the physiological role of heme carrier protein 1 in heme iron uptake is controversial. Within enterocytes, heme iron is released from porphyrin by heme oxygenase.

Ferrous iron is exported through the basolateral membrane by ferroportin, the only known cellular iron exporter in vertebrates.[89] This step requires the cooperation of hephaestin, a multicopper oxidase, structurally homologous with ceruloplasmin, which oxidizes Fe^{+2} to Fe^{+3} for uptake by plasma transferrin.[90] The truncated hephaestin expressed in sex-linked anemia (sla) mice impairs basolateral iron transport, leading to iron accumulation in the enterocytes and systemic iron deficiency with microcytic anemia.[91]

Transferrin is the primary iron transport protein in the blood. Each transferrin molecule is capable of binding two atoms of iron in specific binding clefts. Under normal conditions, only 30% of iron-binding sites are saturated, whereas this number increases to 55–100% in iron overload. Diferric transferrin has a 10-fold higher affinity for its receptor (TFR1, transferrin receptor 1, now known as TFRC) than monoferric transferrin and 2000-fold higher affinity than apotransferrin.[92] The excess in iron-binding capacity by blood transferrin prevents the formation of potentially toxic nontransferrin-bound iron, which can cross cellular membranes in a nonregulated way and induce iron-related toxicity.[93]

Most cell types acquire iron from plasma transferrin, though less well-defined alternative mechanisms of cellular iron acquisition exist. This is evident in humans suffering from the rare genetic disorder hypotransferrinemia and in hpx mice, in which increased iron absorption and liver iron overload coexist with microcytic anemia.

Plasma iron is utilized mainly by erythroid progenitors in the bone marrow. To maintain the total daily output of >2 billion erythrocytes, 20–30 mg of iron is required. Because this amount is about 10-fold greater than the total amount in the plasma transferrin iron pool ($\sim$3 mg), this pool turns over $\geq$10 times a day.[94] Absorbed iron is limited to only 1–2 mg; therefore, the iron demands in bone marrow are met from the recycling of senescent erythrocytes by reticuloendothelial macrophages. Iron released from senescent red cells is exported from macrophages in the ferrous form via ferroportin with the assistance of the multicopper oxidase ceruloplasmin. Humans with aceruloplasminemia have anemia, retinal degeneration, iron overload in liver and spleen, diabetes, and iron accumulation in basal ganglia.

Erythroblasts acquire iron through TFR1-mediated endocytosis. Inside the erythroid endosomes, iron is reduced by a ferrireductase (STEAP3) and exported to the cytoplasm by DMT1. *Tfr1*$^{-/-}$ mice are characterized by embryonic lethality, and *Tfr1* haploinsufficiency causes microcytic anemia. The murine mutant of *Steap3* nm1054 suffers from iron-deficiency anemia.[95] Developing erythroblasts use iron for hemoglobin synthesis, a process involving a fine coordination between globin and heme synthesis with iron supply.

The majority of cellular iron is utilized in the mitochondria, where parts of porphyrin and ISC biosynthesis take place. Defects of the heme synthesis pathway in humans cause porphyrias or sideroblastic anemia, while defects of the ISC biosynthesis pathway cause sideroblastic anemia, Friedreich's ataxia, and hereditary myopathy with severe exercise intolerance.

The human body contains 3–5 g of iron distributed in various tissues. Hemoglobin iron constitutes approximately 60–70% of total-body iron, while other iron-rich organs are the liver, the spleen, and skeletal muscles (myoglobin iron).

The expression of cellular proteins participating in iron uptake and storage, such as TFR1 and ferritin, is posttranscriptionally regulated through mRNA interactions with IRPs. The mRNAs of TFR1 and ferritin harbor IREs in their 3′ and 5′untranslated regions, respectively. IREs are hairpin structures of 30–40 nucleotides, which provide binding sites for IRPs. In iron-depleted cells, IRPs bind with high affinity to IREs, stabilize the otherwise unstable mRNA of TFR1, and repress the translation of ferritin. Increased levels of TFR1 promote iron acquisition from plasma while unnecessary synthesis of ferritin is prevented. Conversely, in iron-rich cells, the IRPs are unavailable for IRE binding, leading consequently to TFR1 mRNA degradation and increased ferritin translation.[90,96] Other genes directly involved in iron trafficking and utilization, such as solute carrier family 40, member 1 (*SLC40A1*, encoding ferroportin), solute carrier family 11, member 2 (*SLC11A2*, encoding *DMT1*), and aminolevulinate, delta-, synthase 2 (*ALAS2*), contain IREs in their mRNAs (Fig. 1).

Systemic iron absorption is known to be increased in iron deficiency and markedly increased in iron-deficiency anemia and in pregnancy, as well as in genetic hemochromatosis and other rare genetic diseases with primary iron overload. Iron absorption is decreased in healthy subjects with iron-repleted stores and patients with chronic inflammatory diseases suffering from "anemia

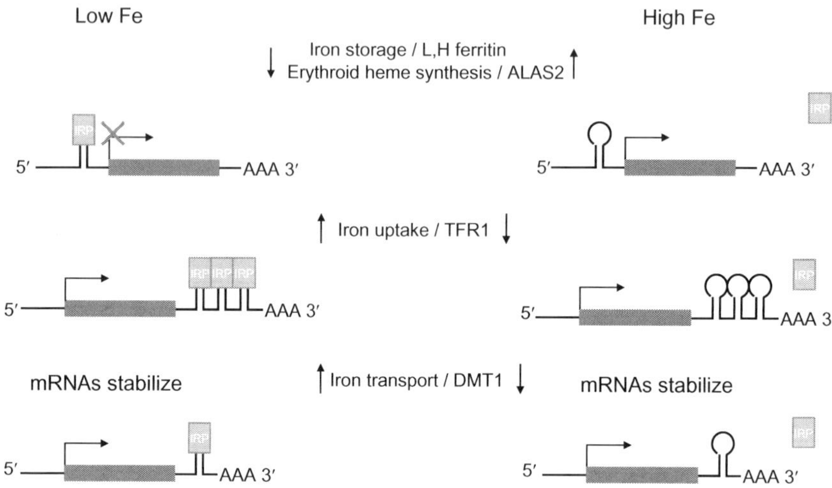

FIG. 1. Posttranscriptional regulation of gene expression in response to intracellular iron. IRPs bind to IREs in iron-depleted cells and stabilize transcripts with IREs in their 3′UTR, increasing the expression of proteins of iron uptake (TFR1) and iron transport (DMT1) and repressing the translation of the iron storage protein ferritin. Conversely, in iron-repleted cells, TFR1 transcripts degrade, and ferritin is readily translated. ALAS2, aminolevulinate, delta-, synthase 2; DMT1, divalent metal transporter 1; IREs, iron-responsive elements; IRPs, iron-regulating proteins; TFR1, transferrin receptor 1. (For color version of this figure, the reader is referred to the online version of this chapter.)

of chronic disease." Anemias characterized by ineffective erythropoiesis (in which immature erythroid progenitors lysate within the bone marrow as in thalassemias, sideroblastic, and congenital dyserythropoietic anemias) show pronounced increases in iron absorption despite systemic iron overload.

Humans do not have a physiological mechanism to increase iron excretion, as they do with other metals and nutrients. Therefore, maintenance of iron homeostasis depends upon tight regulation of iron absorption. This is achieved by hepcidin, a small 25-amino acid peptide hormone produced by liver hepatocytes. Plasma hepcidin binds to ferroportin and induces its internalization and subsequent degradation, acting as a negative regulator of cellular iron export. Increased plasma hepcidin inhibits iron absorption, iron release from macrophages, and liver iron storage. Suppressed hepcidin levels lead to increased iron release from macrophages and intestinal enterocytes. Humans with mutations in the gene encoding for hepcidin antimicrobial peptide (*HAMP*) suffer from juvenile hemochromatosis, a severe iron-overload disease. Acute and chronic inflammation is associated with increased hepcidin and reduced plasma iron.

Hepcidin is the final mediator in a complex pathway that regulates iron homeostasis in response to body iron stores, erythropoietic demand for iron, inflammation, and hypoxia. At the systemic level, diverse signals that affect hepcidin expression converge in liver hepatocytes and determine the final amount of excreted hepcidin. The exact molecular mechanisms underlying this regulation are not yet fully understood.

Iron-dependent hepcidin expression is modulated by liver iron stores and iron concentration in plasma. Current evidence suggests that hereditary hemochromatosis protein (HFE), transferrin receptor 2 (TFR2) (proteins mutated in human hemochromatosis), and TFR1 form a complex in the membrane of hepatic cells capable of sensing the transferrin iron. High concentrations of diferric transferrin interact with this "iron-sensing complex" and activate hepcidin transcription.[90] Bone morphogenetic protein 6 (BMP6), whose mRNA levels are positively correlated with iron stores, is thought to induce signaling in an autocrine manner through hemojuvelin (HJV)/BMP receptors. HJV, a protein mutated in human juvenile hemochromatosis, forms a complex with BMP receptors and functions as a coreceptor to BMP receptor signaling, which increases hepcidin transcription via phosphorylation of SMAD 1, 5, 8 proteins.

A membrane-associated protease, transmembrane protease serine 6 (matriptase-2) (TMPRSS6), plays an important role in the regulation of hepcidin expression. This protease is thought to cleave HJV; thus, its physiological role is to act as a negative regulator of hepcidin expression. Patients with mutations in *TMPRSS6* have iron-deficiency anemia unresponsive to oral iron administration and high serum hepcidin, while Tmprss6$^{-/-}$ and mask mice exhibit growth retardation, alopecia, and severe anemia associated with hepcidin upregulation. The phenotype is partially suppressed by parenteral iron administration.[97] Genome-wide association studies have shown an association between *TMPRSS6* variants with hemoglobin levels, serum iron, transferrin saturation, and mean erythrocyte cell volume.[98,99]

Experimental evidence indicates that hepcidin downregulation in response to anemia requires ongoing erythropoietic activity. It has been hypothesized that a circulating humoral factor, the so-called erythroid regulator, signals bone marrow iron needs. Growth differentiation factor 15 and twisted gastrulation protein homolog 1, members of the transforming growth factor-β superfamily, are overexpressed in patients with thalassemia and suppress hepcidin expression *in vitro*. Both factors are released by erythroid precursors and are found in high concentrations in patients with ineffective erythropoiesis. They also interfere with BMP-mediated hepcidin expression and dysregulate iron homeostasis in thalassemic syndromes.[100,101]

Inflammatory cytokines IL-1 and IL-6 are well-established inducers of hepcidin production. IL-6 binds to its receptor in hepatic cells and promotes hepcidin transcription through the signal transducer and activator of transcription 3

(STAT3) signaling pathway. This finding explains the molecular mechanism underlying hypoferremia in acute infection and anemia of chronic disease. The latter is characterized by normochromic normocytic anemia, increased macrophage iron stores, decreased serum iron and transferrin saturation, and markedly increased serum hepcidin.

Hypoxia can inhibit hepcidin expression in hepatocellular cell lines. Inhibition of prolyl hydroxylases reduces hepcidin transcription *in vitro*, while mice with liver-specific conditional inactivation of hypoxia-inducible factor 1-alpha (*Hif1a*) have inappropriately high hepcidin levels when they follow an iron-deficient diet.[102]

D. Iron, Genetics, and Human Diseases

Iron deficiency is the most common nutritional deficiency, estimated to affect 2 billion people worldwide. Its prevalence correlates with the socioeconomic status of the population. The main cause of iron deficiency is an inadequate nutritional supply, resulting in a negative iron balance, especially when iron losses (menstruating women) and systemic requirements (infancy, childhood, and pregnancy) are increased. Iron deficiency impairs cognitive performance and physical growth in children, and its effects are not likely reversible by subsequent iron therapy. During pregnancy, iron deficiency increases perinatal risks for mothers and neonates and overall infant mortality.[103] Preventive iron supplementation has become a standard of practice for children and pregnant women, who represent high-risk population groups. Ferrous sulfate is the more widely used iron preparation in the therapy and prevention of iron deficiency.

Among disorders with Mendelian inheritance involved in iron homeostasis, hereditary hemochromatosis is the more common in populations of European ancestry. Hemochromatosis is a genetically heterogeneous disease characterized by increased iron absorption and progressive iron accumulation predominantly in the liver. Clinical complications of hemochromatosis include fibrosis, cirrhosis, hepatocellular carcinoma, hypogonadism, arthritis, cardiomyopathy, diabetes, and skin hyperpigmentation. The majority of adult cases are attributed to homozygosity for the C282Y mutation of the *HFE* gene. Compound heterozygotes C282Y/H63D are also considered to be at risk. Population studies have shown that the penetrance of the homozygous genotype is relatively low, suggesting that genetic and environmental factors are important for the expression of clinical disease.[104] Rare genetic types of non-HFE hemochromatosis are attributed to mutations of the genes *TFR2*, *HJV*, and *HAMP*. Juvenile hemochromatosis, caused by mutations in the *HJV* and *HAMP* genes, is characterized by early onset of severe iron overload, signs of hypogonadism, and cardiac complications. The autosomal dominant form of hemochromatosis is due to mutations in *SLC40A1*, encoding ferroportin. Interestingly, the clinical phenotype of *SLC40A1* mutations depends on their functional consequences. Certain mutations prevent the cellular expression of the mutant protein and lead to macrophage iron accumulation and iron-restricted erythropoiesis. Other mutations lead to resistance to hepcidin by altering sites of the protein involved in hepcidin binding and protein internalization.[105] The severity of the clinical phenotype in hemochromatosis is directly related to the level of hepcidin insufficiency.

V. Selenium (Se)

A. Dietary Sources

Among all trace elements, selenium has been perhaps the most controversial concerning its nutritional value and pathophysiological effects. The history of its research is a continuous twist between toxicity and essentiality, between great hopes for its putative health benefits and disappointments over the outcomes of clinical studies. Selenium initially attracted scientific interest for its high toxicity until the mid-1970s when two bacterial enzymes and mammalian glutathione peroxidase (GPX) were found to contain selenium.[106] Since then, several selenium-dependent mechanisms of mammalian physiology have been discovered, while selenium intake and status have been correlated with the risk of many chronic diseases such as cancer and CVD. However, the results of the latest clinical trials with selenium supplements have highlighted the complexity of selenium's biological roles and the difficulty of applying the knowledge obtained from cellular and animal models to humans.

Dietary intake of selenium is determined by its content in different foods, the bioavailability of its chemical forms, and the dietary patterns adopted by different populations. The selenium content of foods varies according to the concentration of selenium in the soil, and several physicochemical properties of soil, such as pH and moisture, can affect the entry of this mineral in the food chain. Geochemical mapping has revealed areas poor in selenium (e.g., Scandinavia, New Zealand, certain parts of China) and seleniferous areas such as certain parts of the United States, Canada, China, and Central America. The corresponding dietary intakes vary from 3 μg/day in selenium-poor areas of China to 350 μg/day in Colombia.[107] However, the enrichment of soils with selenium-rich fertilizers and the use of supplements in animal agriculture have partially overcome low intakes due to selenium-poor soils.[108] Nevertheless, determination of the daily selenium intake by food frequency questionnaires or 24-h recalls is almost impossible because the amount of selenium differs greatly among the same foods, and nutritional tables are unreliable for this trace element. Beef, white bread, pork, chicken, eggs, and fish seem to be the main contributors of selenium in a typical Western diet, while Brazil nuts have the highest content of selenium among all foods.[109] Data from the Greek study

ATTICA revealed that red meat was the major dietary determinant of serum selenium in Greek adults and that the adoption of a carnivorous diet correlates with higher levels of total serum selenium.[110]

Intake is determined not only by the concentration of selenium in foods, but also by its bioavailability, which determines the pool of selenium species entering the human body. Selenium is highly bioavailable; however, its bioavailability is affected by its chemical form (organic forms such as selenomethionine and selenocysteine are more bioavailable than inorganic forms), the food matrix, the content of other dietary fibers, and cooking processes. Both the inorganic and organic forms of selenium are converted to hydrogen selenide in the human body, which, in turn, serves as a precursor for selenocysteine and selenoproteins.[111]

Assessment of daily selenium requirements is a difficult task because of the diversity of selenoproteins and complexity of selenium metabolism. Therefore, no consensus has been achieved among the main national and international health organizations for this issue. The main discrepancy is the clinical or biochemical criterion that should be used for the estimation of recommended selenium daily intakes. Until now, most guidelines have been based on the ability of dietary selenium to saturate the plasma or erythrocyte GPX, which is a sensitive marker of selenium exposure. According to this criterion, a daily intake of approximately 55 μg selenium is sufficient to maximize plasma GPX-3. Daily intakes of 350–700 μg selenium may reach toxic levels, while 20 μg/day is thought to be the basal daily requirement for selenium based on the fact that lower intakes led to Keshan disease, an endemic cardiomyopathy that appeared in selenium-poor areas of China.[112] However, it is now obvious that a single biochemical criterion (e.g., activity of GPXs) for the estimation of optimal selenium intake is inadequate and may lead to contradictory results and false expectations in nutritional studies, as recently shown.[113]

B. Biological Roles

Selenium exerts its actions through its incorporation into proteins. Proteins can incorporate selenium in several ways: (a) nonspecifically in the form of selenomethionine, which antagonizes methionine (e.g., selenoalbumin); (b) specifically as a cofactor (e.g., SLP-14, SELENBP2); and (c) as an integral part of their primary structure in the form of genetically encoded selenocysteine (Sec).[114] Sec has its own codon (UGA) in the nuclear genome; therefore, it is characterized as the 21st proteinogenic amino acid.[115] The UGA triplet in mRNA usually codes for the termination of translation; however, in the presence of secondary mRNA structures called selenocysteine-inserting sequences (SECIS), this codon is recognized by the anticodon of tRNA carrying the selenocysteinyl residue tRNA[Ser]Sec.[114] So far, 25 human selenoproteins containing Sec as part of their polypeptide chain have been characterized.

Among them, GPXs (five genes), thioredoxin reductases (TXRs) (three genes), iodothyronine deiodinases (three genes), selenoprotein P (SelP), and selenophosphate synthetase 2 are the best functionally characterized selenoproteins.[116] They catalyze redox reactions, and the presence of Sec instead of cysteine in their active center significantly increases their catalytic potential.[117]

- GPXs, either the intracellular cytosolic isoforms (GPX-1, GPX-2, GPX-4, and GPX-6) or the circulatory isoform (GPX-3), catalyze the reduction of H_2O_2, organic hydroperoxides, and phospholipid hydroperoxides to the corresponding alcohols using reduced glutathione as the electron donor. GPXs reduce ROS levels under conditions of oxidative stress and protect biomolecules and membranes from extensive oxidative damage.[118]
- Thioredoxin reductases (TXRs) are part of the thioredoxin system, along with thioredoxins and NADPH. Thioredoxins facilitate the reduction of other proteins with a concomitant oxidation of two vicinal –SH groups to a –S–S– bridge. TXRs keep thioredoxins in the reduced state in an NADPH-dependent reaction. TXRs participate in several vital cellular processes such as the reduction of nucleotide diphosphates to deoxynucleotide diphosphates, the regeneration of antioxidant enzymatic systems, the regulation of the intracellular redox tone, and redox-dependent signal transduction pathways that lead to gene expression.[119]
- The 5'-deiodinases (Type I and II) are responsible for the peripheral conversion of the thyroid prohormone tetraiodothyronine (thyroxin, T4) to the active hormone triiodothyronine (T3), while type III 5'-deiodinase degrades T4 and T3 to inactive metabolites. In this way, deiodinases regulate the intracellular levels and consequently the actions of thyroid hormones.[120]
- Selenoprotein P (SelP) is the major plasma selenoprotein carrying 10 atoms of selenium per polypeptide chain. Apart from its role as the main selenium carrier in plasma, SelP has ROS- and reactive nitrogen species (RNS) scavenging properties protecting the luminal side of the endothelium from oxidative stress.[121]
- Finally, selenophosphate synthetase 2 is implicated in the biosynthesis of selenocysteyl-tRNA[Ser]Sec by phosphorylating selenide, which is the donor of selenium to selenocysteyl-tRNA[Ser]Sec. Therefore, the biosynthesis of all selenoproteins is dependent on the activity of a protein which is also a selenoprotein.[122]

Novel functions have also been attributed to the less studied members of selenoproteins such as selenoprotein N (SEPN1), which is involved in the redox regulation of calcium channels in the sarcoplasmic reticulum, and selenoprotein R (SELR), which reduces oxidized methionine residues of proteins.[114] Moreover, organic and inorganic metabolites of selenium (e.g., selenomethionine, selenocysteine, sodium selenite, sodium selenate) also have antioxidant properties by scavenging ROS and RNS or binding metals.[123] The evidence clearly shows that selenium status can affect several mechanisms of human pathophysiology and may protect against the initiation and propagation of chronic diseases.

C. Metabolism

Selenide (H_2Se) is the common intermediate for both inorganic and organic dietary selenium sources. The main inorganic selenium species, namely, selenite and selenate, are converted to selenide by glutathione-coupled reactions. On the other hand, the dietary selenocysteine or the endogenously produced selenocysteine (through transsulfuration of dietary selenomethionine) is metabolized to selenide by beta-lyases. Selenide, in turn, is the precursor of proteinogenic selenocysteine and selenophosphate, both of which are indispensable for selenoprotein synthesis. Alternatively, selenomethionine antagonizes methionine for its incorporation into proteins. The excess of selenide is transformed to methylselenol (CH_3SeH), dimethylselenide ($(CH_3)_2Se$), and trimethylselenonium ion ($(CH3)_3Se^+$), or to selenosugars that are either excreted in urine or exhaled by breath. The tissues with the highest concentrations of selenium are the kidney cortex, pituitary gland, thyroid gland, liver, spleen, and cerebral cortex. However, the largest pool of body selenium is the skeletal muscle, where almost 50% of the whole-body selenium is stored.[124]

D. Selenium, Genetics, and Human Diseases

Despite the promising results of the Nutritional Prevention of Cancer (NPC) study on the effect of selenium supplementation on prevention of nonmelanoma skin cancers,[125] the outcomes of the Selenium and Vitamin E Cancer Prevention Trial (SELECT) were disappointing. Selenomethionine supplementation (200 μg/day) failed to prevent cancer in almost 33,000 healthy U.S. residents.[126] Moreover, the NPC study demonstrated a pro-diabetic role of selenium supplementation, especially in the volunteers with high baseline selenium levels. Other population studies (NHANES III, EVA Study) also showed a positive relationship between selenium levels and prevalence of diabetes or fasting glucose. However, it should be mentioned that these studies were conducted in populations with high selenium levels. It is possible that further administration of selenium to such populations may modify the homeostasis of selenium and lead to overexpression of metabolites/proteins (not known yet) and an unfavorable clinical phenotype. Prospective epidemiological and intervention studies on the effect of selenium levels and supplementation on cardiovascular risk have demonstrated that the protective effects of selenium lie within a narrow therapeutic range. The inverse relationship between selenium levels and the risk for coronary heart disease appears only in populations with low selenium levels. Moreover, selenium supplementation may have an adverse effect on CVD risk factors because it worsens dyslipidemia (especially hypercholesterolemia) and hypertension.[127]

Based on these results, detailed dose–response intervention studies of selenium supplementation are necessary, and several additional parameters should be taken into account, such as the gender, age, race, lifestyle (smoking, amount and type of exercise), clinical characteristics, different distribution of selenium to selenoproteins according to genetic differences, and selenium levels for saturation of other selenoproteins. Several cross-sectional studies have demonstrated the dependence of selenium status on age. Most studies have revealed a decline of serum selenium and selenoprotein levels in older ages. Data from the ATTICA study also demonstrated a significant decline of selenium levels with age (18–75). This decline was more obvious between the first quartile of age (18–31 years) and the other quartiles, and it was independent of anthropometric, lifestyle, biochemical, and nutritional indices.[128] Concerning the effect of gender on selenium status, most epidemiological studies have not revealed differences of serum selenium between males and females. However, nutritional studies in wild-type and genetically modified rodents clearly show a relationship between sex and selenium status since there is sex-specific expression of selenoproteins and sexual dimorphism of selenium on clinical outcomes.[129] Another lifestyle characteristic that should be taken into account is physical activity. People who exercise, especially athletes, require a higher antioxidant capacity because of exercise-induced ROS production and have increased needs for dietary antioxidants. From this perspective, a study from our group showed that beyond overt selenium deficiency, people with suboptimal selenium status had worse muscle functional decrements subsequent to eccentric muscle contractions.[130]

Two known genetic defects in selenoprotein synthesis can lead to an impaired clinical phenotype. The first defect is characterized by mutations in the SelN (*SEPN1*) gene, leading to lower levels of the protein and subsequently to a congenital muscular dystrophy. The second defect is a missense mutation of the gene encoding selenium-binding protein 2 (SBP2; also known as SECIS-binding protein 2 [SECISBP2]), which results in impaired biosynthesis of all selenoproteins, leading to thyroid dysfunctions. Both mutations affect the SECIS element, thus demonstrating its importance for selenoprotein biosynthesis. However, dietary selenium could not reverse the phenotype of these mutations. On the other hand, SNPs in the genes encoding selenoproteins or enzymes of their biosynthetic machinery could alter the functionality and requirements for dietary selenium. Only a few polymorphisms of selenoprotein genes (encoding GPX-1, GPX-3, GPX-4, SelP, SelS) have shown functional effects in cellular studies. Among the most interesting polymorphisms is the C/T at position 718 of the 3'UTR, which alters the expression of GPX-1, GPX-3, and GPX-4 in response to selenium supplementation and affects the lipoxygenase activity of lymphocytes. Population studies have linked this polymorphism with susceptibility to breast cancer and ulcerative colitis.[131] Similarly, SNPs of the SelP (*SEPP1*) gene also affect several markers of selenium status and the response of SelP to selenium supplementation, emphasizing the central regulatory role of SelP in selenium metabolism.[131]

Despite the aforementioned studies, the field of selenium nutrigenomics is actually unexplored, and further studies investigating how multiple SNPs affect selenium's biological roles and dietary requirements are needed.[131,132] Moreover, the genomic analysis should be combined with other –omics approaches (proteomics, lipidomics, metabolomics) to clarify the key players in selenium metabolism and actions. With this goal, members of our team applied a selenomics analysis to determine the levels of the three major selenoproteins of plasma (GPX-3, SelP, and selenoalbumin) in the ATTICA cohort.[110] The first results show that the distribution of serum selenium to the selenoproteins can vary markedly even between people with the same levels of total serum selenium and that the ratios of the selenoproteins correlate with several risk factors for CVD (unpublished results). Whether an impaired clinical profile alters selenium distribution to selenoproteins, or a genetically determined variation in selenoprotein synthesis modulates CVD risk factors, is not known yet.

VI. Zinc (Zn)

A. Dietary Sources

Zinc is an essential trace element for the organism, and its content in the human body is 2–3 g. So far, more than 300 enzymes have been identified that contain zinc, in which the mineral acts either as a cofactor or as a structural modulator.[133] Transcriptional factors are another group of proteins that contain zinc in the form of zinc fingers. It is estimated that approximately 10% of the human genome encodes zinc-binding proteins.[134]

Zinc is present in all food groups, yet the main dietary sources of zinc include oysters, red meat and poultry, fish and seafood, legumes, nuts, whole grains, and dairy products. The recommended dietary allowance for zinc is 8 and 11 mg for adult women and men, respectively, while it ranges from 12 to 14 mg for women during pregnancy and breastfeeding and from 2 to 9 mg during childhood and adolescence.[135]

Zinc is absorbed in the small intestine, mainly in the jejunum, through a transcellular process. This process is saturable, with an increase in transport velocity as zinc is depleted. Furthermore, low dietary zinc intake upregulates transporter expression in the small intestine, resulting in decreased intestinal loss of zinc. Overall intestinal zinc absorption ranges from 12% to 59%[136,137] and is modulated by several dietary factors.[137] The total amount of protein in a meal seems to improve zinc absorption, which is relevant given that foods rich in protein are also good sources of zinc.[138] The type of protein influences zinc bioavailability, as most animal proteins improve intestinal absorption of the mineral, probably because of the release of amino acids that keep zinc in solution.[137,138] On the contrary, casein in milk has a negative effect on zinc absorption, probably because zinc binds to phosphorylated serine and threonine residues of undigested casein subunits.[137] Dietary fiber and phytate, in particular, have an inhibitory effect on zinc absorption, which is mostly driven by the formation of insoluble complexes between zinc and the phosphate groups in inositol hexaphosphate.[137] Dietary iron is not known to affect zinc absorption, while the administration of both micronutrients in a single supplement is not inhibitory for either's absorption, as long as the iron-to-zinc ratio is not too high. Chelators and ligands (e.g., EDTA) and some amino acids (e.g., histidine and methionine), as well as organic acids (e.g., citrate), improve the absorption and bioavailability of zinc, mostly through increased solubility of the mineral.[137]

Zinc homeostasis is regulated via the gastrointestinal track, including the coordinated functions of many transporters. All zinc transporters have transmembrane domains and are encoded by two gene families: *SLC30* encoding ZnT and *SLC39* encoding Zip.[139] In humans, the 9 ZnT and 15 Zip transporters have opposite functions.[140] ZnT transporters decrease the intracellular zinc content via either zinc efflux from cells or zinc movement into intracellular vesicles. Zip transporters display the opposite effect, leading to an increase in intracellular zinc content. ZnT1, the first zinc transporter identified, is located in the small intestine, renal tubular epithelium, and placenta, mainly transporting zinc from enterocytes to circulation.[48] Its expression is regulated by dietary zinc intake, with increased intake leading to upregulation of ZnT1 mRNA levels. The ZnT4 transporter is mainly expressed in the mammary gland and Zip4 in the small intestine and liver, and Zip4 expression increases under zinc depletion.[141,142]

B. Biological Roles

The basic characteristics of zinc include a highly concentrated charge, a small radius (0.65 Å), no variable valence (low risk of free radical production), transformation from one symmetry to another without exchange, rapid exchange of ligands, and binding mostly to S and N donors in biological systems.[143] These properties enable zinc to play a major biological role as a catalyst. Zinc is essential for the biological function of more than 300 enzymes, while it is important for the structure, function, stability, and flexibility of biomembranes because of its ability to bind to the sulfhydryl domain and form mercaptides.[133] Zinc also regulates the balance between gene expression of metalloproteinases and the tissue inhibitors of matrix metalloproteinases, and this balance is necessary for the optimal function of many biological systems.[133,143] Zinc is present in zinc-finger domains of many proteins, peptides, enzymes, hormones, transcriptional factors, and cytokines, which act in maintaining body homeostasis. Zinc also regulates mRNA stability and extracellular matrix.[133,143,144]

In addition, zinc is required to maintain the enzymatic activity of nitric-oxide synthase (iNOS), through its binding to two cysteine residues in the heme domain of iNOS. Because nitric oxide (NO) is involved in MT mRNA expression and prevention of PARP (poly [ADP-ribose] polymerase) activation, the structural role of the mineral in NO production is important. Furthermore, NO is implicated in zinc release from MT to antioxidant enzymes.[145–149]

C. Zinc and Metallothioneins

MTs are a group of low-molecular-weight metal-binding proteins with a high affinity for zinc and copper.[150] Each MT molecule contains 20 cysteines and 7 binding sites for zinc atoms through mercaptide bonds.[151] There are at least 10 genes (in chromosome 16) identified in humans that encode four subfamilies of MTs. MT1 and MT2 are expressed in all tissues, MT3 is mainly expressed in the brain, and MT4 is expressed in the squamous epithelium.[152] A pivotal role of MT1 and MT2 is that of homeostasis regulation and intracellular limitation of oxidative stress.

The antioxidant properties of MTs, via the regulation of the intracellular zinc concentration, are important for their protective role in transient stress in both the young and the adults. However, the physiological function of MTs seems to change with advancing age, with subsequent modulation of zinc homeostasis and bioavailability in the immune system. In particular, in young adults, the presence of oxidative stress or inflammation leads to increased MT production, with a subsequent release of zinc promoting an optimal immune response. Zinc release from MTs ensures adequate NO production from iNOS and PARP activation toward DNA repair. On the contrary, zinc deficiency in old age, due to inadequate dietary zinc intake or/and decreased intestinal absorption, combined with increased MT production due to chronic exposure to oxidative stress, leads to overproduction of MTs with diminished ability to release zinc. In the elderly, exposure to oxidative stress or inflammation leads to overproduction of MTs, which sequester zinc. The diminished MT-induced zinc release results in impaired immune responses and the activation of PARP toward cell apoptosis.[143,153]

D. Metabolism

The greatest amount of zinc in the body is stored in skeletal muscle and bones; 11% of the total-body zinc is found in liver and skin, while plasma zinc accounts for only 0.1% of the total zinc content.[140,154] Plasma zinc levels range between 10 and 15 μmol/L and are tightly regulated even under fluctuations in dietary zinc intake, unless the latter are severe or/and prolonged.[155] Zinc has a high affinity to bind with protein at neutral pH and mostly (80%) binds to albumin in plasma.[140] The loss of zinc occurs mainly through the pancreatic secretions in the intestines, which ranges from 27 to 90 μmol/day. Urinary losses of zinc are significantly smaller (8–11 μmol/day) because the kidney reabsorbs the mineral. However, it is estimated that the use of thiazide diuretics increases the urinary loss of zinc up to 60%.[140,154]

E. Zinc, Genetics, and Human Diseases

Severe zinc deficiency is manifested by growth retardation, skin lesions, impaired wound healing, anemia, anorexia, hypogonadism, and impaired immune response.[156,157]

1. Zinc and the Immune System

Zinc is an essential mineral for the optimal structure and function of the immune system.[148] The production, maturation, and activation of lymphocytes depend on adequate zinc concentration. Zinc binds to DNA transcription and translation enzymes during the cell cycle of lymphocytes, is part of structural proteins of the lymphocytes' cytomembrane, and activates thymuline. Thymuline is a hormone secreted from the epithelial cells of the thymus gland and is zinc bound when activated. Thymuline is essential for the maturation, differentiation, and activation of T-cells, for cytokine production, and for the optimal function of natural killer cells. Zinc deficiency is manifested by thymus gland atrophy, lymphopenia, impaired cytotoxic function of the immune system, and decreased thymuline activity.[153,158]

T-lymphocytes are especially sensitive to adequate zinc concentration. Zinc deficiency leads to a smaller peripheral T-lymphocyte population, impaired T-helper function, and decreased cytotoxicity. Zinc is also an important element of the major histocompatibility class I and II, and the manifestations of its deficiency include impaired antibody production.[146,158] Other manifestations include imbalanced cytokine production by T-helpers, leading to decreased IL-2, IL-12, IFN-α, and IFN-γ levels but increased TNF-α, IL-1, and IL-6 levels.[147,152,159] The zinc-induced cytokine imbalance causes decreased activation and function of natural killer cells.[148,158] Results from supplementation trials in elderly populations suggest that dietary zinc intake interacts with genetic variants in *IL6* and *MT1*, with implications for chronic diseases and inflammatory biomarkers.[160–163]

Apart from the T-lymphocytes, zinc is also important for the maturation of the B-lymphocytes, and zinc deficiency is related to a decreased population of immature B-lymphocytes.[148,158] Furthermore, zinc deficiency is manifested in the immune system by impaired chemotaxis and phagocytosis in neutrophils, monocytes, and tissue macrophages. In humans, the most characteristic manifestation of zinc deficiency is acrodermatitis enteropathica, a rare autosomal recessive metabolic disorder that causes thymic atrophy and increased vulnerability to bacterial, fungal, and virus infections. The pathology of the disease includes zinc malabsorption due to mutations in *SLC39A4*, which encodes the intestinal zinc transporter protein Zip4.[158]

2. Zinc and Diabetes

Zinc is necessary in β-cells for insulin crystallization in hexamers.[164] Moreover, it is cosecreted with insulin, exerts insulinomimetic and antioxidant actions, and participates in the regulation of β-cell mass.[165,166] Zinc homeostasis is impaired in diabetic animals and humans of both types, while type 2 diabetes (T2D) is associated with decreased plasma zinc levels.[167] Consistent findings from several zinc supplementation trials in animal models support the protective effects of zinc against T2D.[167] Despite this evidence, the link between zinc and T2D in humans is not well established. There are only a few interventional trials investigating the effect of zinc supplementation on glucose metabolism, insulin homeostasis, and T2D risk, and even fewer report consistent findings.[167,168] Limited data from population studies provide evidence that dietary and total (food sources and supplements) zinc intake could reduce T2D risk.[169–171] Type 2 diabetes is characterized by increased urinary zinc excretion, leading to decreased plasma zinc levels. In supplementation trials with rodent models, zinc consistently appears to have inverse effects on hyperglycemia, hyperinsulinemia, and islet destruction, the latter mainly attributed to zinc-stimulated expression of MTs.[167] Decreased HbA1c and glucose levels also have been reported in T2D patients after zinc supplementation[172,173]; however, these findings failed to replicate, though under different experimental settings (supplementation period and dose).[174,175] In nondiabetic, obese Brazilian women, insulin sensitivity improved after a 4-week zinc supplementation trial.[176]

The results from a recent meta-analysis suggest that higher total zinc intake may attenuate the glucose-raising effect of the rs11558471 *SLC30A8* variant.[177] The *SLC30A8* gene encodes the newly characterized ZnT8 zinc transporter, while rs1158471 is associated with glycemic traits.[178,179] The ZnT8 β-cell-specific knockout (Znt8KO) mice are glucose intolerant and have reduced β-cell zinc accumulation, atypical insulin granules, reduced first-phase glucose-stimulated insulin secretion, reduced insulin-processing enzyme transcripts, and increased proinsulin levels.[180] Furthermore, a genetic variant (rs13266634; in strong linkage disequilibrium with rs11558471) in *SLC30A8* has been reliably associated with fasting glucose levels and T2D risk in several genome-wide association studies.[181,182] Interestingly, observations suggest that this variant impairs islet ZnT8 expression, insulin secretion, or glucose homeostasis and is associated with the production of a less active zinc transporter protein.[183,184] However, more research is needed to further elucidate the potential link between zinc, genetic variants, and diabetes.

VII. Conclusion

A description of the current knowledge on selected minerals and their relations with genetics and diseases has been presented. Figure 2 shows a synoptic diagram of dietary sources, metabolism, and implicated genes for

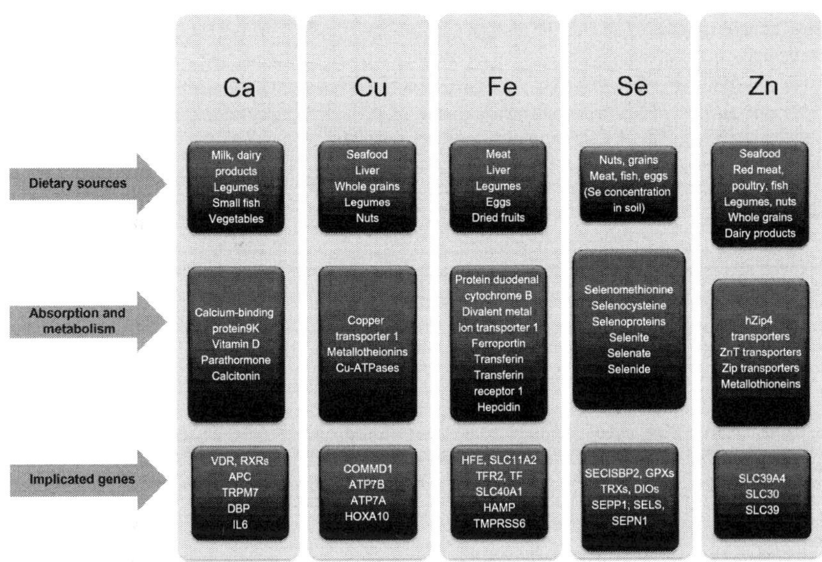

FIG. 2. Synoptic diagram of the dietary sources, metabolism, and implicated genes of the selected minerals. APC, adenomatous polyposis coli; ATP7A, ATPase Cu^{2+} transporting, alpha polypeptide; ATP7B, ATPase, Cu^{2+} transporting, beta polypeptide; COMMD1, copper metabolism (Murr1) domain containing 1; DBP, D site of albumin promoter-binding protein; DIOs, deiodinases, iodothyronine; GPXs, glutathione peroxidises; HAMP, hepcidin antimicrobial peptide; HFE, hemochromatosis; HOXA10, homeobox A10; IL6, interleukin 6; RXRs, retinoid X receptors; SECISBP2, SECIS-binding protein 2; SEPN1, selenoprotein N, 1; SEPP1, selenoprotein P, plasma, 1; SELS, selenoprotein S; SLC11A2, solute carrier family 11 (proton-coupled divalent metal ion transporters), member 2 (encodes DMT1); SLC30, solute carrier family 30 (encodes ZnT); SLC39, solute carrier family 39 (encodes Zip); SLC39A4, solute carrier family 39 (zinc transporter, member 4); SLC40A1, solute carrier family 40 (iron-regulated transporter), member 1 (ferroportin); TF, transferrin; TFR2, transferrin receptor 2; TMPRSS6, transmembrane protease, serine 6; TRPM7, transient receptor potential cation channel, subfamily M, member 7; TRXs, thioredoxin reductases; VDR, vitamin D receptor. (For color version of this figure, the reader is referred to the online version of this chapter.)

each mineral discussed. Although in a preliminary stage, the study of minerals and their interaction with genetic variants is expected to increase our knowledge of human physiology and to lead to the realization of the personalized nutrition concept. In addition, the combinatory use of –omics approaches has broadened the range of research tools for the study of mineral metabolism and its involvement in health and disease.

References

1. Gropper SS, Smith JL, Groff JL. *Advanced nutrition and human metabolism.* 5th ed. Wadswoth: Cengage Learning; 2008.
2. National Academy of Science Institute of Medicine Food and Nutritional Board. *Dietary reference intakes: recommended intakes for individuals.* The National Academies Press, Washington, D.C. 2010.
3. National Institutes of Health . NIH state-of-the-science conference statement on multivitamin/mineral supplements and chronic disease prevention. *NIH Consens State Sci Statements* 2006;**23**:1–30.
4. Chung M, Balk EM, Brendel M, Ip S, Lau J, Lee J, et al. Vitamin D and calcium: a systematic review of health outcomes. *Evid Rep Technol Assess (Full Rep)* 2009;**183**:1–420.
5. Dennehy C, Tsourounis C. A review of select vitamins and minerals used by postmenopausal women. *Maturitas* 2010;**66**:370–80.
6. Christakos S, Dhawan P, Liu Y, Peng X, Porta A. New insights into the mechanisms of vitamin D action. *J Cell Biochem* 2003;**88**:695–705.
7. Dusso AS, Brown AJ, Slatopolsky E. Vitamin D. *Am J Physiol Renal Physiol* 2005;**289**: F8–F28.
8. St-Arnaud R. The direct role of vitamin D on bone homeostasis. *Arch Biochem Biophys* 2008;**473**:225–30.
9. Norman AW. Minireview: vitamin D receptor: new assignments for an already busy receptor. *Endocrinology* 2006;**147**:5542–8.
10. Uitterlinden AG, Fang Y, Bergink AP, van Meurs JB, van Leeuwen HP, Pols HA. The role of vitamin D receptor gene polymorphisms in bone biology. *Mol Cell Endocrinol* 2002;**197**:15–21.
11. Krall EA, Parry P, Lichter JB, Dawson-Hughes B. Vitamin D receptor alleles and rates of bone loss: influences of years since menopause and calcium intake. *J Bone Miner Res* 1995;**10**:978–84.
12. Ferrari S, Rizzoli R, Chevalley T, Slosman D, Eisman JA, Bonjour JP. Vitamin-D-receptor-gene polymorphisms and change in lumbar-spine bone mineral density. *Lancet* 1995;**345**:423–4.
13. Salamone LM, Glynn NW, Black DM, Ferrell RE, Palermo L, Epstein RS, et al. Determinants of premenopausal bone mineral density: the interplay of genetic and lifestyle factors. *J Bone Miner Res* 1996;**11**:1557–65.
14. Kiel DP, Myers RH, Cupples LA, Kong XF, Zhu XH, Ordovas J, et al. The BsmI vitamin D receptor restriction fragment length polymorphism (bb) influences the effect of calcium intake on bone mineral density. *J Bone Miner Res* 1997;**12**:1049–57.
15. Ferrari SL, Rizzoli R, Slosman DO, Bonjour JP. Do dietary calcium and age explain the controversy surrounding the relationship between bone mineral density and vitamin D receptor gene polymorphisms? *J Bone Miner Res* 1998;**13**:363–70.
16. Brown MA, Haughton MA, Grant SF, Gunnell AS, Henderson NK, Eisman JA. Genetic control of bone density and turnover: role of the collagen 1alpha1, estrogen receptor, and vitamin D receptor genes. *J Bone Miner Res* 2001;**16**:758–64.

17. Stathopoulou MG, Dedoussis GV, Trovas G, Theodoraki EV, Katsalira A, Dontas IA, et al. The role of vitamin D receptor gene polymorphisms in the bone mineral density of Greek postmenopausal women with low calcium intake. *J Nutr Biochem* 2011;**22**:752–7.
18. Fang Y, van Meurs JB, Bergink AP, Hofman A, van Duijn CM, van Leeuwen JP, et al. Cdx-2 polymorphism in the promoter region of the human vitamin D receptor gene determines susceptibility to fracture in the elderly. *J Bone Miner Res* 2003;**18**:1632–41.
19. Ferrari S, Rizzoli R, Manen D, Slosman D, Bonjour JP. Vitamin D receptor gene start codon polymorphisms (FokI) and bone mineral density: interaction with age, dietary calcium, and 3′-end region polymorphisms. *J Bone Miner Res* 1998;**13**:925–30.
20. Jehan F, Voloc A, Esterle L, Walrant-Debray O, Nguyen TM, Garabedian M. Growth, calcium status and vitamin D receptor (VDR) promoter genotype in European children with normal or low calcium intake. *J Steroid Biochem Mol Biol* 2010;**121**:117–20.
21. Macdonald HM, McGuigan FE, Stewart A, Black AJ, Fraser WD, Ralston S, et al. Large-scale population-based study shows no evidence of association between common polymorphism of the VDR gene and BMD in British women. *J Bone Miner Res* 2006;**21**:151–62.
22. Lau EM, Lam V, Li M, Ho K, Woo J. Vitamin D receptor start codon polymorphism (Fok I) and bone mineral density in Chinese men and women. *Osteoporos Int* 2002;**13**:218–21.
23. Stathopoulou MG, Dedoussis GV, Trovas G, Katsalira A, Hammond N, Deloukas P, et al. Low-density lipoprotein receptor-related protein 5 polymorphisms are associated with bone mineral density in Greek postmenopausal women: an interaction with calcium intake. *J Am Diet Assoc* 2010;**110**:1078–83.
24. Fang Y, van Meurs JB, Arp P, van Leeuwen JP, Hofman A, Pols HA, et al. Vitamin D binding protein genotype and osteoporosis. *Calcif Tissue Int* 2009;**85**:85–93.
25. Li X, He GP, Zhang B, Chen YM, Su YX. Interactions of interleukin-6 gene polymorphisms with calcium intake and physical activity on bone mass in pre-menarche Chinese girls. *Osteoporos Int* 2008;**19**:1629–37.
26. Ferrari SL, Karasik D, Liu J, Karamohamed S, Herbert AG, Cupples LA, et al. Interactions of interleukin-6 promoter polymorphisms with dietary and lifestyle factors and their association with bone mass in men and women from the Framingham Osteoporosis Study. *J Bone Miner Res* 2004;**19**:552–9.
27. Karami S, Brennan P, Navratilova M, Mates D, Zaridze D, Janout V, et al. Vitamin d pathway genes, diet, and risk of renal cell carcinoma. *Int J Endocrinol* 2010;**2010**:879362.
28. McCullough ML, Stevens VL, Diver WR, Feigelson HS, Rodriguez C, Bostick RM, et al. Vitamin D pathway gene polymorphisms, diet, and risk of postmenopausal breast cancer: a nested case-control study. *Breast Cancer Res* 2007;**9**:R9.
29. Wong HL, Seow A, Arakawa K, Lee HP, Yu MC, Ingles SA. Vitamin D receptor start codon polymorphism and colorectal cancer risk: effect modification by dietary calcium and fat in Singapore Chinese. *Carcinogenesis* 2003;**24**:1091–5.
30. Guerreiro CS, Cravo ML, Brito M, Vidal PM, Fidalgo PO, Leitao CN. The D1822V APC polymorphism interacts with fat, calcium, and fiber intakes in modulating the risk of colorectal cancer in Portuguese persons. *Am J Clin Nutr* 2007;**85**:1592–7.
31. Dai Q, Shrubsole MJ, Ness RM, Schlundt D, Cai Q, Smalley WE, et al. The relation of magnesium and calcium intakes and a genetic polymorphism in the magnesium transporter to colorectal neoplasia risk. *Am J Clin Nutr* 2007;**86**:743–51.
32. Jenab M, McKay J, Bueno-de-Mesquita HB, van Duijnhoven FJ, Ferrari P, Slimani N, et al. Vitamin D receptor and calcium sensing receptor polymorphisms and the risk of colorectal cancer in European populations. *Cancer Epidemiol Biomarkers Prev* 2009;**18**:2485–91.
33. Dong LM, Ulrich CM, Hsu L, Duggan DJ, Benitez DS, White E, et al. Genetic variation in calcium-sensing receptor and risk for colon cancer. *Cancer Epidemiol Biomarkers Prev* 2008;**17**:2755–65.

34. Peters U, Chatterjee N, Yeager M, Chanock SJ, Schoen RE, McGlynn KA, et al. Association of genetic variants in the calcium-sensing receptor with risk of colorectal adenoma. *Cancer Epidemiol Biomarkers Prev* 2004;**13**:2181–6.
35. Poynter JN, Jacobs ET, Figueiredo JC, Lee WH, Conti DV, Campbell PT, et al. Genetic variation in the vitamin D receptor (VDR) and the vitamin D-binding protein (GC) and risk for colorectal cancer: results from the Colon Cancer Family Registry. *Cancer Epidemiol Biomarkers Prev* 2010;**19**:525–36.
36. Zillikens MC, van Meurs JB, Rivadeneira F, Hofman A, Oostra BA, Sijbrands EJ, et al. Interactions between dietary vitamin E intake and SIRT1 genetic variation influence body mass index. *Am J Clin Nutr* 2010;**91**:1387–93.
37. Pilvi TK, Storvik M, Louhelainen M, Merasto S, Korpela R, Mervaala EM. Effect of dietary calcium and dairy proteins on the adipose tissue gene expression profile in diet-induced obesity. *J Nutrigenet Nutrigenomics* 2008;**1**:240–51.
38. van der Meer-van Kraaij C, Kramer E, Jonker-Termont D, Katan MB, van der Meer R, Keijer J. Differential gene expression in rat colon by dietary heme and calcium. *Carcinogenesis* 2005;**26**:73–9.
39. Yang K, Lipkin M, Newmark H, Rigas B, Daroqui C, Maier S, et al. Molecular targets of calcium and vitamin D in mouse genetic models of intestinal cancer. *Nutr Rev* 2007;**65**: S134–7.
40. Nittke T, Selig S, Kallay E, Cross HS. Nutritional calcium modulates colonic expression of vitamin D receptor and pregnane X receptor target genes. *Mol Nutr Food Res* 2008;**52**(Suppl. 1):S45–51.
41. Turnlund JR, Keyes WR, Anderson HL, Acord LL. Copper absorption and retention in young men at three levels of dietary copper by use of the stable isotope 65Cu. *Am J Clin Nutr* 1989;**49**:870–8.
42. Velasco-Reynold C, Navarro-Alarcon M, López-Ga De La Serrana H, Lopez-Martinez MC. Copper in foods, beverages and waters from South East Spain: influencing factors and daily dietary intake by the Andalusian population. *Food Addit Contam Part A Chem Anal Control Expo Risk Assess* 2008;**25**:937–45.
43. Voskaki I, Arvanitidou V, Athanasopoulou H, Tzagkaraki A, Tripsianis G, Giannoulia-Karantana A. Serum copper and zinc levels in healthy greek children and their parents. *Biol Trace Elem Res* 2010;**134**:136–45.
44. Sanchez C, Lopez-Jurado M, Aranda P, Llopis J. Plasma levels of copper, manganese and selenium in an adult population in southern Spain: influence of age, obesity and lifestyle factors. *Sci Total Environ* 2010;**408**:1014–20.
45. Pennington JT, Calloway DH. Copper content of foods. Factors affecting reported values. *J Am Diet Assoc* 1973;**63**:143–53.
46. Patterson KY, Holbrook JT, Bodner JE, Kelsay JL, Smith Jr. JC, Veillon C. Zinc, copper, and manganese intake and balance for adults consuming self-selected diets. *Am J Clin Nutr* 1984;**40**:1397–403.
47. Pennington JA, Wilson DB. Daily intakes of nine nutritional elements: analyzed vs. calculated values. *J Am Diet Assoc* 1990;**90**:375–81.
48. Liuzzi JP, Blanchard RK, Cousins RJ. Differential regulation of zinc transporter 1, 2, and 4 mRNA expression by dietary zinc in rats. *J Nutr* 2001;**131**:46–52.
49. European Commission Publications Office. *Tolerable upper intake levels for vitamins and minerals.* Scientific Committee on Food, EFSA; 2006.
50. Ashwell M, Lambert JP, Alles MS, Branca F, Bucchini L, Brzozowska A, et al. How we will produce the evidence-based EURRECA toolkit to support nutrition and food policy. *Eur J Nutr* 2008;**47**(Suppl. 1):2–16.

51. de Romana DL, Olivares M, Uauy R, Araya M. Risks and benefits of copper in light of new insights of copper homeostasis. *J Trace Elem Med Biol* 2011;**25**:3–13.
52. Kim BE, Nevitt T, Thiele DJ. Mechanisms for copper acquisition, distribution and regulation. *Nat Chem Biol* 2008;**4**:176–85.
53. Lee J, Pena MM, Nose Y, Thiele DJ. Biochemical characterization of the human copper transporter Ctr1. *J Biol Chem* 2002;**277**:4380–7.
54. Freedman JH, Ciriolo MR, Peisach J. The role of glutathione in copper metabolism and toxicity. *J Biol Chem* 1989;**264**:5598–605.
55. Wijmenga C, Klomp LW. Molecular regulation of copper excretion in the liver. *Proc Nutr Soc* 2004;**63**:31–9.
56. Petris MJ, Smith K, Lee J, Thiele DJ. Copper-stimulated endocytosis and degradation of the human copper transporter, hCtr1. *J Biol Chem* 2003;**278**:9639–46.
57. Muller P, van Bakel H, van de Sluis B, Holstege F, Wijmenga C, Klomp LW. Gene expression profiling of liver cells after copper overload in vivo and in vitro reveals new copper-regulated genes. *J Biol Inorg Chem* 2007;**12**:495–507.
58. La Fontaine S, Mercer JF. Trafficking of the copper-ATPases, ATP7A and ATP7B: role in copper homeostasis. *Arch Biochem Biophys* 2007;**463**:149–67.
59. Vulpe C, Levinson B, Whitney S, Packman S, Gitschier J. Isolation of a candidate gene for Menkes disease and evidence that it encodes a copper-transporting ATPase. *Nat Genet* 1993;**3**:7–13.
60. Klomp AE, van de Sluis B, Klomp LW, Wijmenga C. The ubiquitously expressed MURR1 protein is absent in canine copper toxicosis. *J Hepatol* 2003;**39**:703–9.
61. Tao TY, Liu F, Klomp L, Wijmenga C, Gitlin JD. The copper toxicosis gene product Murr1 directly interacts with the Wilson disease protein. *J Biol Chem* 2003;**278**:41593–6.
62. Burstein E, Ganesh L, Dick RD, van De Sluis B, Wilkinson JC, Klomp LW, et al. A novel role for XIAP in copper homeostasis through regulation of MURR1. *EMBO J* 2004;**23**:244–54.
63. Pyle GG, Kamunde CN, McDonald DG, Wood CM. Dietary sodium inhibits aqueous copper uptake in rainbow trout (Oncorhynchus mykiss). *J Exp Biol* 2003;**206**:609–18.
64. Handy RD, Eddy FB, Baines H. Sodium-dependent copper uptake across epithelia: a review of rationale with experimental evidence from gill and intestine. *Biochim Biophys Acta* 2002;**1566**:104–15.
65. Canessa CM, Schild L, Buell G, Thorens B, Gautschi I, Horisberger JD, et al. Amiloride-sensitive epithelial Na+ channel is made of three homologous subunits. *Nature* 1994;**367**:463–7.
66. Scheinberg IH, Sternlieb I. Wilson disease and idiopathic copper toxicosis. *Am J Clin Nutr* 1996;**63**:842S–5S.
67. Gitlin JD. Wilson disease. *Gastroenterology* 2003;**125**:1868–77.
68. Menkes JH, Richardson F, Verplanck S. Program for the detection of metabolic diseases. Chromatographic screening to find metabolic diseases, especially those involving the nervous system. *Arch Neurol* 1962;**6**:462–70.
69. Mercer JF, Livingston J, Hall B, Paynter JA, Begy C, Chandrasekharappa S, et al. Isolation of a partial candidate gene for Menkes disease by positional cloning. *Nat Genet* 1993;**3**:20–5.
70. Chelly J, Tumer Z, Tonnesen T, Petterson A, Ishikawa-Brush Y, Tommerup N, et al. Isolation of a candidate gene for Menkes disease that encodes a potential heavy metal binding protein. *Nat Genet* 1993;**3**:14–9.
71. Wilson FP. A preliminary study of the bio-chemical relations of various lipoid substances in the liver. *Biochem J* 1912;**6**:100–5.
72. Yamaguchi Y, Heiny ME, Gitlin JD. Isolation and characterization of a human liver cDNA as a candidate gene for Wilson disease. *Biochem Biophys Res Commun* 1993;**197**:271–7.
73. Tanzi RE, Petrukhin K, Chernov I, Pellequer JL, Wasco W, Ross B, et al. The Wilson disease gene is a copper transporting ATPase with homology to the Menkes disease gene. *Nat Genet* 1993;**5**:344–50.

74. Bull PC, Thomas GR, Rommens JM, Forbes JR, Cox DW. The Wilson disease gene is a putative copper transporting P-type ATPase similar to the Menkes gene. *Nat Genet* 1993; **5**:327–37.
75. Dedoussis GV, Genschel J, Sialvera TE, Bochow B, Manolaki N, Manios Y, et al. Wilson disease: high prevalence in a mountainous area of Crete. *Ann Hum Genet* 2005;**69**:268–74.
76. Wijmenga C, Muller T, Murli IS, Brunt T, Feichtinger H, Schonitzer D, et al. Endemic Tyrolean infantile cirrhosis is not an allelic variant of Wilson's disease. *Eur J Hum Genet* 1998;**6**:624–8.
77. Muller T, Feichtinger H, Berger H, Muller W. Endemic Tyrolean infantile cirrhosis: an ecogenetic disorder. *Lancet* 1996;**347**:877–80.
78. Tanner MS. Role of copper in Indian childhood cirrhosis. *Am J Clin Nutr* 1998;**67**: 1074S–81S.
79. Xia X, Xie C, Wang Y, Cai S, Zhu C, Yang X. The forces imposed by the novel T-shape Cu/ LDPE nanocomposite intrauterine devices on the simulated uterine cavity. *Contraception* 2007;**76**:326–30.
80. Krumlauf R. Hox genes in vertebrate development. *Cell* 1994;**78**:191–201.
81. Tetrault AM, Richman SM, Fei X, Taylor HS. Decreased endometrial HOXA10 expression associated with use of the copper intrauterine device. *Fertil Steril* 2009;**92**:1820–4.
82. Squitti R, Ghidoni R, Scrascia F, Benussi L, Panetta V, Pasqualetti P, et al. Free copper distinguishes mild cognitive impairment subjects from healthy elderly individuals. *J Alzheimers Dis* 2011;**23**:239–48.
83. Northrop-Clewes CA, Thurnham DI. Monitoring micronutrients in cigarette smokers. *Clin Chim Acta* 2007;**377**:14–38.
84. Planells E, Sanchez C, Montellano MA, Mataix J, Llopis J. Vitamins B6 and B12 and folate status in an adult Mediterranean population. *Eur J Clin Nutr* 2003;**57**:777–85.
85. Mataix J, Lopez-Frias M, Martinez-de-Victoria E, Lopez-Jurado M, Aranda P, Llopis J. Factors associated with obesity in an adult Mediterranean population: influence on plasma lipid profile. *J Am Coll Nutr* 2005;**24**:456–65.
86. Johnson PE, Milne DB, Lykken GI. Effects of age and sex on copper absorption, biological half-life, and status in humans. *Am J Clin Nutr* 1992;**56**:917–25.
87. Song IS, Chen HH, Aiba I, Hossain A, Liang ZD, Klomp LW, et al. Transcription factor Sp1 plays an important role in the regulation of copper homeostasis in mammalian cells. *Mol Pharmacol* 2008;**74**:705–13.
88. Hunt JR, Zito CA, Johnson LK. Body iron excretion by healthy men and women. *Am J Clin Nutr* 2009;**89**:1792–8.
89. De Domenico I, Ward DM, Langelier C, Vaughn MB, Nemeth E, Sundquist WI, et al. The molecular mechanism of hepcidin-mediated ferroportin down-regulation. *Mol Biol Cell* 2007; **18**:2569–78.
90. Hentze MW, Muckenthaler MU, Galy B, Camaschella C. Two to tango: regulation of Mammalian iron metabolism. *Cell* 2010;**142**:24–38.
91. Chen H, Attieh ZK, Su T, Syed BA, Gao H, Alaeddine RM, et al. Hephaestin is a ferroxidase that maintains partial activity in sex-linked anemia mice. *Blood* 2004;**103**:3933–9.
92. Anderson GJ, Vulpe CD. Mammalian iron transport. *Cell Mol Life Sci* 2009;**66**:3241–61.
93. Papanikolaou G, Pantopoulos K. Iron metabolism and toxicity. *Toxicol Appl Pharmacol* 2005;**202**:199–211.
94. Cavill I. Erythropoiesis and iron. *Best Pract Res Clin Haematol* 2002;**15**:399–409.
95. Ohgami RS, Campagna DR, Greer EL, Antiochos B, McDonald A, Chen J, et al. Identification of a ferrireductase required for efficient transferrin-dependent iron uptake in erythroid cells. *Nat Genet* 2005;**37**:1264–9.
96. Wang J, Pantopoulos K. Regulation of cellular iron metabolism. *Biochem J* 2011;**434**:365–81.

97. Folgueras AR, de Lara FM, Pendas AM, Garabaya C, Rodriguez F, Astudillo A, et al. Membrane-bound serine protease matriptase-2 (Tmprss6) is an essential regulator of iron homeostasis. *Blood* 2008;**112**:2539–45.
98. Benyamin B, Ferreira MA, Willemsen G, Gordon S, Middelberg RP, McEvoy BP, et al. Common variants in TMPRSS6 are associated with iron status and erythrocyte volume. *Nat Genet* 2009;**41**:1173–5.
99. Chambers JC, Zhang W, Li Y, Sehmi J, Wass MN, Zabaneh D, et al. Genome-wide association study identifies variants in TMPRSS6 associated with hemoglobin levels. *Nat Genet* 2009; **41**:1170–2.
100. Tanno T, Noel P, Miller JL. Growth differentiation factor 15 in erythroid health and disease. *Curr Opin Hematol* 2010;**17**:184–90.
101. Tanno T, Porayette P, Sripichai O, Noh SJ, Byrnes C, Bhupatiraju A, et al. Identification of TWSG1 as a second novel erythroid regulator of hepcidin expression in murine and human cells. *Blood* 2009;**114**:181–6.
102. Fleming MD. The regulation of hepcidin and its effects on systemic and cellular iron metabolism. *Hematology Am Soc Hematol Educ Program* 2008;**2**:151–8.
103. WHO. Iron deficiency anaemia: assessment, prevention and control. A guide for programme managers; 2001. A (C282Y) HFE hereditary haemochromatosis mutation in the USA. Lancet 359, 211–218.
104. Beutler E, Felitti VJ, Koziol JA, Ho NJ, Gelbart T. Penetrance of 845G–> A (C282Y) HFE hereditary haemochromatosis mutation in the USA. *Lancet* 2002;**359**:211–8.
105. De Domenico I, Ward DM, Musci G, Kaplan J. Iron overload due to mutations in ferroportin. *Haematologica* 2006;**91**:92–5.
106. Flohe L, Andreesen JR, Brigelius-Flohe R, Maiorino M, Ursini F. Selenium, the element of the moon, in life on earth. *IUBMB Life* 2000;**49**:411–20.
107. Rayman MP. Food-chain selenium and human health: emphasis on intake. *Br J Nutr* 2008; **100**:254–68.
108. Combs Jr. GF. Selenium in global food systems. *Br J Nutr* 2001;**85**:517–47.
109. Navarro-Alarcon M, Cabrera-Vique C. Selenium in food and the human body: a review. *Sci Total Environ* 2008;**400**:115–41.
110. Letsiou S, Nomikos T, Panagiotakos D, Pergantis SA, Fragopoulou E, Antonopoulou S, et al. Dietary habits of Greek adults and serum total selenium concentration: the ATTICA study. *Eur J Nutr* 2010;**49**:465–72.
111. Finley JW. Bioavailability of selenium from foods. *Nutr Rev* 2006;**64**:146–51.
112. Thomson CD. Assessment of requirements for selenium and adequacy of selenium status: a review. *Eur J Clin Nutr* 2004;**58**:391–402.
113. Ashton K, Hooper L, Harvey LJ, Hurst R, Casgrain A, Fairweather-Tait SJ. Methods of assessment of selenium status in humans: a systematic review. *Am J Clin Nutr* 2009;**89**:2025S–39S.
114. Papp LV, Lu J, Holmgren A, Khanna KK. From selenium to selenoproteins: synthesis, identity, and their role in human health. *Antioxid Redox Signal* 2007;**9**:775–806.
115. Lee BJ, Worland PJ, Davis JN, Stadtman TC, Hatfield DL. Identification of a selenocysteyl-tRNA(Ser) in mammalian cells that recognizes the nonsense codon, UGA. *J Biol Chem* 1989; **264**:9724–7.
116. Lobanov AV, Hatfield DL, Gladyshev VN. Eukaryotic selenoproteins and selenoproteomes. *Biochim Biophys Acta* 2009;**1790**:1424–8.
117. Arner ES. Selenoproteins—what unique properties can arise with selenocysteine in place of cysteine? *Exp Cell Res* 2010;**316**:1296–303.
118. Arthur JR. The glutathione peroxidases. *Cell Mol Life Sci* 2000;**57**:1825–35.
119. Arner ES. Focus on mammalian thioredoxin reductases—important selenoproteins with versatile functions. *Biochim Biophys Acta* 2009;**1790**:495–526.

120. Kohrle J, Jakob F, Contempre B, Dumont JE. Selenium, the thyroid, and the endocrine system. *Endocr Rev* 2005;**26**:944–84.
121. Burk RF, Hill KE. Selenoprotein P-expression, functions, and roles in mammals. *Biochim Biophys Acta* 2009;**1790**:1441–7.
122. Ganichkin OM, Xu XM, Carlson BA, Mix H, Hatfield DL, Gladyshev VN, et al. Structure and catalytic mechanism of eukaryotic selenocysteine synthase. *J Biol Chem* 2008;**283**:5849–65.
123. Ramoutar RR, Brumaghim JL. Antioxidant and anticancer properties and mechanisms of inorganic selenium, oxo-sulfur, and oxo-selenium compounds. *Cell Biochem Biophys* 2010; **58**:1–23.
124. Suzuki KT, Ogra Y. Metabolic pathway for selenium in the body: speciation by HPLC-ICP MS with enriched Se. *Food Addit Contam* 2002;**19**:974–83.
125. Clark LC, Combs Jr. GF, Turnbull BW, Slate EH, Chalker DK, Chow J, et al. Effects of selenium supplementation for cancer prevention in patients with carcinoma of the skin. A randomized controlled trial. Nutritional Prevention of Cancer Study Group. *JAMA* 1996; **276**:1957–63.
126. Lippman SM, Klein EA, Goodman PJ, Lucia MS, Thompson IM, Ford LG, et al. Effect of selenium and vitamin E on risk of prostate cancer and other cancers: the Selenium and Vitamin E Cancer Prevention Trial (SELECT). *JAMA* 2009;**301**:39–51.
127. Stranges S, Navas-Acien A, Rayman MP, Guallar E. Selenium status and cardiometabolic health: state of the evidence. *Nutr Metab Cardiovasc Dis* 2010;**20**:754–60.
128. Letsiou S, Nomikos T, Panagiotakos D, Pergantis SA, Fragopoulou E, Antonopoulou S, et al. Serum total selenium status in Greek adults and its relation to age. The ATTICA study cohort. *Biol Trace Elem Res* 2009;**128**:8–17.
129. Schomburg L, Schweizer U. Hierarchical regulation of selenoprotein expression and sex-specific effects of selenium. *Biochim Biophys Acta* 2009;**1790**:1453–62.
130. Milias GA, Nomikos T, Fragopoulou E, Athanasopoulos S, Antonopoulou S. Effects of baseline serum levels of Se on markers of eccentric exercise-induced muscle injury. *Biofactors* 2006;**26**:161–70.
131. Hesketh J. Nutrigenomics and selenium: gene expression patterns, physiological targets, and genetics. *Annu Rev Nutr* 2008;**28**:157–77.
132. Schomburg L. Genetics and phenomics of selenoenzymes–how to identify an impaired biosynthesis? *Mol Cell Endocrinol* 2010;**322**:114–24.
133. Vallee BL, Falchuk KH. The biochemical basis of zinc physiology. *Physiol Rev* 1993; **73**:79–118.
134. Andreini C, Banci L, Bertini I, Rosato A. Counting the zinc-proteins encoded in the human genome. *J Proteome Res* 2006;**5**:196–201.
135. Trumbo P, Yates AA, Schlicker S, Poos M. Dietary reference intakes: vitamin A, vitamin K, arsenic, boron, chromium, copper, iodine, iron, manganese, molybdenum, nickel, silicon, vanadium, and zinc. *J Am Diet Assoc* 2001;**101**:294–301.
136. Lee HH, Prasad AS, Brewer GJ, Owyang C. Zinc absorption in human small intestine. *Am J Physiol* 1989;**256**:G87–91.
137. Lonnerdal B. Dietary factors influencing zinc absorption. *J Nutr* 2000;**130**:1378S–83S.
138. Sandstrom B, Cederblad A. Zinc absorption from composite meals. II. Influence of the main protein source. *Am J Clin Nutr* 1980;**33**:1778–83.
139. Liuzzi JP, Cousins RJ. Mammalian zinc transporters. *Annu Rev Nutr* 2004;**24**:151–72.
140. Tuerk MJ, Fazel N. Zinc deficiency. *Curr Opin Gastroenterol* 2009;**25**:136–43.
141. Kelleher SL, Lonnerdal B. Zinc transporters in the rat mammary gland respond to marginal zinc and vitamin A intakes during lactation. *J Nutr* 2002;**132**:3280–5.
142. Wang K, Zhou B, Kuo YM, Zemansky J, Gitschier J. A novel member of a zinc transporter family is defective in acrodermatitis enteropathica. *Am J Hum Genet* 2002;**71**:66–73.

143. Mocchegiani E, Costarelli L, Giacconi R, Cipriano C, Muti E, Tesei S, et al. Nutrient-gene interaction in ageing and successful ageing. A single nutrient (zinc) and some target genes related to inflammatory/immune response. *Mech Ageing Dev* 2006;**127**:517–25.
144. Taylor CM, Bacon JR, Aggett PJ, Bremner I. Homeostatic regulation of zinc absorption and endogenous losses in zinc-deprived men. *Am J Clin Nutr* 1991;**53**:755–63.
145. Bogdan C, Rollinghoff M, Diefenbach A. The role of nitric oxide in innate immunity. *Immunol Rev* 2000;**173**:17–26.
146. Mocchegiani E, Giacconi R, Muzzioli M, Cipriano C. Zinc, infections and immunosenescence. *Mech Ageing Dev* 2000;**121**:21–35.
147. Mocchegiani E, Muzzioli M, Giacconi R. Zinc, metallothioneins, immune responses, survival and ageing. *Biogerontology* 2000;**1**:133–43.
148. Mocchegiani E, Muzzioli M, Giacconi R. Zinc and immunoresistance to infection in aging: new biological tools. *Trends Pharmacol Sci* 2000;**21**:205–8.
149. Zangger K, Oz G, Haslinger E, Kunert O, Armitage IM. Nitric oxide selectively releases metals from the amino-terminal domain of metallothioneins: potential role at inflammatory sites. *FASEB J* 2001;**15**:1303–5.
150. Kagi JH, Schaffer A. Biochemistry of metallothionein. *Biochemistry* 1988;**27**:8509–15.
151. Maret W, Jacob C, Vallee BL, Fischer EH. Inhibitory sites in enzymes: zinc removal and reactivation by thionein. *Proc Natl Acad Sci USA* 1999;**96**:1936–40.
152. Mocchegiani E, Muzzioli M, Cipriano C, Giacconi R. Zinc, T-cell pathways, aging: role of metallothioneins. *Mech Ageing Dev* 1998;**106**:183–204.
153. Mocchegiani E, Muzzioli M, Giacconi R, Cipriano C, Gasparini N, Franceschi C, et al. Metallothioneins/PARP-1/IL-6 interplay on natural killer cell activity in elderly: parallelism with nonagenarians and old infected humans. Effect of zinc supply. *Mech Ageing Dev* 2003;**124**:459–68.
154. King JC, Shames DM, Woodhouse LR. Zinc homeostasis in humans. *J Nutr* 2000; **130**:1360S–6S.
155. Cousins RJ. Theoretical and practical aspects of zinc uptake and absorption. *Adv Exp Med Biol* 1989;**249**:3–12.
156. Ibs KH, Rink L. Zinc-altered immune function. *J Nutr* 2003;**133**:1452S–6S.
157. McClain CJ, McClain M, Barve S, Boosalis MG. Trace metals and the elderly. *Clin Geriatr Med* 2002;**18**:801–18 vii–viii.
158. Haase H, Rink L. The immune system and the impact of zinc during aging. *Immun Ageing* 2009;**6**:9.
159. Bao B, Prasad AS, Beck FW, Godmere M. Zinc modulates mRNA levels of cytokines. *Am J Physiol Endocrinol Metab* 2003;**285**:E1095–102.
160. Kanoni S, Dedoussis GV, Herbein G, Fulop T, Varin A, Jajte J, et al. Assessment of gene-nutrient interactions on inflammatory status of the elderly with the use of a zinc diet score—ZINCAGE study. *J Nutr Biochem* 2010;**21**:526–31.
161. Mariani E, Neri S, Cattini L, Mocchegiani E, Malavolta M, Dedoussis GV, et al. Effect of zinc supplementation on plasma IL-6 and MCP-1 production and NK cell function in healthy elderly: interactive influence of +647 MT1a and −174 IL-6 polymorphic alleles. *Exp Gerontol* 2008;**43**:462–71.
162. Mocchegiani E, Giacconi R, Costarelli L, Muti E, Cipriano C, Tesei S, et al. Zinc deficiency and IL-6-174G/C polymorphism in old people from different European countries: effect of zinc supplementation. ZINCAGE study. *Exp Gerontol* 2008;**43**:433–44.
163. Mocchegiani E, Malavolta M. Zinc-gene interaction related to inflammatory/immune response in ageing. *Genes Nutr* 2008;**3**:61–75.
164. Scott DA. Crystalline insulin. *Biochem J* 1934;**28**:1592–602 1.
165. Rungby J. Zinc, zinc transporters and diabetes. *Diabetologia* 2010;**53**:1549–51.

166. Wijesekara N, Chimienti F, Wheeler MB. Zinc, a regulator of islet function and glucose homeostasis. *Diabetes Obes Metab* 2009;**11**(Suppl. 4):202–14.
167. Jansen J, Karges W, Rink L. Zinc and diabetes—clinical links and molecular mechanisms. *J Nutr Biochem* 2009;**20**:399–417.
168. Haase H, Overbeck S, Rink L. Zinc supplementation for the treatment or prevention of disease: current status and future perspectives. *Exp Gerontol* 2008;**43**:394–408.
169. Shi Z, Yuan B, Qi L, Dai Y, Zuo H, Zhou M. Zinc intake and the risk of hyperglycemia among Chinese adults: the prospective Jiangsu Nutrition Study (JIN). *J Nutr Health Aging* 2010; **14**:332–5.
170. Singh RB, Niaz MA, Rastogi SS, Bajaj S, Gaoli Z, Shoumin Z. Current zinc intake and risk of diabetes and coronary artery disease and factors associated with insulin resistance in rural and urban populations of North India. *J Am Coll Nutr* 1998;**17**:564–70.
171. Sun Q, van Dam RM, Willett WC, Hu FB. Prospective study of zinc intake and risk of type 2 diabetes in women. *Diabetes Care* 2009;**32**:629–34.
172. Al-Maroof RA, Al-Sharbatti SS. Serum zinc levels in diabetic patients and effect of zinc supplementation on glycemic control of type 2 diabetics. *Saudi Med J* 2006;**27**:344–50.
173. Gupta R, Garg VK, Mathur DK, Goyal RK. Oral zinc therapy in diabetic neuropathy. *J Assoc Physicians India* 1998;**46**:939–42.
174. Anderson RA, Roussel AM, Zouari N, Mahjoub S, Matheau JM, Kerkeni A. Potential antioxidant effects of zinc and chromium supplementation in people with type 2 diabetes mellitus. *J Am Coll Nutr* 2001;**20**:212–8.
175. Brandao-Neto J, da Silva CA, Figueiredo NB, Shuhama T, da Cunha NF, Dourado FB, et al. Lack of acute zinc effects in glucose metabolism in healthy and insulin-dependent diabetes mellitus patients. *Biometals* 1999;**12**:161–5.
176. Marreiro DN, Geloneze B, Tambascia MA, Lerario AC, Halpern A, Cozzolino SM. Effect of zinc supplementation on serum leptin levels and insulin resistance of obese women. *Biol Trace Elem Res* 2006;**112**:109–18.
177. Kanoni S, Nettleton JA, Hivert MF, Ye Z, Rooij FJA, Shungin D, et al. Total zinc intake may modify the glucose-raising effect of a zinc transporter (SLC30A8) variant: a 14-cohort meta-analysis. *Diabetes* 2011;**60**:2407–16.
178. Chimienti F, Devergnas S, Favier A, Seve M. Identification and cloning of a beta-cell-specific zinc transporter, ZnT-8, localized into insulin secretory granules. *Diabetes* 2004;**53**:2330–7.
179. Dupuis J, Langenberg C, Prokopenko I, Saxena R, Soranzo N, Jackson AU, et al. New genetic loci implicated in fasting glucose homeostasis and their impact on type 2 diabetes risk. *Nat Genet* 2010;**42**:105–16.
180. Wijesekara N, Dai FF, Hardy AB, Giglou PR, Bhattacharjee A, Koshkin V, et al. Beta cell-specific Znt8 deletion in mice causes marked defects in insulin processing, crystallisation and secretion. *Diabetologia* 2010;**53**:1656–68.
181. Saxena R, Voight BF, Lyssenko V, Burtt NP, de Bakker PI, Chen H, et al. Genome-wide association analysis identifies loci for type 2 diabetes and triglyceride levels. *Science* 2007; **316**:1331–6.
182. Sladek R, Rocheleau G, Rung J, Dina C, Shen L, Serre D, et al. A genome-wide association study identifies novel risk loci for type 2 diabetes. *Nature* 2007;**445**:881–5.
183. Cauchi S, Del Guerra S, Choquet H, D'Aleo V, Groves CJ, Lupi R, et al. Meta-analysis and functional effects of the SLC30A8 rs13266634 polymorphism on isolated human pancreatic islets. *Mol Genet Metab* 2010;**100**:77–82.
184. Nicolson TJ, Bellomo EA, Wijesekara N, Loder MK, Baldwin JM, Gyulkhandanyan AV, et al. Insulin storage and glucose homeostasis in mice null for the granule zinc transporter ZnT8 and studies of the type 2 diabetes-associated variants. *Diabetes* 2009;**58**:2070–83.

Gene–Sodium Interaction and Blood Pressure: Findings from Genomics Research of Blood Pressure Salt Sensitivity

TANIKA N. KELLY* AND
JIANG HE*,†

*Department of Epidemiology, Tulane University School of Public Health and Tropical Medicine, New Orleans, Louisiana, USA

†Department of Medicine, Tulane University School of Medicine, New Orleans, Louisiana, USA

High blood pressure (BP) is a complex trait determined by both genetic and environmental factors, as well as the interactions between these factors. Over the past few decades, there has been substantial progress in elucidating the genetic determinants underlying the BP response to sodium intake, or BP salt sensitivity. Research of monogenic BP disorders has highlighted the importance of renal salt handling in BP regulation, implicating genes and biological pathways related to salt sensitivity. Candidate gene studies have contributed important information toward understanding the genomic mechanisms underlying the BP response to salt intake, identifying genes in the renin–angiotensin–aldosterone system, renal sodium channels/transporters, and the endothelial system related to this phenotype. Despite these advancements, genome-wide association studies are still needed to uncover novel mechanisms underlying salt sensitivity, while future sequencing efforts promise the discovery of functional variants related to this complex trait. Delineating the genetic

Progress in Molecular Biology
and Translational Science, Vol. 108
DOI: 10.1016/B978-0-12-398397-8.00010-1

1877-1173/12 $35.00

architecture of salt sensitivity will be critical to understanding how genes and dietary sodium interact to influence BP.

I. Introduction

Hypertension is a major public health challenge because of its high prevalence and concomitant increase in the risk for cardiovascular disease and all-cause mortality.[1–3] As a complex trait, hypertension is influenced by multiple environmental and genetic determinants, as well as by the interactions between these factors.[4–6] Among environmental determinants, dietary sodium intake is the most common and important risk factor for hypertension.[7,8] Evidence from animal models,[9,10] observational studies,[11,12] and clinical trials[13,14] has demonstrated a causal relationship between dietary sodium intake and high blood pressure (BP). Epidemiological studies have also indicated that a high dietary sodium intake is associated with an increased risk of cardiovascular disease.[15–18] However, there is substantial evidence suggesting variability in individuals' BP responses to dietary sodium intake, a phenomenon described as BP salt sensitivity.[19–21]

Guided by known physiological mechanisms and findings from early research of monogenic BP disorders, investigators have begun to examine genes in biological pathways implicated in the BP response to sodium intake. As an example, the Genetic Epidemiology Network of Salt Sensitivity (GenSalt) study, which is the largest dietary sodium feeding study to date, was designed to examine how genes and sodium interact to influence BP.[22] While the results from genetics studies of BP salt sensitivity have sometimes been inconsistent, in total, there is strong and accumulating evidence for the role of genes in the determination of this phenotype.

II. Renal Sodium Handling and Blood Pressure Regulation

The physiological mechanisms that link renal sodium handling to BP regulation have been well described and are presented in Fig. 1. Briefly, sodium homeostasis (and BP control) is dependent on the kidney's ability to reabsorb salt, which is accomplished by exchangers, transporters, and ion channels located in the kidney's nephron.[24] The epithelial sodium channel (ENaC), located in the nephron's cortical collecting tubule, represents the principal site for the determination of sodium balance. ENaC is highly regulated by the renin–angiotensin–aldosterone system (RAAS), which exerts its effects through a complex cascade of events beginning with the kidney's

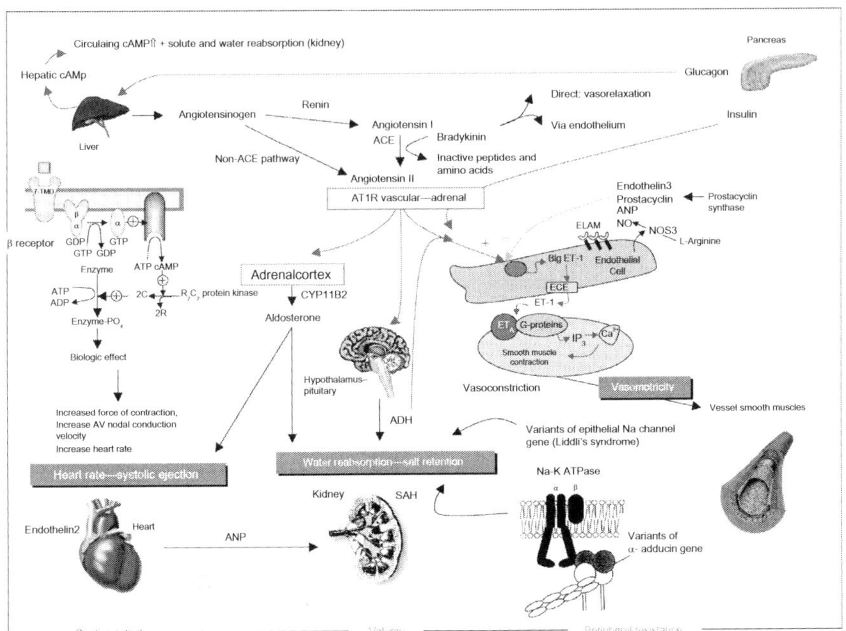

FIG. 1. Network of pathways and genes postulated to be associated with BP regulation. ACE, angiotensin-converting enzyme; ADH, antidiuretic hormone (vasopressin); ANP, atrial natriuretic peptide; AT1R, angiotensin II type 1 receptor; AV, atrioventricular; cAMP, cyclic adenosine monophosphate; CYP11B2, ECE, endothelin-converting enzyme; ELAM, endothelial leukocyte adhesion molecule 1 (E-selectin); ET-1, endothelin-1; ET_A, endothelin receptor type A; GDP, guanine diphosphate; GTP, guanine triphosphate; IP3, inositol tris-phosphate; NO, nitric oxide; NOS3, nitric oxide synthase; R_2C_2 protein kinase, protein kinase complex of two catalytic subunits and the regulatory dimer; SAH, SA hypertension-associated homolog (rat); 7-TMD, seven-transmembrane domain (from Marteau *et al.*[23]).

release of renin in response to decreased renal perfusion.[24,25] Renin (REN), a proteolytic enzyme, cleaves angiotensinogen (AGT) to angiotensin I, which is in turn cleaved by the angiotensin-converting enzyme (ACE) to form angiotensin II, the main effector protein of the RAAS.[25] Angiotensin II raises BP by direct vasoconstriction, increasing both sympathetic nerve activity and myocardial contractility.[26] In addition, angiotensin II binds to a G protein-coupled membrane receptor on the surface of cells in the zona glomerulosa, setting off a chain of events that rapidly activates aldosterone biosynthesis and secretion.[25] It should be noted that the action of potassium on adrenal glomerulosa cells is also an important regulator of aldosterone secretion.[25] Through the interaction of aldosterone with the mineralocorticoid receptor (MR) in renal tubular cells, ENaC activity is increased, which, in turn, increases sodium and water reabsorption and potassium excretion.[26] This ultimately results in expanded plasma volume, increased cardiac output, and increased BP.[27]

Because of the clear role of renal sodium handling in BP regulation, genes encoding components of these pathways have frequently been targeted in genomics research of BP salt sensitivity. However, other biological pathways also play important roles in sodium homeostasis and BP regulation (Fig. 1).

III. Monogenic Blood Pressure Disorders

Some of the earliest progress in genomics research of salt sensitivity was in the identification of genes responsible for severe inherited BP disorders.[24,25,27] While many physiological mechanisms are involved in BP regulation, the vast majority of genes identified for Mendelian forms of hypertension and hypotension have been shown to exert their effects by influencing renal salt handling (Table I).[24,25,27–60] Monogenic hypertensive disorders such as apparent mineralocorticoid excess,[24,25,27–31] congenital adrenal hyperplasia,[24,25] familial hyperaldosteronism, type I (glucocorticoid remediable aldosteronism),[24,25,32,33] and early onset hypertension with severe exacerbation during pregnancy[24,39] are all caused by mutations in genes encoding components of the RAAS. Conversely, components of the RAAS are also responsible for monogenic hypotensive disorders including corticosterone methyloxidase type I deficiency,[24,54] pseudohypoaldosteronism, type Ia,[56] and salt-wasting congenital adrenal hyperplasia.[58,59] The majority of other monogenic BP disorders, including familial hyperaldosteronism, type III,[37,38] Liddle's syndrome,[40–42] pseudohyperaldosteronism type II,[43–46] Bartter syndrome,[24,47–53] Gitelman syndrome,[24,55] and pseudohypoaldosteronism, type Ib,[57] are caused by mutations in genes encoding either ion channels, exchangers and transporters, or enzymes involved in renal ion transport.

Research on monogenic BP disorders provides important lessons for investigation of the complex salt-sensitivity phenotype. First, these findings indicate that interaction between dietary sodium intake and genotype exists among these rare Mendelian conditions. For example, a low-sodium diet reduces BP in carriers of the Liddle's syndrome gene mutation.[61] The fact that environmental factors can influence the phenotypic expression of a monogenic disorder lends credence to the idea that gene–diet interaction could affect complex traits that are more strongly determined by environmental factors. In addition, with the majority of monogenic BP disorders clearly elucidated,[24] the finding that nearly all of them result from deficits in renal sodium handling highlights the crucial role of sodium homeostasis in BP regulation, strongly implicating these genes and biological pathways in the etiology of BP salt sensitivity.

TABLE I

MENDELIAN INHERITED BLOOD PRESSURE DISORDERS RESULTING FROM MUTATIONS IN GENES INVOLVED IN RENAL SODIUM HANDLING

Syndrome	Gene(s)	Chromosome	Gene location (bp)	Inheritance	Mechanism	References
Hypertension disorders						
Apparent mineralocorticoid excess	*HSD11B2*	16	67465036–67471456	Autosomal recessive	Inhibits conversion of cortisol to cortisone (increased MR activity)	24,25,27–31
Congenital adrenal hyperplasia	*CYP11B1*	8	143953773–143961236	Autosomal recessive	Accumulation of 21-hydroxylated steroids (increased MR activity)	24,25
	CYP17A1	10	104590288–104597290			
Familial hyperaldosteronism, type I	*CYP11B1/CYP11B2*	8	143953773–143999259	Autosomal dominant	Constitutive expression of aldosterone synthase (increased aldosterone biosynthesis and MR activity)	24,25,32,33
Familial hyperaldosteronism, type II	Unknown	7	Unknown	Autosomal dominant	Unknown	34–36
Familial hyperaldosteronism, type III	*KCNJ5*	11	128761313–128787964	Unknown	Constitutive expression of aldosterone synthase (increased aldosterone synthesis and MR activity)	37,38
Hypertension, early onset, with severe exacerbation in pregnancy	*NR3C2*	4	148999915– 149363672	Autosomal dominant	Constitutive activation of MR	24,39
Liddle's syndrome	*SCNN1B*	16	23313591–23392620	Autosomal dominant	Constitutive activation of ENaC	40–42
	SCNN1G	16	23194040–23228200			
Pseudohyperaldosteronism, type II	Unknown	1	Unknown	Autosomal dominant	Dysregulation of sodium chloride transporter and ENaC	43–46
	WNK1	12	862089–1020618			
	WNK4	17	40932649–40949084			

(Continues)

TABLE I (*Continued*)

Syndrome	Gene(s)	Chromosome	Gene location (bp)	Inheritance	Mechanism	References
Hypotension disorders						
Bartter syndrome	*SLC12A1*	15	48498498–48596275	Autosomal recessive	Loss of function of the renal ion transporters, channels, and their subunits	24,47–53
	KCNJ1	11	128707915–128737268			
	CLCNKB	1	16370247–16383803			
	BSND	1	55464617–55474465			
	CLCNKB +*CLCNKA*	1	16348486–16383803			
Corticosterone methyloxidase type I deficiency	*CYP11B2*	8	143991975–143999259	Autosomal recessive	Impaired aldosterone biosynthesis	24,54
Gitelman syndrome	*SLC12A3*	16	56899119–56949762	Autosomal recessive	Loss of function of the renal thiazide-sensitive sodium chloride cotransporter	24,55
Pseudohypoaldosteronism, type I	*NR3C2*	4	148999915–149363672	Autosomal dominant	Loss of MR function	56
	SCNN1A	12	6456009–6486523	Autosomal recessive	Loss of ENaC function	
	SCNN1B	16	23313591–23392620			57
	SCNN1G	16	23194040–23228200			
Salt-wasting congenital adrenal hyperplasia	*CYP21A2*	6	32006093–32009447	Autosomal recessive	Aldosterone deficiency (decreased MR activity)	58,59
	HSD3B2	1	119957554–119965662			
	CYP11A1	15	74630103–74660081			

IV. The Salt-Sensitivity Phenotype

Kawasaki *et al.*[19] and later on Weinberger[20] were among the first to recognize the heterogeneity of the BP response to sodium and to develop the concept of salt sensitivity in humans. In their initial study, Kawasaki and colleagues defined salt sensitivity arbitrarily as an increase in mean arterial pressure (MAP) $\geq$10% with a high-sodium diet (250 mmol/day) compared with a low-sodium diet (10 mmol/day) over a 7-day period.[19] Since then, a variety of protocols have been used to test for the salt sensitivity of BP in humans, including the examination of the BP response to an acute protocol in which patients are salt-loaded with an intravenous infusion of saline and salt-depleted by administration of furosemide[20,21]; BP responses to chronic low- and high-sodium dietary interventions, usually lasting 5–14 days, also have been measured.[22] For example, in the GenSalt study, participants received a low-sodium diet (3 g of salt or 51.3 mmol of sodium per day) for 7 days followed by a high-sodium diet (18 g of salt or 307.8 mmol of sodium per day) for 7 days, with BP measured three times during the last 3 days of each intervention phase (Fig. 2).

Despite the variety of protocols used to test for salt sensitivity, certain findings have been consistently observed. First, salt-induced changes in BP are normally distributed in populations, and there is no evidence for a bimodal distribution.[62] As when defining hypertension, using a cut-point to categorize subjects into salt-sensitive and nonsensitive groups is arbitrary. In the GenSalt study, salt sensitivity is typically defined continuously as the percent or absolute change in systolic BP (SBP), diastolic BP (DBP), or MAP when switching from baseline to low-sodium or low-sodium to high-sodium interventions. In addition, up to 75% of the population may experience BP changes in response to sodium intake.[62] Thus, salt sensitivity is a common complex trait in the general population.

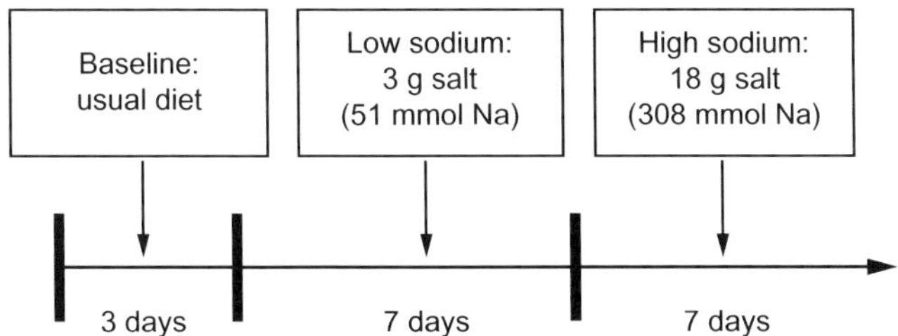

FIG. 2. GenSalt dietary sodium feeding (from The GenSalt Collaborative Research Group[22]).

V. Genomic Etiology of Salt Sensitivity

To date, substantial progress has been made in identifying genomic factors underlying the BP response to sodium intake. With BP salt sensitivity confirmed as an inheritable trait, investigators have expended much energy examining candidate genes implicated by past studies of renal physiology and monogenic BP disorders. Although their findings have at times been inconsistent, insights into the salt-sensitivity phenotype have emerged, revealing important genetic mechanisms influencing this trait and paving a clear direction for future research efforts.

A. Heritability of Salt Sensitivity

Heritability analyses have consistently identified a significant contribution of genetic factors to salt sensitivity, documenting the BP response to sodium as a moderately heritable trait. Among the 1906 Han Chinese participants of the GenSalt dietary feeding study, heritabilities of salt sensitivity, defined as the percent changes in SBP, DBP, and MAP when switching from a low-sodium to high-sodium intervention, were 22%, 33%, and 33%, respectively.[63] In a study of the BP response to a 12-week sodium-restricted diet in Caucasian families, heritable factors accounted for 64% of the SBP response.[64] Finally, using an intravenous sodium-loading and furosemide volume-depletion protocol, Svetkey *et al.* examined 20 African-American families and calculated heritability estimates of 26–84% for MAP and 26–74% for SBP responses to the salt-sensitivity maneuver.[65]

B. Whole-Genome Linkage Analyses of Salt Sensitivity

Animal studies have reported several QTLs on chromosomes 1 and 17 linked to salt sensitivity and salt resistance of BP among spontaneously hypertensive rats or salt-sensitive Sabra hypertension-prone rats.[66,67] In the only human genome-wide linkage scan of BP salt sensitivity, GenSalt detected strong linkage signals at 33–42 cM of chromosome 2 (2p24.3–2p24.1) with maximum limit LOD scores of 3.33 for DBP and 2.91 for MAP responses to dietary sodium intervention.[68] Follow-up of this linkage signal implicated the novel melatonin receptor 1B (*MTNR1B*) gene in BP salt sensitivity.[68] Although these findings are promising and represent some of the only non-hypothesis-based research of the salt-sensitivity phenotype in human populations, replication and fine-mapping studies of the *MNTR1B* gene are warranted.

C. Genetic Association Studies of Salt Sensitivity

Genetic association studies offer a powerful approach for detecting genetic variants that influence common complex traits. The candidate gene studies of BP salt sensitivity have been relatively successful because the biological pathways of sodium homeostasis and BP regulation have been well established.[63,69–117] Table II summarizes genes that have been associated with salt-sensitivity phenotypes in previous studies. Many of these findings are described in detail below, according to their biological pathway.

1. Renin–Angiotensin–Aldosterone System

Because the RAAS plays a central role in renal sodium handling and BP regulation, genes encoding its components have been studied extensively for their association with BP salt sensitivity. Most reports have focused on a handful of variants in select RAAS genes, including the well-known *ACE* I/D polymorphism,[69–71,73,83,118] as well as the *AGT* M235T and G6A polymorphisms[71,73–78]; hydroxysteroid (11-β) dehydrogenase 2 (*HSD11B2*) G534A and CA-repeat polymorphisms[71,85,86]; cytochrome P450, family 11, subfamily B, polypeptide 2 (*CYP11B2*) C-344T and intron 2 conversion polymorphisms[71,82,83,119]; and angiotensin II receptor, type 1 (*AGTR1*) A1166C polymorphism.[71,83] However, studies of these genetic variants have yielded inconsistent results.[97] As with candidate gene studies of other complex traits, the inconsistency in findings has been attributed mainly to methodological limitations including poorly measured and inconsistently defined phenotypes, small sample sizes, permissive *p*-values, and lack of replication evidence.[79,97]

Using clearly defined and well-measured phenotypes, as well as stringent significance thresholds, the GenSalt study comprehensively explored the RAAS, examining the association of salt sensitivity with 191 single-nucleotide polymorphisms (SNPs) covering 12 RAAS genes, including *ACE*; *ACE2*; *AGT*; *AGTR1*; *AGTR2*; *CYP11B1*; *HSD11B1*; *HSD11B2*; nuclear receptor subfamily 3, group C, member 2 (*NR3C2*); *REN*; and REN-binding protein (*RENBP*).[72,79] GenSalt findings failed to replicate the previously reported associations but identified novel genetic variants in the *ACE2* (rs1514283, rs1514282, rs2074192, rs714205, rs4646176, and rs2285666), *AGTR1* (rs4524238 and rs3772616), *HSD11B2* (rs5479), and *RENBP* (rs1557501 and rs2269372) genes which were significantly associated with BP responses to dietary sodium interventions.[72,79] Further follow-up work is ongoing in the GenSalt study, including replication of these findings in independent samples and a sequencing study to pinpoint the causal variants involved in BP salt sensitivity.

2. Ion and Water Channels, Transporters, and Exchangers

Important insights into the etiology of BP salt sensitivity may come from genomics studies of renal ion and water channels, transporters, and exchangers, which help regulate sodium balance, blood volume, and BP. Of particular importance, the ENaC represents the primary site of renal sodium reabsorption and has been definitively linked to the monogenic BP disorders Liddle's syndrome and pseudohypoaldosteronism, type Ib.[24] Recently, GenSalt investigators

TABLE II

GENES REPORTED TO ASSOCIATE WITH BP SALT SENSITIVITY, ACCORDING TO BIOLOGICAL PATHWAY

Gene symbol	Gene	Chromosome	Gene location (bp)	References
Renin–angiotensin–aldosterone system				
ACE	Angiotensin I-converting enzyme 1	17	61554422–61575741	69–71
ACE2	Angiotensin I-converting enzyme 2	X	15579156–15620192	72
AGT	Angiotensinogen	1	230838269–230850336	73–78
AGTR1	Angiotensin II receptor, type 1	3	148415658–148460790	79
AGTR2	Angiotensin II receptor, type 2	X	115301958–115306225	80
CYP11B1[a]	Cytochrome P450, family 11, subfamily B, polypeptide 1	8	143953773–143961236	81
CYP11B2	Cytochrome P450, family 11, subfamily B, polypeptide 2	8	143991975–143999259	82–84
HSD11B2	Hydroxysteroid (11-β) dehydrogenase 2	16	67465036–67471456	71,79,85,86
RENBP	Renin-binding protein	X	153200722–153210232	79
Ion and water channels, transporters, and exchangers				
AQP5	Aquaporin 5	12	50355279–50359461	87
CLCNKA	Chloride channel Ka	1	16348486–16360545	63
SCNN1G	Sodium channel, nonvoltage-gated 1, γ	16	23194040–23228200	88
SLC8A1	Solute carrier family 8 (sodium/calcium exchanger), member 1	2	40339286–40739575	89
SLC24A3	Solute carrier family 24 (sodium/potassium/calcium exchanger), member 3	20	19193290–19703541	89
WNK1	WNK lysine deficient protein kinase 1	12	862089–1020618	90,91
Endothelial system				
CYBA	Cytochrome *b*-245, α polypeptide	16	88709697–88717457	92
EDN1[a]	Endothelin 1	6	12290529–12297427	93
EDNRB	Endothelin receptor type B	13	78469616–78493903	94
NOS3	Nitric oxide synthase 3	7	150688144–150711687	95,96
Intracellular messengers				
ADD1	Adducin 1 (α)	4	2845584–2931789	90,97–105
GNB3	Guanine nucleotide-binding protein, β polypeptide 3	22	6949375–6956557	105
Sympathetic nervous system				
ADRB2	β-2-Adrenergic receptor	5	148206156–148208197	75,106,107

GRK4	G protein-coupled receptor kinase 4	4	2965343–3042474	108
Apelin-APJ system				
APLNR	Apelin receptor	11	57001051–57004927	72
Atrial natriuretic peptides				
NPPA	Natriuretic peptide A	1	11905766–11907840	109
Kallikrein–kinin system				
KLK1	Kallikrein 1	19	51322404–51327043	75
Others				
ABCB1	ATP-binding cassette, subfamily B (MDR/TAP), member 1	7	87133179–87342639	110
CYP3A5	Cytochrome P450, family 3, subfamily A, polypeptide 5	7	99245813–99277621	110–112
DYNLT1[a]	Dynein, light chain, Tctex-type 1	6	159057506–159065804	113
FAH[a]	Fumarylacetoacetate hydrolase	15	80445233–80478924	113
HNRNPK[a]	Heterogeneous nuclear ribonucleoprotein K	9	86582998–86595569	114
HP	Haptoglobin	16	72088508–72094955	115
LSR[a]	Lipolysis-stimulated lipoprotein receptor	19	35739559–35758867	113
MTNR1B	Melatonin receptor 1B	11	92702789–92715948	68
MYADM[a]	Myeloid-associated differentiation marker	19	54369611–54379689	113
NEDD4L	Neural precursor cell expressed, developmentally downregulated 4-like	18	55711619–56068772	90
NOS1[a]	Neuronal nitric oxide synthase 1	12	117645947–117799607	116
SERPINH1[a]	Serpin peptidase inhibitor, clade H (heat-shock protein 47), member 1	11	75273101–75283849	113
STK39	Serine–threonine kinase 39	2	168810530–169104105	117

[a]Findings exclusively from studies of animal models.

comprehensively examined common variants in genes encoding the α-, β-, and γ-subunits of the ENaC (sodium channel, nonvoltage-gated 1, alpha [*SCNN1A*], beta [*SCNN1B*], and gamma [*SCNN1G*], respectively) for their association with BP responses to sodium intake.[88] In this study, Zhao and colleagues identified six *SCNN1G* SNPs strongly associated with salt sensitivity, including two variants predicted to enhance the expression of this gene.[88] Although future works to confirm these findings are needed, along with a functional study to determine the causal variants, this research provides early evidence that common variants in ENaC may play an important role in BP salt sensitivity.

3. Endothelial System

It is well known that the release of nitric oxide from the endothelium increases vasodilation and regional blood flow while lowering BP.[120,121] Acting as "salt sensors," endothelial cells respond to changes in extracellular concentrations of sodium, with the increase in sodium related to decreased endothelial nitric oxide release and increased endothelial cell stiffness.[120,121] For this reason, the endothelial system is implicated in the pathogenesis of salt-sensitive hypertension. Surprisingly, only a few studies have examined the relationship between genes in this biological pathway and the BP response to sodium.[92,94–96] Animal models have suggested a role for the gene encoding the potent vasoconstrictor endothelin-1 (EDN1) in salt sensitivity.[93] In human studies, Miyaki *et al.* identified a significant interaction between the T-786C polymorphism of the endothelial nitric oxide synthase 3 (*NOS3*) gene and sodium intake on BP.[96] Dengel and colleagues later identified a similar association between this variant and BP responses to sodium intake.[95] In contrast, examination of the *EDN1*, *NOS3*, and selectin E (*SELE*) genes in the GenSalt study did not identify positive associations with this phenotype.

In 2008, Caprioli and colleagues examined the endothelin receptor type B (*EDNRB*) gene, which encodes a receptor that mediates vasodilation via nitric oxide. A significant association between the synonymous *EDNRB* G1065A variant and salt sensitivity was identified.[94] Earlier, Castejon and colleagues identified a mutation in the cytochrome *b*-245, α polypeptide (*CYBA*) gene associated with salt sensitivity in women.[92] *CYBA* encodes the superoxide-generating NADPH oxidase light chain subunit, a key player in nitric oxide availability and vascular dysfunction.[92] Thus, current evidence supports a role for the endothelial system in BP salt sensitivity. Further work in this area may provide important insights into the genetic underpinnings of this complex trait.

4. Intracellular Messengers

Both the guanine nucleotide-binding protein (G protein), β polypeptide 3 (*GNB3*) and α-adducin 1 (*ADD1*) genes have been implicated in BP salt sensitivity because of their biological effects on sodium homeostasis via sodium–proton exchanger activity and renal tubular sodium reabsorption, respectively.[97] Many reports have investigated the association between the *GNB3* C825T and *ADD1* Gly460Trp polymorphisms and BP responses to sodium.[83,97–104,122] While there is little evidence for a role of the *GNB3* C825T variant in salt sensitivity, many studies have identified increased salt sensitivity among those with the *ADD1* 460Trp allele compared to those with the 460Gly allele.[97–104] However, the majority of these studies used an acute sodium-loading protocol[98,99,102,103] or the BP response to diuretic therapy[100,104] to determine salt sensitivity. Among the few studies that examined the BP response to a dietary sodium intervention, findings have been conflicting, with a positive association reported by Grant *et al.*[101] and negative findings by Castejon *et al.*[123] and Ciechanowicz *et al.*[124]

The GenSalt study comprehensively examined common variants in the *GNB3* and *ADD1* genes for their association with BP salt sensitivity.[105] After adjustment for multiple comparisons, neither the proxy *GNB3* C825T marker (which was in perfect LD with C825T in the Han Chinese population) nor the *ADD1* Gly460Trp variant were significantly associated with BP responses to dietary sodium interventions among the GenSalt participants. Despite negative findings for these variants, results from other markers implicated both the *GNB3* and *ADD1* genes in BP salt sensitivity. Participants carrying the minor allele of the novel *GNB3* marker rs1129649 had a significantly decreased MAP response to a low-sodium intervention. Similarly, strong associations between the novel rs17833172 variant of the *ADD1* gene and SBP, DBP, and MAP responses to a high-sodium intervention and the DBP response to a low-sodium intervention were observed. Future studies aimed at replicating these novel findings in other populations are warranted.

5. Sympathetic Nervous System

One of the main mechanisms by which the sympathetic nervous system exerts its influence on BP is through its interaction with the kidney and the RAAS. Studies have demonstrated that increased renal sympathetic nerve activity results in renal vasoconstriction, with decreased glomerular filtration rate and renal blood flow, increased renal vascular resistance, increased renal tubular reabsorption of sodium and water, and increased renal release of REN and norepinephrine.[125,126] As part of this system, the genes encoding the G protein-coupled receptor kinase 4 (*GRK4*) and β-2-adrenergic receptor (*ADRB2*) have been targeted as important candidates for BP salt sensitivity.[75,106–108] Although not replicated by the GenSalt study, associations between the *ADRB2* G46A and C79A variants and BP salt sensitivity have been identified in several reports.[75,106,107] Both the G46A (Gly16Arg) and C79G (Gln27Glu) polymorphisms represent nonsynonymous coding variants of the *ADRB2* gene, which could influence the structure and function of its encoded protein. In fact, functional studies have revealed attenuated downregulation of the 46AA/79GG diplotype in response to agonist exposure compared to other diplotype combinations.[106] In total, these findings implicate ADRB2 as a potential genetic mechanism underlying salt sensitivity. Further characterization of the *ADRB2* gene, as well as other genes in this pathway, could yield additional insights into this complex phenotype.

6. Apelin–APJ System

The apelin (APLN) system, including its receptor APJ, is a peptidic signaling pathway that is implicated in the regulation of cardiovascular function and fluid homeostasis.[72,127] While SNPs in the *APLN* and apelion receptor (*APLNR*) genes have been associated with hypertension,[128] the GenSalt study was the first to examine their association with BP responses to dietary sodium intervention.[72] In this study, there was no association between variants in the *APLN* gene and salt sensitivity. However, two SNPs in the 3′UTR (untranslated region) region of *APLNR*, namely, rs2282623 and rs746886, were significantly associated with the BP response to a low-sodium intervention. Although replication work is still needed, these results suggest that the APLN–APJ system might be mechanistically involved in BP salt sensitivity.

7. Atrial Natriuretic Peptides

Genes encoding components of the natriuretic peptide system may be important candidates for BP salt sensitivity because of their BP-lowering effect and interaction with the RAAS. Studies have demonstrated a role for these peptides in the stimulation of diuresis and natriuresis via direct inhibition of sodium reabsorption in the medullary collecting duct, alteration of renal hemodynamics, and inhibition of REN and aldosterone activity.[129,130] Not surprisingly, evidence for a relationship with salt sensitivity has already been suggested by animal models.[129,131] In addition, a genome-wide association study (GWAS) meta-analysis conducted in European populations identified a robust signal for SBP at a gene cluster that includes the atrial natriuretic peptide A (*NPPA*) and B (*NPPB*) genes,[6] with further exploration of these genes providing strong evidence of a role for the *NPPA/NPPB* locus in BP regulation and hypertension.[132] In addition, a recent GWAS meta-analysis in East Asians demonstrated a genome-wide significant association signal at the natriuretic peptide receptor C/guanylate cyclase C (*NPR3*) gene.[4] Still, there is a paucity of data examining the effects of genes encoding the natriuretic peptides (*NPPA*, *NPPB*, and *NPPC*) and their receptors (*NPR1*, *NPR2*, and *NPR3*, respectively) on BP responses to sodium in human populations. To date, there has been only one report identifying an association between an *NPPA* variant and salt sensitivity, a finding that was not replicated by GenSalt.[109] Based on biological plausibility, animal evidence, and genomics studies of human hypertension, further research in this area could provide important clues for unraveling the genetic architecture of BP salt sensitivity.

8. Kallikrein–Kinin System

Derived from the enzymatic action of kallikrein (KLK) on kininogen, kinins cause vasodilation, diuresis, and natriuresis.[133] Physiological studies of mouse models, as well as epidemiological research in human populations, have linked salt sensitivity to decreased urinary KLK levels.[134,135] In addition, studies of the low-urinary-KLK rat model showed that exogenous KLK replacement prevented the exaggerated BP increase in response to a high-sodium diet.[133] These data provide compelling evidence that the KLK–kinin system is mechanistically linked to BP salt sensitivity. Svetkey and colleagues examined the Q121E variant of the *KLK1* gene, identifying a significant association between this polymorphism and the BP response to sodium intake.[75] Although GenSalt did not examine this particular polymorphism, other common *KLK1* variants as well as several additional genes in this pathway were targeted, including those encoding the B1 and B2 bradykinin receptors (*BDKRB1* and *BDKRB2*), kininogen-1 (*KNG1*), plasma KLK (*KLKB1*), KLK inhibitor (*SERPINA4*), and major metabolic enzymes involved in the conversion and degradation of kinins (*CPN1*, *CPN2*, *CPM*, *ECE1*, and *MME*). While confirmatory work is still warranted, preliminary analyses have identified variants in the *BDKRB1* and endothelin-converting enzyme 1 (*ECE1*) genes that were strongly associated with BP salt sensitivity.

D. Sequencing Studies of Salt Sensitivity

Because of the early success of Cohen and colleagues in identifying rare variants with a large influence on lipid phenotypes, sequencing studies have become popular for examining the effect of rare genetic variants on complex traits such as salt sensitivity.[136–139] Rare variants in renal salt-handling genes have already been associated with BP regulation.[140] In a 2008 report, Ji and colleagues showed that carriers of rare functional mutations in three renal salt-handling genes (solute carrier family 12 (sodium/chloride transporters), member 3 (*SLC12A3*); solute carrier family 12 (sodium/potassium/chloride transporters), member 1 (*SLC12A1*); and potassium inwardly rectifying channel, subfamily J, member 1 (*KCNJ1*)) had significantly reduced SBP (mean reduction = 9.0 mmHg, $p = 0.0002$) and DBP (mean reduction = 5.0 mmHg, $p = 0.003$) compared to noncarriers.[140] Although they did not examine the salt-sensitivity phenotype specifically, this early evidence strongly suggests that sequencing studies will be important to uncover rare variants that may confer large effects on BP responses to dietary sodium intake.

E. Future Directions in Genomics Studies of Salt Sensitivity

While candidate gene studies of salt sensitivity have yielded promising findings, we are still in the early phases of understanding the genetic etiology of this complex trait. Further candidate gene studies of key biological pathways, particularly the endothelial system, natriuretic peptide system, and KLK–kinin system, are critically needed. Moreover, GWAS will be extremely valuable for uncovering novel biological pathways and unknown mechanisms related to BP salt sensitivity. In addition, sequencing studies will help to identify functional genetic variants and pinpoint causal variants related to this trait. Finally, functional studies will be needed to confirm findings from genomic epidemiology studies.

VI. Conclusions

Although the genomic architecture of BP salt sensitivity has yet to be fully elucidated, important contributions to the field have already been made. Findings from research on monogenic BP disorders that highlight the importance of renal salt handling have paved the way for genomics studies of the salt-sensitivity phenotype, implicating genes and biological pathways likely related to this complex trait. Candidate gene studies have made great strides in delineating the genomic mechanisms underlying the BP response to salt intake. Extensive research efforts have identified genes in the RAAS; ion and water channels, transporters, and exchangers; endothelial system; intracellular messengers; sympathetic nervous system; APLN–APJ system; atrial natriuretic peptides; the KLK–kinin system; and many others related to this complex phenotype. In addition, inconsistent findings of early genomics research have taught investigators to clearly define the salt-sensitivity phenotype, use stringent significance thresholds for statistical testing, and replicate study findings in order to identify true genotype–phenotype associations. Still, much work will be necessary to unravel the genomic etiology of salt sensitivity. Further exploration of genes encoding components of the multiple systems involved in renal salt handling is warranted. In addition, GWAS will be vital to uncover novel biological pathways related to salt sensitivity, while sequencing efforts will be important for discovering functional variants related to this phenotype.

Delineating the genetic architecture of salt sensitivity will help build an understanding of how genes and dietary sodium interact to influence BP. This knowledge may assist in the development of new antihypertensive medications targeting the molecular mechanisms related to sodium homeostasis and BP regulation. In addition, establishing a relationship between genetic variants and salt sensitivity could help to identify individuals who are at high risk for hypertension and would receive maximum benefit from a low-sodium dietary intervention. Such advancements will have important public health and clinical implications, helping to curb the growing hypertension epidemic at a national and global level.

References

1. Kearney PM, Whelton M, Reynolds K, Muntner P, Whelton PK, He J. Global burden of hypertension: analysis of worldwide data. *Lancet* 2005;**365**:217–23.
2. Lopez AD, Mathers CD, Ezzati M, Jamison DT, Murray CJL. Global and regional burden of disease and risk factors, 2001: systematic analysis of population health data. *Lancet* 2006; **367**:1747–57.
3. Danaei G, Finucane MM, Lin JK, Singh GM, Paciorek CJ, Cowan MJ, et al. National, regional, and global trends in systolic blood pressure since 1980: systematic analysis of health examination surveys and epidemiological studies with 786 country-years and 5.4 million participants. *Lancet* 2011;**377**:568–77.
4. Kato N, Takeuchi F, Tabara Y, Kelly TN, Go MJ, Sim X, et al. Meta-analysis identifies five novel loci associated with blood pressure in East Asians. *Nat Genet* 2011;**43**:531–8.
5. Levy D, Ehret GB, Rice K, Verwoert GC, Launer LJ, Dehghan A, et al. Genome-wide association study of blood pressure and hypertension. *Nat Genet* 2009;**41**:677–87.
6. Newton-Cheh C, Johnson T, Gateva V, Tobin MD, Bochud M, Coin L, et al. Genome-wide association study identifies eight loci associated with blood pressure. *Nat Genet* 2009;**41**:666–76.
7. He J, Whelton PK. Salt intake, hypertension and risk of cardiovascular disease: an important public health challenge. *Int J Epidemiol* 2002;**31**:327–31 discussion 331–332.
8. Chobanian AV, Hill M. National Heart, Lung, and Blood Institute Workshop on Sodium and Blood Pressure : a critical review of current scientific evidence. *Hypertension* 2000;**35**:858–63.
9. Tobian L. Salt and hypertension. Lessons from animal models that relate to human hypertension. *Hypertension* 1991;**17**:I52–8.
10. Denton D, Weisinger R, Mundy NI, Wickings EJ, Dixson A, Moisson P, et al. The effect of increased salt intake on blood pressure of chimpanzees. *Nat Med* 1995;**1**:1009–16.
11. Elliott P, Stamler J, Nichols R, Dyer AR, Stamler R, Kesteloot H, et al. Intersalt revisited: further analyses of 24 hour sodium excretion and blood pressure within and across populations. Intersalt Cooperative Research Group. *BMJ* 1996;**312**:1249–53 [Erratum appears in *BMJ* 1997 Aug 23;315(7106):458].
12. He J, Tell GS, Tang YC, Mo PS, He GQ. Relation of electrolytes to blood pressure in men. The Yi people study. *Hypertension* 1991;**17**:378–85.
13. Cutler JA, Follmann D, Allender PS. Randomized trials of sodium reduction: an overview. *Am J Clin Nutr* 1997;**65**:643S–51S.
14. Sacks FM, Svetkey LP, Vollmer WM, Appel LJ, Bray GA, Harsha D, et al. Effects on blood pressure of reduced dietary sodium and the Dietary Approaches to Stop Hypertension (DASH) diet. DASH-Sodium Collaborative Research Group. *N Engl J Med* 2001;**344**:3–10.
15. He J, Ogden LG, Vupputuri S, Bazzano LA, Loria C, Whelton PK. Dietary sodium intake and subsequent risk of cardiovascular disease in overweight adults. *JAMA* 1999;**282**:2027–34.
16. Tuomilehto J, Jousilahti P, Rastenyte D, Moltchanov V, Tanskanen A, Pietinen P, et al. Urinary sodium excretion and cardiovascular mortality in Finland: a prospective study. *Lancet* 2001; **357**:848–51.

17. Cook NR, Cutler JA, Obarzanek E, Buring JE, Rexrode KM, Kumanyika SK, et al. Long term effects of dietary sodium reduction on cardiovascular disease outcomes: observational follow-up of the trials of hypertension prevention (TOHP). *BMJ* 2007;**334**:885–8.
18. Strazzullo P, D'Elia L, Kandala N-B, Cappuccio FP. Salt intake, stroke, and cardiovascular disease: meta-analysis of prospective studies. *BMJ* 2009;**339**:b4567.
19. Kawasaki T, Delea CS, Bartter FC, Smith H. The effect of high-sodium and low-sodium intakes on blood pressure and other related variables in human subjects with idiopathic hypertension. *Am J Med* 1978;**64**:193–8.
20. Weinberger MH. Salt sensitivity of blood pressure in humans. *Hypertension* 1996; **27**:481–90.
21. Luft FC, Weinberger MH. Heterogeneous responses to changes in dietary salt intake: the salt-sensitivity paradigm. *Am J Clin Nutr* 1997;**65**:612S–7S.
22. GenSalt Collaborative Research G. GenSalt: rationale, design, methods and baseline characteristics of study participants. *J Hum Hypertens* 2007;**21**:639–46.
23. Marteau J-B, Zaiou M, Siest G, Visvikis-Siest S. Genetic determinants of blood pressure regulation. *J Hypertens* 2005;**23**:2127–43.
24. Lifton RP, Gharavi AG, Geller DS. Molecular mechanisms of human hypertension. *Cell* 2001;**104**:545–56.
25. White PC. Disorders of aldosterone biosynthesis and action. *N Engl J Med* 1994;**331**:250–8.
26. Connell JM, MacKenzie SM, Freel EM, Fraser R, Davies E. A lifetime of aldosterone excess: long-term consequences of altered regulation of aldosterone production for cardiovascular function. *Endocr Rev* 2008;**29**:133–54.
27. Lifton RP. Molecular genetics of human blood pressure variation. *Science* 1996;**272**:676–80.
28. Ferrari P, Lovati E, Frey FJ. The role of the 11beta-hydroxysteroid dehydrogenase type 2 in human hypertension. *J Hypertens* 2000;**18**:241–8.
29. Funder JW, Pearce PT, Smith R, Smith AI. Mineralocorticoid action: target tissue specificity is enzyme, not receptor, mediated. *Science* 1988;**242**:583–5.
30. Funder JW. Mineralocorticoids, glucocorticoids, receptors and response elements. *Science* 1993;**259**:1132–3.
31. Stewart PM, Wallace AM, Valentino R, Burt D, Shackleton CH, Edwards CR. Mineralocorticoid activity of liquorice: 11-beta-hydroxysteroid dehydrogenase deficiency comes of age. *Lancet* 1987;**2**:821–4.
32. Lifton RP, Dluhy RG, Powers M, Rich GM, Cook S, Ulick S, et al. A chimaeric 11 beta-hydroxylase/aldosterone synthase gene causes glucocorticoid-remediable aldosteronism and human hypertension. *Nature* 1992;**355**:262–5.
33. Luft FC. Mendelian forms of human hypertension and mechanisms of disease. *Clin Med Res* 2003;**1**:291–300.
34. Lafferty AR, Torpy DJ, Stowasser M, Taymans SE, Lin JP, Huggard P, et al. A novel genetic locus for low renin hypertension: familial hyperaldosteronism type II maps to chromosome 7 (7p22). *J Med Genet* 2000;**37**:831–5.
35. So A, Duffy DL, Gordon RD, Jeske YWA, Lin-Su K, New MI, et al. Familial hyperaldosteronism type II is linked to the chromosome 7p22 region but also shows predicted heterogeneity. *J Hypertens* 2005;**23**:1477–84.
36. Stowasser M, Gordon RD, Tunny TJ, Klemm SA, Finn WL, Krek AL. Familial hyperaldosteronism type II: five families with a new variety of primary aldosteronism. *Clin Exp Pharmacol Physiol* 1992;**19**:319–22.
37. Geller DS, Zhang J, Wisgerhof MV, Shackleton C, Kashgarian M, Lifton RP. A novel form of human mendelian hypertension featuring nonglucocorticoid-remediable aldosteronism. *J Clin Endocrinol Metab* 2008;**93**:3117–23.

38. Choi M, Scholl UI, Yue P, Bjorklund P, Zhao B, Nelson-Williams C, et al. K+ channel mutations in adrenal aldosterone-producing adenomas and hereditary hypertension. *Science* 2011;**331**:768–72.
39. Geller DS, Farhi A, Pinkerton N, Fradley M, Moritz M, Spitzer A, et al. Activating mineralocorticoid receptor mutation in hypertension exacerbated by pregnancy. *Science* 2000; **289**:119–23.
40. Shimkets RA, Lifton RP, Canessa CM. The activity of the epithelial sodium channel is regulated by clathrin-mediated endocytosis. *J Biol Chem* 1997;**272**:25537–41.
41. Snyder PM, Price MP, McDonald FJ, Adams CM, Volk KA, Zeiher BG, et al. Mechanism by which Liddle's syndrome mutations increase activity of a human epithelial Na+ channel. *Cell* 1995;**83**:969–78.
42. Staub O, Dho S, Henry P, Correa J, Ishikawa T, McGlade J, et al. WW domains of Nedd4 bind to the proline-rich PY motifs in the epithelial Na+ channel deleted in Liddle's syndrome. *EMBO J* 1996;**15**:2371–80.
43. Gordon RD, Geddes RA, Pawsey CG, O'Halloran MW. Hypertension and severe hyperkalaemia associated with suppression of renin and aldosterone and completely reversed by dietary sodium restriction. *Australas Ann Med* 1970;**19**:287–94.
44. Kahle KT, Ring AM, Lifton RP. Molecular physiology of the WNK kinases. *Annu Rev Physiol* 2008;**70**:329–55.
45. Mansfield TA, Simon DB, Farfel Z, Bia M, Tucci JR, Lebel M, et al. Multilocus linkage of familial hyperkalaemia and hypertension, pseudohypoaldosteronism type II, to chromosomes 1q31-42 and 17p11-q21. *Nat Genet* 1997;**16**:202–5.
46. Wilson FH, Disse-Nicodeme S, Choate KA, Ishikawa K, Nelson-Williams C, Desitter I, et al. Human hypertension caused by mutations in WNK kinases. *Science* 2001;**293**:1107–12.
47. Birkenhager R, Otto E, Schurmann MJ, Vollmer M, Ruf EM, Maier-Lutz I, et al. Mutation of BSND causes Bartter syndrome with sensorineural deafness and kidney failure. *Nat Genet* 2001;**29**:310–4.
48. Nozu K, Inagaki T, Fu XJ, Nozu Y, Kaito H, Kanda K, et al. Molecular analysis of digenic inheritance in Bartter syndrome with sensorineural deafness. *J Med Genet* 2008; **45**:182–6.
49. Riazuddin S, Anwar S, Fischer M, Ahmed ZM, Khan SY, Janssen AGH, et al. Molecular basis of DFNB73: mutations of BSND can cause nonsyndromic deafness or Bartter syndrome. *Am J Hum Genet* 2009;**85**:273–80.
50. Schlingmann KP, Konrad M, Jeck N, Waldegger P, Reinalter SC, Holder M, et al. Salt wasting and deafness resulting from mutations in two chloride channels. *N Engl J Med* 2004; **350**:1314–9.
51. Simon DB, Bindra RS, Mansfield TA, Nelson-Williams C, Mendonca E, Stone R, et al. Mutations in the chloride channel gene, CLCNKB, cause Bartter's syndrome type III. *Nat Genet* 1997;**17**:171–8.
52. Simon DB, Karet FE, Hamdan JM, DiPietro A, Sanjad SA, Lifton RP. Bartter's syndrome, hypokalaemic alkalosis with hypercalciuria, is caused by mutations in the Na-K-2Cl cotransporter NKCC2. *Nat Genet* 1996;**13**:183–8.
53. Simon DB, Karet FE, Rodriguez-Soriano J, Hamdan JH, DiPietro A, Trachtman H, et al. Genetic heterogeneity of Bartter's syndrome revealed by mutations in the K+ channel, ROMK. *Nat Genet* 1996;**14**:152–6.
54. Mitsuuchi Y, Kawamoto T, Miyahara K, Ulick S, Morton DH, Naiki Y, et al. Congenitally defective aldosterone biosynthesis in humans: inactivation of the P-450C18 gene (CYP11B2) due to nucleotide deletion in CMO I deficient patients. *Biochem Biophys Res Commun* 1993;**190**:864–9.

55. Simon DB, Nelson-Williams C, Bia MJ, Ellison D, Karet FE, Molina AM, et al. Gitelman's variant of Bartter's syndrome, inherited hypokalaemic alkalosis, is caused by mutations in the thiazide-sensitive Na-Cl cotransporter. *Nat Genet* 1996;**12**:24–30.
56. Geller DS, Rodriguez-Soriano J, Vallo Boado A, Schifter S, Bayer M, Chang SS, et al. Mutations in the mineralocorticoid receptor gene cause autosomal dominant pseudohypoaldosteronism type I. *Nat Genet* 1998;**19**:279–81.
57. Chang SS, Grunder S, Hanukoglu A, Rosler A, Mathew PM, Hanukoglu I, et al. Mutations in subunits of the epithelial sodium channel cause salt wasting with hyperkalaemic acidosis, pseudohypoaldosteronism type 1. *Nat Genet* 1996;**12**:248–53.
58. Speiser PW, White PC. Congenital adrenal hyperplasia. *N Engl J Med* 2003;**349**:776–88.
59. White PC, New MI, Dupont B. Congenital adrenal hyperplasia (2). *N Engl J Med* 1987; **316**:1580–6.
60. Caulfield M, Newhouse S, Munroe P. Genetics of essential hypertension. In: Khumar D, Elliot P, editors. *Clinical cardiovascular genetics*. New York: Oxford University Press; 2010. pp. 329–36.
61. Botero-Velez M, Curtis JJ, Warnock DG. Brief report: Liddle's syndrome revisited—a disorder of sodium reabsorption in the distal tubule. *N Engl J Med* 1994;**330**:178–81.
62. He J, Gu D, Chen J, Jaquish CE, Rao DC, Hixson JE, et al. Gender difference in blood pressure responses to dietary sodium intervention in the GenSalt study. *J Hypertens* 2009;**27**:48–54.
63. Barlassina C, Dal Fiume C, Lanzani C, Manunta P, Guffanti G, Ruello A, et al. Common genetic variants and haplotypes in renal CLCNKA gene are associated to salt-sensitive hypertension. *Hum Mol Genet* 2007;**16**:1630–8.
64. Miller JZ, Weinberger MH, Christian JC, Daugherty SA. Familial resemblance in the blood pressure response to sodium restriction. *Am J Epidemiol* 1987;**126**:822–30.
65. Svetkey LP, McKeown SP, Wilson AF. Heritability of salt sensitivity in black Americans. *Hypertension* 1996;**28**:854–8.
66. Yagil C, Sapojnikov M, Kreutz R, Katni G, Lindpaintner K, Ganten D, et al. Salt susceptibility maps to chromosomes 1 and 17 with sex specificity in the Sabra rat model of hypertension. *Hypertension* 1998;**31**:119–24.
67. Cowley Jr. AW, Stoll M, Greene AS, Kaldunski ML, Roman RJ, Tonellato PJ, et al. Genetically defined risk of salt sensitivity in an intercross of Brown Norway and Dahl S rats. *Physiol Genomics* 2000;**2**:107–15.
68. Mei H, Gu D, Hixson JE, Rice TK, Chen J, Shimmin LC, Schwander K, Kelly TN, Liu D, Chen S, Huang J, Jaquish CE, Rao DC, He J. Genome-wide linkage and association study of blood pressure response to dietary sodium intervention: The GenSalt Study. *Am J Epidemiol*. [Accepted].
69. Dengel DR, Brown MD, Ferrell RE, Supiano MA. Role of angiotensin converting enzyme genotype in sodium sensitivity in older hypertensives. *Am J Hypertens* 2001;**14**:1178–84.
70. Hiraga H, Oshima T, Watanabe M, Ishida M, Ishida T, Shingu T, et al. Angiotensin I-converting enzyme gene polymorphism and salt sensitivity in essential hypertension. *Hypertension* 1996;**27**:569–72.
71. Poch E, Gonzalez D, Giner V, Bragulat E, Coca A, de La Sierra A. Molecular basis of salt sensitivity in human hypertension. Evaluation of renin-angiotensin-aldosterone system gene polymorphisms. *Hypertension* 2001;**38**:1204–9.
72. Zhao Q, Hixson JE, Rao DC, Gu D, Jaquish CE, Rice TK, et al. Genetic variants in the apelin system and blood pressure responses to dietary sodium interventions: a family-based association study. *J Hypertens* 2010;**28**:756–63.
73. Johnson AG, Nguyen TV, Davis D. Blood pressure is linked to salt intake and modulated by the angiotensinogen gene in normotensive and hypertensive elderly subjects. *J Hypertens* 2001;**19**:1053–60.

74. Schorr U, Blaschke K, Beige J, Distler A, Sharma AM. Angiotensinogen M235T variant and salt sensitivity in young normotensive Caucasians. *J Hypertens* 1999;**17**:475–9.
75. Svetkey LP, Harris EL, Martin E, Vollmer WM, Meltesen GT, Ricchiuti V, et al. Modulation of the BP response to diet by genes in the renin-angiotensin system and the adrenergic nervous system. *Am J Hypertens* 2011;**24**:209–17.
76. Hunt SC, Cook NR, Oberman A, Cutler JA, Hennekens CH, Allender PS, et al. Angiotensinogen genotype, sodium reduction, weight loss, and prevention of hypertension: trials of hypertension prevention, phase II. *Hypertension* 1998;**32**:393–401.
77. Hunt SC, Geleijnse JM, Wu LL, Witteman JC, Williams RR, Grobbee DE. Enhanced blood pressure response to mild sodium reduction in subjects with the 235T variant of the angiotensinogen gene. *Am J Hypertens* 1999;**12**:460–6.
78. Norat T, Bowman R, Luben R, Welch A, Khaw KT, Wareham N, et al. Blood pressure and interactions between the angiotensin polymorphism AGT M235T and sodium intake: a cross-sectional population study. *Am J Clin Nutr* 2008;**88**:392–7.
79. Gu D, Kelly TN, Hixson JE, Chen J, Liu D, Chen J-c, et al. Genetic variants in the renin-angiotensin-aldosterone system and salt sensitivity of blood pressure. *J Hypertens* 2010;**28**:1210–20.
80. Miyaki K, Hara A, Araki J, Zhang L, Song Y, Kimura T, et al. C3123A polymorphism of the angiotensin II type 2 receptor gene and salt sensitivity in healthy Japanese men. *J Hum Hypertens* 2006;**20**:467–9.
81. Cicila GT, Garrett MR, Lee SJ, Liu J, Dene H, Rapp JP. High-resolution mapping of the blood pressure QTL on chromosome 7 using Dahl rat congenic strains. *Genomics* 2001;**72**:51–60.
82. Iwai N, Kajimoto K, Tomoike H, Takashima N. Polymorphism of CYP11B2 determines salt sensitivity in Japanese. *Hypertension* 2007;**49**:825–31.
83. Pamies-Andreu E, Ramirez-Lorca R, Stiefel Garcia-Junco P, Muniz-Grijalbo O, Vallejo-Maroto I, Garcia Morillo S, et al. Renin-angiotensin-aldosterone system and G-protein beta-3 subunit gene polymorphisms in salt-sensitive essential hypertension. *J Hum Hypertens* 2003;**17**:187–91.
84. Wrona A, Widecka K, Adler G, Czekalski S, Ciechanowicz A. Promoter variants of aldosterone synthase gene (CYP11B2) and salt-sensitivity of blood pressure. *Pol Arch Med Wewn* 2004;**111**:191–7.
85. Lovati E, Ferrari P, Dick B, Jostarndt K, Frey BM, Frey FJ, et al. Molecular basis of human salt sensitivity: the role of the 11beta-hydroxysteroid dehydrogenase type 2. *J Clin Endocrinol Metab* 1999;**84**:3745–9.
86. Agarwal AK, Giacchetti G, Lavery G, Nikkila H, Palermo M, Ricketts M, et al. CA-Repeat polymorphism in intron 1 of HSD11B2: effects on gene expression and salt sensitivity. *Hypertension* 2000;**36**:187–94.
87. Adamzik M, Frey UH, Bitzer K, Jakob H, Baba HA, Schmieder RE, et al. A novel-1364A/C aquaporin 5 gene promoter polymorphism influences the responses to salt loading of the renin-angiotensin-aldosterone system and of blood pressure in young healthy men. *Basic Res Cardiol* 2008;**103**:598–610.
88. Zhao Q, Gu D, Hixson JE, Liu D, Rao DC, Jaquish CE, et al. Common variants in epithelial sodium channel genes contribute to salt-sensitivity of blood pressure: The GenSalt Study. *Circ Cardiovasc Genet* 2011;**4**:375–80.
89. Citterio L, Simonini M, Zagato L, Salvi E, Delli Carpini S, Lanzani C, et al. Genes involved in vasoconstriction and vasodilation system affect salt-sensitive hypertension. *PLoS One* 2011;**6**:e19620.
90. Manunta P, Lavery G, Lanzani C, Braund PS, Simonini M, Bodycote C, et al. Physiological interaction between alpha-adducin and WNK1-NEDD4L pathways on sodium-related blood pressure regulation. *Hypertension* 2008;**52**:366–72.

91. Osada Y, Miyauchi R, Goda T, Kasezawa N, Horiike H, Iida M, et al. Variations in the WNK1 gene modulates the effect of dietary intake of sodium and potassium on blood pressure determination. *J Hum Genet* 2009;**54**:474–8.
92. Castejon AM, Bracero J, Hoffmann IS, Alfieri AB, Cubeddu LX. NAD(P)H oxidase p22phox gene C242T polymorphism, nitric oxide production, salt sensitivity and cardiovascular risk factors in Hispanics. *J Hum Hypertens* 2006;**20**:772–9.
93. Shindo T, Kurihara H, Maemura K, Kurihara Y, Ueda O, Suzuki H, et al. Renal damage and salt-dependent hypertension in aged transgenic mice overexpressing endothelin-1. *J Mol Med* 2002;**80**:105–16.
94. Caprioli J, Mele C, Mossali C, Gallizioli L, Giacchetti G, Noris M, et al. Polymorphisms of EDNRB, ATG, and ACE genes in salt-sensitive hypertension. *Can J Physiol Pharmacol* 2008; **86**:505–10.
95. Dengel DR, Brown MD, Ferrell RE, Reynolds TH, Supiano MA. A preliminary study on T-786C endothelial nitric oxide synthase gene and renal hemodynamic and blood pressure responses to dietary sodium. *Physiol Res* 2007;**56**:393–401.
96. Miyaki K, Tohyama S, Murata M, Kikuchi H, Takei I, Watanabe K, et al. Salt intake affects the relation between hypertension and the T-786C polymorphism in the endothelial nitric oxide synthase gene. *Am J Hypertens* 2005;**18**:1556–62.
97. Beeks E, Kessels AGH, Kroon AA, van der Klauw MM, de Leeuw PW. Genetic predisposition to salt-sensitivity: a systematic review. *J Hypertens* 2004;**22**:1243–9.
98. Barlassina C, Schork NJ, Manunta P, Citterio L, Sciarrone M, Lanella G, et al. Synergistic effect of alpha-adducin and ACE genes causes blood pressure changes with body sodium and volume expansion. *Kidney Int* 2000;**57**:1083–90.
99. Cusi D, Barlassina C, Azzani T, Casari G, Citterio L, Devoto M, et al. Polymorphisms of alpha-adducin and salt sensitivity in patients with essential hypertension. *Lancet* 1997;**349**:1353–7 [Erratum appears in *Lancet* 1997 Aug 16;350(9076):524].
100. Glorioso N, Manunta P, Filigheddu F, Troffa C, Stella P, Barlassina C, et al. The role of alpha-adducin polymorphism in blood pressure and sodium handling regulation may not be excluded by a negative association study. *Hypertension* 1999;**34**:649–54.
101. Grant FD, Romero JR, Jeunemaitre X, Hunt SC, Hopkins PN, Hollenberg NH, et al. Low-renin hypertension, altered sodium homeostasis, and an alpha-adducin polymorphism. *Hypertension* 2002;**39**:191–6.
102. Manunta P, Cusi D, Barlassina C, Righetti M, Lanzani C, D'Amico M, et al. Alpha-adducin polymorphisms and renal sodium handling in essential hypertensive patients. *Kidney Int* 1998;**53**:1471–8.
103. Manunta P, Maillard M, Tantardini C, Simonini M, Lanzani C, Citterio L, et al. Relationships among endogenous ouabain, alpha-adducin polymorphisms and renal sodium handling in primary hypertension. *J Hypertens* 2008;**26**:914–20.
104. Sciarrone MT, Stella P, Barlassina C, Manunta P, Lanzani C, Bianchi G, et al. ACE and alpha-adducin polymorphism as markers of individual response to diuretic therapy. *Hypertension* 2003;**41**:398–403.
105. Kelly TN, Rice TK, Gu D, Hixson JE, Chen J, Liu D, et al. Novel genetic variants in the alpha-adducin and guanine nucleotide binding protein beta-polypeptide 3 genes and salt sensitivity of blood pressure. *Am J Hypertens* 2009;**22**:985–92.
106. Pojoga L, Kolatkar NS, Williams JS, Perlstein TS, Jeunemaitre X, Brown NJ, et al. Beta-2 adrenergic receptor diplotype defines a subset of salt-sensitive hypertension. *Hypertension* 2006;**48**:892–900.
107. Sun B, Williams JS, Svetkey LP, Kolatkar NS, Conlin PR. Beta2-adrenergic receptor genotype affects the renin-angiotensin-aldosterone system response to the Dietary Approaches to Stop Hypertension (DASH) dietary pattern. *Am J Clin Nutr* 2010;**92**:444–9.

108. Sanada H, Yatabe J, Midorikawa S, Hashimoto S, Watanabe T, Moore JH, et al. Single-nucleotide polymorphisms for diagnosis of salt-sensitive hypertension. *Clin Chem* 2006;**52**:352–60.
109. Widecka K, Ciechanowicz A, Adler G, Szychot E, Wodecki M, Czekalski S. Analysis of polymorphisms Sma (Hpa II) and Sca I gene precursors of atrial natriuretic peptide (ANP) in patients with essential hypertension. *Pol Arch Med Wewn* 1998;**100**:27–34.
110. Eap CB, Bochud M, Elston RC, Bovet P, Maillard MP, Nussberger J, et al. CYP3A5 and ABCB1 genes influence blood pressure and response to treatment, and their effect is modified by salt. *Hypertension* 2007;**49**:1007–14.
111. Bochud M, Eap CB, Elston RC, Bovet P, Maillard M, Schild L, et al. Association of CYP3A5 genotypes with blood pressure and renal function in African families. *J Hypertens* 2006;**24**:923–9.
112. Ho H, Pinto A, Hall SD, Flockhart DA, Li L, Skaar TC, et al. Association between the CYP3A5 genotype and blood pressure. *Hypertension* 2005;**45**:294–8.
113. Yagil C, Hubner N, Monti J, Schulz H, Sapojnikov M, Luft FC, et al. Identification of hypertension-related genes through an integrated genomic-transcriptomic approach. *Circ Res* 2005;**96**:617–25.
114. Tian Z, Greene AS, Usa K, Matus IR, Bauwens J, Pietrusz JL, et al. Renal regional proteomes in young Dahl salt-sensitive rats. *Hypertension* 2008;**51**:899–904.
115. Havlik RJ. Predictors of hypertension. Population studies. *Am J Hypertens* 1991;**4**:586S–9S.
116. Castrop H, Kurtz A. Differential nNOS gene expression in salt-sensitive and salt-resistant Dahl rats. *J Hypertens* 2001;**19**:1223–31.
117. Fava C, Danese E, Montagnana M, Sjogren M, Almgren P, Engstrom G, et al. Serine/threonine kinase 39 is a candidate gene for primary hypertension especially in women: results from two cohort studies in Swedes. *J Hypertens* 2011;**29**:484–91.
118. Kojima S, Inenaga T, Matsuoka H, Kuramochi M, Omae T, Nara Y, et al. The association between salt sensitivity of blood pressure and some polymorphic factors. *J Hypertens* 1994;**12**:797–801.
119. Brand E, Schorr U, Ringel J, Beige J, Distler A, Sharma AM. Aldosterone synthase gene (CYP11B2) C-344T polymorphism in Caucasians from the Berlin Salt-Sensitivity Trial (BeSST). *J Hypertens* 1999;**17**:1563–7.
120. Oberleithner H, Kusche-Vihrog K, Schillers H. Endothelial cells as vascular salt sensors. *Kidney Int* 2010;**77**:490–4.
121. Toda N, Arakawa K. Salt-induced hemodynamic regulation mediated by nitric oxide. *J Hypertens* 2011;**29**:415–24.
122. Schorr U, Beige J, Ringel J, Turan S, Kreutz R, Distler A, et al. Hpa II polymorphism of the atrial natriuretic peptide gene and the blood pressure response to salt intake in normotensive men. *J Hypertens* 1997;**15**:715–8.
123. Castejon AM, Alfieri AB, Hoffmann IS, Rathinavelu A, Cubeddu LX. Alpha-adducin polymorphism, salt sensitivity, nitric oxide excretion, and cardiovascular risk factors in normotensive Hispanics. *Am J Hypertens* 2003;**16**:1018–24.
124. Ciechanowicz A, Widecka K, Drozd R, Adler G, Cyrylowski L, Czekalski S. Lack of association between Gly460Trp polymorphism of alpha-adducin gene and salt sensitivity of blood pressure in Polish hypertensives. *Kidney Blood Press Res* 2001;**24**:201–6.
125. DiBona GF. Sympathetic nervous system and the kidney in hypertension. *Curr Opin Nephrol Hypertens* 2002;**11**:197–200.
126. Stella A, Zanchetti A. Interactions between the sympathetic nervous system and the kidney: experimental observations. *J Hypertens Suppl* 1985;**3**:S19–25.
127. Reaux A, De Mota N, Skultetyova I, Lenkei Z, El Messari S, Gallatz K, et al. Physiological role of a novel neuropeptide, apelin, and its receptor in the rat brain. *J Neurochem* 2001;**77**:1085–96.

128. Li W-W, Niu W-Q, Zhang Y, Wu S, Gao P-J, Zhu D-L. Family-based analysis of apelin and AGTRL1 gene polymorphisms with hypertension in Han Chinese. *J Hypertens* 2009; **27**:1194–201.
129. Melo LG, Veress AT, Chong CK, Pang SC, Flynn TG, Sonnenberg H. Salt-sensitive hypertension in ANP knockout mice: potential role of abnormal plasma renin activity. *Am J Physiol* 1998;**274**:R255–61.
130. Ellis KL, Newton-Cheh C, Wang TJ, Frampton CM, Doughty RN, Whalley GA, et al. Association of genetic variation in the natriuretic peptide system with cardiovascular outcomes. *J Mol Cell Cardiol* 2011;**50**:695–701.
131. John SW, Krege JH, Oliver PM, Hagaman JR, Hodgin JB, Pang SC, et al. Genetic decreases in atrial natriuretic peptide and salt-sensitive hypertension. *Science* 1995;**267**:679–81 [Erratum appears in *Science* 1995 Mar 24;267(5205):1753].
132. Newton-Cheh C, Larson MG, Vasan RS, Levy D, Bloch KD, Surti A, et al. Association of common variants in NPPA and NPPB with circulating natriuretic peptides and blood pressure. *Nat Genet* 2009;**41**:348–53.
133. Madeddu P, Varoni MV, Demontis MP, Chao J, Simson JA, Glorioso N, et al. Kallikrein-kinin system and blood pressure sensitivity to salt. *Hypertension* 1997;**29**:471–7.
134. Madeddu P, Varoni MV, Demontis MP, Pinna-Parpaglia P, Glorioso N, Anania V. Urinary kallikrein: a marker of blood pressure sensitivity to salt. *Kidney Int* 1996;**49**:1422–7.
135. Bonner G, Thieven B, Rutten H, Chrosch R, Krone W. Renal kallikrein is a determinant of salt sensitivity. *J Hypertens Suppl* 1993;**11**:S210–1.
136. Cohen J, Pertsemlidis A, Kotowski IK, Graham R, Garcia CK, Hobbs HH. Low LDL cholesterol in individuals of African descent resulting from frequent nonsense mutations in PCSK9. *Nat Genet* 2005;**37**:161–5 [Erratum appears in *Nat Genet* 2005 Mar;37(3):328].
137. Cohen JC, Kiss RS, Pertsemlidis A, Marcel YL, McPherson R, Hobbs HH. Multiple rare alleles contribute to low plasma levels of HDL cholesterol. *Science* 2004;**305**:869–72.
138. Cohen JC, Pertsemlidis A, Fahmi S, Esmail S, Vega GL, Grundy SM, et al. Multiple rare variants in NPC1L1 associated with reduced sterol absorption and plasma low-density lipoprotein levels. *Proc Natl Acad Sci USA* 2006;**103**:1810–5.
139. Cohen JC, Boerwinkle E, Mosley Jr. TH, Hobbs HH. Sequence variations in PCSK9, low LDL, and protection against coronary heart disease. *N Engl J Med* 2006;**354**:1264–72.
140. Ji W, Foo JN, O'Roak BJ, Zhao H, Larson MG, Simon DB, et al. Rare independent mutations in renal salt handling genes contribute to blood pressure variation. *Nat Genet* 2008;**40**:592–9.

Alcohol Intake

Dolores Corella

Genetic and Molecular Epidemiology Unit, School of Medicine, University of Valencia, Valencia, Spain

CIBER Fisiopatología de la Obesidad y Nutrición, Instituto de Salud Carlos III, Madrid, Spain

Alcohol consumption and its association with health or illness states are of great interest from the nutritional genomics point of view. This interest is centered not only on investigating the genetic variants that can modulate the effects of alcoholic beverages on different intermediate and final disease phenotypes (mainly cardiovascular diseases and cancer), but also on finding out how the genome influences the amount of alcohol consumed and consumption habits. This chapter reviews the latest findings on alcohol consumption trends, the methodological limitations in the analysis of alcohol consumption, and the main genes and polymorphisms related to alcohol intake, including the inconsistent results from genome-wide association studies (GWASs). It also reviews the effects of alcohol consumption on cardiovascular diseases and cancer and the studies analyzing the interactions between different genetic polymorphisms and alcohol in phenotypes related to these diseases, discussing the studies' advantages and limitations as well as future research perspectives.

Progress in Molecular Biology
and Translational Science, Vol. 108
DOI: 10.1016/B978-0-12-398397-8.00011-3

1877-1173/12 $35.00

I. Introduction

Alcohol intake can be studied from a number of viewpoints. When viewed from a nutritional genomics point of view, variations in the genome can be influential in determining the total amount of alcohol consumed per person, the preferences of a consumer for a certain kind of alcoholic beverage, and the different toxicological effects of the same quantity of alcohol. Likewise, variations in the genome may explain why the effects of alcohol consumption on health are different in some people than in others. Numerous gene–alcohol interactions influencing both intermediate (plasma lipid concentrations, glucose, anthropometric measures, inflammatory markers, etc.) and final disease phenotypes (cardiovascular diseases, cancer, neurodegenerative diseases, etc.) have been described.[1–3] Moreover, nutrigenomics studies have shown that alcohol may have an influence on the expression of multiple genes in different tissues, causing either an underexpression or overexpression depending on the gene.[4] This chapter analyzes the latest evidence on different nutrigenetics aspects of alcohol intake and health. It begins with an overview of the data and figures available on alcohol consumption and general information on the association of alcohol consumption with different intermediate and final disease phenotypes. The next sections discuss the studies that have analyzed the variability of these responses depending on genotype as well as the main genes associated with alcohol consumption. Finally, the chapter concludes with remarks regarding the evidence obtained.

II. General Information on Alcohol Consumption

Throughout the world, numerous drinking cultures exist, and attitudes toward alcohol vary. Alcohol has attracted general interest for thousands of years, and the scientific debate about the risks and benefits of alcohol consumption, which started several centuries ago, continues today.

A. Evolution of Alcohol Consumption and Types of Alcoholic Beverages Consumed in Various Countries

According to the Organisation for Economic Co-operation and Development (OECD) *Health Data 2010*, released on 29 June 2010 (www.oecd.org/health/healthdata), the weighted yearly average alcohol consumption in countries of the OECD, per population aged above 15 years, was 9.6 l/capita. This figure is for the year 2003, which was the year when complete data were available for most countries. In that year, the country with the greatest alcohol consumption was Luxemburg with 15.5 l/capita, followed by France (14.8 l),

Ireland (13.5 l), Hungary (13.4 l), the Czech Republic (12.1 l), and Spain (11.7 l). The United States ranked 20th with 8.3 l/capita, followed by Poland (8.1 l), Italy (8.0), Canada (7.8 l), and Japan (7.6 l). The country with the lowest consumption was Turkey with 1.5 l/capita. According to OECD data, alcohol consumption has, in general, been falling slightly. Figure 1 shows the alcohol-consumption data per capita in 16 selected countries from 1960 to 2008. Geographic variability is observed, and it appears to be associated with culture, customs, religion, eating habits, and perhaps other factors.

Besides the large differences in overall alcohol consumption, there is also great variability in the geographic distribution and evolution of the types of alcoholic beverages consumed. Hence, according to data obtained with the same methodology in various countries by the Food and Agriculture Organization of the United Nations, published in the *World Drink Trends* report of 2003, different patterns of consumption of wine, beer, and spirits can be observed depending on the country analyzed. As an example, Fig. 2 shows those trends in three selected countries: France, the United States, and Japan. In France, just as in most Mediterranean countries, there has been a sharp drop in wine consumption, the alcoholic beverage that in previous decades occupied the number one position. In some countries such as Spain, consumption of beer now exceeds that of wine. In the United States, the leading

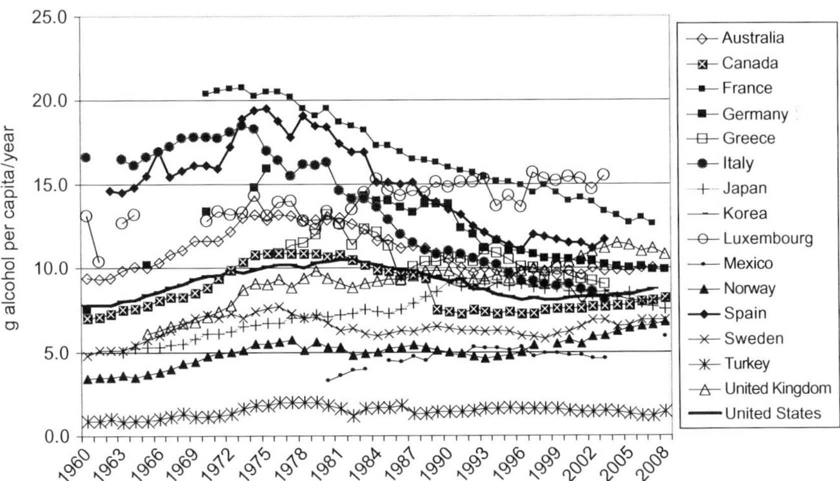

FIG. 1. Trends in alcohol consumption (1960–2008) in selected countries expressed in grams of ethanol per capita per year, according to the Organisation for Economic Co-operation and Development (OECD) *Health Data 2010*. (For color version of this figure, the reader is referred to the online version of this chapter.)

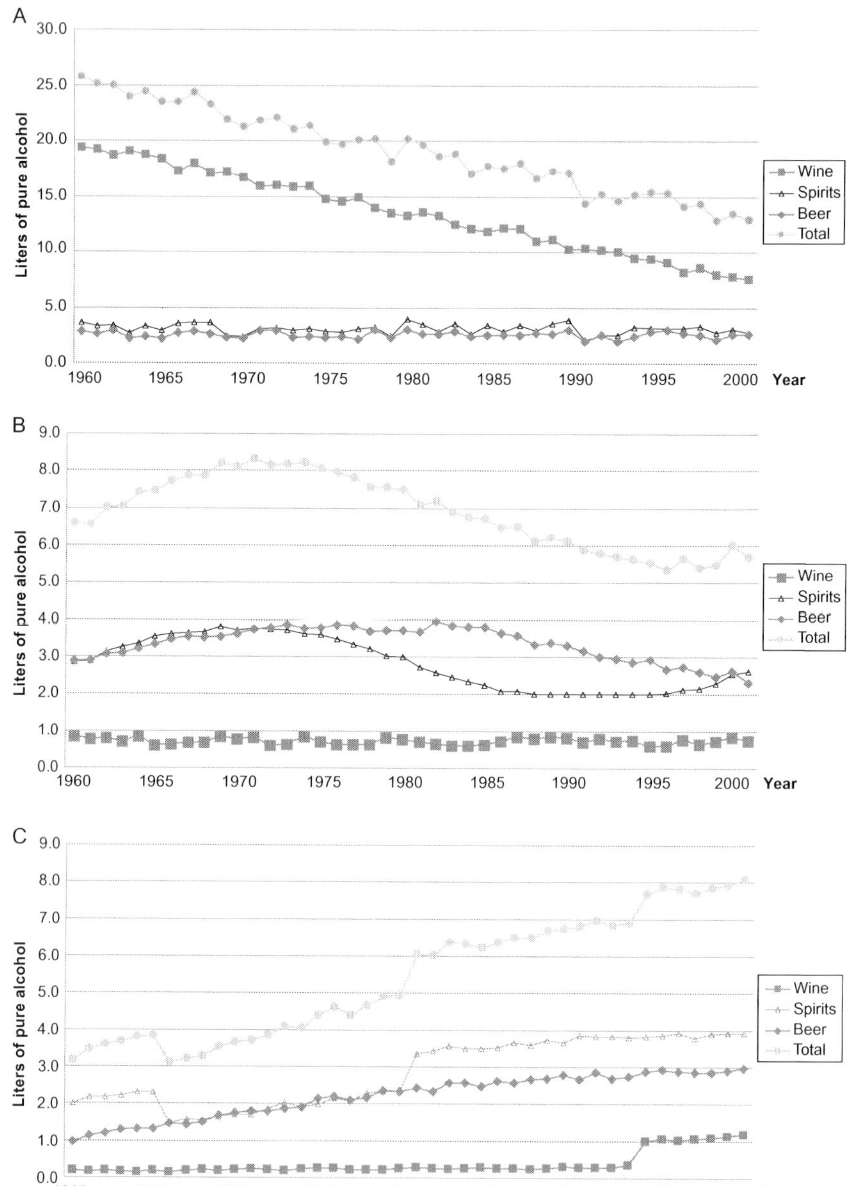

FIG. 2. Trends in the consumption of wine, beer, or spirits in three selected countries: France (A), the United States (B), and Japan (C), according to the data of the Food and Agriculture Organization of the United Nations, published in the *World Drink Trends* report of 2003.

alcoholic beverage consumed is beer, while the consumption of spirits has been falling in recent decades. In Japan, the main alcoholic beverages consumed are spirits, followed by beer, whereas wine consumption is of limited relevance, quantitatively speaking.

B. Patterns of Alcohol Consumption and Tools for Measuring Them

Alcohol consumption varies widely between individuals and even throughout the life of an individual. The variability in consumption patterns hinders evaluation of the effects of alcohol consumption on health.[5] Another factor adding to the complexity of accurately measuring alcohol intake is that consumption is conventionally expressed in grams of ethanol. However, participants in studies cannot be asked "how many grams of ethanol you consume" because people do not know this information and they remember only the type of drink consumed and the approximate amount. For this reason, researchers have to ask about the number of "drinks," "units," "bottles," or "cans" a person typically consumes, depending on the national culture. Once the information on the number of "drinks" has been obtained, it is necessary to transform that number of drinks into the quantity of ethanol contained in order to generate estimates of alcohol consumption. However, once again, definitions describing how much alcohol is there in a drink vary remarkably.[5–10] Although there are differences among countries,[9] the National Institute on Alcohol Abuse and Alcoholism definition stipulates that a standard drink is 17.74 ml of pure alcohol. This amount is equivalent to 354.8 ml of beer, 147.9 ml of wine, or 44.4 ml of (40% alcohol by volume) hard liquor.[10]

There is wide variation in the types of questionnaires used to measure alcohol consumption. In the large-scale studies, detailed questions are not usually asked about alcohol, as their inclusion would increase the time involved in administering the questionnaire. Bearing in mind that assessment time is limited but that interest in alcohol consumption is high, a task force of the National Institute on Alcohol Abuse and Alcoholism recommended that all studies include at least three questions in order to capture patterns of alcohol consumption over a period of 12 months.[11] These include questions about the frequency of drinking, the number of drinks consumed in a typical drinking day, and the frequency of binge drinking (5 or more standard drinks within a 2-h period for men; 4 or more for women). This three-question set is similar to the three alcohol consumption questions from the Alcohol Use Disorders Identification Test (AUDIT).[12] Another of the questionnaires most frequently used as a screening tool for alcohol problems in clinical practice, clinical studies, and in general population studies is the CAGE questionnaire.[13,14] This questionnaire contains four questions each of which contributes toward the acronym CAGE: (1) Have you ever tried to *c*ut down on your drinking, (2) do you get *a*nnoyed when people talk about your drinking, (3) do you feel *g*uilty about your drinking, and (4) have you ever had an *e*ye-opener (a drink first thing in the morning) to steady your nerves or get rid of a hangover?

There are various types of questionnaires validated for each country and population type studied.[15–17] Depending on the questionnaire, an over or underestimate of the amount of alcohol consumed may be produced. To minimize the use of questionnaires, research has focused on discovering the biomarkers of alcohol consumption. However, at present, there is no good biomarker of average alcohol intake.[18]

Although alcohol concentrations in breath or blood are indicators of recent alcohol intake, a person's blood-alcohol or breath-alcohol concentrations are constantly changing, both during and after drinking, and the final result depends on many environmental and genetic factors. Among the biomarkers most used in epidemiological studies as approximations to alcohol consumption are carbohydrate-deficient transferrin (CDT) and gamma-glutamyltransferase (GGT) levels in plasma.[18] The World Health Organization (WHO)/International Society for Biomedical Research on Alcoholism Study on State and Trait Markers of Alcohol Use and Dependence was designed to assess and compare alcohol use markers and traits of alcohol dependence in a multicenter trial. This study included subjects aged 18 and over, recruited in Australia, Brazil, Canada, Finland, and Japan. The alcohol consumption categories were nondrinkers, current drinkers, and persons receiving treatment for alcohol dependence. The authors concluded that CDT and GGT had comparable performance as biomarkers. They also found that variables such as gender, age, body mass index, and quantity of alcohol consumed influenced the test results.[19] Thus, CDT was a better marker of high-risk consumption in men. All biomarkers were more effective for detection of high-risk rather than intermediate-risk drinking. More recent studies have arrived at similar conclusions.[20,21] In addition to CDT and GGT, commonly used biomarkers include mean corpuscular volume, aspartate aminotransferase, alanine aminotransferase, sialylation of apolipoprotein, ethyl glucuronide, and 5-hydroxytryptophol.[21]

In addition, there are no standardized criteria for classifying the amount of alcohol consumption. Thus, terms such as light, moderate, and heavy drinking are variably defined.[6] There is also no general consensus on when to consider an individual a moderate or excessive drinker, with different cut-off points proposed with different criteria.[5–7]

The number of abstainers also varies by country. WHO defines abstainers as people who abstain from drinking alcohol, either over the year preceding the survey (last-year abstainers) or throughout their life (lifetime abstainers). Religion can play an important role in determining the number of abstainers. Muslim countries have the highest levels of abstinence. According to WHO data, the percentage of last-year abstainers ranged from 2.5% in Luxembourg to 99.5% in Egypt.[22] Other relevant factors that are involved in determining the number of abstainers are gender (across cultures, traditionally more women abstain from alcohol than men), age, and the integration of alcohol use in the customs of each region (celebrations, wine consumption at meals, etc.).

C. Alcohol Metabolism

To understand alcohol metabolism, it is important not only to know the metabolites that are produced and their possible toxic action, but also to investigate the candidate genes codifying the main enzymes involved in alcohol metabolism and their influence on alcohol consumption and on the health effects of that consumption.

Alcohol is the favorite mood-altering drug in most parts of the world, and its effects, both pleasant and unpleasant, are well known. Each individual can have a different perception of the pleasant and unpleasant effects of alcohol, and that conditions the amount of alcohol consumed and the consumption pattern. The alcohol in a beverage, on consumption, passes from the stomach and intestines into the blood, a process referred to as absorption. Elimination of alcohol is achieved mostly through metabolism (95–98%), with a small fraction being excreted through exhalation, sweating, or urine. In the liver, ethanol is metabolized to acetaldehyde by alcohol dehydrogenase (ADH), cytochrome P450 2E1 (CYP2E1), and, to a much lesser extent, catalase.[23,24] ADH1 (class I ADH) is the key enzyme in alcohol metabolism in the body. Catalase and CYP2E1 form part of what is called "the non-ADH pathway," and its contribution is, generally speaking, minor. However, this pathway has a greater metabolic role when the level of blood alcohol is high or when the intake of alcohol is chronic.[25]

The human ADH family has been grouped into five classes (I–V). Class I ADH has three separate gene loci (*ADH1A*, *ADH1B*, and *ADH1C*). The genes of the human ADH family cluster in a region of chromosome 4q23, spanning 370 kb, in a tandem array with the same transcriptional orientation in the order 5′-ADH4–ADH1C–ADH1B–ADH1A–ADH5–ADH2–ADH3-3′.[26] Class I and class II ADH contribute to hepatic metabolism of ethanol, whereas class I ADH1C and class IV ADH participate in gastric metabolism. Class III ADH seems to be more involved in gastrointestinal metabolism at very high ethanol concentrations. The role of class V ADH remains unknown.[26] Interestingly, the prevalence of the different polymorphisms in the ADH family genes varies among populations of the world. Thus, the ADH1B*1 allele is prevalent among Caucasians and American Indians. ADH1B*2 is predominant among East Asians, ADH1B*3 is more prevalent in African populations, and ADH1C*1 is predominant among East Asians and African populations.[27]

Acetaldehyde is further oxidized to acetate, mainly by acetaldehyde dehydrogenase (ALDH), before being cleared into the systemic circulation. The human ALDH superfamily comprises 10 families which have been mapped to 11 chromosomes.[28] The superfamily contains related enzymes that metabolize a wide spectrum of aldehydes. Cytosolic ALDH1A1 and mitochondrial ALDH2 primarily contribute to the oxidation of acetaldehyde. The *ALDH1A1* and *ALDH2* genes have been mapped to chromosomes 9q21 and 12q24, respectively.[29]

The liver can metabolize only a certain amount of alcohol per hour. The rate of alcohol metabolism depends on the activity of metabolizing enzymes, which varies among individuals. Many studies have analyzed the influence on alcohol consumption of polymorphisms of the two key enzymes involved in ethanol metabolism (ADH and ALDH). Certain ADH1B and ADH1C alleles encode particularly active ADH enzymes, resulting in more rapid conversion of ethanol to acetaldehyde, with a protective effect on the risk of high alcohol consumption.[30] Similarly, there is a polymorphism in *ALDH2* (consisting of an amino acid substitution from glutamic acid to lysine at the 504th position), highly prevalent in populations from East Asia, that is associated with a less active ALDH2 and, hence, with high blood acetaldehyde levels upon ethanol intake.[30,31] In these individuals, alcohol intake leads to a number of unpleasant effects, including nausea, dizziness, and a characteristic facial flushing response.[31] These unfavorable effects appear to be protective against high alcohol consumption, reducing the risk of alcoholism by a very high percentage[32,33] (see also Chapter "Adaptive Genetic Variation and Population Differences"). This mechanism has led to the development of a drug called disulfiram which inhibits hepatic ADH. Acetaldehyde is responsible for the unpleasant effects that remain until the alcohol consumed is metabolized. The effects of disulfiram last for 1 or 2 weeks after taking the last dosage. This drug has been successfully used for treating alcoholism.[34]

III. Alcohol and Health

Alcoholic drinks not only contain different amounts of ethanol per volume consumed but also contain different quantities of carbohydrates, micronutrients, and other nonnutritional components. These nonnutritional components can have protective effects on health, possibly by confounding the effect of ethanol and thus reducing the risk of disease. In the case of red wine, the ethanol content is accompanied by resveratrol, flavonoids, and other antioxidant compounds that can have a significant anti-inflammatory and cardioprotective effect.[35] Thus, numerous studies have shown that, to determine the effects of alcohol consumption on health, one has to take into account the quantity of ethanol consumed, the type of drink consumed, and the drinking pattern.[36,37] There is, therefore, some evidence that wine may have more beneficial effects than beer and distilled spirits and that the cardioprotective effect of alcohol is generally higher for steady versus binge drinking.[37] Data also suggest that some of the benefits of alcoholic beverages in general, and wine in particular, are the result of socioeconomic confounders.[37] It is difficult to find studies that analyze in detail the effect of different alcoholic beverages on health and the impact of socioeconomic factors on these effects.

Before reviewing genetic variants that modulate the association of alcohol consumption with various health problems, we shall first briefly summarize the evidence on the effect of alcohol consumption on different diseases. Although high alcohol consumption is an important public health problem, potentially resulting in violent behavior, absence from work, increased risk of accidents, hepatic diseases, mental disorders, alterations in pregnancy, addiction, and other health problems,[38–40] this review focuses only on the effects that alcohol consumption has on cancer and cardiovascular diseases.

For the most prevalent diseases, alcohol consumption appears to have opposing effects.[41–43] In general, most studies find a direct association between alcohol consumption and a greater risk of cancer,[41,42] whereas for cardiovascular diseases, the overall effect of alcohol consumption may be protective, although this association is not linear, as it is more favorable only for moderate alcohol consumption.[43]

A. Alcohol Consumption and Cancer

Studies of the associations between alcohol consumption and cancer have reported conflicting results. For some types of cancer, there are still inconsistent associations, whereas for other types the results are more convincing. Currently, there is strong evidence that alcohol consumption increases the risks of cancers of the liver, oral cavity, pharynx, larynx, and esophagus.[44–47] The risks tend to increase with the amount of ethanol drunk. Nevertheless, there is still great controversy over whether alcohol consumption increases or decreases the risk of gastric cancer, breast cancer, prostate cancer, bladder cancer, or lung cancer, among others.[44] There are also disagreements and little information on the specific association of the kinds of alcoholic drinks with these cancers, given that several studies have observed a protective effect of red wine consumption but a harmful effect of spirits consumption.

1. Liver Cancer

Heavy, long-term alcohol use, together with hepatitis B or C virus infections, is the main risk factor for hepatocellular carcinoma.[48] In around 80% of cases, hepatocellular carcinoma is associated with cirrhosis or advanced fibrosis, inflammation, and oxidative stress, and high alcohol consumption contributes to the development of these conditions. A recent study combined the results from various prospective cohort studies with representative population-based data on alcohol exposure in eight countries (France, Italy, Spain, United Kingdom, the Netherlands, Greece, Germany, and Denmark) participating in the European Prospective Investigation into Cancer and Nutrition (EPIC) study,[44] including 109,118 men and 254,870 women. Alcohol consumption (former or current) was associated with a significant 17% (10–25%) increase in the rate of liver cancer.

2. Cancer of the Oral Cavity, Pharynx, Larynx, and Esophagus

Evidence for the human carcinogenic effects of alcohol consumption on the risk of oral cavity and pharynx cancers was considered sufficient by the International Agency for Research on Cancer (IARC) in 1988.[49] Turati and colleagues[45] updated the estimation of alcohol consumption risk associated with oral cavity cancer by combining findings from all case–control and cohort studies published up to September 2009, using a meta-analytical approach. According to their estimates, the risk of oral cancer is statistically significant in light drinkers. Thus, compared to nondrinkers or occasional drinkers, the overall relative risks (RRs) for light drinkers were 1.17 (95% confidence interval [CI]: 1.01–1.35) for oral and 1.23 (95% CI: 0.87–1.73) for pharyngeal cancer. This risk increased considerably in heavy drinkers: RRs for heavy drinkers were 4.64 (95% CI: 3.78–5.70) for oral and 6.62 (95% CI: 4.72–9.29) for pharyngeal cancer. Islami *et al.*[47] conducted a meta-analysis to estimate the association between alcohol consumption and the risk of laryngeal cancer; 38 case–control and 2 cohort studies reporting on at least three levels of alcohol consumption were included. Overall, alcohol drinking versus nondrinking was associated with an approximately twofold increase in risk of laryngeal cancer (RR = 1.90; 95% CI: 1.59–2.28), while light alcohol drinking ($\geq$1 drink/day) did not show any significant association with risk of laryngeal cancer. Moderate drinking (>1 to <4 drinks/day) was associated with a 1.5-fold increase in risk (RR = 1.47; 95% CI: 1.25–1.72), and heavy drinking ($\geq$4 drinks/day) was associated with a 2.5-fold increased risk (RR = 2.62; 95% CI: 2.13–3.23).

A recent review and meta-analysis[50] focused on the association between esophageal squamous cell carcinoma (ESCC) and alcohol drinking. The authors included 40 case–control and 13 cohort studies in the analysis. A significant association between alcohol consumption and a greater risk of ESCC, even with light alcohol drinking ($\leq$12.5 g/day), RR = 1.38 (95% CI: 1.14–1.67), was found with adjustment for age, sex, and tobacco smoking. These estimates were also significant for moderate drinkers (>12.5 to <50 g/day), RR = 2.15 (95% CI: 1.55–2.98), and for high alcohol intake ($\geq$50 g/day), RR = 3.35 (95% CI: 2.06–5.46). The association was slightly stronger in Asian countries than in other populations, and the authors suggested a possible role for genetic susceptibility factors modulating the association between alcohol consumption and ESCC.

3. Gastric Cancer

In 1988, IARC concluded that there was inadequate evidence for the carcinogenicity of alcohol on gastric cancer.[49] However, the suggestion that alcohol consumption might be associated with an increased risk of gastric cancer has become stronger, as there may have been confounding because of tobacco smoking and diet in the results of earlier studies.[41] We shall here consider only the data from a recent meta-analysis published by Tramacere *et al.*[51] These authors identified 44 case–control and 15 cohort studies, including a total of 34,557 gastric cancer cases. Although the overall RR of gastric cancer and alcohol drinking (drinkers vs. nondrinkers) was statistically significant (RR = 1.07; 95% CI: 1.01–1.13), some inconsistencies were observed. The risk estimate was nonsignificant for cohort studies (RR = 1.04; 95% CI: 0.97–1.11). When the authors considered the dose–response relationship, they found no association between moderate alcohol drinking and gastric cancer risk. Interestingly, the results for heavy drinkers varied by geographic area, depending on where the study was carried out. Thus, the risk of gastric cancer for heavy alcohol drinking was nonsignificant, RR = 0.90 (95% CI: 0.65–1.25), among Asians but statistically significant, RR = 1.39 (95% CI: 1.14–1.69), among non-Asian populations. To explain this heterogeneity, the authors suggested the possible role of *ADH* and *ALDH* polymorphisms, which have a different prevalence between Asian and non-Asian populations.[30–33]

4. Colorectal Cancer

For many years, studies on the association between alcohol consumption and the risk of colorectal cancer have reported contradictory results, although many have concluded that there is a greater risk of colon cancer. Hence, in a 2007 review of the associations between alcohol and different types of cancer, the IARC included colorectal cancer as a cancer localization related to alcohol consumption.[52] Fedirko *et al.*, in a recent meta-analysis, included 27 cohort and 34 case–control studies with results for at least three categories of alcohol intake published before May 2010.[53] They found a strong association between alcohol consumption and colorectal cancer, which followed a dose–response relationship. The greater the alcohol consumption, the greater the risk of colorectal cancer. Thus, compared with nondrinkers or occasional drinkers, the RR for moderate drinkers was 1.21 (95% CI: 1.13–1.28) and that for heavy drinkers was 1.52 (95% CI: 1.27–1.81). As with gastric cancer, the authors found heterogeneity depending on the geographical area. However, in contrast to gastric cancer, the colorectal cancer risk for heavy drinkers was greater in studies carried out in Asian populations.

5. Breast Cancer

As with gastric cancer, there is great heterogeneity in the results published on the association between alcohol consumption and breast cancer. In the IARC study published in 2007,[52] the breast was included as a cancer localization related to alcohol consumption. A recent meta-analysis[54] identified 35 articles that had studied the relationship between dietary patterns and breast cancer, of which 16 were included. Dietary patterns were classified as "prudent/healthy," "Western/unhealthy," and "drinker." The combined analysis showed an increase in the risk of breast cancer for the highest compared to the lowest categories of the drinker dietary pattern (odds ratio [OR] = 1.21; 95% CI: 1.04–1.41; $P = 0.01$). The authors hypothesized that increased alcohol consumption can lead to higher estrogen concentrations and that these estrogens contribute to an increased risk of cancer.

6. Other Cancer Sites

For prostate cancer, a recent review[55] concluded that there is no evidence that moderate alcohol consumption of up to about 3 drinks/day increases prostate cancer risk; however, heavy consumption of about 7 or more drinks/day may be associated with a statistically significant increase in risk.

For bladder cancer, a recent meta-analysis[56] based on 19 studies did not observe any association between alcohol consumption and bladder cancer (OR = 1.00; 95% CI: 0.89–1.10). Curiously, in the specific beverage analysis, beer consumption was associated with a lower risk of bladder cancer (OR = 0.86; 95% CI: 0.76–0.96). Wine consumption presented a similar association pattern on the limit of statistical significance (OR = 0.85; 95% CI: 0.71–1.00).

For endometrial cancer, another recent meta-analysis[57] concluded that alcohol intake was not significantly associated with the risk of this cancer among prospective studies, RR = 1.04 (95% CI: 0.91–1.18), or among case–control studies, OR = 0.89 (95% CI: 0.76–1.05). However, as in the previous case, heterogeneity was found for the type of alcohol analyzed. A greater consumption of spirits did show an association with a higher risk of endometrial cancer (RR = 1.22; 95% CI: 1.03–1.45), but not a greater consumption of wine or beer.

Finally, the association between alcohol consumption and lung cancer is also controversial, mainly because of the confounding effect resulting from the frequent association between tobacco smoking and alcohol. A recent meta-analysis estimated the association between alcohol intake and lung cancer risk in individuals who have never smoked.[58] The conclusion was that alcohol consumption does not increase the risk of lung cancer in nonsmokers; RR for drinkers versus nondrinkers was 1.21 (95% CI: 0.95–1.55).

B. Alcohol Consumption and Cardiovascular Diseases

For several decades now, epidemiological studies have suggested that moderate alcohol consumption reduces the risk of cardiovascular diseases.[59] In the absence of large prospective clinical trials, dozens of studies, both ecological and case–control and cohort, have shown an inverse relationship between moderate alcohol consumption and several outcomes related to cardiovascular diseases.[43,59,60] A meta-analysis[43] of prospective cohort studies has recently been conducted on the association between alcohol consumption and overall mortality from cardiovascular diseases, the incidence of and mortality from coronary heart disease, and the incidence of and mortality from stroke. Following a selection process of the more than 4000 studies found, 84 were included in the final analysis. The duration of follow-up for study end points ranged from 2.5 to 35 years, with a mean follow-up of 11 years. The primary exposure variable was the presence of active alcohol drinking at baseline compared with a reference group of nondrinkers. Alcohol intake was categorized as <2.5 (<0.5 drink), 2.5–14.9 (about 0.5–1 drink), 15–29.9 (about 1–2.5 drinks), 30–60 (about 2.5–5 drinks), and >60 g/day (≥ 5 drinks). Overall, for cardiovascular disease mortality and for incidence and mortality for coronary heart disease, alcohol consumption was associated with lower risk. These figures were RR = 0.75 (95% CI: 0.70–0.80) for cardiovascular disease mortality, RR = 0.71 (95% CI: 0.66–0.77) for incident coronary heart disease, and RR = 0.75 (95% CI: 0.68–0.81) for coronary heart disease mortality. However, for stroke, no statistically significant protective associations were found: RR = 0.98 (95% CI: 0.91–1.06) for incident stroke and RR = 1.06 (95% CI: 0.91–1.23) for stroke mortality.

Analyses of the doses of alcohol consumed showed that approximately 1 drink/day was protective for all five outcomes compared with no alcohol intake. An alcohol consumption of >60 g/day (≥ 5 drinks) was not, however, significantly associated with a protective effect against overall mortality for cardiovascular diseases, although it was for coronary heart disease mortality, RR = 0.75 (95% CI: 0.63–0.89). In contrast, this high level of alcohol consumption was significantly associated with a higher incidence of stroke, RR = 1.62 (95% CI: 1.32–1.98).

The meta-analysis carried out by Ronksley *et al.*[43] evaluated the effects of alcohol consumption on individuals free from cardiovascular diseases at baseline. However, moderate alcohol consumption also presented favorable effects on the secondary prevention of cardiovascular diseases in another study.[61] Costanzo *et al.*[62] undertook a meta-analysis of the articles published up to October 2009 and examined the relationship between alcohol and cardiovascular mortality and total mortality in patients with a history of cardiovascular events. After identifying 54 publications, only 8 publications were included in the meta-analysis for reasons of quality control. These eight publications included 16,351 patients with a history of cardiovascular diseases. All selected studies were prospective.

Results showed that moderate alcohol consumption presented significant protection against cardiovascular mortality in these patients. Researchers found a J-shaped pooled curve with a significant maximal protection (average 22%) by alcohol at approximately 26 g/day (about 1.5 drinks/day). Total mortality was also inversely associated with moderate alcohol consumption.

The results of these recent meta-analyses have helped consolidate the protective role of moderate alcohol consumption in both primary and secondary outcomes of cardiovascular diseases. However, the mechanisms that give rise to this protection are not well understood. For several decades, there has been evidence that alcohol consumption increases the plasma concentrations of high-density lipoprotein cholesterol (HDL-C) as well as apolipoprotein A-I, these two biomarkers being associated with lower cardiovascular risk.[63] Apart from a favorable change in plasma lipid concentrations, recent studies have observed a favorable effect on other markers of inflammation and endothelial damage, as well as on hemostatic factors.

A recent meta-analysis reviewed the results from interventional studies on the effects of alcohol consumption on 21 biological markers associated with coronary heart disease risk in adults without known cardiovascular diseases.[64] The markers studied were lipids (triglycerides, total cholesterol, HDL-C, low-density lipoprotein cholesterol [LDL-C], lipoprotein(a), and apolipoprotein A-I), inflammatory markers (C-reactive protein, leukocytes, interleukin-6, and tumor necrosis factor α), hemostatic factors (plasminogen activator inhibitor 1, von Willebrand factor, tissue plasminogen activator, fibrinogen, and e-selectin), endothelial cell function markers (intracellular adhesion molecule 1 and vascular cell adhesion molecule), and hormones (leptin and adiponectin). After undertaking a screening of thousands of articles, 63 relevant articles were finally selected in accordance with the selection criteria. Of these, 44 articles on 13 biomarkers were meta-analyzed in fixed or random effects models. Most of these studies analyzed plasma lipid biomarkers. Based on the pooled results, the authors found that alcohol consumption significantly increased HDL-C and apolipoprotein A-I concentrations, with linear dose–response relationships. In contrast, alcohol did not significantly change the levels of total cholesterol, LDL-C, triglycerides, or lipoprotein(a). Researchers also observed that a very high dose of alcohol (>60 g/day) caused an increase in plasma triglyceride concentrations, but these results come from only two studies.

Regarding inflammatory markers, the associations of alcohol with concentrations of C-reactive protein, interleukin-6, and tumor necrosis factor α were not significant. Neither were the results significant for hemostatic factors, apart from fibrinogen, whose plasma concentrations significantly decreased after alcohol consumption. For the other markers analyzed, significant results were found only for adiponectin, whose concentrations increased following intervention with alcohol.

IV. Main Genes Related to Alcohol Intake

It has been previously observed that many of the effects of alcohol consumption on health are dose-dependent. It is now not only a question of whether one consumes alcohol or not, but that greater or lesser amounts of alcohol intake have notably different effects on health. Thus, it is important to know what factors determine an individual's consumption of more or less alcohol and the different consumption patterns that arise. In addition, the types of alcoholic beverages consumed are also important.

The determinants of alcoholic beverage consumption represent a complex network of genetic and environmental factors that is still not well understood. Multiple studies have been undertaken to characterize the environmental determinants of alcohol consumption.[65–67] Outstanding among these determinants are the socioeconomic level, the alcohol consumption behavior of relatives, a person's social network, and immediate neighbors and co-workers, a permissive environment, availability of cheap alcohol, loud music, tobacco smoking, low levels of social support, dysfunctional coping strategies, social myths, social disturbances, violence, and drug abuse, among others. Importance is also given to the genetic factors determining alcohol consumption, although they have only recently begun to be identified.

A. Candidate Genes

The first investigations of genetic factors determining alcohol consumption focused on the so-called candidate genes related to the different pathways on which alcohol acted or was metabolized.[30,31,68] Hence, as ethanol interacts with several neurotransmitters (dopaminergic, glutamatergic, opioidergic, and cannabinoid) and neuromodulators and these interactions are involved in the development and maintenance of alcohol self-administration, the genes related to these pathways were among the first candidates to be investigated.[68–71] Variants in neuropeptide Y, a major endogenous regulator of anxiety-related behaviors and emotions, have been related to alcohol consumption.[72,73] Similarly, a central serotonin (5-HT) deficit is thought to be involved in the pathogenesis of alcohol dependence by modulating motivational behavior.[74,75] Variations in many genes that encode receptors, enzymes, and transporters of the 5-HT system have been tested as risk factors for alcohol dependence with discordant results.[75–77] Dozens of polymorphisms in other neurotransmitters and drug-related receptors (e.g., gamma-aminobutyric acid-A, glutamate, cannabinoid and opioid receptors) have also been studied for their relationships with alcohol.[78–81] The studies that have analyzed these polymorphisms and their associations with alcohol consumption have been reviewed recently by Kimura and Higuchi.[82]

Continuing with this research approach into certain genes that are candidates because their peptide products are functionally related to ethanol, the polymorphisms in alcohol-metabolizing enzymes,[26–33] including *ALDH*, *ADH*, and *CYP2E1*, were also considered in the recent review of Kimura and Higuchi.[82] Other candidate genes that may be important in determining the amount of alcohol consumed and the types of beverages consumed are those that codify the taste receptors.[83] Initial studies have suggested that individuals have different capacities for perceiving the taste of ethanol. Thus, in an experiment that administered 10% ethanol in which participants were asked to indicate its taste, the majority of subjects described 10% ethanol as bitter, and 30% of the subjects described it as sweet and/or sour.[84] In mammals, sweetness perception is initiated when sweeteners interact with taste receptor proteins from the taste receptor, type 1 (*TAS1R*) family expressed in taste receptor cells in taste buds of the oral cavity. This family includes three genes: *TAS1R1*, *TAS1R2*, and *TAS1R3*. Considerable detail on these receptors, their variations, and their associations with the consumption of certain foods has recently been provided by Bachmanov *et al.*[85] (also see Chapter "Taste Preferences"). The mechanism that may link variants in these genes with alcohol consumption is based on the observation that lingual application of ethanol activates gustatory nerves.[86] Moreover, central mechanisms that determine hedonic responses to ethanol and sweeteners also overlap and involve opioidergic, serotonergic, and dopaminergic brain neurotransmitter systems.[87] Furthermore, variations in genes related to the perception of bitter taste, fundamentally in the bitter receptors TAS2R (TAS2R16 and TAS2R38), have been associated in some studies with the amount of alcohol consumed, as well as with the preference for the consumption of bitter drinks.[88,89] However, the results of other studies have not been consistent.[90,91]

One of the factors that may affect the consistency of these study results is the definition of "alcohol consumption" itself, as there is great heterogeneity in the phenotypes considered. Thus, many studies have analyzed the association of these polymorphisms with alcohol dependence, while others have considered regular intake versus nonconsumption of alcohol and only a few have studied the association of polymorphisms with amount of alcohol consumed. It is not, therefore, surprising that the results are heterogeneous.

B. Genome-Wide Association Studies

In addition to the classic approach of searching for polymorphisms in candidate genes, hypothesis-free studies have been undertaken to find new genes related to alcohol consumption. The first genome-wide linkage studies were carried out in the Collaborative Study on the Genetics of Alcoholism

(COGA), a large family-based data set including more than 300 extended families densely affected by alcoholism.[92] Linkage studies have allowed identification of chromosomal regions associated with dependence.[93,94]

Later, as array technology advanced, genome-wide association studies (GWASs) have been undertaken to identify new loci related to alcohol consumption. The first such GWAS was published in 2006[95] and was undertaken with participants in the COGA study. Researchers analyzed about 100,000 single-nucleotide polymorphisms (SNPs) in 120 alcohol-dependence cases and 160 controls. Results showed that 188 SNPs clustering in 51 regions satisfied the criteria of genome-wide significance and suggested enrichment in genes related to gene regulation, cellular signaling, and development. However, this study was limited by its small sample size and relatively small number of SNPs analyzed. Another GWAS was undertaken on a larger sample size (487 German cases, 1358 controls) with a greater number of measured SNPs (500,000 SNPs).[96]

Following these initial GWASs, others have been carried out but have provided inconsistent information. Thus, Bierut *et al.*[97] genotyped 1 million polymorphisms in 1897 European-American and African-American subjects with alcohol dependence and in 1932 unrelated, alcohol-exposed, nondependent controls. None of the polymorphisms met genome-wide significance; 15 polymorphisms yielded $P < 10^{-5}$, but in two independent replication series, no polymorphism passed the replication threshold ($P < 0.05$). Likewise, in the COGA participants, no single polymorphism met the genome-wide criteria for significance.[98] However, a cluster of genes on chromosome 11 (*SLC22A18*, *PHLDA2*, *NAP1L4*, *SNORA54*, *CARS*, and *OSBPL5*) that presented low *P*-values and replicated previous findings with alcohol dependence was observed. More significant results have been obtained in a GWAS undertaken in 1064 unrelated individuals drawn from a study of alcohol dependence.[99] The most relevant SNP was rs7916403 in the serotonin receptor 7 gene (*HTR7*) on chromosome 10q23.

Other interesting results were obtained in 1721 Korean male drinkers in which alcohol consumption was investigated as a quantitative trait.[100] Researchers found that 12 SNPs on chromosome 12q24 had genome-wide significant associations with alcohol consumption (*P*-value range: 1×10^{-7} to 3.7×10^{-48}). In a sample of 1113 male drinkers from an independent cohort, SNPs in or near chromosome 12 open reading frame 51 (*C12orf51*), coiled-coil domain containing 63 (*CCDC63*), and myosin, light chain 2, regulatory cardiac, slow (*MYL2*) on chromosome 12q24 were successfully replicated. Carriers with minor alleles of these SNPs showed reduced alcohol consumption. Interestingly, rs2074356 in *C12orf51* was in high linkage disequilibrium with SNPs in *ALDH2*, a gene for which the variant allele has less activity to eliminate acetaldehyde, increasing the concentrations of this compound in blood and resulting in adverse effects.[30–33] It would now be of great interest to undertake a meta-analysis of these GWASs to verify whether greater consistency can be attained.

V. Gene–Alcohol Interactions in Determining Cardiovascular Diseases

Although moderate alcohol consumption has been associated with a lower risk of cardiovascular disease incidence and mortality, these estimations do not take into account possible genetic heterogeneity. From a nutrigenetics point of view, it is interesting to find out whether the effects of alcohol consumption on intermediate and disease phenotypes of cardiovascular diseases are the same for all individuals, or whether there are important differences in the effects of alcohol consumption (or of the different types of alcoholic beverages) depending on the genotype. It would also be interesting to find out which genes are the most relevant in determining these effects and whether a number of them act in a synergetic or antagonistic way. Over the past 10 years, dozens of reports on different populations and focusing on different genes have been published.[1,101,102] However, the results are heterogeneous, and we still do not have a sufficiently high level of confidence to conclude that specific genetic polymorphisms modify the effects of alcohol consumption on intermediate and disease phenotypes. Because of the heterogeneity of the results, this section focuses on only two classic polymorphisms in cardiovascular epidemiology.

One of these is the common polymorphism in the apolipoprotein E (*APOE*) gene (alleles E2, E3, and E4). In 2001, we published one of the first gene–alcohol interactions determining LDL-C concentrations in the Framingham Heart Study.[103] We showed that alcohol consumption acts by diminishing plasma LDL-C concentrations in E2 individuals, whereas it increases LDL-C in E4 subjects. Later, Djoussé and colleagues[104] reported an interaction between alcohol consumption and the *APOE* polymorphism on HDL-C concentrations. They found that the increase in HDL-C associated with alcohol consumption is stronger in subjects without the E4 allele. In carriers of the E4 allele, HDL-C concentrations did not significantly increase with increasing alcohol consumption.

The apparently deleterious effect of alcohol consumption in carriers of the E4 allele and higher protective effects in carriers of the E2 allele have been supported by recent results from a nested case–control study in the Spanish EPIC cohort.[105] In this study, healthy men and women (41,440 participants) were followed up over a 10-year period for incident coronary heart disease. A significant interaction between alcohol consumption and the *APOE* genotype in determining LDL-C concentrations was found; E2 individuals had lower LDL-C concentrations if they were alcohol consumers than if they were not. In agreement with Djoussé *et al.*,[104] the *APOE* polymorphism was not associated with HDL-C concentrations in nondrinkers, while in drinkers, the *APOE* polymorphism was associated with HDL-C ($P = 0.013$) with a decreasing effect from E2 to E4. Moreover, in the same Spanish EPIC study,[105] an interaction between alcohol consumption and the *APOE* genotype was observed on in determining the incidence of coronary heart disease. In the nondrinker category, the *APOE* polymorphism was not significantly associated with coronary heart disease incidence. However, in drinkers, E2 carriers did present significantly lower coronary heart disease incidence, with the risk in E4 carriers being higher. These differences were magnified with higher alcohol consumption. These opposite effects observed in E2 and E4 carriers need to be confirmed in other studies.

A second polymorphism of interest is the cholesterol ester transfer protein, plasma (*CETP*)-TaqIB (alleles B1 and B2). An initial study found an interaction between this polymorphism and alcohol consumption in determining HDL-C concentrations and cardiovascular disease risk[106] in men participating in the Etude Cas Témoins de l'Infarctus du Myocarde (ECTIM) study. The increasing effect of B2 on HDL-C concentrations was absent in subjects drinking < 25 g/day of alcohol but markedly increased with higher values of alcohol consumption (P for interaction < 0.001). The risk for myocardial infarction in B2 homozygotes decreased from 1.0 in nondrinkers to 0.34 in those drinking ≥ 75 g/day. Following this observation, there have been divergent results published on this apparent interaction.[107–112] In the Framingham Heart Study[101,107] and in the Prevención con Dieta Mediterránea (PREDIMED) Study,[108] no statistically significant alcohol-TaqIB interaction in determining HDL-C concentrations was found. Similarly, in the meta-analysis of Boekholdt *et al.*[113] in about 13,000 individuals, no statistically significant interaction was observed for the *CETP*-TaqIB polymorphism and alcohol consumption (drinkers vs. nondrinkers) on HDL-C concentrations. Interestingly, in a recent study undertaken on the participants of the EPIC-Spain cohort,[112] we also did not find a significant interaction between alcohol consumption and the CETP-TaqIB polymorphism in determining HDL-C, but we detected a greater risk of coronary heart disease incidence in homozygous subjects for the B2 allele who drink. In nondrinkers, the B2B2 genotype was associated with a nonsignificant lower coronary heart disease risk, whereas in drinkers it was associated with a higher risk (OR = 1.55; 95% CI: 1.05–2.29). These results illustrate how misleading it can be to translate the results from a single study to clinical practice; replication of results is absolutely essential and in various populations.

Table I summarizes the results of other relevant studies that have focused on genes related to alcohol metabolism[114–122] and that have reported gene–alcohol interactions on intermediate and cardiovascular disease phenotypes. Despite the interesting results on gene–alcohol interactions and cardiovascular disease phenotypes,[123–127] the replication record of such studies is very low.

TABLE I

RESULTS OF SELECTED STUDIES ANALYZING GENE–ALCOHOL INTERACTION ON CARDIOVASCULAR DISEASE PHENOTYPES

Reference	Study and population	Outcome	Exposure	Genes	Main results
Hines *et al.*[114]	Nested case–control study in the prospective Physicians' Health Study	Newly diagnosed myocardial infarction (396 cases and 2 matched controls) HDL-C in this study and in a replication sample of women	Alcohol consumption assessed by questionnaire	*ADH3* (or *ADH1C*) genotype (rs698: gamma1/2 alleles)	The *ADH3* genotype significantly modified the effect of alcohol on the risk of myocardial infarction. Homozygous subjects for the gamma2 allele who consumed at least 1 drink/day, had the greatest reduction in risk. These subjects (men and women in the replication sample) also had the highest HDL-C concentrations
Younis *et al.*[115]	Study in middle-aged men participating in the prospective Second Northwick Park Heart Study (NPHS II)	Incidence of coronary heart disease (220 cases) and HDL-C concentrations	Alcohol consumption assessed by questionnaire	*ADH1C* genotype (gamma1/2 alleles)	A statistically significant interaction between the *ADH1C* genotype and alcohol on coronary heart disease risk (higher reduction in risk in gamma2 homozygotes with moderate alcohol drinking). No significant interaction on HDL-C
Djoussé *et al.*[116]	Study on unrelated subjects in the Framingham Offspring Study	Prevalent cardiovascular disease and HDL-C concentrations	Alcohol consumption assessed by questionnaire	*ADH1C* rs1693482 and rs698	Borderline significant interactions between *ADH1C* polymorphisms and alcohol on cardiovascular disease. No significant interactions on HDL-C
Marques-Vidal *et al.*[117]	Cross-sectional study in a representative population sample from France	Intima-media thickness and HDL-C concentrations	Alcohol consumption assessed by questionnaire	*ADH1C* genotype (gamma1/2 alleles)	No significant interactions between the *ADH1C* polymorphism and alcohol on intima-media thickness or HDL-C concentrations

Heidrich *et al.*[118]	Prospective cohort study in the MONICA/ KORA-Augsburg cohort 1994/ 1995–2002	Incidence of coronary heart disease and HDL-C concentrations	Alcohol consumption assessed by questionnaire	*ADH1C* genotype (gamma1/2 alleles)	*ADH1C* modified the effect of alcohol consumption on coronary risk. No statistically significant interactions on HDL-C concentrations
Ebrahim *et al.*[119]	Prospective study in two cohorts: British Women's Heart and Health Study and Caerphilly cohorts	Incidence of coronary heart disease and HDL-C concentrations	Alcohol consumption assessed by questionnaire	*ADH1C* genotype (gamma1/2 alleles)	No evidence of interactions between *ADH1C* variants and alcohol intake on HDL-C or coronary heart disease incidence
Latella *et al.*[120]	Cross-sectional study of 974 healthy European subjects	HDL-C concentrations and other related factors	Alcohol consumption assessed by questionnaire	*ADH1C* genotype	No significant interaction between alcohol consumption and the *ADH1C* polymorphism on HDL-C
Tolstrup *et al.*[121]	Prospective study in a Danish general population in the Copenhagen City Heart Study	Incidence of myocardial infarction (663 cases) and HDL-C concentrations	Alcohol consumption assessed by questionnaire	*ADH1C* and *ADH1B* genotypes	No significant interactions between alcohol consumption and *ADH1C* or *ADH1B* polymorphisms on HDL-C concentrations or myocardial infarction risk
Husemoen *et al.*[122]	Cross-sectional study in Northern European men and women from Denmark	HDL-C concentrations and other plasma lipids and metabolic syndrome measurements	Alcohol consumption assessed by questionnaire	*ADH1B*, *ADH1C*, *ADH7*, *ALDH2*, *ALDH1B1*, and *ALDH1B1*	No statistically significant interactions between alcohol consumption and the *ADH1C* or the other genotypes in determining HDL-C concentrations. Significant interactions between alcohol and *ADH1B* (rs1229984) on LDL-C and between alcohol and *ALDH2* (rs886205) on impaired glucose/diabetes

VI. Gene–Alcohol Interactions in Determining Cancer Risk

There are a variety of studies that have investigated the association between polymorphisms in enzymes involved in alcohol metabolism and alcohol-related cancers.[128–131] Outstanding among them is the work conducted by Hashibe *et al.*[128] analyzing six *ADH* polymorphisms in over 3800 aerodigestive cancer cases (and 5200 controls) from three individual studies. They found that polymorphisms in the *ADH1B* (rs1229984) and in the *ADH7* (rs1573496) genes were significantly associated with a lower risk of cancer. The main metabolite of ethanol, namely, acetaldehyde, can induce DNA lesions, which if left unrepaired can initiate carcinogenesis.[132] Thus, polymorphisms that have functional differences in enzyme activity lead to differences in acetaldehyde exposure among drinkers which may result in variable cancer risk. When the authors[128] analyzed these associations taking into account the amount of alcohol consumed, these associations were more apparent with increasing alcohol intake. Druesne-Pecollo and colleagues[2] undertook a review of several studies and concluded that the data lend support to the existence of some interactions between polymorphisms in the *ADH1B* and *ALDH2* genes and alcohol consumption increased the risk of cancer. For other polymorphisms, the results were insufficient or inconclusive. Table II summarizes some of the most relevant studies focused on gene–alcohol interactions on cancer risk.[129,133–137]

VII. Concluding Remarks

Despite the difficulties encountered in measuring alcohol consumption and in classifying consumption patterns, there is substantial evidence on the protective effect of moderate alcohol consumption in the prevention of cardiovascular diseases. This evidence is corroborated by various meta-analyses. However, there are no long-term intervention trials to provide the highest level of evidence on this apparent protective effect of alcohol or on the type of alcoholic beverages consumed. This protection appears to be higher for coronary heart disease than for stroke. A higher risk of stroke was actually associated with high levels of alcohol consumption. Alcohol consumption also appears to be associated with a higher risk of cancer, with the most consistent associations found in liver, upper aerodigestive tract (i.e., oral cavity, pharynx, or esophagus), colorectal, and breast cancers.

There are numerous studies showing that some genetic polymorphisms modulate the effects of alcohol consumption on intermediate and final disease phenotypes by increasing or reducing risk in certain individuals. However, the results of these studies are very heterogeneous. Reports are often based on different variants, and the level of consistency among studies that analyzed the

TABLE II

Results of Selected Studies Analyzing Gene–Alcohol Interaction on Cancer

Reference	Study and population	Outcome	Exposure	Genes	Main results
Oze *et al.*[129]	Case–control study in Japanese	585 Upper aerodigestive tract cancer cases and 1170 controls	Alcohol consumption assessed by questionnaire	*ADH4*, *ADH7*, *ADH1B*, *ADH1C*, and *ALDH2* genotypes	Polymorphisms in *ADH4*, *ADH1B*, *ADH1C*, and *ADH7* were associated with risk of aerodigestive tract cancers. This association was greater in subjects with a high alcohol consumption
Hashibe *et al.*[133]	Multicenter case–control study in European countries	811 Upper aerodigestive tract cancer cases and 1083 controls	Alcohol consumption assessed by questionnaire	*ADH1B*, *ADH1C*, and *ALDH2*	Polymorphisms in the *ADH1B* and *ALDH2* genes were associated with upper aerodigestive tract cancer and interact with alcohol consumption, increasing the risk
Platek *et al.*[134]	Case–control study in the Western New York Exposures and Breast Cancer study	Breast cancer cases (1063) and controls (1890)	Alcohol consumption assessed by questionnaire	*MTHFR* and *MTR* genotypes	Among postmenopausal women, there was an increase in breast cancer risk for women who were homozygote TT for *MTHFR* C677T and had high lifetime alcohol intake
Benzon Larsen *et al.*[135]	Nested case–control study in the prospective Diet, Cancer and Health study	Breast cancer cases (809 postmenopausal) and 809 controls	Alcohol consumption assessed by questionnaire	*ADH1C* polymorphism (Arg272Gln)	In carriers of the variant allele, alcohol intake increased the risk of breast cancer 14% (95% CI: 1.04–1.24) per 10 g alcohol/day, but not among wild-type subjects

(*Continues*)

TABLE II (*Continued*)

Reference	Study and population	Outcome	Exposure	Genes	Main results
Zhang *et al.*[136]	Meta-analysis in Chinese Han population	Esophageal cancer (1450 cases and 2459 controls)	Alcohol consumption assessed by questionnaire	*ADH1B* and *ALDH2* genotypes	Polymorphisms *ADH1B* His47Arg and *ALDH2* Glu487Lys were significantly associated with esophageal cancer and interacted with alcohol consumption, increasing the risk
Shin *et al.*[137]	Case–control study in Korean population	445 Patients with gastric cancer and 370 controls	Alcohol consumption assessed by questionnaire	*ALDH2* genotype	Alcohol consumption interacted with the *ALDH2* polymorphism in determining gastric cancer risk. Among heavy drinkers, carriers of the variant allele at *ALDH2* had a fourfold increased risk compared with wild-type homozygotes

same genetic variant is very low. It would therefore be useful if the methodology for studying gene–alcohol interactions was standardized in any new study undertaken in order to improve the consistency level among studies. Finally, although real progress has been made in the identification of variant genes that have an influence on alcohol consumption, the methodology employed for studying their influence still needs to be improved. The same is true for the study of interaction between genetic variants, alcohol consumption, and complex social and environmental factors.

Acknowledgments

This work was supported by grants from the Ministerio de Ciencia e Innovación, Spain, and Fondo Europeo de Desarrollo Regional: CIBER CB06/03/0035, PI07-0954, CNIC-06, AGL2006-14228-C03-03 and AGL2010-22319-C03-03, and the Generalitat Valenciana (GVACOMP2011-151, AP111/10, CS2011-AP-042, ACOMP2012-190 and BEST11/263), Valencia, Spain.

References

1. Corella D. Gene-alcohol interactions in the metabolic syndrome. *Nutr Metab Cardiovasc Dis* 2007;**17**:140–7.
2. Druesne-Pecollo N, Tehard B, Mallet Y, Gerber M, Norat T, Hercberg S, et al. Alcohol and genetic polymorphisms: effect on risk of alcohol-related cancer. *Lancet Oncol* 2009;**10**:173–80.
3. Young-Wolff KC, Enoch MA, Prescott CA. The influence of gene-environment interactions on alcohol consumption and alcohol use disorders: a comprehensive review. *Clin Psychol Rev* 2011;**31**:800–16.
4. Farris SP, Wolen AR, Miles MF. Using expression genetics to study the neurobiology of ethanol and alcoholism. *Int Rev Neurobiol* 2010;**91**:95–128.
5. Bloomfield K, Stockwell T, Gmel G, Rehn N. International comparisons of alcohol consumption. *Alcohol Res Health* 2003;**27**:95–109.
6. Kloner RA, Rezkalla SH. To drink or not to drink? That is the question. *Circulation* 2007; **116**:1306–17.
7. Leeman RF, Heilig M, Cunningham CL, Stephens DN, Duka T, O'Malley SS. Ethanol consumption: how should we measure it? Achieving consilience between human and animal phenotypes. *Addict Biol* 2010;**15**:109–24.
8. Greenfield TK, Kerr WC, Bond J, Ye Y, Stockwell T. Improving graduated frequencies alcohol measures for monitoring consumption patterns: results from an Australian national survey and a US diary validity study. *Contemp Drug Probl* 2009;**36**:75056015.
9. Devos-Comby L, Lange JE. "My drink is larger than yours"? A literature review of self-defined drink sizes and standard drinks. *Curr Drug Abuse Rev* 2008;**1**:162–76.
10. National Institute on Alcohol Abuse and Alcoholism. . *Helping patients who drink too much: a clinician's guide, updated 2005 edition.*. Bethesda, MD: National Institute on Alcohol Abuse and Alcoholism; 2005 Reprinted May 2007. NIH Publication No. 07-3769.
11. National Institute on Alcohol Abuse and Alcoholism. . *Task Force on Recommended Alcohol Questions—National Council on Alcohol Abuse and Alcoholism, recommended sets of alcohol consumption questions*. October 15–16, 2003. http://www.niaaa.nih.gov/Resources/ResearchResources/TaskForce.htm Accessed 16 January 2012.

12. Berner MM, Kriston L, Bentele M, Härter M. The alcohol use disorders identification test for detecting at-risk drinking: a systematic review and meta-analysis. *J Stud Alcohol Drugs* 2007;**68**:461–73.
13. Dhalla S, Kopec JA. The CAGE questionnaire for alcohol misuse: a review of reliability and validity studies. *Clin Invest Med* 2007;**30**:33–41.
14. Skogen JC, Overland S, Knudsen AK, Mykletun A. Concurrent validity of the CAGE questionnaire. The Nord-Trøndelag Health Study. *Addict Behav* 2011;**36**:302–7.
15. Heeb JL, Gmel G. Measuring alcohol consumption: a comparison of graduated frequency, quantity frequency, and weekly recall diary methods in a general population survey. *Addict Behav* 2005;**30**:403–13.
16. Friesema IH, Veenstra MY, Zwietering PJ, Knottnerus JA, Garretsen HF, Lemmens PH. Measurement of lifetime alcohol intake: utility of a self-administered questionnaire. *Am J Epidemiol* 2004;**159**:809–17.
17. Giovannucci E, Colditz G, Stampfer MJ, Rimm EB, Litin L, Sampson L, et al. The assessment of alcohol consumption by a simple self-administered questionnaire. *Am J Epidemiol* 1991; **133**:810–7.
18. Niemelä O, Alatalo P. Biomarkers of alcohol consumption and related liver disease. *Scand J Clin Lab Invest* 2010;**70**:305–12.
19. Conigrave KM, Degenhardt LJ, Whitfield JB, Saunders JB, Helander A, Tabakoff B, et al. CDT, GGT, and AST as markers of alcohol use: the WHO/ISBRA collaborative project. *Alcohol Clin Exp Res* 2002;**26**:332–9.
20. Bergström JP, Helander A. Clinical characteristics of carbohydrate-deficient transferrin (% disialotransferrin) measured by HPLC: sensitivity, specificity, gender effects, and relationship with other alcohol biomarkers. *Alcohol Alcohol* 2008;**43**:436–41.
21. Freeman WM, Vrana KE. Future prospects for biomarkers of alcohol consumption and alcohol-induced disorders. *Alcohol Clin Exp Res* 2010;**34**:946–54.
22. Glanz J, Grant B, Monteiro M, Tabakoff B, WHO/ISBRA Study on State and Trait Markers of Alcohol Use and Dependence Investigators . WHO/ISBRA Study on State and Trait Markers of Alcohol Use and Dependence: analysis of demographic, behavioral, physiologic, and drinking variables that contribute to dependence and seeking treatment. International Society on Biomedical Research on Alcoholism. *Alcohol Clin Exp Res* 2002;**26**:1047–61.
23. Seitz HK, Stickel F. Molecular mechanisms of alcohol-mediated carcinogenesis. *Nat Rev Cancer* 2007;**7**:599–612.
24. Zhang Y, Ren J. ALDH2 in alcoholic heart diseases: molecular mechanism and clinical implications. *Pharmacol Ther* 2011;**132**:86–95.
25. Lieber CS, DeCarli LM. The role of the hepatic microsomal ethanol oxidizing system (MEOS) for ethanol metabolism in vivo. *J Pharmacol Exp Ther* 1972;**181**:279–87.
26. Chen YC, Peng GS, Wang MF, Tsao TP, Yin SJ. Polymorphism of ethanol-metabolism genes and alcoholism: correlation of allelic variations with the pharmacokinetic and pharmacodynamic consequences. *Chem Biol Interact* 2009;**178**:2–7.
27. Liu J, Zhou Z, Hodgkinson CA, Yuan Q, Shen PH, Mulligan CJ, et al. Haplotype-based study of the association of alcohol-metabolizing genes with alcohol dependence in four independent populations. *Alcohol Clin Exp Res* 2011;**35**:304–16.
28. Zakhari S. Overview: how is alcohol metabolized by the body? *Alcohol Res Health* 2006;**29**:245–54.
29. Vasiliou V, Bairoch A, Tipton KF, Nebert DW. Eukaryotic aldehyde dehydrogenase (ALDH) genes: human polymorphisms, and recommended nomenclature based on divergent evolution and chromosomal mapping. *Pharmacogenetics* 1999;**9**:421–34.
30. Edenberg HJ. The genetics of alcohol metabolism: role of alcohol dehydrogenase and aldehyde dehydrogenase variants. *Alcohol Res Health* 2007;**30**:5–13.

31. Pautassi RM, Camarini R, Quadros IM, Miczek KA, Israel Y. Genetic and environmental influences on ethanol consumption: perspectives from preclinical research. *Alcohol Clin Exp Res* 2010;**34**:976–87.
32. Luczak SE, Elvine-Kreis B, Shea SH, Carr LG, Wall TL. Genetic risk for alcoholism relates to level of response to alcohol in Asian-American men and women. *J Stud Alcohol* 2002;**63**:74–82.
33. Peng GS, Yin SJ. Effect of the allelic variants of aldehyde dehydrogenase ALDH2°2 and alcohol dehydrogenase ADH1B°2 on blood acetaldehyde concentrations. *Hum Genomics* 2009;**3**:121–7.
34. Barth KS, Malcolm RJ. Disulfiram: an old therapeutic with new applications. *CNS Neurol Disord Drug Targets* 2010;**9**:5–12.
35. Rodrigo R, Miranda A, Vergara L. Modulation of endogenous antioxidant system by wine polyphenols in human disease. *Clin Chim Acta* 2011;**412**:410–24.
36. Rehm J, Sempos CT, Trevisan M. Alcohol and cardiovascular disease—more than one paradox to consider. Average volume of alcohol consumption, patterns of drinking and risk of coronary heart disease—a review. *J Cardiovasc Risk* 2003;**10**:15–20.
37. Tolstrup J, Grønbaek M. Alcohol and atherosclerosis: recent insights. *Curr Atheroscler Rep* 2007;**9**:116–24.
38. McCambridge J, McAlaney J, Rowe R. Adult consequences of late adolescent alcohol consumption: a systematic review of cohort studies. *PLoS Med* 2011;**8**:e1000413.
39. Taylor B, Irving HM, Kanteres F, Room R, Borges G, Cherpitel C, et al. The more you drink, the harder you fall: a systematic review and meta-analysis of how acute alcohol consumption and injury or collision risk increase together. *Drug Alcohol Depend* 2010;**110**:108–16.
40. Mukamal KJ, Rimm EB. Alcohol consumption: risks and benefits. *Curr Atheroscler Rep* 2008;**10**:536–43.
41. Li Y, Yang H, Cao J. Association between alcohol consumption and cancers in the Chinese population—a systematic review and meta-analysis. *PLoS One* 2011;**6**:e18776.
42. Inoue M, Nagata C, Tsuji I, Sugawara Y, Wakai K, Tamakoshi A, et al. Impact of alcohol intake on total mortality and mortality from major causes in Japan: a pooled analysis of six large-scale cohort studies. *J Epidemiol Community Health* 2012;**66**:448–56.
43. Ronksley PE, Brien SE, Turner BJ, Mukamal KJ, Ghali WA. Association of alcohol consumption with selected cardiovascular disease outcomes: a systematic review and meta-analysis. *BMJ* 2011;**342**:d671.
44. Schütze M, Boeing H, Pischon T, Rehm J, Kehoe T, Gmel G, et al. Alcohol attributable burden of incidence of cancer in eight European countries based on results from prospective cohort study. *BMJ* 2011;**342**:d1584.
45. Turati F, Garavello W, Tramacere I, Bagnardi V, Rota M, Scotti L, et al. A meta-analysis of alcohol drinking and oral and pharyngeal cancers. Part 2: results by subsites. *Oral Oncol* 2010;**46**:720–6.
46. Tramacere I, Negri E, Bagnardi V, Garavello W, Rota M, Scotti L, et al. A meta-analysis of alcohol drinking and oral and pharyngeal cancers. Part 1: overall results and dose-risk relation. *Oral Oncol* 2010;**46**:497–503.
47. Islami F, Tramacere I, Rota M, Bagnardi V, Fedirko V, Scotti L, et al. Alcohol drinking and laryngeal cancer: overall and dose-risk relation—a systematic review and meta-analysis. *Oral Oncol* 2010;**46**:802–10.
48. Schütte K, Bornschein J, Malfertheiner P. Hepatocellular carcinoma—epidemiological trends and risk factors. *Dig Dis* 2009;**27**:80–92.
49. Schütte . *Alcohol drinking: summary of data reported and evaluation.*, vol. 44. Lyon:International Agency for Research on Cancer; 1998.

50. Islami F, Fedirko V, Tramacere I, Bagnardi V, Jenab M, Scotti L, et al. Alcohol drinking and esophageal squamous cell carcinoma with focus on light-drinkers and never-smokers: a systematic review and meta-analysis. *Int J Cancer* 2011;**129**:2473–84.
51. Tramacere I, Negri E, Pelucchi C, Bagnardi V, Rota M, Scotti L, et al. A meta-analysis on alcohol drinking and gastric cancer risk. *Ann Oncol* 2012;**23**:28–36.
52. Baan R, Straif K, Grosse Y, Secretan B, El Ghissassi F, Bouvard F, et al. Carcinogenicity of alcoholic beverages. *Lancet Oncol* 2007;**8**:292–3.
53. Fedirko V, Tramacere I, Bagnardi V, Rota M, Scotti L, Islami F, et al. Alcohol drinking and colorectal cancer risk: an overall and dose-response meta-analysis of published studies. *Ann Oncol* 2011;**22**:1958–72.
54. Brennan SF, Cantwell MM, Cardwell CR, Velentzis LS, Woodside JV. Dietary patterns and breast cancer risk: a systematic review and meta-analysis. *Am J Clin Nutr* 2010;**91**:1294–302.
55. Rizos C, Papassava M, Golias C, Charalabopoulos K. Alcohol consumption and prostate cancer: a mini review. *Exp Oncol* 2010;**32**:66–70.
56. Mao Q, Lin Y, Zheng X, Qin J, Yang K, Xie L. A meta-analysis of alcohol intake and risk of bladder cancer. *Cancer Causes Control* 2010;**21**:1843–50.
57. Sun Q, Xu L, Zhou B, Wang Y, Jing Y, Wang B. Alcohol consumption and the risk of endometrial cancer: a meta-analysis. *Asia Pac J Clin Nutr* 2011;**20**:125–33.
58. Bagnardi V, Rota M, Botteri E, Scotti L, Jenab M, Bellocco R, et al. Alcohol consumption and lung cancer risk in never smokers: a meta-analysis. *Ann Oncol* 2011;**22**:2631–9.
59. Di Castelnuovo A, Rotondo S, Iacoviello L, Donati MB, De Gaetano G. Meta-analysis of wine and beer consumption in relation to vascular risk. *Circulation* 2002;**105**:2836–44.
60. Mukamal KJ, Conigrave KM, Mittleman MA, Camargo Jr. CA, Stampfer MJ, Willett WC, et al. Roles of drinking pattern and type of alcohol consumed in coronary heart disease in men. *N Engl J Med* 2003;**348**:109–18.
61. Costanzo S, Di Castelnuovo A, Donati MB, Iacoviello L, de Gaetano G. Cardiovascular and overall mortality risk in relation to alcohol consumption in patients with cardiovascular disease. *Circulation* 2010;**121**:1951–9.
62. Costanzo S, Di Castelnuovo A, Donati MB, Iacoviello L, de Gaetano G. Alcohol consumption and mortality in patients with cardiovascular disease: a meta-analysis. *J Am Coll Cardiol* 2010;**55**:1339–47.
63. Rimm EB, Williams P, Fosher K, Criqui M, Stampfer MJ. Moderate alcohol intake and lower risk of coronary heart disease: meta-analysis of effects on lipids and haemostatic factors. *BMJ* 1999;**319**:1523–8.
64. Brien SE, Ronksley PE, Turner BJ, Mukamal KJ, Ghali WA. Effect of alcohol consumption on biological markers associated with risk of coronary heart disease: systematic review and meta-analysis of interventional studies. *BMJ* 2011;**342**:d636.
65. Hughes K, Quigg Z, Eckley L, Bellis M, Jones L, Calafat A, et al. Environmental factors in drinking venues and alcohol-related harm: the evidence base for European intervention. *Addiction* 2011;**106**(Suppl. 1):37–46.
66. Batty GD, Lewars H, Emslie C, Benzeval M, Hunt K. Problem drinking and exceeding guidelines for "sensible" alcohol consumption in Scottish men: associations with life course socioeconomic disadvantage in a population-based cohort study. *BMC Public Health* 2008;**8**:302.
67. Wicki M, Kuntsche E, Gmel G. Drinking at European universities? A review of students' alcohol use. *Addict Behav* 2010;**35**:913–24.
68. van der Zwaluw CS, Engels RC. Gene-environment interactions and alcohol use and dependence: current status and future challenges. *Addiction* 2009;**104**:907–14.
69. Enoch MA. The role of GABA(A) receptors in the development of alcoholism. *Pharmacol Biochem Behav* 2008;**90**:95–104.

70. Whitfield JB. Alcohol and gene interactions. *Clin Chem Lab Med* 2005;**43**:480–7.
71. Schuckit MA, Smith TL, Kalmijn J. The search for genes contributing to the low level of response to alcohol: patterns of findings across studies. *Alcohol Clin Exp Res* 2004;**28**:1449–58.
72. Thiele TE, Badia-Elder NE. A role for neuropeptide Y in alcohol intake control: evidence from human and animal research. *Physiol Behav* 2003;**79**:95–101.
73. Francès F, Guillen M, Verdú F, Portolés O, Castelló A, Sorlí JV, et al. The 1258 G>A polymorphism in the neuropeptide Y gene is associated with greater alcohol consumption in a Mediterranean population. *Alcohol* 2011;**45**:131–6.
74. Kirby LG, Zeeb FD, Winstanley CA. Contributions of serotonin in addiction vulnerability. *Neuropharmacology* 2011;**61**:421–32.
75. Sari Y, Johnson VR, Weedman JM. Role of the serotonergic system in alcohol dependence: from animal models to clinics. *Prog Mol Biol Transl Sci* 2011;**98**:401–43.
76. Johnson BA, Ait-Daoud N, Seneviratne C, Roache JD, Javors MA, Wang XQ, et al. Pharmacogenetic approach at the serotonin transporter gene as a method of reducing the severity of alcohol drinking. *Am J Psychiatry* 2011;**168**:265–75.
77. Cao JX, Hu J, Ye XM, Xia Y, Haile CA, Kosten TR, et al. Association between the 5-HTR1B gene polymorphisms and alcohol dependence in a Han Chinese population. *Brain Res* 2011;**1376**:1–9.
78. Zuo L, Kranzler HR, Luo X, Covault J, Gelernter J. CNR1 variation modulates risk for drug and alcohol dependence. *Biol Psychiatry* 2007;**62**:616–26.
79. Terranova C, Tucci M, Forza G, Barzon L, Palù G, Ferrara SD. Alcohol dependence and glutamate decarboxylase gene polymorphisms in an Italian male population. *Alcohol* 2010;**44**:407–13.
80. Enoch MA, Hodgkinson CA, Yuan Q, Albaugh B, Virkkunen M, Goldman D. GABRG1 and GABRA2 as independent predictors for alcoholism in two populations. *Neuropsychopharmacology* 2009;**34**:1245–54.
81. López-Moreno JA, López-Jiménez A, Gorriti MA, de Fonseca FR. Functional interactions between endogenous cannabinoid and opioid systems: focus on alcohol, genetics and drug-addicted behaviors. *Curr Drug Targets* 2010;**11**:406–28.
82. Kimura M, Higuchi S. Genetics of alcohol dependence. *Psychiatry Clin Neurosci* 2011;**65**:213–25.
83. Temussi PA. Sweet, bitter and umami receptors: a complex relationship. *Trends Biochem Sci* 2009;**34**:296–302.
84. Scinska A, Koros E, Habrat B, Kukwa A, Kostowski W, Bienkowski P. Bitter and sweet components of ethanol taste in humans. *Drug Alcohol Depend* 2000;**60**:199–206.
85. Bachmanov AA, Bosak NP, Floriano WB, Inoue M, Li X, Lin C, et al. Genetics of sweet taste preferences. *Flavour Fragr J* 2011;**26**:286–94.
86. Sako N, Yamamoto T. Electrophysiological and behavioral studies on taste effectiveness of alcohols in rats. *Am J Physiol* 1999;**276**:R388–96.
87. Fortuna JL. Sweet preference, sugar addiction and the familial history of alcohol dependence: shared neural pathways and genes. *J Psychoactive Drugs* 2010;**42**:147–51.
88. Wang JC, Hinrichs AL, Bertelsen S, Stock H, Budde JP, Dick DM, et al. Functional variants in TAS2R38 and TAS2R16 influence alcohol consumption in high-risk families of African-American origin. *Alcohol Clin Exp Res* 2007;**31**:209–15.
89. Hayes JE, Wallace MR, Knopik VS, Herbstman DM, Bartoshuk LM, Duffy VB. Allelic variation in TAS2R bitter receptor genes associates with variation in sensations from and ingestive behaviors toward common bitter beverages in adults. *Chem Senses* 2011;**36**:311–9.

90. Eny KM, Wolever TM, Corey PN, El-Sohemy A. Genetic variation in TAS1R2 (Ile191Val) is associated with consumption of sugars in overweight and obese individuals in 2 distinct populations. *Am J Clin Nutr* 2010;**92**:1501–10.
91. El-Sohemy A, Stewart L, Khataan N, Fontaine-Bisson B, Kwong P, Ozsungur S, et al. Nutrigenomics of taste—impact on food preferences and food production. *Forum Nutr* 2007;**60**:176–82.
92. Reich T. A genomic survey of alcohol dependence and related phenotypes: results from the Collaborative Study on the Genetics of Alcoholism (COGA). *Alcohol Clin Exp Res* 1996;**20**:133A–7A.
93. Bergen AW, Korczak JF, Weissbecker KA, Goldstein AM. A genome-wide search for loci contributing to smoking and alcoholism. *Genet Epidemiol* 1999;**17**(Suppl. 1):S55–60.
94. Cantor RM, Lanning CD. Comparison of evidence supporting a chromosome 6 alcoholism gene. *Genet Epidemiol* 1999;**17**(Suppl. 1):S91–6.
95. Johnson C, Drgon T, Liu QR, Walther D, Edenberg H, Rice J, et al. Pooled association genome scanning for alcohol dependence using 104,268 SNPs: validation and use to identify alcoholism vulnerability loci in unrelated individuals from the collaborative study on the genetics of alcoholism. *Am J Med Genet B Neuropsychiatr Genet* 2006;**141B**:844–53.
96. Treutlein J, Cichon S, Ridinger M, Wodarz N, Soyka M, Zill P, et al. Genome-wide association study of alcohol dependence. *Arch Gen Psychiatry* 2009;**66**:773–84.
97. Bierut LJ, Agrawal A, Bucholz KK, Doheny KF, Laurie C, Pugh E, et al. A genome-wide association study of alcohol dependence. *Proc Natl Acad Sci USA* 2010;**107**:5082–7.
98. Edenberg HJ, Koller DL, Xuei X, Wetherill L, McClintick JN, Almasy L, et al. Genome-wide association study of alcohol dependence implicates a region on chromosome 11. *Alcohol Clin Exp Res* 2010;**34**:840–52.
99. Zlojutro M, Manz N, Rangaswamy M, Xuei X, Flury-Wetherill L, Koller D, et al. Genome-wide association study of theta band event-related oscillations identifies serotonin receptor gene HTR7 influencing risk of alcohol dependence. *Am J Med Genet B Neuropsychiatr Genet* 2011;**156B**:44–58.
100. Baik I, Cho NH, Kim SH, Han BG, Shin C. Genome-wide association studies identify genetic loci related to alcohol consumption in Korean men. *Am J Clin Nutr* 2011;**93**:809–16.
101. Corella D, Ordovas JM. Single nucleotide polymorphisms that influence lipid metabolism: interaction with dietary factors. *Annu Rev Nutr* 2005;**25**:341–90.
102. Andreassi MG. Metabolic syndrome, diabetes and atherosclerosis: influence of gene-environment interaction. *Mutat Res* 2009;**667**:35–43.
103. Corella D, Tucker K, Lahoz C, Coltell O, Cupples LA, Wilson PW, et al. Alcohol drinking determines the effect of the APOE locus on LDL-cholesterol concentrations in men: the Framingham Offspring Study. *Am J Clin Nutr* 2001;**73**:736–45.
104. Djoussé L, Pankow JS, Arnett DK, Eckfeldt JH, Myers RH, Ellison RC. Apolipoprotein E polymorphism modifies the alcohol-HDL association observed in the National Heart, Lung, and Blood Institute Family Heart Study. *Am J Clin Nutr* 2004;**80**:1639–44.
105. Corella D, Portolés O, Arriola L, Chirlaque MD, Barrricarte A, Francés F, et al. Saturated fat intake and alcohol consumption modulate the association between the APOE polymorphism and risk of future coronary heart disease: a nested case-control study in the Spanish EPIC cohort. *J Nutr Biochem* 2011;**22**:487–94.
106. Fumeron F, Betoulle D, Luc G, Behague I, Ricard S, Poirier O, et al. Alcohol intake modulates the effect of a polymorphism of the cholesteryl ester transfer protein gene on plasma high density lipoprotein and the risk of myocardial infarction. *J Clin Invest* 1995; **96**:1664–71.
107. Ordovas JM, Cupples LA, Corella D, Otvos JD, Osgood D, Martinez A, et al. Association of cholesteryl ester transfer protein-TaqIB polymorphism with variations in lipoprotein

subclasses and coronary heart disease risk: the Framingham study. *Arterioscler Thromb Vasc Biol* 2000;**20**:1323–9.
108. Corella D, Carrasco P, Fitó M, Martínez-González MA, Salas-Salvadó J, Arós F, et al. Gene-environment interactions of CETP gene variation in a high cardiovascular risk Mediterranean population. *J Lipid Res* 2010;**51**:2798–807.
109. Corella D, Sáiz C, Guillén M, Portolés O, Mulet F, González JI, et al. Association of TaqIB polymorphism in the cholesteryl ester transfer protein gene with plasma lipid levels in a healthy Spanish population. *Atherosclerosis* 2000;**152**:367–76.
110. Tsujita Y, Nakamura Y, Zhang Q, Tamaki S, Nozaki A, Amamoto K, et al. The association between high-density lipoprotein cholesterol level and cholesteryl ester transfer protein TaqIB gene polymorphism is influenced by alcohol drinking in a population-based sample. *Atherosclerosis* 2007;**191**:199–205.
111. Jensen MK, Mukamal KJ, Overvad K, Rimm EB. Alcohol consumption, TaqIB polymorphism of cholesteryl ester transfer protein, high-density lipoprotein cholesterol, and risk of coronary heart disease in men and women. *Eur Heart J* 2008;**29**:104–12.
112. Corella D, Carrasco P, Amiano P, Arriola L, Chirlaque MD, Huerta JM, et al. Common cholesteryl ester transfer protein gene variation related to high-density lipoprotein cholesterol is not associated with decreased coronary heart disease risk after a 10-year follow-up in a Mediterranean cohort: modulation by alcohol consumption. *Atherosclerosis* 2010;**211**:531–8.
113. Boekholdt SM, Sacks FM, Jukema JW, Shepherd J, Freeman DJ, McMahon AD, et al. Cholesteryl ester transfer protein TaqIB variant, high-density lipoprotein cholesterol levels, cardiovascular risk, and efficacy of pravastatin treatment: individual patient meta-analysis of 13,677 subjects. *Circulation* 2005;**111**:278–87.
114. Hines LM, Stampfer MJ, Ma J, Gaziano JM, Ridker PM, Hankinson SE, et al. Genetic variation in alcohol dehydrogenase and the beneficial effect of moderate alcohol consumption on myocardial infarction. *N Engl J Med* 2001;**344**:549–55.
115. Younis J, Cooper JA, Miller GJ, Humphries SE, Talmud PJ. Genetic variation in alcohol dehydrogenase 1C and the beneficial effect of alcohol intake on coronary heart disease risk in the Second Northwick Park Heart Study. *Atherosclerosis* 2005;**180**:225–32.
116. Djoussé L, Levy D, Herbert AG, Wilson PW, D'Agostino RB, Cupples LA, et al. Influence of alcohol dehydrogenase 1C polymorphism on the alcohol-cardiovascular disease association (from the Framingham Offspring Study). *Am J Cardiol* 2005;**96**:227–32.
117. Marques-Vidal P, Bal Dit Sollier C, Drouet L, Boccalon H, Ruidavets JB, Ferrières J. Lack of association between ADH3 polymorphism, alcohol intake, risk factors and carotid intima-media thickness. *Atherosclerosis* 2006;**184**:397–403.
118. Heidrich J, Wellmann J, Döring A, Illig T, Keil U. Alcohol consumption, alcohol dehydrogenase and risk of coronary heart disease in the MONICA/KORA-Augsburg cohort 1994/1995-2002. *Eur J Cardiovasc Prev Rehabil* 2007;**14**:769–74.
119. Ebrahim S, Lawlor DA, Shlomo YB, Timpson N, Harbord R, Christensen M, et al. Alcohol dehydrogenase type 1C (ADH1C) variants, alcohol consumption traits, HDL-cholesterol and risk of coronary heart disease in women and men: British Women's Heart and Health Study and Caerphilly cohorts. *Atherosclerosis* 2008;**196**:871–8.
120. Latella MC, Di Castelnuovo A, de Lorgeril M, Arnout J, Cappuccio FP, Krogh V, et al. Genetic variation of alcohol dehydrogenase type 1C (ADH1C), alcohol consumption, and metabolic cardiovascular risk factors: results from the IMMIDIET study. *Atherosclerosis* 2009;**207**:284–90.
121. Tolstrup JS, Grønbaek M, Nordestgaard BG. Alcohol intake, myocardial infarction, biochemical risk factors, and alcohol dehydrogenase genotypes. *Circ Cardiovasc Genet* 2009;**2**:507–14.
122. Husemoen LL, Jørgensen T, Borch-Johnsen K, Hansen T, Pedersen O, Linneberg A. The association of alcohol and alcohol metabolizing gene variants with diabetes and coronary heart disease risk factors in a white population. *PLoS One* 2010;**5**:e11735.

123. Jerrard-Dunne P, Sitzer M, Risley P, Steckel DA, Buehler A, von Kegler S, et al. Interleukin-6 promoter polymorphism modulates the effects of heavy alcohol consumption on early carotid artery atherosclerosis: the Carotid Atherosclerosis Progression Study (CAPS). *Stroke* 2003; **34**:402–7.
124. Mukamal KJ, Pai JK, Jensen MK, Rimm EB. Paraoxonase 1 polymorphisms and risk of myocardial infarction in women and men. *Circ J* 2009;**73**:1302–7.
125. Yin RX, Li YY, Liu WY, Zhang L, Wu JZ. Interactions of the apolipoprotein A5 gene polymorphisms and alcohol consumption on serum lipid levels. *PLoS One* 2011;**6**:e17954.
126. Ruixing Y, Yiyang L, Meng L, Kela L, Xingjiang L, Lin Z, et al. Interactions of the apolipoprotein C-III 3238C>G polymorphism and alcohol consumption on serum triglyceride levels. *Lipids Health Dis* 2010;**9**:86.
127. Vogel U, Segel S, Dethlefsen C, Tjønneland A, Saber AT, Wallin H, et al. PPARgamma Pro12Ala polymorphism and risk of acute coronary syndrome in a prospective study of Danes. *BMC Med Genet* 2009;**10**:52.
128. Hashibe M, McKay JD, Curado MP, Oliveira JC, Koifman S, Koifman R, et al. Multiple ADH genes are associated with upper aerodigestive cancers. *Nat Genet* 2008;**40**:707–9.
129. Oze I, Matsuo K, Suzuki T, Kawase T, Watanabe M, Hiraki A, et al. Impact of multiple alcohol dehydrogenase gene polymorphisms on risk of upper aerodigestive tract cancers in a Japanese population. *Cancer Epidemiol Biomarkers Prev* 2009;**18**:3097–102.
130. Brocic M, Supic G, Zeljic K, Jovic N, Kozomara R, Zagorac S, et al. Genetic polymorphisms of ADH1C and CYP2E1 and risk of oral squamous cell carcinoma. *Otolaryngol Head Neck Surg* 2011;**145**:586–93.
131. Li DP, Dandara C, Walther G, Parker MI. Genetic polymorphisms of alcohol metabolising enzymes: their role in susceptibility to oesophageal cancer. *Clin Chem Lab Med* 2008; **46**:323–8.
132. Yu HS, Oyama T, Isse T, Kitagawa K, Pham TT, Tanaka M, et al. Formation of acetaldehyde-derived DNA adducts due to alcohol exposure. *Chem Biol Interact* 2010;**188**:367–75.
133. Hashibe M, Boffetta P, Zaridze D, Shangina O, Szeszenia-Dabrowska N, Mates D, et al. Evidence for an important role of alcohol- and aldehyde-metabolizing genes in cancers of the upper aerodigestive tract. *Cancer Epidemiol Biomarkers Prev* 2006;**15**:696–703.
134. Platek ME, Shields PG, Marian C, McCann SE, Bonner MR, Nie J, et al. Alcohol consumption and genetic variation in methylenetetrahydrofolate reductase and 5-methyltetrahydrofolate-homocysteine methyltransferase in relation to breast cancer risk. *Cancer Epidemiol Biomarkers Prev* 2009;**18**:2453–9.
135. Benzon Larsen S, Vogel U, Christensen J, Hansen RD, Wallin H, Overvad K, et al. Interaction between ADH1C Arg(272)Gln and alcohol intake in relation to breast cancer risk suggests that ethanol is the causal factor in alcohol related breast cancer. *Cancer Lett* 2010;**295**:191–7.
136. Zhang GH, Mai RQ, Huang B. Meta-analysis of ADH1B and ALDH2 polymorphisms and esophageal cancer risk in China. *World J Gastroenterol* 2010;**16**:6020–5.
137. Shin CM, Kim N, Cho SI, Kim JS, Jung HC, Song IS. Association between alcohol intake and risk for gastric cancer with regard to ALDH2 genotype in the Korean population. *Int J Epidemiol* 2011;**40**:1047–55.

Coffee Intake

Marilyn C. Cornelis

Department of Nutrition, Harvard School of Public Health, Boston, Massachusetts, USA

Coffee is one of the most widely consumed beverages in the world. Its widespread popularity and availability has fostered public health concerns of the potential health consequences of regular coffee consumption. Epidemiological studies of coffee intake and certain health outcomes have been inconsistent. The precise component of coffee potentially contributing to development of these conditions also remains unclear. One step toward addressing the challenges in studying the impact coffee has on health is a better understanding of the factors contributing to its consumption and physiological effects. This chapter focuses on those factors that are genetically determined and briefly summarizes progress in applying this knowledge to epidemiological studies of coffee and disease.

I. Introduction

Coffee is one of the most widely consumed beverages in the world.[1] North American coffee drinkers typically consume ~2 cups per day while the norm is at least 4 cups in many European countries.[1] Coffee production has grown by nearly 200% since 1950, and it is currently among the most important traded commodities in the world. Its popularity and availability have fostered public health concerns regarding the potential consequences of regular coffee consumption. There is convincing epidemiological evidence that regular

Progress in Molecular Biology
and Translational Science, Vol. 108
DOI: 10.1016/B978-0-12-398397-8.00012-5

consumption decreases the risk of Parkinson's disease, Alzheimer's disease,[2] and type 2 diabetes,[3] but coffee's role in other conditions remains inconclusive.[4–7] Shortcomings in study design, including exposure misclassification and confounding, have largely been to blame for inconsistencies in the literature.[8] The precise component of coffee potentially contributing to the development of these conditions also remains unclear.

One step toward addressing these research challenges would be a better understanding of the factors contributing to coffee's consumption and physiological effects. This chapter focuses on those factors that are genetically determined and briefly summarizes progress in applying this knowledge to epidemiological studies of coffee and disease.

II. A Cup of Coffee: A Complex Mixture of Protective and Harmful Components

Roasted coffee is a complex mixture of more than 1000 chemicals. Although a brief review of its composition follows, the precise chemical composition of the beverage depends on multiple factors, from bean species selection to the method of beverage preparation.[9–12]

Coffee is considered one of the richest sources of natural phenolics in the Western diet.[13–16] These compounds contribute to the final acidity, astringency, and bitterness of the beverage[17,18] but have drawn particular attention due, in part, to their antioxidant activities[19] and impact on glucose and insulin homeostasis.[20,21] An array of lipid-soluble heterocyclic compounds (i.e., furans, pyrroles, and maltol) and melanoidins also contribute to coffee's antioxidant content.[11,22–30] These biological properties may convey protection against the development of cardiovascular and metabolic disease and certain cancers.[31–33]

Cafestol and kahweol are diterpenoid alcohols released from roasted and ground coffee beans by hot water, but they are largely trapped by the use of a paper filter in coffee preparation.[34] Consequently, higher levels of diterpenes are present in boiled coffee while lower levels are found in filtered coffee.[35,36] These components have been identified as hypercholesterolemic and so may negatively impact cardiovascular health.[37] These diterpenoids may also produce anticarcinogenic effects including the inhibition of the activity of phase I enzymes (involved in carcinogen activation) and induction of phase II enzymes (involved in carcinogen detoxification), as well as the stimulation of intracellular antioxidant defense mechanisms.[33,38–42]

Heterocyclic amines and polycyclic aromatic hydrocarbons are present in roasted coffee but at only low μg/kg concentrations. Nevertheless, they may be significant contributors to the total mutagen content of the diet when coffee is consumed in large amounts.[11] Indeed, coffee drinking has been classified as a possible human carcinogen (group 2B) by the International Agency for Research on Cancer.[43]

Caffeine (1,3,7-trimethylxanthine) is by far the best characterized naturally occurring component of coffee and is the most widely consumed stimulant in the world. In North American and European countries, over 75% of the caffeine consumed by adults daily comes from coffee.[11,44] Decaffeinated coffee makes up only ~10% of the coffee market.[45,46] The British obtain most of their caffeine from tea, while mate is the primary source in South American countries.[44] Other sources of caffeine include cocoa products, cola, and "energy" beverages.[47,48] Caffeine elicits a variety of physiological effects that may potentially impact health in different ways. The compound has been reported to both stimulate and suppress tumors[49] and to enhance the activation and carcinogenic potential of environmental mutagens.[50] Caffeine has been linked to adverse effects on blood pressure[51] and calcium and glucose homeostasis.[52] It may also modulate endogenous stress and sex hormones.[53–55] Conversely, caffeine may attenuate 1-methyl-4-phenyl-1,2,3,6-tetrahydropyridine (MPTP)-induced dopaminergic toxicity, a neuroprotective property with special relevance to Parkinson's disease development.[56]

While each of the above coffee constituents has been implicated in disease development, the persistence and magnitude of their individual effects in the context of the beverage as a whole remain unclear and further underscore the complexity of coffee's potential impact on health.

III. Factors Contributing to Coffee Intake

Knowledge of the external and internal cues for coffee intake may inform the causal role this beverage has in health and the potential population subgroups most susceptible to the health consequences of regular consumption. Thus far, demographic, social, and health-related factors have been the focus of most research.[57–61] Daily consumption tends to be positively correlated with age, smoking, and alcohol consumption.[59] Current health or perceived health consequences of coffee may also be a factor in certain population settings. Coffee is naturally bitter tasting. Although this bitterness is easily offset by additives, some individuals may avoid or prefer coffee because of personal taste preferences. Individuals may also learn to associate this sensory cue with social context or postingestive signals elicited by biologically active constituents of coffee.[62–66]

The acute behavioral effects and reinforcing properties of coffee's caffeine component are especially important in determining coffee drinking patterns.[44,67,68] In humans, a low to moderate dose of caffeine increases alertness, energetic arousal, and motivation and improves cognitive performance.[44,69–72]

High intakes produce anxiety, nervousness, sleep disturbances, tremor, and tension.[44,70,72,73] The precise doses defining “low” and “high,” however, vary substantially between individuals.[58,69,73,74] The degree of tolerance to caffeine’s acute effects and the withdrawal symptoms caused by abrupt cessation of use in habituated individuals also vary between individuals.[58,75,76] Consequently, many consumers modulate their dietary caffeine intake in order to obtain the desired pleasurable effects and to avoid any unpleasant symptoms.[44,57,58,68,73,77,78] The intake of the most concentrated dietary sources of caffeine (i.e., coffee) may be especially self-monitored.

Many of the above factors have a genetic component, which could therefore indirectly influence coffee intake. Indeed, there is growing evidence that habitual coffee consumption has a significant genetic underpinning.[79] As summarized in the following section, this evidence also corroborates the notion that response to caffeine largely determines coffee drinking behavior in a population.

IV. Heritability of Coffee Intake

Twin studies provide powerful evidence for the heritability of coffee intake and responsive traits related to its caffeine content.[79] These studies and their limitations have recently been reviewed by Yang *et al.*[79] Most studies assessed the level of total caffeine consumption derived from self-reported intakes of caffeinated coffee, tea, and soda. Heritability estimates ranged between 0.30 and 0.58, with higher estimates reported for heavy use (up to 0.77).[79] The genetic component of intake appears specific to caffeine, rather than to a predisposition to stimulant use in general,[80–83] and increases from late adolescence until middle adulthood and then stabilizes thereafter.[84,85] Self-reported symptoms of withdrawal or tolerance also have an important heritable component (0.34–0.40).[80,84] Two studies that separated heritability estimates by caffeine source each reported higher heritability for coffee relative to other sources.[86,87] A strong heritability for the preference of coffee over tea has also been observed (up to 0.62).[86,88] This source specificity may be due in part to differences in caffeine content. That is, if caffeine is the key factor underlying the heritable component of caffeinated beverage intake, the richest caffeine source (i.e., coffee) should demonstrate higher heritability relative to the other sources. Alternatively, a genetic factor underlying taste perception or preferences may contribute to this source-specific heritability, although the taste of each source is easily manipulated.

While twin studies support an important role of genetics in coffee consumption, they do not provide information on the precise molecular or physiological mechanisms at work. The latter have largely been addressed by genetic association studies that have aimed at identifying and characterizing specific genes contributing to the heritability of coffee intake, as well as responses to its individual constituents.

V. Genetic Association Studies of Coffee Intake

A. Candidate Gene Approach

Traditional genetic association studies of coffee intake have been hypothesis-driven, focusing on candidate gene pathways mediating the metabolism of and response to coffee constituents. This section provides a brief review of these pathways and their corresponding genetic association studies. A detailed summary of the literature has recently been provided by Yang *et al.*[79]

1. Caffeine Metabolism and Mechanism of Action

At doses typically consumed in the diet, caffeine is rapidly and completely absorbed from the gastrointestinal tract.[89] The elimination half-life of caffeine in plasma ranges from 2.5 to 12 h in adults.[90–92] Large intra- and interindividual variability in caffeine elimination is mainly due to the variable efficiency in metabolizing and eliminating the compound rather than absorbing it.[93,94] In humans, hepatic cytochrome P450 1A2 (CYP1A2) catalyzes the demethylation of caffeine to its three related dimethylxanthines (paraxanthine > theobromine > theophylline), accounting for more than 95% of caffeine metabolism and making it the most quantitatively important metabolic pathway.[95–97] Each of these caffeine metabolites is subjected to further demethylation into monomethylxanthines.[98] Other enzymes such as CYP2A6, CYP2E1, *N*-acetyltransferase 2 (NAT2), and xanthine oxidase are also involved in caffeine metabolism but have less prominent roles than CYP1A2.[99,100] CYP1A2 expression and activity vary 10- to 60-fold between individuals.[101,102] The well-documented variation in CYP1A2 activity, both within and between individuals, represents a major source of variability in the pharmacokinetics of caffeine. Numerous pharmaceuticals, oral contraceptives, pregnancy, and caffeine itself are among the factors altering the activity of this enzyme.[101,103] Cigarette smoking is an especially potent inducer of CYP1A2 and provides a biological basis for the strong correlation between smoking and coffee intake.[104]

Genetic factors may also contribute to interindividual CYP1A2 variability. More than 150 single-nucleotide polymorphisms (SNPs) have been identified in the *CYP1A2* gene (dbSNP database: www.ncbi.nlm.nih.gov/SNP). However, the functional significance and proportion of CYP1A2 phenotype variability explained by these SNPs remain unclear.[105] The intronic *CYP1A2**1F* (*C734A*, rs762551) polymorphism has gained considerable attention in genetic epidemiological studies. The more common *A* variant (often referred to as the "rapid" metabolizing allele) has been associated with higher inducibility compared to the *C* ("slow") variant, particularly among smokers.[106–108] One study specifically examined the association between genetic variation in *CYP1A2* and coffee intake.[109] Among a population of nonhypertensive Costa Ricans, caffeinated coffee and total dietary caffeine intake were not associated with *CYP1A2*1F* genotype.[109]

The main psychological effects of caffeine in humans are due to competitive inhibition of adenosine (A) receptors, namely the A_1 and A_{2A} receptors (ADORA1, ADORA2A), in the central neurotransmitter system.[44,110,111] The motor and reinforcing effects of caffeine result from the compound's ability to release pre- and postsynaptic brakes that adenosine imposes on the dopaminergic system.[110] The arousing effects of caffeine depend on the blockade of multiple inhibitory mechanisms that adenosine exerts on multiple interconnected arousal systems.[110]

The 1976 $C \rightarrow T$ polymorphism (rs5751876) in *ADORA2A* has been a candidate in several clinical and population studies of caffeine and caffeine-related traits. Retey and colleagues[112] observed greater self-reported caffeine sensitivity and caffeine-induced sleep impairment associated with the *C* variant compared to the *T* variant. Conversely, in two studies of light caffeine consumers, those with the rs5751876 *T/T* genotype reported greater anxiety after acute caffeine administration compared to the other two genotypes.[113,114] In a population-based study of nonhypertensive Costa Ricans, subjects with the rs5751876 *T/T* genotype were likely to habitually consume less caffeine than those with the *C/C* genotype.[109] Interestingly, the rs5751876 *T* allele has also been associated with increased anxiety in response to amphetamine[115] and sympathetic indicators of anxiety-related arousal in blood-injury phobia,[116] as well as increased risk of panic disorder in Caucasian populations[117–119] (although not replicated in Asian populations[120,121]). The rs5751876 SNP may therefore be a general susceptibility locus for anxiety-related disorders, rather than caffeine-induced anxiety specifically. Since this SNP does not cause an amino acid exchange, other variants in strong linkage disequilibrium are presumably the underlying functional variants, which may also explain inconsistencies across traits and populations. The dopamine receptor D_2 (DRD2) and ADORA2A are coexpressed on striatopallidal neurons of the basal ganglia mediating control of locomotor activity, motivation, and addiction.[122,123] Childs *et al.*[114] reported an interaction between *ADORA2A* rs5751876 and *DRD2* rs1079597, as well as a main effect of a second *DRD2* polymorphism (rs1110976) on caffeine-induced anxiety in Caucasians. Finally, in a recent study of sleep strategies, chronotype, and variation in circadian clock genes, a nonsynonymous exonic SNP (rs228669) in period homolog 3 (*PER3*) significantly predicted caffeine consumption.[124]

2. CHEMOSENSORY

Sensitivity to the characteristic bitter taste of coffee may also have a genetic component, which may in turn influence consumption. Hayes *et al.*[125] recently reported that, while a haploblock of SNPs in the taste receptor genes *TAS2R3* (rs765007), *TAS2R4* (rs2234001), and *TAS2R5* (rs2227264, rs2234012) explained *perceived* espresso coffee bitterness, it did not predict coffee *liking* among a group of adults. It is possible that an innate dislike of coffee bitterness is masked by other factors including coffee additives and prior positive experience learned with regular exposure.

Candidate gene studies of coffee intake and response have thus far focused on pathways relevant to caffeine content and bitter properties, and of these pathways only a few genes and a limited number of SNPs have been considered. Relative to our knowledge of caffeine, the absorption, metabolism, chemosensory detection, and physiological effect of other coffee constituents are only now being realized.[126–130] Genetic variation in these other pathways (either independently or via complex interactions) could potentially influence response and intake but has yet to be explored.

B. Genome-Wide Approach

Unlike the candidate gene approach, a genome-wide association approach enables a comprehensive investigation of common genetic variation underlying a specific trait. This approach may therefore yield new insight into trait development, revealing novel molecular pathways worthy of further investigation. In 2011, this approach was applied to habitual dietary caffeine intake and coffee consumption.[131–133] A genome-wide association study (GWAS) of habitual caffeine intake that included over 47,000 individuals sourced from five US population-based studies identified two genome-wide significant loci. The first (rs4410790) was located just upstream of the aryl hydrocarbon receptor (*AHR*) gene, and the second (rs2470893) mapped to the bidirectional promoter of the *CYP1A1–CYP1A2* locus. A GWAS of caffeinated coffee intake in this study sample produced the same loci but with even stronger effects.[131] These latter findings were consistent with a GWAS of predominately caffeinated coffee consumption conducted by Sulem and colleagues.[132] These researchers conducted a meta-analysis of four GWAS of coffee consumption among coffee drinkers from Iceland, the Netherlands, Germany, and the United States ($N=6611$) followed by replication in a sample set from Iceland and Denmark ($N=4050$). Two loci reached genome-wide significance: rs2472297 and rs6968865. These two SNPs are in moderate to high linkage disequilibrium with rs2470893 ($r^2=0.70$) and rs4410790 ($r^2=0.87$), respectively. In another GWAS of coffee intake among predominately European populations (discovery $N=18{,}176$, replication $N=7929$), Amin *et al.*[133] also uncovered rs2472297 and rs2470893, as well as a novel locus (rs382140) near *NRCAM* encoding the neuronal cell adhesion molecule. Genome-wide significant SNPs mapping near *AHR* were not identified.[133]

The protein product of *AHR* is a ligand-activated transcription factor that plays a key role in regulating the expression of several genes including *CYP1A1* and *CYP1A2*.[134,135] Within the common promoter region of *CYP1A1* and *CYP1A2*, rs2472297 and rs2470893 are located in *AHR* response elements that correlate with transcriptional activation of *CYP1A1* and/or *CYP1A2*.[136,137] *CYP1A1* expression in the liver (the target tissue for caffeine metabolism) is low and has a minor, if any, role in caffeine metabolism. *CYP1A2* has already been discussed as a potential candidate but has not previously been linked to consumption behavior. A role for NRCAM in caffeine or coffee consumption is unclear. *NRCAM* has not been implicated in caffeine or other coffee-constituent metabolism; however, it is expressed in the brain, and variation in this gene has previously been associated with autism[138] and addiction.[139,140]

The results of these three GWAS add to the growing evidence that psychological responses to caffeine are key drivers for coffee consumption at the population level. Although functional studies are warranted, individuals with SNPs near *CYP1A2* and *AHR* related to increased intake may seek more caffeine to compensate for their increased metabolism in order to maintain caffeine levels that elicit the most desirable effects. The mean difference in coffee (caffeine) intake between homozygote genotypes was ~0.2 cups per day (40 mg per day) for each of the SNPs near *AHR* and *CYP1A2*.[131,132] The two SNPs together, however, explained no more than 1% of the total variation in coffee (caffeine) intake,[131,132] suggesting additional variants remain to be discovered.

VI. A Nutrigenomics Approach to Studies of Coffee and Human Health

Shortcomings in study design including exposure misclassification and confounding by unhealthy lifestyle factors have largely been to blame for inconsistent relationships between coffee intake and disease found in epidemiological studies.[8] Even if the link between coffee and health is *causal*, traditional epidemiological studies will be limited in providing mechanistic insight to the relationship. Moreover, the ultimate impact that coffee has on health will also vary from person to person based on background risk factor profile, including genetic constitution. The latter is not accounted for in traditional epidemiological studies of unrelated individuals.

Accounting for genetic determinants of coffee intake and response can be a powerful way to study coffee's impact on heath. First, evidence for individual variability may explain inconsistencies in the literature. Genetic variants associated with coffee intake or response may better reflect long-term physiological adaptations to its use. Secondly, evidence of modification would provide insight on the mechanisms of action and causal role of coffee in health. For example, if caffeine is the component of coffee responsible for increased disease risk, individuals with genotypes corresponding to impaired caffeine metabolism and/or enhanced sensitivity to the adverse effects of caffeine should be at a particularly increased risk if they consume coffee. Thirdly, the approach may identify subgroups of the population most predisposed to adverse consequences of coffee intake.

Table I provides a summary of epidemiological studies specifically investigating the role that genetic variation plays in modifying the associations between coffee or total caffeine intake and various conditions. The caffeine and mutagen components of coffee have been the subject of interest in most genetic epidemiological studies of disease outcomes. Caffeine or xenobiotic pathway genes have, therefore, been the primary candidates selected for analysis. An alternative approach taken by a few investigators has been to focus on disease-related loci and examine how coffee intake might modify disease development in susceptible individuals.[169–171] These latter studies are not specifically discussed.

A. Coffee and Cancer

A recent meta-analysis of 40 independent cohorts examining the association between coffee intake and cancer did not support a positive association between this beverage and cancer overall.[172] Rather, evidence showed that regular coffee intake may reduce the risk of certain cancers. There is ongoing debate about whether these findings are causal and, if so, which constituents of coffee may have cancer-specific relevance.

Kotsopoulos and colleagues[141] reported a reduced risk of *breast cancer* in those who "ever" consumed coffee only among *CYP1A2 C* (slow, rs762551) carriers, suggesting that the decreased risk may be attributable to the prolonged exposure of caffeine among slow metabolizers. Although CYP1A2 also activates potential mutagens in coffee, this role is less likely to explain the protective effect of a slow *CYP1A2* genotype, since no interaction was observed between *CYP1A2* and smoking, the latter being a more concentrated source of mutagens. In a case-only study,[142] women with the *CYP1A2 AA* (rapid, rs762551) genotype consumed more coffee than *C* carriers, a finding consistent with the interaction reported by Kotsopoulos *et al.*[141] A third study explored the association between coffee (and other potential sources of mutagens), acetylation status (defined by *NAT2* haplotypes), and receptor-defined breast cancer.[143] An increased risk of breast cancer associated with coffee was

TABLE I

Genetic Epidemiological Studies of Coffee or Caffeine Intake and Selected Health Outcomes

Study reference	Primary outcome	Study sample and setting	Gene (SNP)	Exposure	Significant findings[a]
Cancer					
141	Breast cancer	170 ca 241 co All BRCA1+ F Canada (mostly Caucasian)	*CYP1A2* (rs762551)	Ever vs. never initiated coffee intake prior to age 35	Coffee ↓ risk of cancer Coffee ↓ risk among *C* carriers but not among *A/A* carriers Similar results when restricted to caffeinated coffee
142	Breast cancer: AOD, ER status	458 ca F Sweden	*CYP1A2* (rs762551)	Coffee intake in the past week (preoperative)	Moderate/high coffee intake associated with a later AOD and ER tumors vs. low intake only among *CYP1A2 A/A* carriers CYP1A2 *A/A* associated with ↑ coffee intake vs. *C* carriers
143	Breast cancer	1020 ca 1047 co F Germany	*NAT2* haplotypes (6 SNPs): "slow" and "fast" acetylators	Coffee and tea intake the year before onset (ca) or interview (co)	Coffee ↑ risk for ER- and PR tumors Coffee ↑ cancer risk among "slow" acetylators and ↓ risk among "fast" acetylators
144	Ovarian cancer	164 ca 194 co F USA (Hawaii, Caucasian)	*CYP1A2* (rs762551)	Coffee, tea, and soda intake in the past year	High caffeine intake ↑ risk of cancer vs. low intake Caffeinated coffee intake ↑ risk of cancer vs. no intake These risks were enhanced among *CYP1A2 A/A* carriers

145	Ovarian cancer	445 ca 472 co F USA (Caucasian)	*CYP1A1* (rs1048943, rs4646903)	Caffeine intake the year before onset (ca) or interview (co)	rs1048943 Val variant ↑ risk of cancer among high caffeine consumers
146	Ovarian cancer	1354 ca 1851 co F USA (Caucasian)	*CYP1A1* (rs4646903) *CYP1A2* (rs762551) *CYP2A6* (rs1801272) *CYP19* (rs2446405, rs2445765, rs2470144, rs1004984, rs1902584, rs28566535, rs2445759, rs936306, rs1902586, rs749292, rs1008805, rs4646, rs700519, rs10046, rs727479, rs2414096, rs17601241, rs6493494, rs28757184, rs2445762, rs3751591)	Coffee, tea, and total caffeine intake the year before onset (ca) or questionnaire (co)	Main effect of *CYP19* variation on cancer risk
147	Colon cancer	1579 ca 1898 co M, F USA (mostly Caucasian)	*GSTM1* (Del)	Coffee intake 2 years prior to selection	Null
148	Bladder cancer	197 ca 211 co F Northern Italy	*GSTM1* (Del) *GSTT1* (Del) *GSTP1* (rs947894, Ile105Val) *NAT1* ("fast," "slow") *NAT2* ("fast," "slow") *SULT1A1* (rs9282861) *XRCC1* (rs25487) *XRCC3* (rs861539) *XPD* (rs1052559)	Lifetime coffee intake	Heavy coffee intake ↑ risk of cancer among *GSTP1 105Val* carriers
149	Bladder cancer	1136 ca 1138 co M, F Spain	*NAT2* ("fast," "slow") *CYP1A2* (rs762551) *CYP1A1* (rs4646421, rs2198843, rs2472299) *CYP2E1* (rs2070676, rs8192766)	Lifetime coffee intake	Null

(*Continues*)

TABLE I (*Continued*)

Study reference	Primary outcome	Study sample and setting	Gene (SNP)	Exposure	Significant findings[a]
150	Bladder cancer	185 ca 180 co M Italy	*CYP1A2* (rs35694136, rs762551)	Lifetime coffee intake	Coffee ↑ risk of cancer
151	Acute childhood leukemia	280 ca 288 co M, F (children) France	*CYP1A1* (rs4646903) *GSTM1* (Del) *GSTP1* (rs947894, Ile105Val) *GSTT1* (Del) *NQO1* (rs1800566) *EPHX1* (rs1051740, rs2234922)	Maternal coffee intake during pregnancy and breastfeeding	Coffee intake ↑ risk Risk associated with coffee intake attenuated among children with *NQO1* variant
Cardiovascular disease					
152	Incidence of coronary heart disease events	78 Events over ~13-year follow-up of 773 disease-free at baseline M Finland	*COMT* (rs4680, Met158Val) Met: low activity Val: high activity	Coffee and tea intake based on 4-day food records (recent intake)	Heavy coffee intake ↑ risk of CHD among *COMT* Met but not *COMT* Val variant carriers
153	MI	2014 ca 2014 co M, F Costa Rica	*CYP1A2* (rs762551)	Caffeinated coffee intake the year before onset (ca) or interview (co)	Coffee ↑ risk of MI Coffee ↑ risk among *C* "slow" carriers but not among *A/A* "rapid" carriers Trend toward coffee ↓ risk among *A/A* carriers

154	Hypertension	323 Cases over ~8.2-year follow-up of 553 disease-free at baseline M, F Italy, Italians	*CYP1A2* (rs762551)	Caffeinated coffee intake at baseline	Coffee ↑ risk of hypertension Coffee ↑ risk of hypertension among *C* "slow" carriers but not among *A/A* "rapid" carriers Similar interaction observed for blood pressure measures Urinary epinephrine was higher in coffee drinkers than abstainers but only among *C* carriers
Reproductive health					
155	Stillbirth	142 ca 157 co F (history of pregnancy) Denmark	*CYP1A2* (rs762551) *NAT2* ("fast," "slow") *GSTA1* (rs3957357)	Caffeine intake at ~16 weeks gestation	*CYP1A2* (*C* "slow") + *NAT2* ("slow") + *GSTA1* ("low") ↑ risk vs. other genotype combinations
156	Recurrent pregnancy loss	187 ca 109 co F (history of pregnancy) The Netherlands	*GSTP1* (rs947894, Ile105Val) *GSTM1* (Del) *GSTT1* (Del) *CYP1A1* (rs1048943)	Coffee intake (time of exposure not reported)	*GSTP1 Val/Val* ↑ risk (enhanced risk among coffee drinkers or smokers)
157	Recurrent pregnancy loss	58 ca 147 co F (history of pregnancy) Japan	*CYP1A2* (rs762551)	Coffee, tea, and soda intake during pregnancy	Total caffeine ↑ risk of recurrent pregnancy loss only among *A/A* "rapid" carriers
158	Small for gestational age	493 ca 472 co M, F (infants + mother) Canada (mostly Caucasian)	*CYP1A2* (rs2069514) *CYP2E1* (*5A) Maternal and infant	Coffee, tea, and cola intake during each trimester and 1 month before delivery	Null

(*Continues*)

TABLE I (*Continued*)

Study reference	Primary outcome	Study sample and setting	Gene (SNP)	Exposure	Significant findings[a]
159	Neural tube defects	768 ca 4143 co Genotyped: 306 ca 669 co M, F (infants +parents) USA (mostly Caucasian)	*CYP1A2* (rs762551) *NAT2* (rs1799929, rs1799930: "fast," "slow")	Coffee, tea, soda, and chocolate intake 1 year preceding pregnancy and during pregnancy	Infant *NAT2* "slow" ↑ risk of neural tube defects Maternal *CYP1A2 A/A* ("rapid") ↑ risk of neural tube defects Maternal caffeine intake (year preceding pregnancy) ↑ risk of neural tube defects only among infants with *CYP1A2 A/A* ("rapid")
160	Unexplained recurrent miscarriage	103 ca 101 co F (history of pregnancy) Japan	*GSTP1* (rs947894, Ile105Val) *GSTM1* (Del) *GSTT1* (Del) *CYP1A1* (rs1048943)	Coffee intake (ever vs. never, time of exposure not reported)	Among coffee drinkers, *GSTM1* (Del) more frequent in ca vs. co
161	Fecundability (probability of pregnancy per time unit)	319 USA (mostly Caucasian)	*NAT2* (rs1799929, rs1799930, rs1208)	Coffee, tea, and soda intake 1 year preceding pregnancy	Null
Parkinson's disease					
162	Parkinson's disease	418 ca 468 co M, F Singapore	*CYP1A2* (rs762551)	Lifetime coffee and tea intake	Total caffeine intake ↓ risk of Parkinson's disease
163	Parkinson's disease	222 ca 219 co M, F Singapore	*ADORA2A* (rs35320474)	Lifetime coffee and tea intake	Coffee ↓ risk of Parkinson's disease

164	Parkinson's disease	Siblings:446 ca 446 co Unrelated: 158 ca 159 co M, F USA	*ADORA2A* (rs5751876, rs3032740) *CYP1A2* (rs35694136, rs762551)	Lifetime coffee, soda, and tea intake	*ADORA2A* (rs3032740, Ins ↓ expression) ↓ risk of Parkinson's disease in M only
165	Parkinson's disease	159 F ca 724 F co 139 M ca 561 M co USA	*NAT2* (rs1801280, rs1799930, rs1799931) *CYP1A2* (rs762551) *ESR1* (rs2077647, rs2228480, rs379857, rs1801132) *ESR2* (rs1256049, rs1256030, rs928554, rs1255998, rs1152579)	Cumulative average coffee, tea, chocolate, and soda intake up to 2 years before disease onset	*CYP1A2* rs762551 *C* "slow" ↑ risk of Parkinson's disease in F only
166	Parkinson's disease	1325 ca 1735 co 925 ca 1249 co M, F USA (mostly Caucasian)	*ADORA2A* (rs5751876*, rs71651683, rs3032740*, rs5996696, *LD) *CYP1A2* (rs762551, rs2472304*, rs2470890*, *LD)	Caffeinated coffee, tea, and soda intake (time of exposure varied by study: lifetime or past week intake)	Coffee ↓ risk of Parkinson's disease Protective effect of coffee strongest among subjects with *CYP1A2* rs762551 *C/C* or rs2470890 *C/C* genotypes *ADORA2A* rs71651683 *C* ↓ risk of Parkinson's disease *ADORA2A* rs5996696 *C* ↓ risk of Parkinson's disease
167	Parkinson's disease	1458 ca 931 co M, F Caucasian	*CYP1A2* (rs762551, rs2472304*, rs2470890*, *LD)	Coffee intake (ever vs. never and high vs. low, time of exposure not reported)	Coffee ↓ risk of Parkinson's disease

(Continues)

TABLE I (*Continued*)

Study reference	Primary outcome	Study sample and setting	Gene (SNP)	Exposure	Significant findings[a]
168	Parkinson's disease	Discovery: 1458 ca 931 co Replication: 1014 ca 1917 co M, F USA (Caucasian)	Genome-wide	Lifetime caffeinated coffee intake	Joint test of SNP and SNP–coffee revealed SNPs (index: rs4998386) neighboring *GRIN2A* as associated with Parkinson's disease

ca, cases; co, controls; M, males; F, females; AOD, age of diagnosis; ER, estrogen receptor; PR, progesterone receptor; CHD, coronary heart disease; MI, myocardial infarction; LD, linkage disequilibrium; Del, deletion; Ins, insertion.

[a]Table presents results relevant to this review.

observed among slow acetylators, with a decreased risk among fast acetylators. An adverse constituent of coffee potentially metabolized by NAT2 and thus *directly* mediating this interaction remains uncertain.

Terry *et al.*[145] examined two SNPs in *CYP1A1* (encoding a phase 1 activating enzyme) and their interaction with various potential mutagens, including caffeine, on the risk of *ovarian cancer*. An elevated risk associated with an Ile to Val substitution (rs1048943), conferring increased activity, was observed only among high caffeine consumers. No interaction was observed between the second SNP (rs4646903) and caffeine intake. Goodman and colleagues[144] evaluated the association between *CYP1A2* (rs762551), coffee, and ovarian cancer among a population of predominately Asian and Pacific Islanders. They observed an increased risk of ovarian cancer with coffee and caffeine intake among women with the *AA* (rapid) genotype but not *C* carriers (slow). A more recent and larger study of coffee and ovarian cancer among Caucasians considered variation in *CYP1A2* as well as other genes implicated in metabolism of caffeine, estrogen, or both.[146] No effect modification by genotype was reported.

Three studies investigated SNP–coffee interactions and the risk of *bladder cancer*. The first investigated whether variants in genes encoding xenobiotic metabolizing and DNA repair enzymes interacted with total coffee intake on the risk of bladder cancer.[148] Heavy coffee consumption enhanced the risk associated with a more active glutathione *S*-transferase pi 1 (*GSTP1*) variant, but whether this effect equated to a significant statistical interaction was not indicated. Villanueva *et al.*[149] and Pavanello *et al.*[150] focused on variation in caffeine metabolism genes (*CYP1A2*, *CYP1A1*, and *CYP2E1*) and reported no significant SNP–coffee interactions with bladder cancer. SNP–coffee interactions have also been studied in the context of colon cancer[147] and childhood leukemia,[151] but no clear associations have been reported.

B. Reproductive Health and Fetal Development

Evidence for an effect of caffeine on human reproductive health and fetal development is currently limited by the inability to rule out confounding by pregnancy symptoms and smoking, as well as by exposure measurement error.[173,174] Four studies investigated whether genetic variation in xenobiotic metabolism interacts with coffee and/or total caffeine on the risk of recurrent pregnancy loss[156,157,160] or fecundability,[161] but no consistent and/or interpretable results were reported. Four studies specifically examined whether reproductive effects of coffee intake are modified by variation in caffeine metabolism genes.

Sata and colleagues[157] reported an increased risk of recurrent pregnancy loss associated with caffeine only among women with the *CYP1A2 AA* (rapid, rs762551) genotype, suggesting that a metabolite of caffeine adversely impacts pregnancy. Bech *et al.*[155] studied the same *CYP1A2* variant (as well as SNPs in *NAT2* and *GSTA1*), caffeine intake, and risk of stillbirth. Only a three-genotype signature corresponding to slow/low activity was associated with stillbirth, regardless of caffeine exposure. Infante-Rivard[158] examined SNPs in *CYP1A2* and *CYP2E1* and their relationships with small-for-gestational-age babies, taking into account caffeine and other exposures at pregnancy. No main effect of caffeine or SNP–caffeine interactions was observed. Finally, Schmidt and coworkers[159] recently examined the association between maternal exposure to caffeine and the risk of neural tube defects, taking into consideration maternal and child *CYP1A2* and *NAT2* genotype. Infant *NAT2* slow or maternal *CYP1A2 AA* rapid genotypes increased the risk of neural tube defects, regardless of caffeine intake. Maternal prepregnancy exposure to caffeine increased the risk among infants with *CYP1A2 AA* genotype. However, the number of tests performed warrants caution when interpreting results. It is also unclear how variation in an infant's *CYP1A2* gene might impact predisposition, given that the expression of this enzyme is very low relative to adults. Based on the studies conducted to date, it appears that *CYP1A2* genotype may have an important role in reproductive health that may be enhanced with dietary caffeine consumption.

C. Parkinson's Disease

Epidemiological studies support an inverse association between caffeine/coffee consumption and the risk of developing Parkinson's disease that is unlikely explained by bias or uncontrolled confounding.[175] Experimental studies support these findings and further implicate caffeine's target receptors in mediating this inverse relationship.[176] Studies of human genetic variation provide another way in which to gain mechanistic insight to the role that caffeine processing and caffeine's targets of action play in Parkinson's disease development. Early studies by Tan *et al.*[162,163] and Facheris *et al.*[164] investigated variation in *ADORA2A* and *CYP1A2* but did not report significant main effects of the genes or interactions with coffee/caffeine intake on the risk of Parkinson's. More recently, Palacios and colleagues[165] observed an increased risk of Parkinson's associated with the *CYP1A2 C* (slow, rs762551) variant among females (but not males), regardless of caffeine consumption. *CYP1A2* has a role in metabolism of MPTP,[177] providing one potential mechanism underlying this association. The sex-specific effect, however, is unclear. The largest study conducted to date reported that two SNPs in *ADORA2A* (rs71651683, rs5996696) each independently decreased the risk of Parkinson's disease.[166] Further, the strongest inverse association between coffee and Parkinson's was observed among individuals homozygous for the slow *CYP1A2 C* allele or the rs2470890 *C* allele. The former association is consistent with the protective effect of coffee/caffeine exposure on Parkinson's. The rs2470890 variant has already been discussed with respect to its significant association with coffee intake, possibly mediated by impaired caffeine metabolism.[131–133] These *CYP1A2*–coffee interactions, however, were not replicated by the

NeuroGenetics Research Consortium.[167] More recently, this consortium embarked on a genome-wide gene–coffee interaction analysis in order to exploit the role caffeine plays in Parkinson's disease development as a means for disease gene discovery.[168] In a joint test of SNP main effect and SNP–coffee interactions, the most significant signals came from rs4998386 and neighboring SNPs in *GRIN2A*. This gene encodes the NMDA glutamate receptor subunit 2A, which regulates excitatory neurotransmission in the brain and is a plausible candidate in Parkinson's disease etiology. GRIN2A might also be a target for caffeine and thus a novel candidate worthy of further investigation.

D. Cardiovascular Disease

Randomized controlled trials have shown a significant but modest increase in blood pressure with regular coffee intake.[51,178] However, prospective cohort studies of longer coffee drinking duration generally do not support an elevated risk of hypertension associated with this beverage.[7] Palatini *et al.*[154] tested the *CYP1A2*–coffee interaction among Italians and reported an increased risk of hypertension among *CYP1A2 C* (slow, rs762551) carriers but not among those homozygous for the rapid *A* allele.

The role of coffee in the development of coronary heart disease is also controversial,[179–183] with some studies suggesting that coffee consumption may actually reduce coronary heart disease risk.[7,183] Two case–control studies considered genetic variation in metabolism of or response to caffeine in associations of coffee intake and coronary heart disease. The first reported an increased risk of nonfatal myocardial infarction associated with caffeinated coffee intake only among carriers of the *CYP1A2 C* (slow, rs762551) allele,[153] a pattern consistent with that reported by Palatini *et al.* for hypertension.[154] In the second study, the risk of acute myocardial infarction in heavy coffee drinkers was found to be higher in subjects possessing the *COMT* rs4680 allele conferring lower catechol *O*-methyltransferase (COMT) activity.[152] COMT is the main enzyme responsible for metabolism of catecholamines. Elevated catecholamine levels have been linked to both coffee intake and increased risk of coronary heart disease.[184]

VII. Conclusions and Future Studies of Coffee Intake

Twin studies underscore a strong heritable component to coffee consumption behavior. Further evidence suggests that this component is largely mediated by the caffeine content of the beverage. GWASs of coffee and caffeine intake provide convincing support for the latter and have thus far identified *at least* two robust loci associated with habitual consumption behavior. Knowledge of the genetic determinants of coffee intake and response may provide ways to study the potential health effects of coffee more comprehensively by using genetic markers as surrogate variables for coffee intake or by accounting for gene–coffee interactions. Indeed, this knowledge has recently been applied to epidemiological studies of coffee intake and disease, but efforts have been limited and findings inconclusive. Larger sample sizes, adequate replication, and more exhaustive SNP panels are needed. Genetic factors that impact the exposure as well as the acute and chronic response pose a considerable challenge in genetic epidemiology. This notion clearly applies to coffee intake and, along with other well-known determinants of coffee intake, should be carefully considered in future study design.

As clearly emphasized throughout this chapter, *caffeine* has been an overwhelming theme in genetic studies of coffee intake. However, this does not eliminate the possibility that genes involved in the bioavailability of coffee phytochemicals or mediating coffee's array of caffeine-independent effects could have implications in modifying effects on certain health conditions; future studies of the latter are warranted. Finally, besides our growing knowledge and particular interest in the role the human genome has on consumption of or response to coffee intake, our understanding of the reverse relationship—the impact coffee has on the human genome—is rather limited.[39,185–192] Nevertheless, with the continued advancements in nutrigenomics, further progress in this area is highly anticipated.

References

1. International Coffee Organization. *Annual review*; 2009–2010.
2. Rosso A, Mossey J, Lippa CF. Caffeine: neuroprotective functions in cognition and Alzheimer's disease. *Am J Alzheimers Dis Other Demen* 2008;**23**:417–22.
3. van Dam RM, Hu FB. Coffee consumption and risk of type 2 diabetes: a systematic review. *JAMA* 2005;**294**:97–104.
4. Cnattingius S, Signorello LB, Anneren G, Clausson B, Ekbom A, Ljunger E, et al. Caffeine intake and the risk of first-trimester spontaneous abortion. *N Engl J Med* 2000;**343**:1839–45.
5. van Dam RM. Coffee consumption and risk of type 2 diabetes, cardiovascular diseases, and cancer. *Appl Physiol Nutr Metab* 2008;**33**:1269–83.
6. Cornelis MC, El-Sohemy A. Coffee, caffeine, and coronary heart disease. *Curr Opin Lipidol* 2007;**18**:13–9.
7. Zhang Z, Hu G, Caballero B, Appel L, Chen L. Habitual coffee consumption and risk of hypertension: a systematic review and meta-analysis of prospective observational studies. *Am J Clin Nutr* 2011;**93**:1212–9.
8. Schreiber GB, Robins M, Maffeo CE, Masters MN, Bond AP, Morganstein D. Confounders contributing to the reported association of coffee or caffeine with disease. *Prev Med* 1988;**17**:295–309.
9. D'Amicis A, Viani R. The consumption of coffee. In: Garattini S, editor. *Caffeine, coffee, and health*. New York: Raven Press; 1993. pp. 1–16.
10. Varnam AH, Sutherland JP. *Beverages: technology, chemistry, and microbiology*. London: Chapman and Hall; 1994.

11. Spiller MA. The chemical components of coffee. In: Spiller GA, editor. *Caffeine*. Boca Raton: CRC; 1998. pp. 97–161.
12. Gilbert RM. Caffeine consumption. In: Spiller GA, editor. *The methylxanthine beverages and foods: chemistry, consumption, and health effects*. New York: Alan R. Liss Inc; 1984. pp. 185–213.
13. Higdon JV, Frei B. Coffee and health: a review of recent human research. *Crit Rev Food Sci Nutr* 2006;**46**:101–23.
14. Scalbert A, Williamson G. Dietary intake and bioavailability of polyphenols. *J Nutr* 2000;**130**:2073S–NaN.
15. Clifford MN. Chlorogenic acids and other cinnamates, nature, occurrence, dietary burden, absorption and metabolism. *J Sci Food Agric* 2000;**80**:1033–43.
16. Hervert-Hernandez D, Goni I. Contribution of beverages to the intake of polyphenols and antioxidant capacity in obese women from rural Mexico. *Public Health Nutr* 2012;**15**:6–12.
17. Turgo LC, Macrae R. A study of the effect of roasting on the chlorogenic acid composition of coffee using HPLC. *Food Chem* 1984;**15**:219–27.
18. Variyar PS, Ahmed R, Bhat R, Niyas Z, Sharma A. Flavoring components of raw monsooned arabica coffee and their changes during radiation processing. *J Agric Food Chem* 2003;**51**: 7945–50.
19. Natella F, Nardini M, Giannetti I, Dattilo C, Scaccini C. Coffee drinking influences plasma antioxidant capacity in humans. *J Agric Food Chem* 2002;**50**:6211–6.
20. Rodriguez de Sotillo DV, Hadley M, Sotillo JE. Insulin receptor exon 11+/- is expressed in Zucker (fa/fa) rats, and chlorogenic acid modifies their plasma insulin and liver protein and DNA. *J Nutr Biochem* 2006;**17**:63–71.
21. Johnston KL, Clifford MN, Morgan LM. Coffee acutely modifies gastrointestinal hormone secretion and glucose tolerance in humans: glycemic effects of chlorogenic acid and caffeine. *Am J Clin Nutr* 2003;**78**:728–33.
22. Chuyen NV. Maillard reaction and food processing. Application aspects. *Adv Exp Med Biol* 1998;**434**:213–35.
23. Daglia M, Papetti A, Gregotti C, Berte F, Gazzani G. In vitro antioxidant and ex vivo protective activities of green and roasted coffee. *J Agric Food Chem* 2000;**48**:1449–54.
24. Borrelli RC, Visconti A, Mennela C, Anese M, Fogliano V. Chemical characterization and antioxidant properties of coffee melanoidins. *J Agric Food Chem* 2002;**50**:6527–33.
25. Iwai K, Kishimoto N, Kakino Y, Mochida K, Fujita T. In vitro antioxidative effects of tyrosinase inhibitory activities of seven hydroxycinnamoyl derivatives in green coffee beans. *J Agric Food Chem* 2004;**52**:4893–8.
26. Kono Y, Kobayashi K, Tagawa S, Adachi K, Ueda A, Sawa Y, et al. Antioxidant activity of polyphenolics in diets. Rate constants of reactions of chlorogenic acid and caffeic acid with reactive species of oxygen and nitrogen. *Biochem Biophys Acta* 1997;**1335**:335–42.
27. Mazur WM, Wahala K, Rasku S, Salakka A, Hase T, Adlercreutz H. Lignan and isoflavonoid concentrations in tea and coffee. *Br J Nutr* 1998;**79**:37–45.
28. Pellegrini N, Serafini M, Colombi B, Del Rio D, Salvatore S, Bianchi M, et al. Total antioxidant capacity of plant foods, beverages and oils consumed in Italy assessed by three different in vitro assays. *J Nutr* 2003;**133**:2812–9.
29. Yanagimoto K, Ochi H, Lee KG, Shibamoto T. Antioxidative activities of fractions obtained from brewed coffee. *J Agric Food Chem* 2004;**52**:592–6.
30. Fogliano V, Morales FJ. Estimation of dietary intake of melanoidins from coffee and bread. *Food Funct* 2011;**2**:117–23.
31. Shearer J, Farah A, de Paulis T, Bracy DP, Pencek RR, Graham TE, et al. Quinides of roasted coffee enhance insulin action in conscious rats. *J Nutr* 2003;**133**:3529–32.
32. Renehan AG, Roberts DL, Dive C. Obesity and cancer: pathophysiological and biological mechanisms. *Arch Physiol Biochem* 2008;**114**:71–83.

33. Ramos S. Cancer chemoprevention and chemotherapy: dietary polyphenols and signalling pathways. *Mol Nutr Food Res* 2008;**52**:507–26.
34. Viani R. Composition of coffee. In: Garattini S, editor. *Caffeine, coffee and health*. New York: Raven Press; 1993. pp. 17–41.
35. Ranheim T, Halvorsen B. Coffee consumption and human health—beneficial or detrimental?—Mechanisms for effects of coffee consumption on different risk factors for cardiovascular disease and type 2 diabetes mellitus. *Mol Nutr Food Res* 2005;**49**:274–84.
36. Urgert R. Levels of the cholesterol-elevating diterpenes cafestol and kahweol in various coffee brews. *J Agric Food Chem* 1995;**43**:2167–72.
37. Urgert R, Katan MB. The cholesterol-raising factor from coffee beans. *Annu Rev Nutr* 1997;**17**:305–24.
38. Huber WW, Scharf G, Nagel G, Prustomersky S, Schulte-Hermann R, Kaina B. Coffee and its chemopreventive components Kahweol and Cafestol increase the activity of O6-methylguanine-DNA methyltransferase in rat liver—comparison with phase II xenobiotic metabolism. *Mutat Res* 2003;**522**:57–68.
39. Huber WW, Rossmanith W, Grusch M, Haslinger E, Prustomersky S, Peter-Vorosmarty B, et al. Effects of coffee and its chemopreventive components kahweol and cafestol on cytochrome P450 and sulfotransferase in rat liver. *Food Chem Toxicol* 2008;**46**:1230–8.
40. Majer BJ, Hofer E, Cavin C, Lhoste E, Uhl M, Glatt HR, et al. Coffee diterpenes prevent the genotoxic effects of 2-amino-1-methyl-6-phenylimidazo[4,5-b]pyridine (PhIP) and N-nitrosodimethylamine in a human derived liver cell line (HepG2). *Food Chem Toxicol* 2005;**43**:433–41.
41. Okamura S, Suzuki K, Yanase M, Koizumi M, Tamura HO. The effects of coffee on conjugation reactions in human colon carcinoma cells. *Biol Pharm Bull* 2005;**28**:271–4.
42. Vucic EA, Brown CJ, Lam WL. Epigenetics of cancer progression. *Pharmacogenomics* 2008; **9**:215–34.
43. IARC Working Group on the Evaluation of Carcinogenic Risks to Humans. Coffee, tea, mate, methylxanthines and methylglyoxal. Lyon, 27 February to 6 March 1990. *IARC Monogr Eval Carcinog Risks Hum* 1991;**51**:1–513.
44. Fredholm BB, Battig K, Holmen J, Nehlig A, Zvartau EE. Actions of caffeine in the brain with special reference to factors that contribute to its widespread use. *Pharmacol Rev* 1999;**51**:83–133.
45. Silvarola MB, Mazzafera P, Fazioli LC. A naturally decaffeinated arabic coffee. *Nature* 2004; **249**:826.
46. Ramalakshmi K, Raghavan B. Caffeine in coffee: its removal. Why and how? *Crit Rev Food Sci Nutr* 1999;**39**:441–56.
47. Barone JJ, Roberts HR. Caffeine consumption. *Food Chem Toxicol* 1996;**34**:119–29.
48. Mandel HG. Update on caffeine consumption, disposition, and action. *Food Chem Toxicol* 2002;**40**:1231–4.
49. American Institute for Cancer Research. . *Food, nutrition, physical activity and the prevention of cancer: a global perspective.* World Cancer Research Fund/American Institute for Cancer Research;; 2007.
50. Zhou T, Chen Y, Huang C, Chen G. Caffeine induction of sulfotransferases in rat liver and intestine. *J Appl Toxicol* 2011.
51. Jee SH, He J, Whelton PK, Suh I, Klag MJ. The effect of chronic coffee drinking on blood pressure: a meta-analysis of controlled clinical trials. *Hypertension* 1999;**33**:647–52.
52. Beaudoin MS, Graham TE. Methylxanthines and human health: epidemiological and experimental evidence. *Handb Exp Pharmacol* 2011;**200**:509–48.
53. Lee AJ, Cai MX, Thomas PE, Conney AH, Zhu BT. Characterization of the oxidative metabolites of 17beta-estradiol and estrone formed by 15 selectively expressed human cytochrome p450 isoforms. *Endocrinology* 2003;**144**:3382–98.

54. Ferrini RL, Barrett-Connor E. Caffeine intake and endogenous sex steroid levels in postmenopausal women. The Rancho Bernardo Study. *Am J Epidemiol* 1996;**144**:642–4.
55. Nagata C, Kabuto M, Shimizu H. Association of coffee, green tea, and caffeine intakes with serum concentrations of estradiol and sex hormone-binding globulin in premenopausal Japanese women. *Nutr Cancer* 1998;**30**:21–4.
56. Chen JF, Xu K, Petzer JP, Staal R, Xu YH, Beilstein M. Neuroprotection by caffeine and A(2A) adenosine receptor inactivation in a model of Parkinson's disease. *J Neurosci* 2001;**21**:RC143.
57. Brice CF, Smith AP. Factors associated with caffeine consumption. *Int J Food Sci Nutr* 2002;**53**:55–64.
58. Evans SM, Griffiths RR. Caffeine tolerance and choice in humans. *Psychopharmacology (Berl)* 1992;**108**:51–9.
59. Hewlett P, Smith A. Correlates of daily caffeine consumption. *Appetite* 2006;**46**:97–9.
60. Jones HA, Lejuez CW. Personality correlates of caffeine dependence: the role of sensation seeking, impulsivity, and risk taking. *Exp Clin Psychopharmacol* 2005;**13**:259–66.
61. Swift CG, Tiplady B. The effects of age on the response to caffeine. *Psychopharmacology (Berl)* 1988;**94**:29–31.
62. Birch LL. Development of food preferences. *Annu Rev Nutr* 1999;**19**:41–62.
63. Mattes RD. Influences on acceptance of bitter foods and beverages. *Physiol Behav* 1994;**56**:1229–36.
64. Guinard JX, Zoumas-Morse C, Dietz J, Goldberg S, Holz M, Heck E, et al. Does consumption of beer, alcohol, and bitter substances affect bitterness perception? *Physiol Behav* 1996;**59**:625–31.
65. Rozin P, Vollmecke TA. Food likes and dislikes. *Annu Rev Nutr* 1986;**6**:433–56.
66. Cines BM, Rozin P. Some aspects of the liking for hot coffee and coffee flavor. *Appetite* 1982;**3**:23–34.
67. Nehlig A. Are we dependent upon coffee and caffeine? A review on human and animal data. *Neurosci Biobehav Rev* 1999;**23**:563–76.
68. Griffiths RR, Mumford J. Caffeine reinforcement, discrimination, tolerance and physical dependence in laboratory animals and humans. In: Schuster CR, Kuhar MJ, editors. *Pharmacological aspects of drug dependence: toward an integrated neurobehavioral approach*. Berlin: Springer Verlag; 1996. pp. 315–41 Chapter 9.
69. Lieberman HR, Wurtman RJ, Emde GG, Roberts C, Coviella IL. The effects of low doses of caffeine on human performance and mood. *Psychopharmacology (Berl)* 1987;**92**:308–12.
70. Kaplan GB, Greenblatt DJ, Ehrenberg BL, Goddard JE, Cotreau MM, Harmatz JS, et al. Dose-dependent pharmacokinetics and psychomotor effects of caffeine in humans. *J Clin Pharmacol* 1997;**37**:693–703.
71. Lorist MM, Tops M. Caffeine, fatigue and cognition. *Brain Cogn* 2003;**53**:82–94.
72. Smith A. Effects of caffeine on human behavior. *Food Chem Toxicol* 2002;**40**:1243–55.
73. Evans SM, Griffiths RR. Dose-related caffeine discrimination in normal volunteers: individual differences in subjective and self-reported cues. *Behav Pharmacol* 1991;**2**:345–56.
74. Daly JW, Fredholm BB. Caffeine—an atypical drug of dependence. *Drug Alcohol Depend* 1998;**51**:199–206.
75. Juliano LM, Griffiths RR. A critical review of caffeine withdrawal: empirical validation of symptoms and signs, incidence, severity, and associated features. *Psychopharmacology (Berl)* 2004;**176**:1–29.
76. Griffiths RR, Woodson PP. Reinforcing effects of caffeine in humans. *J Pharmacol Exp Ther* 1988;**246**:21–9.
77. Soroko S, Chang J, Barrett-Connor E. Reasons for changing caffeinated coffee consumption: the Rancho Bernardo Study. *J Am Coll Nutr* 1996;**15**:97–101.

78. Stern KN, Chait LD, Johansson CE. Reinforcing and subjective effects of caffeine in normal human volunteers. *Psychopharmacology (Berl)* 1989;**98**:81–8.
79. Yang A, Palmer AA, de Wit H. Genetics of caffeine consumption and responses to caffeine. *Psychopharmacology (Berl)* 2010;**211**(3):245–57.
80. Kendler KS, Prescott CA. Caffeine intake, tolerance, and withdrawal in women: a population-based twin study. *Am J Psychiatry* 1999;**156**:223–8.
81. Hettema JM, Corey LA, Kendler KS. A multivariate genetic analysis of the use of tobacco, alcohol, and caffeine in a population based sample of male and female twins. *Drug Alcohol Depend* 1999;**57**:69–78.
82. Kendler KS, Myers J, Prescott CA. Specificity of genetic and environmental risk factors for symptoms of cannabis, cocaine, alcohol, caffeine, and nicotine dependence. *Arch Gen Psychiatry* 2007;**64**:1313–20.
83. Swan GE, Carmelli D, Cardon LR. The consumption of tobacco, alcohol, and coffee in Caucasian male twins: a multivariate genetic analysis. *J Subst Abuse* 1996;**8**:19–31.
84. Kendler KS, Schmitt E, Aggen SH, Prescott CA. Genetic and environmental influences on alcohol, caffeine, cannabis, and nicotine use from early adolescence to middle adulthood. *Arch Gen Psychiatry* 2008;**65**:674–82.
85. Laitala VS, Kaprio J, Silventoinen K. Genetics of coffee consumption and its stability. *Addiction* 2008;**103**:2054–61.
86. Luciano M, Kirk KM, Heath AC, Martin NG. The genetics of tea and coffee drinking and preference for source of caffeine in a large community sample of Australian twins. *Addiction* 2005;**100**:1510–7.
87. Teucher B, Skinner J, Skidmore PM, Cassidy A, Fairweather-Tait SJ, Hooper L, et al. Dietary patterns and heritability of food choice in a UK female twin cohort. *Twin Res Hum Genet* 2007;**10**:734–48.
88. Vink JM, Staphorsius AS, Boomsma DI. A genetic analysis of coffee consumption in a sample of Dutch twins. *Twin Res Hum Genet* 2009;**12**:127–31.
89. Bonati M, Latini R, Galletti F, Young JF, Tognoni G, Garattini S. Caffeine disposition after oral doses. *Clin Pharmacol Ther* 1982;**32**:98–106.
90. Benowitz NL. Clinical pharmacology of caffeine. *Annu Rev Med* 1990;**41**:277–88.
91. Birkett DJ, Miners JO. Caffeine renal clearance and urine caffeine concentrations during steady state dosing. Implications for monitoring caffeine intake during sports events. *Br J Clin Pharmacol* 1991;**31**:405–8.
92. Lelo A, Birkett DJ, Robson RA, Miners JO. Comparative pharmacokinetics of caffeine and its primary demethylated metabolites paraxanthine, theobromine, and theophylline in man. *Br J Clin Pharmacol* 1986;**22**:177–82.
93. Kalow W, Tang BK. Caffeine as a metabolic probe: exploration of the enzyme-inducing effect of cigarette smoking. *Clin Pharmacol Ther* 1991;**49**:44–8.
94. Kashuba AD, Bertino Jr. JS, Kearns GL, Leeder JS, James AW, Gotschall R, et al. Quantitation of three-month intraindividual variability and influence of sex and menstrual cycle phase on CYP1A2, N-acetyltransferase-2, and xanthine oxidase activity determined with caffeine phenotyping. *Clin Pharmacol Ther* 1998;**63**:540–51.
95. Ferrero JL, Neims AH. Metabolism of caffeine by mouse liver microsomes: GSH or cytosol causes a shift in products from 1,3,7-trimethylurate to a substituted diaminouracil. *Life Sci* 1983;**33**:1173–8.
96. Latini R, Bonati M, Marzi E, Garattini S. Urinary excretion of an uracilic metabolite from caffeine rat, monkey and man. *Toxicol Lett* 1981;**7**:267–72.
97. Lelo A, Miners JO, Robson R, Birkett DJ. Quantitative assessment of caffeine partial clearance in man. *Br J Clin Pharmacol* 1986;**22**:183–6.

98. Miners JO, Birkett DJ. The use of caffeine as a metabolic probe for human drug metabolizing enzymes. *Gen Pharmacol* 1996;**27**:245–9.
99. Krul C, Hageman G. Analysis of urinary caffeine metabolites to assess biotransformation enzyme activities by reversed-phase high-performance liquid chromatography. *J Chromatogr B Biomed Sci Appl* 1998;**709**:27–34.
100. Rostami-Hodjegan A, Nurminen S, Jackson PR, Tucker GT. Caffeine urinary metabolite ratios as markers of enzyme activity: a theoretical assessment. *Pharmacogenetics* 1996;**6**: 121–49.
101. Gunes A, Dahl ML. Variation in CYP1A2 activity and its clinical implications: influence of environmental factors and genetic polymorphisms. *Pharmacogenomics* 2008;**9**:625–37.
102. Zhou SF, Wang B, Yang LP, Liu JP. Structure, function, regulation and polymorphism and the clinical significance of human cytochrome P450 1A2. *Drug Metab Rev* 2010;**42**:268–354.
103. Berthou F, Goasduff T, Dreano Y, Menez JF. Caffeine increases its own metabolism through cytochrome P4501A induction in rats. *Life Sci* 1995;**57**:541–9.
104. Swanson JA, Lee JW, Hopp JW. Caffeine and nicotine: a review of their joint use and possible interactive effects in tobacco withdrawal. *Addict Behav* 1994;**19**:229–56.
105. Jiang Z, Dragin N, Jorge-Nebert LF, Martin MV, Guengerich FP, Aklillu E, et al. Search for an association between the human CYP1A2 genotype and CYP1A2 metabolic phenotype. *Pharmacogenet Genomics* 2006;**16**:359–67.
106. Sachse C, Brockmoller J, Bauer S, Roots I. Functional significance of a C–>A polymorphism in intron 1 of the cytochrome P450 CYP1A2 gene tested with caffeine. *Br J Clin Pharmacol* 1999;**47**:445–9.
107. Ghotbi R, Christensen M, Roh HK, Ingelman-Sundberg M, Aklillu E, Bertilsson L. Comparisons of CYP1A2 genetic polymorphisms, enzyme activity and the genotype-phenotype relationship in Swedes and Koreans. *Eur J Clin Pharmacol* 2007;**63**:537–46.
108. Gunes A, Ozbey G, Vural EH, Uluoglu C, Scordo MG, Zengil H, et al. Influence of genetic polymorphisms, smoking, gender and age on CYP1A2 activity in a Turkish population. *Pharmacogenomics* 2009;**10**:769–78.
109. Cornelis MC, El-Sohemy A, Campos H. Genetic polymorphism of the adenosine A2A receptor is associated with habitual caffeine consumption. *Am J Clin Nutr* 2007;**86**:240–4.
110. Ferre S. Role of the central ascending neurotransmitter systems in the psychostimulant effects of caffeine. *J Alzheimers Dis* 2010;**20**(Suppl. 1):S35–49.
111. Daly JW, Butts-Lamb P, Padgett W. Subclasses of adenosine receptors in the central nervous system: interaction with caffeine and related methylxanthines. *Cell Mol Neurobiol* 1983;**3**: 69–80.
112. Retey JV, Adam M, Khatami R, Luhmann UF, Jung HH, Berger W, et al. A genetic variation in the adenosine A2A receptor gene (ADORA2A) contributes to individual sensitivity to caffeine effects on sleep. *Clin Pharmacol Ther* 2007;**81**:692–8.
113. Alsene K, Deckert J, Sand P, de Wit H. Association between A2a receptor gene polymorphisms and caffeine-induced anxiety. *Neuropsychopharmacology* 2003;**28**:1694–702.
114. Childs E, Hohoff C, Deckert J, Xu K, Badner J, de Wit H. Association between ADORA2A and DRD2 polymorphisms and caffeine-induced anxiety. *Neuropsychopharmacology* 2008; **33**:2791–800.
115. Hohoff C, McDonald JM, Baune BT, Cook EH, Deckert J, de Wit H. Interindividual variation in anxiety response to amphetamine: possible role for adenosine A2A receptor gene variants. *Am J Med Genet B Neuropsychiatr Genet* 2005;**139B**:42–4.
116. Hohoff C, Domschke K, Schwarte K, Spellmeyer G, Vogele C, Hetzel G, et al. Sympathetic activity relates to adenosine A(2A) receptor gene variation in blood-injury phobia. *J Neural Transm* 2009;**116**:659–62.

117. Deckert J, Nothen MM, Franke P, Delmo C, Fritze J, Knapp M, et al. Systematic mutation screening and association study of the A1 and A2a adenosine receptor genes in panic disorder suggest a contribution of the A2a gene to the development of disease. *Mol Psychiatry* 1998; **3**:81–5.
118. Hamilton SP, Slager SL, De Leon AB, Heiman GA, Klein DF, Hodge SE, et al. Evidence for genetic linkage between a polymorphism in the adenosine 2A receptor and panic disorder. *Neuropsychopharmacology* 2004;**29**:558–65.
119. Hohoff C, Mullings EL, Heatherley SV, Freitag CM, Neumann LC, Domschke K, et al. Adenosine A(2A) receptor gene: evidence for association of risk variants with panic disorder and anxious personality. *J Psychiatr Res* 2010;**44**:930–7.
120. Yamada K, Hattori E, Shimizu M, Sugaya A, Shibuya H, Yoshikawa T. Association studies of the cholecystokinin B receptor and A2a adenosine receptor genes in panic disorder. *J Neural Transm* 2001;**108**:837–48.
121. Lam P, Hong CJ, Tsai SJ. Association study of A2a adenosine receptor genetic polymorphism in panic disorder. *Neurosci Lett* 2005;**378**:98–101.
122. Rosin DL, Robeva A, Woodard RL, Guyenet PG, Linden J. Immunohistochemical localization of adenosine A2A receptors in the rat central nervous system. *J Comp Neurol* 1998;**401**:163–86.
123. Svenningsson P, Le Moin C, Fisone G, Fredholm BB. Distribution, biochemistry and function of striatal adenosine A2a receptors. *Prog Neurobiol* 1999;**59**:355–96.
124. Gamble KL, Motsinger-Reif AA, Hida A, Borsetti HM, Servick SV, Ciarleglio CM, et al. Shift work in nurses: contribution of phenotypes and genotypes to adaptation. *PLoS One* 2011;**6**: e18395.
125. Hayes JE, Wallace MR, Knopik VS, Herbstman DM, Bartoshuk LM, Duffy VB. Allelic variation in TAS2R bitter receptor genes associates with variation in sensations from and ingestive behaviors toward common bitter beverages in adults. *Chem Senses* 2011;**36**:311–9.
126. Ferruzzi MG. The influence of beverage composition on delivery of phenolic compounds from coffee and tea. *Physiol Behav* 2010;**100**:33–41.
127. Gonthier MP, Verny MA, Besson C, Remesy C, Scalbert A. Chlorogenic acid bioavailability largely depends on its metabolism by the gut microflora in rats. *J Nutr* 2003;**133**:1853–9.
128. Bouayed J, Rammal H, Dicko A, Younos C, Soulimani R. Chlorogenic acid, a polyphenol from Prunus domestica (Mirabelle), with coupled anxiolytic and antioxidant effects. *J Neurol Sci* 2007;**262**:77–84.
129. Rosenzweig S, Yan W, Dasso M, Spielman AI. Possible novel mechanism for bitter taste mediated through cGMP. *J Neurophysiol* 1999;**81**:1661–5.
130. Moon SJ, Kottgen M, Jiao Y, Xu H, Montell C. A taste receptor required for the caffeine response in vivo. *Curr Biol* 2006;**16**:1812–7.
131. Cornelis MC, Monda KL, Yu K, Paynter N, Bennett SN, Boerwinkle E, et al. Genome-wide meta-analysis identifies regions on 7p21 (AHR) and 15q24 (CYP1A2) as determinants of habitual caffeine consumption. *PLoS Genet* 2011;**7**(4):e1002033.
132. Sulem P, Gudbjartsson DF, Geller F, Prokopenko I, Feenstra B, Aben KK, et al. Sequence variants at CYP1A1-CYP1A2 and AHR associate with coffee consumption. *Hum Mol Genet* 2011;**20**:2071–7.
133. Amin N, Byrne E, Johnson J, Chenevix-Trench G, Walter S, Nolte IM, et al. Genome-wide association analysis of coffee drinking suggests association with CYP1A1/CYP1A2 and NRCAM. *Mol Psychiatry* 2012.
134. Nukaya M, Bradfield CA. Conserved genomic structure of the Cyp1a1 and Cyp1a2 loci and their dioxin responsive elements cluster. *Biochem Pharmacol* 2009;**77**:654–9.
135. Nukaya M, Moran S, Bradfield CA. The role of the dioxin-responsive element cluster between the Cyp1a1 and Cyp1a2 loci in aryl hydrocarbon receptor biology. *Proc Natl Acad Sci USA* 2009;**106**:4923–8.

136. Jorge-Nebert LF, Jiang Z, Chakraborty R, Watson J, Jin L, McGarvey ST, et al. Analysis of human CYP1A1 and CYP1A2 genes and their shared bidirectional promoter in eight world populations. *Hum Mutat* 2010;**31**:27–40.
137. Ueda R, Iketaki H, Nagata K, Kimura S, Gonzalez FJ, Kusano K, et al. A common regulatory region functions bidirectionally in transcriptional activation of the human CYP1A1 and CYP1A2 genes. *Mol Pharmacol* 2006;**69**:1924–30.
138. Sakurai T, Ramoz N, Reichert JG, Corwin TE, Kryzak L, Smith CJ, et al. Association analysis of the NrCAM gene in autism and in subsets of families with severe obsessive-compulsive or self-stimulatory behaviors. *Psychiatr Genet* 2006;**16**:251–7.
139. Ishiguro H, Liu QR, Gong JP, Hall FS, Ujike H, Morales M, et al. NrCAM in addiction vulnerability: positional cloning, drug-regulation, haplotype-specific expression, and altered drug reward in knockout mice. *Neuropsychopharmacology* 2006;**31**:572–84.
140. Matzel LD, Babiarz J, Townsend DA, Grossman HC, Grumet M. Neuronal cell adhesion molecule deletion induces a cognitive and behavioral phenotype reflective of impulsivity. *Genes Brain Behav* 2008;**7**:470–80.
141. Kotsopoulos J, Ghadirian P, El-Sohemy A, Lynch HT, Snyder C, Daly M, et al. The CYP1A2 genotype modifies the association between coffee consumption and breast cancer risk among BRCA1 mutation carriers. *Cancer Epidemiol Biomarkers Prev* 2007;**16**:912–6.
142. Bageman E, Ingvar C, Rose C, Jernstrom H. Coffee consumption and CYP1A2°1F genotype modify age at breast cancer diagnosis and estrogen receptor status. *Cancer Epidemiol Biomarkers Prev* 2008;**17**:895–901.
143. Rabstein S, Bruning T, Harth V, Fischer HP, Haas S, Weiss T, et al. N-acetyltransferase 2, exposure to aromatic and heterocyclic amines, and receptor-defined breast cancer. *Eur J Cancer Prev* 2010;**19**:100–9.
144. Goodman MT, Tung KH, McDuffie K, Wilkens LR, Donlon TA. Association of caffeine intake and CYP1A2 genotype with ovarian cancer. *Nutr Cancer* 2003;**46**:23–9.
145. Terry KL, Titus-Ernstoff L, Garner EO, Vitonis AF, Cramer DW. Interaction between CYP1A1 polymorphic variants and dietary exposures influencing ovarian cancer risk. *Cancer Epidemiol Biomarkers Prev* 2003;**12**:187–90.
146. Kotsopoulos J, Vitonis AF, Terry KL, De Vivo I, Cramer DW, Hankinson SE, et al. Coffee intake, variants in genes involved in caffeine metabolism, and the risk of epithelial ovarian cancer. *Cancer Causes Control* 2009;**20**:335–44.
147. Slattery ML, Kampman E, Samowitz W, Caan BJ, Potter JD. Interplay between dietary inducers of GST and the GSTM-1 genotype in colon cancer. *Int J Cancer* 2000;**87**:728–33.
148. Covolo L, Placidi D, Gelatti U, Carta A, Scotto Di Carlo A, Lodetti P, et al. Bladder cancer, GSTs, NAT1, NAT2, SULT1A1, XRCC1, XRCC3, XPD genetic polymorphisms and coffee consumption: a case-control study. *Eur J Epidemiol* 2008;**23**:355–62.
149. Villanueva CM, Silverman DT, Murta-Nascimento C, Malats N, Garcia-Closas M, Castro F, et al. Coffee consumption, genetic susceptibility and bladder cancer risk. *Cancer Causes Control* 2009;**20**:121–7.
150. Pavanello S, Mastrangelo G, Placidi D, Campagna M, Pulliero A, Carta A, et al. CYP1A2 polymorphisms, occupational and environmental exposures and risk of bladder cancer. *Eur J Epidemiol* 2010;**25**:491–500.
151. Clavel J, Bellec S, Rebouissou S, Menegaux F, Feunteun J, Bonaiti-Pellie C, et al. Childhood leukaemia, polymorphisms of metabolism enzyme genes, and interactions with maternal tobacco, coffee and alcohol consumption during pregnancy. *Eur J Cancer Prev* 2005;**14**:531–40.
152. Happonen P, Voutilainen S, Tuomainen TP, Salonen JT. Catechol-o-methyltransferase gene polymorphism modifies the effect of coffee intake on incidence of acute coronary events. *PLoS One* 2006;**1**:e117.

153. Cornelis MC, El-Sohemy A, Kabagambe EK, Campos H. Coffee, CYP1A2 genotype, and risk of myocardial infarction. *JAMA* 2006;**295**:1135–41.
154. Palatini P, Ceolotto G, Ragazzo F, Dorigatti F, Saladini F, Papparella I, et al. CYP1A2 genotype modifies the association between coffee intake and the risk of hypertension. *J Hypertens* 2009;**27**:1594–601.
155. Bech BH, Autrup H, Nohr EA, Henriksen TB, Olsen J. Stillbirth and slow metabolizers of caffeine: comparison by genotypes. *Int J Epidemiol* 2006;**35**:948–53.
156. Zusterzeel PL, Nelen WL, Roelofs HM, Peters WH, Blom HJ, Steegers EA. Polymorphisms in biotransformation enzymes and the risk for recurrent early pregnancy loss. *Mol Hum Reprod* 2000;**6**:474–8.
157. Sata F, Yamada H, Suzuki K, Saijo Y, Kato EH, Morikawa M, et al. Caffeine intake, CYP1A2 polymorphism and the risk of recurrent pregnancy loss. *Mol Hum Reprod* 2005;**11**:357–60.
158. Infante-Rivard C. Caffeine intake and small-for-gestational-age birth: modifying effects of xenobiotic-metabolising genes and smoking. *Paediatr Perinat Epidemiol* 2007;**21**:300–9.
159. Schmidt RJ, Romitti PA, Burns TL, Murray JC, Browne ML, Druschel CM, et al. Caffeine, selected metabolic gene variants, and risk for neural tube defects. *Birth Defects Res A Clin Mol Teratol* 2010;**88**:560–9.
160. Nonaka T, Takakuwa K, Tanaka K. Analysis of the polymorphisms of genes coding biotransformation enzymes in recurrent miscarriage in the Japanese population. *J Obstet Gynaecol Res* 2011;**37**:1352–8.
161. Taylor KC, Small CM, Dominguez CE, Murray LE, Tang W, Wilson MM, et al. Alcohol, smoking, and caffeine in relation to fecundability, with effect modification by NAT2. *Ann Epidemiol* 2011;**21**:864–72.
162. Tan EK, Chua E, Fook-Chong SM, Teo YY, Yuen Y, Tan L, et al. Association between caffeine intake and risk of Parkinson's disease among fast and slow metabolizers. *Pharmacogenet Genomics* 2007;**17**:1001–5.
163. Tan EK, Lu ZY, Fook-Chong SM, Tan E, Shen H, Chua E, et al. Exploring an interaction of adenosine A2A receptor variability with coffee and tea intake in Parkinson's disease. *Am J Med Genet B Neuropsychiatr Genet* 2006;**141B**:634–6.
164. Facheris MF, Schneider NK, Lesnick TG, de Andrade M, Cunningham JM, Rocca WA, et al. Coffee, caffeine-related genes, and Parkinson's disease: a case-control study. *Mov Disord* 2008;**23**:2033–40.
165. Palacios N, Weisskopf M, Simon K, Gao X, Schwarzschild M, Ascherio A. Polymorphisms of caffeine metabolism and estrogen receptor genes and risk of Parkinson's disease in men and women. *Parkinsonism Relat Disord* 2010;**16**:370–5.
166. Popat RA, Van Den Eeden SK, Tanner CM, Kamel F, Umbach DM, Marder K, et al. Coffee, ADORA2A, and CYP1A2: the caffeine connection in Parkinson's disease. *Eur J Neurol* 2011;**18**:756–65.
167. Hill-Burns EM, Hamza TH, Zabetian CP, Factor SA, Payami H. An attempt to replicate interaction between coffee and CYP1A2 gene in connection to Parkinson's disease. *Eur J Neurol* 2011;**18**:e107–8.
168. Hamza T, Chen H, Hill-Burns E, Rhodes S, Montimurro JS, et al. Genome-wide gene-environment study identifies glutamate receptor gene GRIN2A as a Parkinson's disease modifier gene via interaction with coffee. *PLoS Genet* 2011;**7**:e1002237.
169. McCulloch CC, Kay DM, Factor SA, Samii A, Nutt JG, Higgins DS, et al. Exploring gene-environment interactions in Parkinson's disease. *Hum Genet* 2008;**123**:257–65.
170. Hancock DB, Martin ER, Vance JM, Scott WK. Nitric oxide synthase genes and their interactions with environmental factors in Parkinson's disease. *Neurogenetics* 2008;**9**:249–62.

171. Kokaze A, Ishikawa M, Matsunaga N, Karita K, Yoshida M, Ohtsu T, et al. NADH dehydrogenase subunit-2 237 Leu/Met polymorphism modulates the effects of coffee consumption on the risk of hypertension in middle-aged Japanese men. *J Epidemiol* 2009;**19**:231–6.
172. Yu X, Bao Z, Zou J, Dong J. Coffee consumption and risk of cancers: a meta-analysis of cohort studies. *BMC Cancer* 2011;**11**:96.
173. Peck JD, Leviton A, Cowan LD. A review of the epidemiologic evidence concerning the reproductive health effects of caffeine consumption: a 2000-2009 update. *Food Chem Toxicol* 2010;**48**:2549–76.
174. Leviton A, Cowan L. A review of the literature relating caffeine consumption by women to their risk of reproductive hazards. *Food Chem Toxicol* 2002;**40**:1271–310.
175. Costa J, Lunet N, Santos C, Santos J, Vaz-Carneiro A. Caffeine exposure and the risk of Parkinson's disease: a systematic review and meta-analysis of observational studies. *J Alzheimers Dis* 2010;**20**(Suppl. 1):S221–38.
176. Chen JF, Chern Y. Impacts of methylxanthines and adenosine receptors on neurodegeneration: human and experimental studies. *Handb Exp Pharmacol* 2011;**200**:267–310.
177. Coleman T, Ellis SW, Martin IJ, Lennard MS, Tucker GT. 1-Methyl-4-phenyl-1,2,3,6-tetrahydropyridine (MPTP) is N-demethylated by cytochromes P450 2D6, 1A2 and 3A4—implications for susceptibility to Parkinson's disease. *J Pharmacol Exp Ther* 1996;**277**:685–90.
178. Noordzij M, Uiterwaal CS, Arends LR, Kok FJ, Grobbee DE, Geleijnse JM. Blood pressure response to chronic intake of coffee and caffeine: a meta-analysis of randomized trials. *J Hypertens* 2005;**23**:921–8.
179. Myers MG, Basinski A. Coffee and coronary heart disease. *Arch Intern Med* 1992;**152**:1767–72.
180. Greenland S. A meta-analysis of coffee, myocardial infarction, and coronary death. *Epidemiology* 1993;**4**:366–74.
181. Kawachi I, Colditz GA, Stone CB. Does coffee drinking increase the risk of coronary heart disease? Results from a meta-analysis. *Br Heart J* 1994;**72**:269–75.
182. Sofi F, Conti AA, Gori AM, Eliana Luisi ML, Casini A, Abbate R, et al. Coffee consumption and risk of coronary heart disease: a meta-analysis. *Nutr Metab Cardiovasc Dis* 2007;**17**:209–23.
183. Cornelis MC, El-Sohemy A. Coffee, caffeine, and coronary heart disease. *Curr Opin Clin Nutr Metab Care* 2007;**10**:745–51.
184. Abraham J, Mudd JO, Kapur NK, Klein K, Champion HC, Wittstein IS. Stress cardiomyopathy after intravenous administration of catecholamines and beta-receptor agonists. *J Am Coll Cardiol* 2009;**53**:1320–5.
185. Fisone G, Borgkvist A, Usiello A. Caffeine as a psychomotor stimulant: mechanism of action. *Cell Mol Life Sci* 2004;**61**:857–72.
186. Ferre S. An update on the mechanisms of the psychostimulant effects of caffeine. *J Neurochem* 2008;**105**:1067–79.
187. Sreerama L, Hedge MW, Sladek NE. Identification of a class 3 aldehyde dehydrogenase in human saliva and increased levels of this enzyme, glutathione S-transferases, and DT-diaphorase in the saliva of subjects who continually ingest large quantities of coffee or broccoli. *Clin Cancer Res* 1995;**1**:1153–63.
188. Fukushima Y, Kasuga M, Nakao K, Shimomura I, Matsuzawa Y. Effects of coffee on inflammatory cytokine gene expression in mice fed high-fat diets. *J Agric Food Chem* 2009;**57**:11100–5.
189. Stonehouse AH, Adachi M, Walcott EC, Jones FS. Caffeine regulates neuronal expression of the dopamine 2 receptor gene. *Mol Pharmacol* 2003;**64**:1463–73.
190. Connolly S, Kingsbury TJ. Caffeine modulates CREB-dependent gene expression in developing cortical neurons. *Biochem Biophys Res Commun* 2010;**397**:152–6.

191. Lam LK, Sparnins VL, Wattenberg LW. Isolation and identification of kahweol palmitate and cafestol palmitate as active constituents of green coffee beans that enhance glutathione S-transferase activity in the mouse. *Cancer Res* 1982;**42**:1193–8.
192. Cavin C, Holzhauser D, Constable A, Huggett AC, Schilter B. The coffee-specific diterpenes cafestol and kahweol protect against aflatoxin B1-induced genotoxicity through a dual mechanism. *Carcinogenesis* 1998;**19**:1369–75.

Nutrigenetics and Nutrigenomics of Caloric Restriction

Itziar Abete, Santiago Navas-Carretero, Amelia Marti, and J. Alfredo Martinez

Department of Nutrition, Food Science, Physiology and Toxicology, University of Navarra, Pamplona, Spain

Obesity is a complex disease resulting from a chronic and long-term positive energy balance in which both genetic and environmental factors are involved. Weight-reduction methods are mainly focused on dietary changes and increased physical activity. However, responses to nutritional intervention programs show a wide range of interindividual variation, which is importantly influenced by genetic determinants. In this sense, subjects carrying several obesity-related single-nucleotide polymorphisms (SNPs) show differences in the response to calorie-restriction programs. Furthermore, there is evidence indicating that dietary components not only fuel the body but also participate in the modulation of gene expression. Thus, the expression pattern and nutritional regulation of several obesity-related genes have been studied, as well as those that are differentially expressed by caloric restriction. The responses to caloric restriction linked to the presence of SNPs in obesity-related genes are reviewed in this chapter. Also, the influence of energy restriction on gene expression pattern in different tissues is addressed.

Progress in Molecular Biology
and Translational Science, Vol. 108
DOI: 10.1016/B978-0-12-398397-8.00013-7

I. Introduction

The prevalence of obesity continues to increase in many countries and is rising in parallel with the associated metabolic disorders.[1] It is known that obesity and the accompanying cardiovascular/metabolic complications are conditioned by genetic and environmental interactions.[2]

In fact, the predisposition to common forms of obesity is probably influenced by numerous susceptibility genes, accounting for variations in energy requirements, fuel utilization, tissue metabolic activity, and taste preferences.[1] Likewise, the response or adaptation to a dietary component or pattern depends in part upon the genotype of the individual.[3,4]

The initial step in most obesity treatments is usually the establishment of an adequate caloric restriction to create a negative energy balance favoring weight loss and cardiovascular benefits.[5] Thus, the effects of caloric restriction on human health, mainly on weight management and metabolic profile, have been studied.[2,6–9] It is evident that improved body weight through restriction of energy intake and even modified macronutrient or food distribution leads to a better quality of life,[2,6,7,10] improving values related to lipid and glucose metabolism as well as oxidative status.[11–15]

However, available data suggest that genetic variations related to energy metabolism, physical activity, appetite control, and the utilization of dietary components play an important role in the response to an obesity treatment.[16] Also, it is well known that some obese patients achieve higher weight loss than others in response to the same negative energy balance.[4,17,18] Several polymorphisms have been studied in order to promote personalized nutrition strategies based on the genetic make-up,[19] but the genes that are responsible for individual differences in body-weight regulation and sensitivity to alterations in energy balance have not yet been fully identified.[3,4] Multiple research designs and technologies are available to identify these genes and to delineate the nature and the extent of the genetic mechanisms involved.[20]

On the other hand, in recent years, the term “nutrigenomics” has been coined and the topic is being studied.[21,22] One branch of nutrigenomics focuses on the mechanisms by which energy restriction influences fuel homeostasis and specifically gene expression in humans, in adipocytes, skeletal muscle, or other target organs. In addition to changes observed in mRNA levels as a consequence of weight and fat mass loss,[23,24] energy restriction *per se* influences the expression of a panel of genes related to adipokines, hormones, and metabolic factors (Fig. 1).[25–27]

Hence, the aim of this chapter is to summarize the effects of gene polymorphisms on body weight and fat mass loss after hypocaloric diets, as well as the influence of energy restriction on the expression of genes from specific pathways

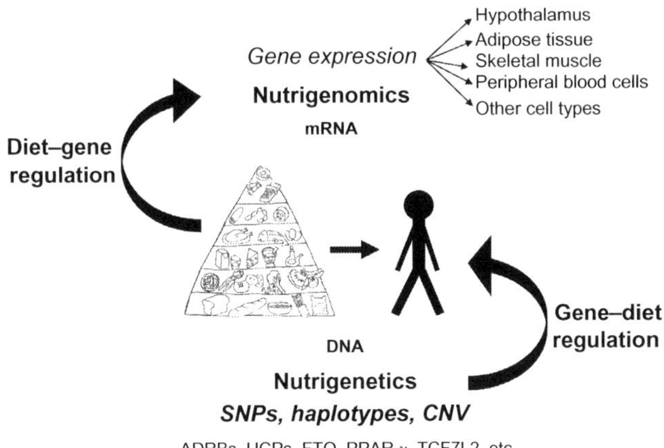

FIG. 1. Nutrigenetics and nutrigenomics. Influence of genotype on the response to a caloric-restriction program (nutrigenetics) and the effect of caloric restriction on gene expression (nutrigenomics).

II. Nutrigenetics and Caloric Restriction

Single-nucleotide polymorphisms (SNPs) in obesity candidate genes have been investigated in relation to weight loss induced by hypocaloric diets.[1,28,29] Numerous SNPs belong to metabolic pathways such as those involved in resting energy expenditure, thermogenic effect of food, and the ability to oxidize fat, as well as psychobehavioral factors that might affect the outcome to a weight-reduction program. This section reviews several gene variants involved in variable outcomes of obesity treatment, and the main findings are summarized in Table I.[30–71]

A. Beta-3 Adrenergic Receptor (*ADRB3*)

The *ADRB3* gene is expressed in adipose tissue and plays a significant role in controlling energy expenditure through its role in the regulation of lipolysis and thermogenesis.[32] A common polymorphism of the *ADRB3* gene is the Trp64Arg variant (rs4994).[32,72] A 2008 meta-analysis including 97 studies involving 44,833 individuals revealed that the Trp64Arg variant was associated with body mass index (BMI), with Arg64 allele carriers having higher BMI compared with noncarriers.[72] Yoshida *et al.*[32] also reported that the Trp64Arg variant of *ADRB3* was related to difficulties in lowering weight; homozygotes for the Arg64 allele were less successful in reducing weight than Trp64 carriers.[32] Likewise, a subsequent study also confirmed that women with the

TABLE I

OBESITY CANDIDATE GENES THAT INFLUENCE WEIGHT LOSS INDUCED BY CALORIC-RESTRICTION THERAPY

Gene	Gene variants	Effect of caloric restriction	References
Genes related to energy metabolism			
Adrenergic receptor	*ADRB2* (Glu27Gln and Arg16Gly)	Glu allele showed greater reduction in body weight	30
	ADRB3 (Trp64Arg)	Arg64 allele carriers lost less weight	31–33
		No influence on BMI reduction	34,35
Uncoupling proteins	*UCP1* (A-3826G)	GG homozygotes resistant to weight loss	36
	UCP2 G-866A	A allele associated with decreased lipid oxidation	37,38
	Ala55Val	ValVal genotype resistant to body fat loss	39
	UCP3 (-55C > T)	haplotype 1 [CGTACC] significantly associated with increased reduction in body weight	40,41
	UCP2–3	Associated with VLCD-induced fat mass reduction	38
Genes related to appetite control			
Leptin	*LEP* (C-2549A) (5′-region)	Carriers of the -254A allele registered less weight loss	42
Leptin receptor	*LEPR* (Ser343Ser) (T/C)	C allele carriers lost more weight	43
	LEPR (Lys109Arg)	AA genotype group showed higher fat loss	44
	LEPR (Lys656Asn)	No influence on weight loss	45
Pro-opiomelanocortin	*POMC* (R236G)	No influence on weight loss	46
5-Hydroxytryptamine (serotonin) receptor 2C	*HTR2C* promoter (C-759T)	Heterozygous more resistant to weight loss	47
Neuromedin beta	*NMB* (Pro73Thr)	Male T allele carriers resistant to weight loss	48
Melanocortin receptor	*MC3R*, *MC4R*	No influence on weight loss	49,50
Fat mass and obesity associated	*FTO*	No effect on weight loss and fat distribution	51–56

Adipogenic genes			
Peroxisome proliferator-activated receptor gamma	*PPARG2* (Pro12Ala)	12Ala allele carriers had greater reduction in body weight and waist circumference	57–59
		Women were nonresponders to VLCD	60
Transcription factor 7-like 2	*TCF7L2* (rs7903146 and others)	Risk allele carriers had smaller reduction in BMI	61
Fatty acid-binding protein 2, intestinal	*FABP2* (Ala54Thr)	Minor allele carriers had greater decrease in fat mass	62
Genes related to insulin resistance			
Insulin receptor substrates/insulin-like growth factor 1 receptor	*IRS1* (Gly972Arg)/*IRS2* (G1057D)/*IGF1R* (GAA1013GAA)	Lifestyle intervention not successful in subjects carrying polymorphisms	63
Insulin-induced gene 2	*INSIG2* (rs7566605)	CC homozygous more resistant to weight loss	64
Adiponectin, C1Q and collagen domain containing	*ADIPOQ* (-11391G/C)	A allele protected against weight regain	65
Genes related to lipid metabolism			
Apolipoprotein A–V	*ApOA5* (T1131C)	Weight reduction higher in C allele carriers	66
Hepatic lipase	*LIPC* (G-250G)	Subjects did not show differences in weight loss	67
Perilipin	*PLIN* (G11482A)	GG homozygotes lost more weight	68
	PLIN (A14995T)	14995A > T had increased free fatty acid levels with a rapid loss in abdominal fat	69
Inflammation genes			
Interleukin-6	*IL6* (-174G > C)	C allele protected against weight regain	70
Tumor necrosis factor	*TNF* (G308A)	No influence on weight loss	71

Arg64 allele lose weight more slowly when enrolled in a weight-loss program.[31] In contrast, some studies have suggested that *ADRB3* Trp64Arg does not influence BMI reduction after an exercise-based intervention program.[34,35]

One trial found a synergistic effect between *ADBR3* Trp64Arg and insulin receptor substrate 1 (*IRS-1*) gene polymorphism (Gly972Agr) on weight loss. Thus, after 13 weeks of a weight-loss program, the group with minor frequency alleles had lost less body weight than obese controls without mutations.[33] Other authors have also shown that the presence of the *ADBR3* polymorphism decreased the changes in the ratio of visceral to subcutaneous abdominal fat during a 3-month weight-reduction program.[73] Overall, the data suggest that the Trp64Arg mutation of *ADBR3* is involved in the resistance to a low-calorie diet, but the evidence remains somewhat controversial.

B. Beta-2 Adrenergic Receptor, Surface (*ADRB2*)

The *ADRB2* gene encodes a key lipolytic receptor in human white adipose tissue.[74] Different polymorphic forms, point mutations, and downregulation of *ADRB2* have been associated with obesity and type 2 diabetes.[74–76] A meta-analysis evaluating the association between two common *ADRB2* polymorphisms and obesity included 10,404 subjects genotyped at Glu27Gln (rs1042714) and 4328 subjects genotyped at Arg16Gly (rs1042713). The frequency of Glu27 allele carriers and Arg16 allele carriers varied between race groups (Europeans, Asians, Pacific Islanders, and American Indians). The presence of the Glu27 allele in the *ADRB2* gene appeared to be a significant risk factor for obesity in Asians, Pacific Islanders, and American Indians, but not in Europeans.[74] The role of the *ADRB2* Gln27Glu and Arg16Gly polymorphisms on body weight and body composition response to energy restriction has also been investigated. In 78 Spanish obese women on a 12-week energy-restricted diet,[30] those carrying the Glu allele had greater reduction in body weight than non-Glu allele carriers, and they also lost more lean mass than the Gln27Gln group. However, a significant interaction effect was not found between the Arg16Gly polymorphism and diet-induced changes.

C. Uncoupling Proteins

The uncoupling proteins (UCPs) are a class of proteins found in the inner mitochondrial membrane, and allelic variants of the associated genes may influence body weight through variation in resting energy expenditure, substrate oxidation, and exercise efficiency.[77] The effects of the Ala55Val (rs660339) genetic polymorphism of *UCP2* were evaluated in 386 subjects after a 1-month caloric-restriction program.[39] The results showed that the ValVal genotype had a smaller decrease in total body fat compared with the other types, whereas changes in lean body mass, protein, mineral, and water contents did not differ on the basis of the Ala55Val polymorphism. Another study examined the effects on weight loss and subsequent maintenance of the Trp64Arg mutation in the *ADBR3* gene and the A > G mutation in the *UCP1* gene.[78] Seventy-seven obese subjects were included in a 12-week weight-loss program and a 40-week weight-maintenance phase. Subjects with both mutations had a lower weight reduction during the very low calorie diet (VLCD) and regained the weight during the maintenance phase. Thus, the presence of UCP polymorphisms may participate in weight and fat mass changes, but the evidence is not conclusive.

D. Fat Mass and Obesity Associated Gene (*FTO*)

The *FTO* gene was identified as an important locus harboring common variants with an unequivocal impact on obesity predisposition and fat mass at the population level.[51,52] Likewise, it was found, in studies concerning large numbers of subjects, that the *FTO* gene polymorphism is related to higher BMI, as well as weight and abdominal circumference.[56] However, the knowledge of a potential modifying effect of the *FTO* gene on changes in body weight achieved by lifestyle interventions is limited. In this context, the Finnish Diabetes Study examined the association of the *FTO* gene variant (rs9939609, T/A) with body weight and BMI and long-term weight changes. Five-hundred and twenty-two subjects with impaired glucose tolerance were randomized to control and lifestyle intervention groups. The results confirmed the association between the common *FTO* variant and BMI; however, there was no association between the *FTO* variant and the magnitude of weight reduction achieved by a long-term lifestyle intervention.[79] In accordance with this, in a case–control study enrolling 519 German overweight and obese children and adolescents and 178 normal-weight adults, no differential effects of the rs9939609 alleles on weight loss were observed after a nutritional intervention program.[80] A more recent study analyzed the influence of *FTO* variants rs17817449 (first intron) and rs17818902 (third intron) on changes in BMI after a short-term lifestyle intervention, and no significant interplay was found between the BMI decrease and *FTO* variants.[53] Furthermore, interventions in children[52] led to a significant decrease in BMI, which was modulated by the *FTO* genotype, whereas carriers of the obesity-related genotype profited more from the intervention.

It is not known whether the association between genetic variation in *FTO* and obesity is mediated through effects on energy intake and energy expenditure. Fischer and coworkers[81] showed that the loss of *FTO* in mice led to a significant reduction of adipose tissue and lean body mass as a consequence of increased energy expenditure and systemic sympathetic activation. On the other hand, data from the Avon Longitudinal Study of Parents and Children showed that subjects carrying minor variants at rs9939609 consumed more fat and total energy than those individuals not carrying such variants.[54] Likewise, other authors have suggested that *FTO* variants that confer predisposition to obesity do not appear to be involved in the regulation of energy expenditure but may have a role in the control of food intake and food choice.[55,82–85] There are few interventional studies that have evaluated the effect of *FTO* SNPs on the response to weight-loss programs, yet such studies are needed to understand the influence of these polymorphisms on weight reduction and other related parameters such as appetite and energy expenditure.

E. Peroxisome Proliferator-Activated Receptor Gamma (*PPARG2*)

The *PPARG2* gene encodes an intracellular transcription factor that plays a role in adipogenesis and glucose and lipid homeostasis, as well as in body fat distribution.[86] A study carried out with the Pro12Ala polymorphism (rs1801282) of the *PPARG2* gene has shown a positive correlation with weight gain; during a 10-year study, the Ala12 carriers gained more weight over time than the Ala12 noncarriers.[57] However, there are also reports suggesting that the Pro12Ala *PPARG2* polymorphism is not associated with BMI or metabolic syndrome parameters in postmenopausal women, but that it does predispose to a less favorable lipid profile in this population.[87] On the other hand, it is unclear whether the *PPARG2* genotype influences weight reduction in response to caloric restriction in humans. Several years ago, Vidal-Puig *et al.*[88] observed that following a low-calorie diet specifically downregulates the expression of PPARG2 mRNA in the adipose tissue of obese humans, which may favor fat mass reduction.[88] In this sense, several studies have been carried out to assess the relationship between *PPARG* genotype and weight loss. An intervention study evaluated the effect of polymorphisms in the *PPARG2* and *ADRB2* genes in 60 obese women who followed a low-calorie diet for 10 weeks. Diet composition was modified to evaluate the influence of macronutrient composition. Results showed that the Pro12Ala polymorphism in the *PPARG2* gene influenced energy metabolism regardless of the genotype of the *ADRB2* gene. Thus, fat oxidation and energy expenditure were lower in Pro12Pro carriers compared to Pro12Ala/Ala12Ala genotypes. Likewise, a higher polyunsaturated fatty acid intake increased fat oxidation in Pro12Ala/Gln27Glu, which could result in greater body weight loss.[89–91]

Another study investigated whether *PPARG* gene variations were associated with weight reduction and changes in coronary heart disease risk factors in 95 middle-aged Japanese women instructed to consume a nutritionally balanced diet of 1200 kcal/day for 14 weeks.[58] Eight SNPs in the *PPARG* gene (rs1801282 [Pro/Ala], rs2292101, rs2959272, rs1386835, rs709158, rs1175540, rs1175544, and 1797912) were screened. The weight reduction was marked during the intervention, with a mean body weight loss of $11.3 \pm 4.4\%$. Six *PPARG* SNPs were significantly associated with weight reduction (rs2959272, rs1386835, rs709158, rs1175540, rs1175544, and rs1797912), with rs1175544 having the strongest association ($P=0.004$).[58] However, no relationship was found between these SNPs and the changes in coronary heart disease risk factors that accompanied weight loss. Results of an interventional study with overweight Korean female subjects suggested that the *PPARG2* Pro12Ala PA/AA genotype was associated with higher subcutaneous and visceral fat areas in comparison to PP genotype subjects, but the genotype did not affect outcomes of a weight-reduction program.[92] Another trial established that the Pro12Ala polymorphism was associated with diet resistance, as women carrying this polymorphism were found to be nonresponders to a VLCD program.[60]

On the other hand, a different frequency distribution for the *PPARG2* genotypes has been observed, with a successful weight-management group ($<10\%$ regain after weight loss) having fewer heterozygous (PA) subjects and more subjects with the homozygous Pro allele (PP) compared to an unsuccessful group ($\geq 10\%$ regain).[86] More recently, in a substudy within the PREDIMED project, the effects of a 2-year nutritional intervention with Mediterranean-style diets were evaluated depending on the Pro12Ala polymorphism of the *PPARG* gene. Thus, carriers of the 12Ala allele allocated to the control group (following a conventional low-fat diet) had higher waist circumference compared with wild-type subjects after 2 years of nutritional intervention. However, subjects with the 12Ala allele in the Mediterranean diet group reduced waist circumference, reversing the negative effect that the *PPARG2* gene appeared to have in the 12Ala allele carriers. The Pro12Ala substitution in this gene has also been widely associated with diabetes.[93] Indeed, a positive association between the 12Ala allele of *PPARG* and obesity-related traits was reported among diabetic patients.[94]

Most associations between dietary intake, risk factors for chronic diseases, and genetics have come from observational epidemiological studies, so in this sense, more dietary intervention trials are needed to provide the strongest evidence of causality.

F. Transcription Factor 7-Like 2

The transcription factor 7-like 2 (*TCF7L2*) gene is expressed in adipose tissue and involved in Wnt-dependent regulation of adipogenesis.[61] Current evidence suggests that the *TCF7L2* gene influences the risk for type 2 diabetes by reducing glucose-induced insulin secretion.[95] Thus, eight subjects with risk-conferring *TCF7L2* genotypes (TT or TC at rs7903146) and 10 matched subjects with wild-type genotype (CC) underwent a 5-h oral glucose tolerance test, which showed that those subjects with the risk-conferring *TCF7L2* genotypes had 50% lower beta-cell responsivity to the oral glucose test.[95] Gene variants of *TCF7L2* have also been associated with less weight loss in response to lifestyle intervention.[61] Thus, 309 German subjects at increased risk for type 2 diabetes were genotyped for SNPs rs7903146, rs12255372, rs11196205, and rs7895340 in *TCF7L2* and instructed to carry out a 2-year exercise and dietary intervention. During the lifestyle intervention, the risk allele carriers of the SNP rs7903146 displayed a smaller reduction in BMI and total body fat. In contrast, SNPS rs11196205 and rs7895340 were not associated with body composition or weight loss during the lifestyle intervention.[61] Thus, the diabetes-associated *TCF7L2* gene variant was found to be a predictor of success of lifestyle intervention in terms of weight reduction.[61]

G. Other Gene Variants

More interactions between relevant gene polymorphisms affecting the amount and composition of weight loss, as well as the changes in obesity-associated risk factors, are under investigation.[1] Polymorphisms of genes related to the control of appetite such as leptin (*LEP*), leptin receptor (*LEPR*), and proopiomelanocortin (*POMC*) have been widely studied.[20] For instance, less weight loss has been described for an *LEP* gene polymorphism in the promoter region 5′ and for the carriers of the -2549A allele at position c-2549A after a low-calorie diet in obese women.[42] Likewise, overweight women carrying the C allele of the Ser343Ser polymorphism of the *LEPR* gene lost more weight than noncarriers, while the Lys656Asn polymorphism in the same gene did not show any influence on weight reduction.[45] Several adipogenic genes and genes related to lipid metabolism such as apolipoproteins (*APOE*, *APOA4*, *APOA5*),[66] perilipins,[68,69] adiponectin, C1Q and collagen domain containing (*ADIPOQ*),[65] and fatty acid-binding protein 2, intestinal (*FABP2*),[62] as well as genes related to the insulin pathway,[63,64] have been investigated for their influence on weight loss during calorie-restriction programs. Genetic polymorphisms that may influence the individual susceptibility to inflammation also have been studied, as the relationship between obesity and low-degree inflammation was established several years ago.[96] Different responses to a weight-reduction program, as well as during a weight-maintenance period, were found in subjects with the -174G > C SNP in the promoter region of the interleukin 6 (*IL6*) gene, with the C allele carriers partially protected against weight regain.[70]

III. Caloric Restriction and Gene Expression Changes

In addition to the statistically significant associations between specific polymorphisms and outcomes of caloric restriction,[21,52,97] there is increasing evidence that energy intake influences gene (mRNA levels) expression.[70] In fact, caloric restriction has been shown to improve longevity and metabolism,[98,99] and one can hypothesize that these changes are mediated in part by a differential gene expression (Table II). In this section, we focus on the impact of caloric restriction on gene expression in adipose tissue, skeletal muscle, and peripheral blood mononuclear cells (PBMCs).

TABLE II

NUTRIGENOMICS IN CALORIC RESTRICTION TRIALS

Type of intervention	Most relevant genes studied	Tissues and main effect observed	References
40 obese women following 600 kcal/day diets either moderate or low in fat content and moderate or high in carbohydrate content	*FASN*, *SCD*, *FADS1*, *FADS2*	Adipose tissue. Coordinated reduction in the expression of genes regulating the production of polyunsaturated fatty acids	23
50 obese women following 30% energy-restricted diets either low or high in fat content	Secreted proteins (*LEP*, *SPARC*), glucose and lipid metabolism (*HSL*, *LPL*, *PDE3B*, *ANPRA*, *CD36*), mitochondrial energy metabolism (*UCP2*), transcription factors and cofactors (*PPARGC1A*, *PPARG2*, *PPARG*t)	Adipose tissue. All genes downregulated except PGC-1α, which was upregulated.	24
Obese subjects following VLCD and healthy subjects following overfeeding	Lipogenesis (*ACLY*, *ACACA*, *FASN*, *SCD*), insulin resistance (*PEDF*, *SPARC*), protein synthesis (*EIF4EBP1*, *EIF4EBP2*), β-oxidation (*CPT1B*)	Adipose tissue. Lipogenesis and insulin resistance gene expression downregulated. Protein synthesis and β-oxidation gene expression upregulated	27,100,101
Obese subjects following VLCD	*LEP*, *RBP4*, *TNF*, *PAI1*	Adipose tissue. Genes downregulated	102–107
12 physically active adults following 20% energy-restricted diet	*AKT1*, *EIF4EBP1*	Skeletal muscle. Protein synthesis downregulated because of a decrease in intracellular signaling	108
Overweight adults following 25% energy-restricted diet either with or without exercise	*TFAM*, *PPARGC1A*	Skeletal muscle. Increased expression of mitochondrial biogenesis activators	109
12 obese subjects following 30% energy-restricted diet	*SIRT1*, *SIRT2*	PBMCs. Increased gene expression and improved antioxidant status	110
Obese subjects following long-term caloric restriction	Anti-inflammatory cytokines (*IL6*, *IL8*), pro-inflammatory cytokines (*TNF*, *IL1B*, *IL1RN*, *TNF/NFKB* signaling cascade)	PBMCs. Anti-inflammatory gene expression upregulated and pro-inflammatory cytokine gene expression downregulated	107,111,112
17 obese subjects following long-term caloric restriction	*COX15*, *MGST2*	PBMC. Attenuation of postprandial downregulation of *COX15* and increase during fasting of *MGST2* gene expression	113

FASN, fatty acid synthase; *SCD*, stearoyl-CoA desaturase; *FADS*, fatty acid desaturase; *LEP*, leptin; *SPARC*, osteonectin; *HSL*, hormone-sensitive lipase; *LPL*, lipoprotein lipase; *PDE3B*, phosphodiesterase 3B, cGMP-inhibited; *ANPRA*, natriuretic peptide receptor A/guanylate cyclase A (now known as *NPR1*); *CD36*, thrombospondin receptor; *UCP2*, uncoupling protein 2; *PPARGC1A*, PPARG, coactivator 1 alpha; *PPARG2*, *PPARG* type 2 isoform; *PPARGt*, *PPARG* total (types 1 and 2); *ACLY*, ATP citrate lyase; *ACACA*, acetyl-CoA carboxylase alpha; *PEDF*, pigment epithelium derived factor; *EIF4EBP* (1,2), eukaryotic translation initiation factor 4E binding proteins 1 and 2; *CPT1B*, carnitine palmitoyl-transferase 1B; *RBP4*, retinol binding protein 4, plasma; *TNF*, tumor necrosis factor; *PAI1*, plasminogen activator inhibitor 1; *AKT1*, v-akt murine thymoma viral oncogene homolog 1; *TFAM*, transcription factor A, mitochondrial; *SIRT* (1 and 2), sirtuin isoforms 1 and 2; *IL6*, interleukin 6; *IL8*, interleukin 8; *IL1B*, interleukin 1, beta; *IL1RN*, interleukin 1 receptor antagonist; *NFKB*, nuclear factor of kappa light polypeptide gene enhancer in B-cells; *COX15*, COX15 homolog, cytochrome *c* oxidase assembly protein; *MGST2*, microsomal glutathione *S*-transferase 2.

A. Adipose Tissue

Body-weight reduction directly affects adipose tissue, influencing not only its activity but also energy storage and adipocyte endocrine functions by improving metabolic disorders commonly associated with overweight.[7] Various studies have documented the gene expression changes in this tissue during weight-loss programs.[23–25] However, it is interesting to distinguish between the consequences of weight reduction and caloric restriction *per se*[26] on gene expression.

A study carried out by Franck *et al.*[27] aimed at identifying those genes affected by caloric intake, regardless of body-weight changes. These investigators analyzed male and female subjects who had been participating in either a weight-loss program[100,101] or a fast-food study.[27] Obese subjects were given a VLCD with gradual reintroduction of ordinary food, while healthy subjects were overfed. Adipose tissue biopsies were taken throughout both nutritional interventions, and gene expression was measured by DNA microarrays.[101] When assessing the outcomes, at least 100 genes were affected by energy intake, with 52 genes upregulated and 50 downregulated because of caloric restriction. Among these genes, some of those involved in lipogenesis (*ACLY*, *ACACA*, *FASN*, *SCD*) and insulin resistance (*PEDF*, now known as *SERPINF1*, *SPARC*) were downregulated, while genes related to protein synthesis (*EIF4BP1*, *EIF4EBP2*) and β-oxidation (*CPT1B*) appeared to be upregulated. Therefore, it seems that caloric restriction *per se* positively influences adipose tissue function, independently of changes in body weight or fat mass. In fact, the decreased *de novo* lipogenesis observed in previous works[114] is attributed mainly to the downregulation of ATP citrate lyase (ACLY), acetyl-coenzyme-A (CoA) carboxylase 1 (ACACA), or fatty acid synthase (FASN). Also, changes in caloric intake could potentially interfere with protein synthesis at the translational level because of higher mRNA levels of eukaryotic translation initiation factor 4E binding protein 1 and 2 (EIF4EBP1, EIF4EBP2).[27,115,116]

In relation to *LEP* gene expression conditioned by caloric restriction, there are some controversial results. While in some studies *LEP* expression has been reported to be influenced by energy intake,[102] some recent reports do not support such an effect.[27] In studies with laboratory animals, both acute and long-term energy deprivations have led to a downregulation in hypothalamic and white adipose tissue *Lep* expression.[117,118] Interestingly, it has also been observed that refeeding after prolonged food restriction reverts this downregulation in white adipose tissue, but the inhibition of *LEP* expression continues in hypothalamus.[118] This observation seems to confirm reports in rats that have associated changes in serum LEP concentration with gene expression in white adipose tissue,[118,119] which is in contrast with other findings in humans.[117] However, in a trial on the effects in obese subjects of two hypocaloric diets differing in fat content (high-fat vs. low-fat), LEP mRNA levels decreased after

8 weeks of energy restriction, regardless of fat content.[24] In the latter nutritional intervention, 10 genes related to glucose and lipid metabolism were reported to have their expression affected by energy restriction, without apparent effects due to macronutrient distribution.[24] In addition to the downregulation of *LEP* expression, osteonectin (*SPARC* [secreted protein, acidic, cysteine-rich]), hormone-sensitive lipase (*HSL*, now known as *LIPE*), lipoprotein lipase (*LPL*), *UCP2*, *PPARG2*, and *PPARG* total (types 1 and 2), among others, were downregulated.[24] Only *PPARG*, coactivator 1 alpha (*PPARGC1A*) expression was upregulated.

The changes in *LEP* gene expression in white adipose tissue also imply changes in the orexigenic neuropeptide-Y (*NPY*)[120]; thus, if *LEP* expression is downregulated, *NPY* gene expression increases in the hypothalamus.[121–123] A study carried out by Sucajtys-Szulc *et al.*[118] tried to discern whether *NPY* expression is regulated by *LEP* expression in the hypothalamus or in white adipose tissue. For this purpose, they studied three groups of rats: control group, energy-restricted group, and energy-restricted with *ad libitum* refeeding afterward. The increase of *Npy* gene expression due to food restriction and the subsequent downregulation of hypothalamic and white adipose tissue *Lep* expression were evident. However, when the animals were refed, only white adipose tissue gene expression increased, while hypothalamic gene expression continued to be downregulated.[118]

In obese human subjects following 3–6 weeks of a VLCD (550–940 kcal/day), adipose tissue *LEP* expression was shown to significantly decrease.[102] However, some of the most recent studies in humans have not been able to establish the relationship between energy restriction and changes in the expression of *LEP* gene.[27] In fact, refeeding and overfeeding in human subjects did not lead to changes in *LEP* and *ADIPOQ* gene expression, with two possible causes: either the refeeding period was too short to induce changes, or most likely, changes in the expression of these two genes are more related to weight changes than energy intake variations.[27,28]

The effect of caloric restriction has also been highlighted in the expression changes of genes encoding for anti-inflammatory markers, both in adipocytes and PBMCs. Thus, it has been observed that obese subjects following VLCD showed a reduction in the expression of retinol binding protein 4, plasma (*RBP4*)[103] in adipose tissue, together with enhanced gene expression of tumor necrosis factor (*TNF*)[102] and plasminogen activator inhibitor 1 (*PAI1*, now known as serpin peptidase inhibitor, clade E, member 1 [*SERPINE1*]).[104] However, in all these studies, it may be difficult to distinguish the effects due to the energy restriction *per se* from those depending on fat mobilization and subsequent fat-mass loss,[24,25,105,106] although most of these markers are not exclusively expressed in adipocytes.[107] Thus, caloric restriction may play a conditional role in expression profile changes secondary to weight loss.

B. Skeletal Muscle

Caloric restriction following weight-loss programs has been shown to influence skeletal muscle loss, whole-body protein metabolism, and energy expenditure.[124,125] The degree and duration of the energy restriction are determinants of these observed effects, as acute energy restriction and fasting will lead to an enhancement of amino acid oxidation and nitrogen excretion as a consequence of the increased proteolysis. These effects are reversed in long-term caloric restriction as a result of adaptive mechanisms.[126–128] Additionally, it has been recently observed that acute energy deprivation leads to a downregulation in skeletal muscle protein synthesis because of the decrease in intracellular signaling.[108] This downregulation appears to be due to a reduction in the phosphorylation of protein kinase B (also known as RAC-alpha serine/threonine-protein kinase, AKT1), which reduces mRNA translation initiation of EIF4EBP1. This leads to a decrease in protein translation initiation.[129]

It has also been reported that caloric restriction improves whole-body energy efficiency by increasing biogenesis in mitochondria, which implies less oxygen utilization and a decreased production of reactive oxygen species.[130,131] In fact, 6 months of caloric restriction in overweight nonobese subjects has been shown to increase the expression levels of transcription factor A, mitochondrial (*TFAM*) and *PPARGC1A*, which suggests an increase in mitochondrial biogenesis in skeletal muscle.[109] Furthermore, evidence shows an increase in the expression of sirtuin 1 (*SIRT1*) after caloric restriction in skeletal muscle, which is thought to induce an increase in the expression of mitochondrial biogenesis activators,[132] resulting in a reduction of oxidative stress and protection against DNA damage.[109,130]

C. Peripheral Blood Mononuclear Cells

Sirtuins are NAD-dependent protein deacetylases, and in recent years, their role in physiological and pathological conditions has been established. They are now considered novel targets for treating some diseases associated with aging and perhaps in extending human life span.[133,134] Although sirtuins are a family with at least seven enzymes (known so far), the most interesting in relation to weight management and metabolic changes in obese/overweight seems to be SIRT1, which protects against cellular oxidative stress and inhibits adipogenesis in adipocytes.[135,136] SIRT2, which is located mainly in the cytoplasm, is involved in the inhibition of cell proliferation in cancer,[137,138] as well as inhibition of adipocyte differentiation and lipid accumulation.[139] It is well known that life span extension is strongly related to caloric restriction, mainly because of the reduction of reactive oxygen species during cellular respiration and the severity of age-related metabolic disorders.[131,140] Sirtuin-related genes have been associated with increased longevity under caloric restriction,[141] and they appear to be implicated in the regulation of different aspects of caloric-restriction responses, such as glucose homeostasis, insulin secretion, fat metabolism, and stress resistance.[142]

It has been shown that, as a consequence of caloric restriction, *SIRT1* and *SIRT2* gene expression in PBMCs is increased,[110] although the mechanisms for each gene are different.[143,144] This increased expression is accompanied by an improvement in antioxidant status in overweight subjects,[110] which suggests an association between both factors. Indeed, Crujeiras *et al.*[110] observed that *SIRT1* and *SIRT2* gene expression is modulated by baseline oxidative status, which reinforces the findings previously linking *SIRT1* with mitochondrial bioenergetics.[109]

Caloric restriction has been found to exert other interesting effects on gene expression in PBMC, and they relate to the modification of inflammatory marker expression.[106,111,112] When obese subjects followed a low-calorie diet for long periods (6–9 months), gene expression was modified in PBMCs, resulting in an overall improvement in low-grade chronic inflammation associated with obesity.[106,111,112] Thus, the gene expression of anti-inflammatory cytokines such as IL6 and IL8 is increased,[112] while there is a decrease in mRNA levels of pro-inflammatory factors such as TNF-α, IL1β, and IL1-receptor antagonist (IL1RN).[106,112] It has also been demonstrated that caloric restriction downregulates gene expression on the *TNF*/nuclear factor of kappa light polypeptide gene enhancer in B-cells (*NFKB*) signaling cascade.[111]

Recent observations have highlighted possible tachyphylactic effects after low-calorie diets.[113] Indeed, the COX15 homolog, cytochrome *c* oxidase assembly protein (*COX15*) and microsomal glutathione *S*-transferase 2 (*MGST2*) gene expression assays in PBMCs have emerged as valuable nutrigenomic biomarkers of the oxidative response. The improvement in antioxidant status induced by energy restriction seems to tone down the postprandial reduction in *COX15* gene expression induced by nutrient intake, while a marked increase in *MGST2* gene expression is observed under fasting conditions after a caloric-restriction period.[113]

IV. Conclusions

Both nutrigenetics and nutrigenomics are emerging scientific areas playing key roles in our understanding of the mechanisms involved in the responses to dietary treatments focused on weight loss induced by caloric restriction. However, it has become evident that the knowledge on polymorphisms and on the regulation of gene expression is not sufficient to reduce the obesity prevalence in our societies. Research focused on the interactions between nutrition and genetics, together with new knowledge areas such as epigenetics, proteomics, and metabolomics, will have the potential to move medical sciences toward practical personalized nutrition.

References

1. Marti A, Goyenechea E, Martinez JA. Nutrigenetics: a tool to provide personalized nutritional therapy to the obese. *World Rev Nutr Diet* 2010;**101**:21–33.
2. Abete I, Astrup A, Martinez JA, Thorsdottir I, Zulet MA. Obesity and the metabolic syndrome: role of different dietary macronutrient distribution patterns and specific nutritional components on weight loss and maintenance. *Nutr Rev* 2010;**68**:214–31.
3. Bouchard C. Gene-environment interactions in the etiology of obesity: defining the fundamentals. *Obesity (Silver Spring)* 2008;**16**(Suppl. 3):S5–S10.
4. Moreno-Aliaga MJ, Santos JL, Marti A, Martinez JA. Does weight loss prognosis depend on genetic make-up? *Obes Rev* 2005;**6**:155–68.
5. Villareal DT, Chode S, Parimi N, Sinacore DR, Hilton T, Armamento-Villareal R, et al. Weight loss, exercise, or both and physical function in obese older adults. *N Engl J Med* 2011;**364**: 1218–29.
6. Abete I, Parra D, Martinez JA. Energy-restricted diets based on a distinct food selection affecting the glycemic index induce different weight loss and oxidative response. *Clin Nutr* 2008;**27**:545–51.
7. Abete I, Parra MD, Zulet MA, Martinez JA. Different dietary strategies for weight loss in obesity: role of energy and macronutrient content. *Nutr Res Rev* 2006;**19**:5–17.
8. Labayen I, Diez N, Parra MD, Gonzalez A, Martinez JA. Time-course changes in macronutrient metabolism induced by a nutritionally balanced low-calorie diet in obese women. *Int J Food Sci Nutr* 2004;**55**:27–35.
9. Larsen TM, Dalskov SM, van Baak M, Jebb SA, Papadaki A, Pfeiffer AF, et al. Diets with high or low protein content and glycemic index for weight-loss maintenance. *N Engl J Med* 2010; **363**:2102–13.
10. Abete I, Parra D, Martinez JA. Legume-, fish-, or high-protein-based hypocaloric diets: effects on weight loss and mitochondrial oxidation in obese men. *J Med Food* 2009; **12**:100–8.
11. Abete I, Parra D, Crujeiras AB, Goyenechea E, Martinez JA. Specific insulin sensitivity and leptin responses to a nutritional treatment of obesity via a combination of energy restriction and fatty fish intake. *J Hum Nutr Diet* 2008;**21**:591–600.
12. Abete I, Parra D, De Morentin BM, Alfredo Martinez J. Effects of two energy-restricted diets differing in the carbohydrate/protein ratio on weight loss and oxidative changes of obese men. *Int J Food Sci Nutr* 2009;**60**(Suppl. 3):1–13.
13. Crujeiras AB, Parra D, Abete I, Martinez JA. A hypocaloric diet enriched in legumes specifically mitigates lipid peroxidation in obese subjects. *Free Radic Res* 2007;**41**:498–506.
14. Hermsdorff HH, Zulet MA, Abete I, Martinez JA. A legume-based hypocaloric diet reduces proinflammatory status and improves metabolic features in overweight/obese subjects. *Eur J Nutr* 2011;**50**:61–9.
15. Hermsdorff HH, Zulet MA, Abete I, Martinez JA. Discriminated benefits of a Mediterranean dietary pattern within a hypocaloric diet program on plasma RBP4 concentrations and other inflammatory markers in obese subjects. *Endocrine* 2009;**36**:445–51.
16. McAllister EJ, Dhurandhar NV, Keith SW, Aronne LJ, Barger J, Baskin M, et al. Ten putative contributors to the obesity epidemic. *Crit Rev Food Sci Nutr* 2009;**49**:868–913.
17. Bouchard C, Tremblay A, Despres JP, Theriault G, Nadeau A, Lupien PJ, et al. The response to exercise with constant energy intake in identical twins. *Obes Res* 1994;**2**:400–10.
18. Hainer V, Stunkard AJ, Kunesova M, Parizkova J, Stich V, Allison DB. Intrapair resemblance in very low calorie diet-induced weight loss in female obese identical twins. *Int J Obes Relat Metab Disord* 2000;**24**:1051–7.

19. Hainer V, Zamrazilova H, Spalova J, Hainerova I, Kunesova M, Aldhoon B, et al. Role of hereditary factors in weight loss and its maintenance. *Physiol Res* 2008;**57**(Suppl. 1):S1–S15.
20. Deram S, Villares SM. Genetic variants influencing effectiveness of weight loss strategies. *Arq Bras Endocrinol Metabol* 2009;**53**:129–38.
21. Bauer M, Hamm A, Pankratz MJ. Linking nutrition to genomics. *Biol Chem* 2004;**385**:593–6.
22. Kaput J, Ordovas JM, Ferguson L, van Ommen B, Rodriguez RL, Allen L, et al. The case for strategic international alliances to harness nutritional genomics for public and personal health. *Br J Nutr* 2005;**94**:623–32.
23. Dahlman I, Linder K, Arvidsson Nordstrom E, Andersson I, Liden J, Verdich C, et al. Changes in adipose tissue gene expression with energy-restricted diets in obese women. *Am J Clin Nutr* 2005;**81**:1275–85.
24. Viguerie N, Vidal H, Arner P, Holst C, Verdich C, Avizou S, et al. Adipose tissue gene expression in obese subjects during low-fat and high-fat hypocaloric diets. *Diabetologia* 2005;**48**:123–31.
25. Capel F, Viguerie N, Vega N, Dejean S, Arner P, Klimcakova E, et al. Contribution of energy restriction and macronutrient composition to changes in adipose tissue gene expression during dietary weight-loss programs in obese women. *J Clin Endocrinol Metab* 2008;**93**: 4315–22.
26. Assali AR, Ganor A, Beigel Y, Shafer Z, Hershcovici T, Fainaru M. Insulin resistance in obesity: body-weight or energy balance? *J Endocrinol* 2001;**171**:293–8.
27. Franck N, Gummesson A, Jernas M, Glad C, Svensson PA, Guillot G, et al. Identification of adipocyte genes regulated by caloric intake. *J Clin Endocrinol Metab* 2011;**96**:E413–8.
28. Sorensen TI, Boutin P, Taylor MA, Larsen LH, Verdich C, Petersen L, et al. Genetic polymorphisms and weight loss in obesity: a randomised trial of hypo-energetic high- versus low-fat diets. *PLoS Clin Trials* 2006;**1**:e12.
29. Nieters A, Becker N, Linseisen J. Polymorphisms in candidate obesity genes and their interaction with dietary intake of n-6 polyunsaturated fatty acids affect obesity risk in a subsample of the EPIC-Heidelberg cohort. *Eur J Nutr* 2002;**41**:210–21.
30. Ruiz JR, Larrarte E, Margareto J, Ares R, Labayen I. Role of beta-adrenergic receptor polymorphisms on body weight and body composition response to energy restriction in obese women: preliminary results. *Obesity (Silver Spring)* 2011;**19**:212–5.
31. Shiwaku K, Nogi A, Anuurad E, Kitajima K, Enkhmaa B, Shimono K, et al. Difficulty in losing weight by behavioral intervention for women with Trp64Arg polymorphism of the beta3-adrenergic receptor gene. *Int J Obes Relat Metab Disord* 2003;**27**:1028–36.
32. Yoshida T, Sakane N, Umekawa T, Sakai M, Takahashi T, Kondo M. Mutation of beta 3-adrenergic-receptor gene and response to treatment of obesity. *Lancet* 1995;**346**:1433–4.
33. Benecke H, Topak H, von zur Muhlen A, Schuppert F. A study on the genetics of obesity: influence of polymorphisms of the beta-3-adrenergic receptor and insulin receptor substrate 1 in relation to weight loss, waist to hip ratio and frequencies of common cardiovascular risk factors. *Exp Clin Endocrinol Diabetes* 2000;**108**:86–92.
34. Fumeron F, Durack-Bown I, Betoulle D, Cassard-Doulcier AM, Tuzet S, Bouillaud F, et al. Polymorphisms of uncoupling protein (UCP) and beta 3 adrenoreceptor genes in obese people submitted to a low calorie diet. *Int J Obes Relat Metab Disord* 1996;**20**:1051–4.
35. Tahara A, Osaki Y, Kishimoto T. Effect of the beta3-adrenergic receptor gene polymorphism Trp64Arg on BMI reduction associated with an exercise-based intervention program in Japanese middle-aged males. *Environ Health Prev Med* 2010;**15**:392–7.
36. Shin HD, Kim KS, Cha MH, Yoon Y. The effects of UCP-1 polymorphisms on obesity phenotypes among Korean female subjects. *Biochem Biophys Res Commun* 2005;**335**: 624–30.

37. Le Fur S, Le Stunff C, Dos Santos C, Bougneres P. The common -866 G/A polymorphism in the promoter of uncoupling protein 2 is associated with increased carbohydrate and decreased lipid oxidation in juvenile obesity. *Diabetes* 2004;**53**:235–9.
38. Yoon Y, Park BL, Cha MH, Kim KS, Cheong HS, Choi YH, et al. Effects of genetic polymorphisms of UCP2 and UCP3 on very low calorie diet-induced body fat reduction in Korean female subjects. *Biochem Biophys Res Commun* 2007;**359**:451–6.
39. Cha MH, Kim KS, Suh D, Yoon Y. Effects of genetic polymorphism of uncoupling protein 2 on body fat and calorie restriction-induced changes. *Hereditas* 2007;**144**:222–7.
40. Cha MH, Shin HD, Kim KS, Lee BH, Yoon Y. The effects of uncoupling protein 3 haplotypes on obesity phenotypes and very low-energy diet-induced changes among overweight Korean female subjects. *Metabolism* 2006;**55**:578–86.
41. Jun HS, Kim IK, Lee HJ, Lee HJ, Kang JH, Kim JR, et al. Effects of UCP2 and UCP3 variants on the manifestation of overweight in Korean children. *Obesity (Silver Spring)* 2009;**17**: 355–62.
42. Mammes O, Betoulle D, Aubert R, Giraud V, Tuzet S, Petiet A, et al. Novel polymorphisms in the 5′ region of the LEP gene: association with leptin levels and response to low-calorie diet in human obesity. *Diabetes* 1998;**47**:487–9.
43. Mammes O, Aubert R, Betoulle D, Pean F, Herbeth B, Visvikis S, et al. LEPR gene polymorphisms: associations with overweight, fat mass and response to diet in women. *Eur J Clin Invest* 2001;**31**:398–404.
44. Abete I, Goyenechea E, Crujeiras AB, Martinez JA. Inflammatory state and stress condition in weight-lowering Lys109Arg LEPR gene polymorphism carriers. *Arch Med Res* 2009;**40**: 306–10.
45. de Luis DA, Aller R, Izaola O, Sagrado MG, Conde R. Influence of Lys656Asn polymorphism of leptin receptor gene on leptin response secondary to two hypocaloric diets: a randomized clinical trial. *Ann Nutr Metab* 2008;**52**:209–14.
46. Santoro N, Perrone L, Cirillo G, Raimondo P, Amato A, Coppola F, et al. Weight loss in obese children carrying the proopiomelanocortin R236G variant. *J Endocrinol Invest* 2006;**29**: 226–30.
47. Pooley EC, Fairburn CG, Cooper Z, Sodhi MS, Cowen PJ, Harrison PJ. A 5-HT2C receptor promoter polymorphism (HTR2C–759C/T) is associated with obesity in women, and with resistance to weight loss in heterozygotes. *Am J Med Genet B Neuropsychiatr Genet* 2004;**126B**:124–7.
48. Spalova J, Zamrazilova H, Vcelak J, Vankova M, Lukasova P, Hill M, et al. Neuromedin beta: P73T polymorphism in overweight and obese subjects. *Physiol Res* 2008;**57**(Suppl. 1): S39–48.
49. Santos JL, De la Cruz R, Holst C, Grau K, Naranjo C, Maiz A, et al. Allelic variants of melanocortin 3 receptor gene (MC3R) and weight loss in obesity: a randomised trial of hypo-energetic high- versus low-fat diets. *PLoS One* 2011;**6**:e19934.
50. Hainerova I, Larsen LH, Holst B, Finkova M, Hainer V, Lebl J, et al. Melanocortin 4 receptor mutations in obese Czech children: studies of prevalence, phenotype development, weight reduction response, and functional analysis. *J Clin Endocrinol Metab* 2007;**92**:3689–96.
51. Frayling TM, Timpson NJ, Weedon MN, Zeggini E, Freathy RM, Lindgren CM, et al. A common variant in the FTO gene is associated with body mass index and predisposes to childhood and adult obesity. *Science* 2007;**316**:889–94.
52. Rendo T, Moleres A, Marti Del Moral A. Effects of the FTO gene on lifestyle intervention studies in children. *Obes Facts* 2009;**2**:393–9.
53. Dlouha D, Suchanek P, Lanskalanskalanska V, Hubacek JA. Body mass index change in females after short-time life style intervention is not dependent on the FTO polymorphisms. *Physiol Res* 2011;**60**:199–202.

54. Timpson NJ, Emmett PM, Frayling TM, Rogers I, Hattersley AT, McCarthy MI, et al. The fat mass- and obesity-associated locus and dietary intake in children. *Am J Clin Nutr* 2008;**88**:971–8.
55. Haupt A, Thamer C, Staiger H, Tschritter O, Kirchhoff K, Machicao F, et al. Variation in the FTO gene influences food intake but not energy expenditure. *Exp Clin Endocrinol Diabetes* 2009;**117**:194–7.
56. Tercjak M, Luczynski W, Wawrusiewicz-Kurylonek N, Bossowski A. The role of FTO gene polymorphism in the pathogenesis of obesity. *Pediatr Endocrinol Diabetes Metab* 2010;**16**: 109–13.
57. Lindi V, Sivenius K, Niskanen L, Laakso M, Uusitupa MI. Effect of the Pro12Ala polymorphism of the PPAR-gamma2 gene on long-term weight change in Finnish non-diabetic subjects. *Diabetologia* 2001;**44**:925–6.
58. Matsuo T, Nakata Y, Katayama Y, Iemitsu M, Maeda S, Okura T, et al. PPARG genotype accounts for part of individual variation in body weight reduction in response to calorie restriction. *Obesity (Silver Spring)* 2009;**17**:1924–31.
59. Razquin C, Alfredo Martinez J, Martinez-Gonzalez MA, Corella D, Santos JM, Marti A. The Mediterranean diet protects against waist circumference enlargement in 12Ala carriers for the PPARgamma gene: 2 years' follow-up of 774 subjects at high cardiovascular risk. *Br J Nutr* 2009;**102**:672–9.
60. Adamo KB, Dent R, Langefeld CD, Cox M, Williams K, Carrick KM, et al. Peroxisome proliferator-activated receptor gamma 2 and acyl-CoA synthetase 5 polymorphisms influence diet response. *Obesity (Silver Spring)* 2007;**15**:1068–75.
61. Haupt A, Thamer C, Heni M, Ketterer C, Machann J, Schick F, et al. Gene variants of TCF7L2 influence weight loss and body composition during lifestyle intervention in a population at risk for type 2 diabetes. *Diabetes* 2010;**59**:747–50.
62. de Luis DA, Aller R, Izaola O, Sagrado MG, Conde R. Influence of Ala54Thr polymorphism of fatty acid-binding protein 2 on weight loss and insulin levels secondary to two hypocaloric diets: a randomized clinical trial. *Diabetes Res Clin Pract* 2008;**82**:113–8.
63. Laukkanen O, Pihlajamaki J, Lindstrom J, Eriksson J, Valle TT, Hamalainen H, et al. Common polymorphisms in the genes regulating the early insulin signalling pathway: effects on weight change and the conversion from impaired glucose tolerance to Type 2 diabetes. The Finnish Diabetes Prevention Study. *Diabetologia* 2004;**47**:871–7.
64. Reinehr T, Hinney A, Nguyen TT, Hebebrand J. Evidence of an influence of a polymorphism near the INSIG2 on weight loss during a lifestyle intervention in obese children and adolescents. *Diabetes* 2008;**57**:623–6.
65. Goyenechea E, Collins LJ, Parra D, Abete I, Crujeiras AB, O'Dell SD, et al. The -11391 G/A polymorphism of the adiponectin gene promoter is associated with metabolic syndrome traits and the outcome of an energy-restricted diet in obese subjects. *Horm Metab Res* 2009; **41**:55–61.
66. Aberle J, Evans D, Beil FU, Seedorf U. A polymorphism in the apolipoprotein A5 gene is associated with weight loss after short-term diet. *Clin Genet* 2005;**68**:152–4.
67. Todorova B, Kubaszek A, Pihlajamaki J, Lindstrom J, Eriksson J, Valle TT, et al. The G-250A promoter polymorphism of the hepatic lipase gene predicts the conversion from impaired glucose tolerance to type 2 diabetes mellitus: the Finnish Diabetes Prevention Study. *J Clin Endocrinol Metabol* 2004;**89**:2019–23.
68. Corella D, Qi L, Sorli JV, Godoy D, Portoles O, Coltell O, et al. Obese subjects carrying the 11482G>A polymorphism at the perilipin locus are resistant to weight loss after dietary energy restriction. *J Clin Endocrinol Metab* 2005;**90**:5121–6.
69. Jang Y, Kim OY, Lee JH, Koh SJ, Chae JS, Kim JY, et al. Genetic variation at the perilipin locus is associated with changes in serum free fatty acids and abdominal fat following mild weight loss. *Int J Obes (Lond)* 2006;**30**:1601–8.

70. Goyenechea E, Dolores Parra M, Alfredo Martinez J. Weight regain after slimming induced by an energy-restricted diet depends on interleukin-6 and peroxisome-proliferator-activated-receptor-gamma2 gene polymorphisms. *Br J Nutr* 2006;**96**:965–72.
71. de Luis DA, Aller R, Izaola O, Gonzalez Sagrado M, Conde R, Romero E. Influence of G308A polymorphism of tumor necrosis factor alpha gene on insulin resistance in obese patients after weight loss. *Med Clin (Barc)* 2007;**129**:401–4.
72. Kurokawa N, Young EH, Oka Y, Satoh H, Wareham NJ, Sandhu MS, et al. The ADRB3 Trp64Arg variant and BMI: a meta-analysis of 44 833 individuals. *Int J Obes (Lond)* 2008;**32**:1240–9.
73. Nakamura M, Tanaka M, Abe S, Itoh K, Imai K, Masuda T, et al. Association between beta 3-adrenergic receptor polymorphism and a lower reduction in the ratio of visceral fat to subcutaneous fat area during weight loss in Japanese obese women. *Nutr Res* 2000;**20**: 25–34.
74. Jalba MS, Rhoads GG, Demissie K. Association of codon 16 and codon 27 beta 2-adrenergic receptor gene polymorphisms with obesity: a meta-analysis. *Obesity (Silver Spring)* 2008;**16**: 2096–106.
75. Al-Rubaish AM. Association of beta(2)-adrenergic receptor gene polymorphisms and nocturnal asthma in Saudi patients. *Ann Thorac Med* 2011;**6**:66–9.
76. Pereira TV, Mingroni-Netto RC, Yamada Y. ADRB2 and LEPR gene polymorphisms: synergistic effects on the risk of obesity in Japanese. *Obesity (Silver Spring)* 2011;**19**:1523–7.
77. Jia JJ, Zhang X, Ge CR, Jois M. The polymorphisms of UCP2 and UCP3 genes associated with fat metabolism, obesity and diabetes. *Obes Rev* 2009;**10**:519–26.
78. Fogelholm M, Valve R, Kukkonen-Harjula K, Nenonen A, Hakkarainen V, Laakso M, et al. Additive effects of the mutations in the beta3-adrenergic receptor and uncoupling protein-1 genes on weight loss and weight maintenance in Finnish women. *J Clin Endocrinol Metab* 1998;**83**:4246–50.
79. Lappalainen TJ, Tolppanen AM, Kolehmainen M, Schwab U, Lindstrom J, Tuomilehto J, et al. The common variant in the FTO gene did not modify the effect of lifestyle changes on body weight: the Finnish Diabetes Prevention Study. *Obesity (Silver Spring)* 2009;**17**:832–6.
80. Muller TD, Hinney A, Scherag A, Nguyen TT, Schreiner F, Schafer H, et al. 'Fat mass and obesity associated' gene (FTO): no significant association of variant rs9939609 with weight loss in a lifestyle intervention and lipid metabolism markers in German obese children and adolescents. *BMC Med Genet* 2008;**9**:85.
81. Fischer J, Koch L, Emmerling C, Vierkotten J, Peters T, Bruning JC, et al. Inactivation of the Fto gene protects from obesity. *Nature* 2009;**458**:894–8.
82. Cecil JE, Tavendale R, Watt P, Hetherington MM, Palmer CN. An obesity-associated FTO gene variant and increased energy intake in children. *N Engl J Med* 2008;**359**:2558–66.
83. Goossens GH, Petersen L, Blaak EE, Hul G, Arner P, Astrup A, et al. Several obesity- and nutrient-related gene polymorphisms but not FTO and UCP variants modulate postabsorptive resting energy expenditure and fat-induced thermogenesis in obese individuals: the NUGENOB study. *Int J Obes (Lond)* 2009;**33**:669–79.
84. Speakman JR. FTO effect on energy demand versus food intake. *Nature* 2010;**464**:E1 discussion E2.
85. Speakman JR, Rance KA, Johnstone AM. Polymorphisms of the FTO gene are associated with variation in energy intake, but not energy expenditure. *Obesity (Silver Spring)* 2008;**16**: 1961–5.
86. Vogels N, Mariman EC, Bouwman FG, Kester AD, Diepvens K, Westerterp-Plantenga MS. Relation of weight maintenance and dietary restraint to peroxisome proliferator-activated receptor gamma2, glucocorticoid receptor, and ciliary neurotrophic factor polymorphisms. *Am J Clin Nutr* 2005;**82**:740–6.

87. Milewicz A, Tworowska-Bardzinska U, Dunajska K, Jedrzejuk D, Lwow F. Relationship of PPARgamma2 polymorphism with obesity and metabolic syndrome in postmenopausal Polish women. *Exp Clin Endocrinol Diabetes* 2009;**117**:628–32.
88. Vidal-Puig AJ, Considine RV, Jimenez-Linan M, Werman A, Pories WJ, Caro JF, et al. Peroxisome proliferator-activated receptor gene expression in human tissues. Effects of obesity, weight loss, and regulation by insulin and glucocorticoids. *J Clin Invest* 1997;**99**: 2416–22.
89. Rosado EL, Bressan J, Hernandez JAM, Martins MF, Cecon PR. Effect of diet and PPAR gamma 2 and beta(2)-adrenergic receptor genes on energy metabolism and body composition in obese women. *Nutr Hosp* 2006;**21**:317–31.
90. Rosado EL, Bressan J, Martinez JA, Marques-Lopes I. Interactions of the PPAR gamma 2 polymorphism with fat intake affecting energy metabolism and nutritional outcomes in obese women. *Ann Nutr Metab* 2011;**57**:242–50.
91. Rosado EL, Bressan J, Martins MF, Cecon PR, Martinez JA. Polymorphism in the PPAR-gamma2 and beta2-adrenergic genes and diet lipid effects on body composition, energy expenditure and eating behavior of obese women. *Appetite* 2007;**49**:635–43.
92. Kim KS, Choi SM, Shin SU, Yang HS, Yoon Y. Effects of peroxisome proliferator-activated receptor-gamma 2 Pro12Ala polymorphism on body fat distribution in female Korean subjects. *Metabolism* 2004;**53**:1538–43.
93. Tonjes A, Stumvoll M. The role of the Pro12Ala polymorphism in peroxisome proliferator-activated receptor gamma in diabetes risk. *Curr Opin Clin Nutr Metab Care* 2007;**10**:410–4.
94. Franks PW, Jablonski KA, Delahanty L, Hanson RL, Kahn SE, Altshuler D, et al. The Pro12Ala variant at the peroxisome proliferator-activated receptor gamma gene and change in obesity-related traits in the Diabetes Prevention Program. *Diabetologia* 2007;**50**:2451–60.
95. Villareal DT, Robertson H, Bell GI, Patterson BW, Tran H, Wice B, et al. TCF7L2 variant rs7903146 affects the risk of type 2 diabetes by modulating incretin action. *Diabetes* 2010; **59**:479–85.
96. Hotamisligil GS. Inflammatory pathways and insulin action. *Int J Obes Relat Metab Disord* 2003;**27**(Suppl. 3):S53–5.
97. Marti A, Martinez-Gonzalez MA, Martinez JA. Interaction between genes and lifestyle factors on obesity. *Proc Nutr Soc* 2008;**67**:1–8.
98. Galikova M, Flatt T. Dietary restriction and other lifespan extending pathways converge at the activation of the downstream effector takeout. *Aging (Albany NY)* 2010;**2**:387–9.
99. Mattson MP. Dietary factors, hormesis and health. *Ageing Res Rev* 2008;**7**:43–8.
100. Lantz H, Peltonen M, Agren L, Torgerson JS. Intermittent versus on-demand use of a very low calorie diet: a randomized 2-year clinical trial. *J Intern Med* 2003;**253**:463–71.
101. Gummesson A, Jernas M, Svensson PA, Larsson I, Glad CA, Schele E, et al. Relations of adipose tissue CIDEA gene expression to basal metabolic rate, energy restriction, and obesity: population-based and dietary intervention studies. *J Clin Endocrinol Metab* 2007;**92**:4759–65.
102. Bastard JP, Hainque B, Dusserre E, Bruckert E, Robin D, Vallier P, et al. Peroxisome proliferator activated receptor-gamma, leptin and tumor necrosis factor-alpha mRNA expression during very low calorie diet in subcutaneous adipose tissue in obese women. *Diabetes Metab Res Rev* 1999;**15**:92–8.
103. Vitkova M, Klimcakova E, Kovacikova M, Valle C, Moro C, Polak J, et al. Plasma levels and adipose tissue messenger ribonucleic acid expression of retinol-binding protein 4 are reduced during calorie restriction in obese subjects but are not related to diet-induced changes in insulin sensitivity. *J Clin Endocrinol Metab* 2007;**92**:2330–5.
104. Bastard JP, Vidal H, Jardel C, Bruckert E, Robin D, Vallier P, et al. Subcutaneous adipose tissue expression of plasminogen activator inhibitor-1 gene during very low calorie diet in obese subjects. *Int J Obes Relat Metab Disord* 2000;**24**:70–4.

105. Salas-Salvado J, Bullo M, Garcia-Lorda P, Figueredo R, Del Castillo D, Bonada A, et al. Subcutaneous adipose tissue cytokine production is not responsible for the restoration of systemic inflammation markers during weight loss. *Int J Obes (Lond)* 2006;**30**:1714–20.
106. Goyenechea E, Parra D, Crujeiras AB, Abete I, Martinez JA. A nutrigenomic inflammation-related PBMC-based approach to predict the weight-loss regain in obese subjects. *Ann Nutr Metab* 2009;**54**:43–51.
107. Clement K, Viguerie N, Poitou C, Carette C, Pelloux V, Curat CA, et al. Weight loss regulates inflammation-related genes in white adipose tissue of obese subjects. *FASEB J* 2004;**18**: 1657–69.
108. Pasiakos SM, Vislocky LM, Carbone JW, Altieri N, Konopelski K, Freake HC, et al. Acute energy deprivation affects skeletal muscle protein synthesis and associated intracellular signaling proteins in physically active adults. *J Nutr* 2010;**140**:745–51.
109. Civitarese AE, Carling S, Heilbronn LK, Hulver MH, Ukropcova B, Deutsch WA, et al. Calorie restriction increases muscle mitochondrial biogenesis in healthy humans. *PLoS Med* 2007;**4**:e76.
110. Crujeiras AB, Parra D, Goyenechea E, Martinez JA. Sirtuin gene expression in human mononuclear cells is modulated by caloric restriction. *Eur J Clin Invest* 2008;**38**:672–8.
111. Crujeiras AB, Parra D, Milagro FI, Goyenechea E, Larrarte E, Margareto J, et al. Differential expression of oxidative stress and inflammation related genes in peripheral blood mononuclear cells in response to a low-calorie diet: a nutrigenomics study. *OMICS* 2008;**12**:251–61.
112. de Mello VD, Kolehmainen M, Schwab U, Mager U, Laaksonen DE, Pulkkinen L, et al. Effect of weight loss on cytokine messenger RNA expression in peripheral blood mononuclear cells of obese subjects with the metabolic syndrome. *Metabolism* 2008;**57**:192–9.
113. Crujeiras AB, Parra D, Goyenechea E, Abete I, Martinez JA. Tachyphylaxis effects on postprandial oxidative stress and mitochondrial-related gene expression in overweight subjects after a period of energy restriction. *Eur J Nutr* 2009;**48**:341–7.
114. O'Dea K, Koletsky S. Effect of caloric restriction on basal insulin levels and the in vivo lipogenesis and glycogen synthesis from glucose in the Koletsky obese rat. *Metabolism* 1977;**26**:763–72.
115. Polak P, Cybulski N, Feige JN, Auwerx J, Ruegg MA, Hall MN. Adipose-specific knockout of raptor results in lean mice with enhanced mitochondrial respiration. *Cell Metab* 2008;**8**: 399–410.
116. Proud CG. Regulation of mammalian translation factors by nutrients. *Eur J Biochem* 2002; **269**:5338–49.
117. Wilkinson M, Brown R, Imran SA, Ur E. Adipokine gene expression in brain and pituitary gland. *Neuroendocrinology* 2007;**86**:191–209.
118. Sucajtys-Szulc E, Goyke E, Korczynska J, Stelmanska E, Rutkowski B, Swierczynski J. Refeeding after prolonged food restriction differentially affects hypothalamic and adipose tissue leptin gene expression. *Neuropeptides* 2009;**43**:321–5.
119. Nogalska A, Stelmanska E, Sledzinski T, Swierczynski J. Surgical removal of perirenal and epididymal adipose tissue decreases serum leptin concentration and increases lipogenic enzyme activities in remnant adipose tissue of old rats. *Gerontology* 2009;**55**:224–8.
120. Margareto J, Aguado M, Oses-Prieto JA, Rivero I, Monge A, Aldana I, et al. A new NPY-antagonist strongly stimulates apoptosis and lipolysis on white adipocytes in an obesity model. *Life Sci* 2000;**68**:99–107.
121. Schwartz MW, Woods SC, Seeley RJ, Barsh GS, Baskin DG, Leibel RL. Is the energy homeostasis system inherently biased toward weight gain? *Diabetes* 2003;**52**:232–8.
122. Kaelin CB, Gong L, Xu AW, Yao F, Hockman K, Morton GJ, et al. Signal transducer and activator of transcription (stat) binding sites but not stat3 are required for fasting-induced

transcription of agouti-related protein messenger ribonucleic acid. *Mol Endocrinol* 2006; **20**:2591–602.

123. Higuchi H, Hasegawa A, Yamaguchi T. Transcriptional regulation of neuronal genes and its effect on neural functions: transcriptional regulation of neuropeptide Y gene by leptin and its effect on feeding. *J Pharmacol Sci* 2005;**98**:225–31.
124. Ravussin E, Lillioja S, Knowler WC, Christin L, Freymond D, Abbott WG, et al. Reduced rate of energy expenditure as a risk factor for body-weight gain. *N Engl J Med* 1988;**318**:467–72.
125. Young VR, Yu YM, Fukagawa NK. Protein and energy interactions throughout life. Metabolic basis and nutritional implications. *Acta Paediatr Scand Suppl* 1991;**373**:5–24.
126. Hoffer LJ, Bistrian BR, Young VR, Blackburn GL, Matthews DE. Metabolic effects of very low calorie weight reduction diets. *J Clin Invest* 1984;**73**:750–8.
127. Hoffer LJ, Forse RA. Protein metabolic effects of a prolonged fast and hypocaloric refeeding. *Am J Physiol* 1990;**258**:E832–40.
128. Stein TP, Rumpler WV, Leskiw MJ, Schluter MD, Staples R, Bodwell CE. Effect of reduced dietary intake on energy expenditure, protein turnover, and glucose cycling in man. *Metabolism* 1991;**40**:478–83.
129. Kimball SR, Jefferson LS. Signaling pathways and molecular mechanisms through which branched-chain amino acids mediate translational control of protein synthesis. *J Nutr* 2006; **136**:227S–31S.
130. Lambert AJ, Wang B, Yardley J, Edwards J, Merry BJ. The effect of aging and caloric restriction on mitochondrial protein density and oxygen consumption. *Exp Gerontol* 2004; **39**:289–95.
131. Lopez-Lluch G, Hunt N, Jones B, Zhu M, Jamieson H, Hilmer S, et al. Calorie restriction induces mitochondrial biogenesis and bioenergetic efficiency. *Proc Natl Acad Sci USA* 2006; **103**:1768–73.
132. Rodgers JT, Lerin C, Haas W, Gygi SP, Spiegelman BM, Puigserver P. Nutrient control of glucose homeostasis through a complex of PGC-1alpha and SIRT1. *Nature* 2005;**434**:113–8.
133. Engel N, Mahlknecht U. Aging and anti-aging: unexpected side effects of everyday medication through sirtuin1 modulation. *Int J Mol Med* 2008;**21**:223–32.
134. Michan S, Sinclair D. Sirtuins in mammals: insights into their biological function. *Biochem J* 2007;**404**:1–13.
135. Haigis MC, Guarente LP. Mammalian sirtuins—emerging roles in physiology, aging, and calorie restriction. *Genes Dev* 2006;**20**:2913–21.
136. Picard F, Kurtev M, Chung N, Topark-Ngarm A, Senawong T, Machado De Oliveira R, et al. Sirt1 promotes fat mobilization in white adipocytes by repressing PPAR-gamma. *Nature* 2004;**429**:771–6.
137. Hiratsuka M, Inoue T, Toda T, Kimura N, Shirayoshi Y, Kamitani H, et al. Proteomics-based identification of differentially expressed genes in human gliomas: down-regulation of SIRT2 gene. *Biochem Biophys Res Commun* 2003;**309**:558–66.
138. North BJ, Marshall BL, Borra MT, Denu JM, Verdin E. The human Sir2 ortholog, SIRT2, is an NAD+-dependent tubulin deacetylase. *Mol Cell* 2003;**11**:437–44.
139. Jing E, Gesta S, Kahn CR. SIRT2 regulates adipocyte differentiation through FoxO1 acetylation/deacetylation. *Cell Metab* 2007;**6**:105–14.
140. Finkel T, Holbrook NJ. Oxidants, oxidative stress and the biology of ageing. *Nature* 2000;**408**:239–47.
141. Guarente L. Calorie restriction and SIR2 genes—towards a mechanism. *Mech Ageing Dev* 2005;**126**:923–8.
142. Chen D, Guarente L. SIR2: a potential target for calorie restriction mimetics. *Trends Mol Med* 2007;**13**:64–71.

143. Lamming DW, Latorre-Esteves M, Medvedik O, Wong SN, Tsang FA, Wang C, et al. HST2 mediates SIR2-independent life-span extension by calorie restriction. *Science* 2005; **309**:1861–4.
144. Wang F, Nguyen M, Qin FX, Tong Q. SIRT2 deacetylates FOXO3a in response to oxidative stress and caloric restriction. *Aging Cell* 2007;**6**:505–14.

Individualized Weight Management: What Can Be Learned from Nutrigenomics and Nutrigenetics?

Iwona Rudkowska[*] and
Louis Pérusse[*,†]

[*]Institute of Nutraceuticals and Functional Foods (INAF), Laval University, Quebec, Canada
[†]Department of Kinesiology, Laval University, Quebec, Canada

The rise in the prevalence of obesity observed over the past decades is taken by many as an indication of the predominance of environmental factors (the so-called obesogenic environment) over genetic factors in explaining why obesity has reached epidemic proportions. While a changing environment favoring increased food intake and decreased physical activity levels has clearly contributed to shifting the distribution of body mass index (BMI) at the population level, not everyone is becoming overweight or obese. This suggests that there are genetic factors interacting with environmental factors to predispose some individuals to obesity. This gene–environment interaction is not only important in determining an individual's susceptibility to obesity but can also influence the outcome of weight-loss programs and weight-management strategies in overweight and obese subjects. This chapter reviews the role of gene–nutrient interactions in the context of weight management. The first section reviews the application of transcriptomics in human nutrition intervention studies on the molecular impact of caloric restriction and macronutrient composition. The second section reviews the effects of various obesity candidate gene

Progress in Molecular Biology
and Translational Science, Vol. 108
DOI: 10.1016/B978-0-12-398397-8.00014-9

1877-1173/12 $35.00

polymorphisms on the response of body weight or weight-related phenotypes to weight-loss programs which include nutritional interventions.

I. Introduction

One of the pressing issues faced by the nutrition research community is the obesity epidemic. The causes of this epidemic are not clearly established, but most agree that our modern lifestyle favoring sedentary behavior coupled with the easy access to low-cost, energy-dense foods is largely responsible for the dramatic increases in the prevalence of obesity observed over the past 40 years. However, not everybody exposed to this "obesogenic" environment becomes obese, suggesting that obesity is likely the result of complex interactions between susceptibility genes and a host of environmental factors promoting increased energy intake and reduced energy expenditure. Although considerable progress has been made in the identification of obesity genes,[1] there has been relatively little progress in the identification of gene–environment interactions relevant to obesity. These gene–environment interactions are not only important in determining an individual's susceptibility to become obese but can also play a role in determining the response to weight-loss programs and the effectiveness of weight-management strategies in overweight and obese subjects.

This chapter provides an overview of nutrigenomics and nutrigenetics in the context of weight management. The first section reviews the impact of caloric restriction and various diets on gene expression profiles. The second section reviews candidate gene polymorphisms that have been shown to influence changes in body weight or body fatness in response to dietary interventions. Only human studies with a nutritional intervention and studies in which at least one of the outcome variables is related to body weight or body fatness are reviewed.

II. Weight Management and Nutrigenomics

Dietary interventions for obesity may include hypocaloric diets, a variety of diets with different macronutrient compositions, and the addition of various functional foods. Dietary components provide energy and essential nutrients in addition to participating in the modulation of gene expression. Gene expression profiling is thought to be more sensitive to nutritional intervention than the traditional biochemical parameters. Here we examine the studies of transcriptomics in human nutrition intervention studies to determine the molecular impact of caloric intake, macronutrient composition, and the addition of functional foods (e.g., olive oil, nuts, $n-3$ polyunsaturated fatty acids (PUFAs), antioxidants) in order to better define weight-management strategies. A summary of the relevant studies is presented in Table I.

A. Energy Intake

Understanding the molecular changes associated with obesity and body-weight homeostasis is a crucial step in the development of effective therapeutic strategies against excess body weight.

Investigating the effects of overfeeding on changes in gene expression is essential to understanding the molecular basis of obesity. Meugnier and colleagues[2] wanted to define the metabolic responses and changes in gene expression in healthy volunteers during fat overfeeding. The results confirmed that fat overfeeding promotes the storage of excess energy. In addition, transcriptomics data identified the key lipid metabolic pathways and suggested the involvement of the sterol regulatory element-binding proteins in the short-term adaptation to fat overfeeding in skeletal muscle.[2] It is well known that lipid accumulation in skeletal muscle is associated with insulin resistance; therefore, these data confirm a detrimental effect of excess fat intake at the molecular level.

Similarly, understanding the molecular basis of hypoenergetic diets commonly used to reduce body fat mass and metabolic risk factors in overweight or obese subjects should help in the establishment of personalized nutrition for maintenance of health and disease prevention. Very low calorie diets (VLCDs) are used to promote short-term weight loss in obese patients. Ong *et al.*[5] demonstrated that genes involved in glycolytic and lipid synthesis pathways were downregulated after a VLCD in overweight and obese women. In other studies, the beneficial effect of weight loss via VLCD was associated with the modification of inflammation-related gene expression.[6,28] Further, gene expression in obese subjects after dietary restriction was closer to the profile of lean subjects than to the pattern of obese subjects before dietary restriction.[6] Thus, obese individuals may have a gene expression profile improved by weight loss.

Long-term weight management usually involves moderate caloric restrictions (–500 kcal/day) over long time periods to achieve sustainable weight loss. A study demonstrated that long-term weight reduction downregulated genes of the extracellular matrix and cell death in calorie-restricted subjects compared to controls.[3] Moreover, the expression of tenomodulin (*TNMD*), an angiogenesis inhibition gene, was downregulated and was correlated with insulin sensitivity and body adiposity.[3] Crujeiras and colleagues[7] showed decreases in oxidative stress and inflammation genes in obese men after an 8-week diet. Similarly, Bouchard *et al.*[4] observed differences in gene expression profiles after dieting, including genes related to angiogenesis. Overall, these studies show that long-term, moderate caloric restriction alters gene expression.

TABLE I

SUMMARY OF WEIGHT-MANAGEMENT STUDIES USING TRANSCRIPTOMICS

Intervention	Study design	Methodology (subjects and tissue type)	Outcome	Reference
Effects of energy intake on gene expression				
Diet rich in energy including SFAs for 4 weeks	Free-living subjects with a supplemented diet	Eight lean young healthy men *Skeletal muscle tissue*	55 Genes modified Changes in genes involved in stimulation of triacylglycerol synthesis, inhibition of lipolysis, reduction in fatty acid oxidation, development of adipocytes Sterol regulatory element-binding proteins play an important role	2
Moderate long-term weight-reduction program for 12 weeks followed by weight maintenance for 21 weeks or control group for 33 weeks	Randomized, parallel-arm dietary advice	46 Subjects with impaired fasting glycemia or impaired glucose tolerance and features of metabolic syndrome *Adipose tissue*	105 Genes, of which 86 were downregulated, including genes involved in the extracellular matrix and cell death	3
Caloric restriction to reduce body weight by 10% over 6 months	Parallel-arm, free-living study with dietary advice	14 Overweight and obese postmenopausal women *Subcutaneous adipose tissue*	644 Genes differentially expressed, including 334 upregulated and 342 downregulated, between the two groups after dieting Including genes involved in metabolic pathways related to angiogenesis and cerebellar long-term depression	4
Dietary energy restriction or normal eating patterns for one menstrual cycle	Randomized, parallel-arm study	19 Overweight and obese women at moderately increased risk of breast cancer *Breast and abdominal fat tissues*	161 Genes changed after dietary energy restriction, including 113 genes downregulated Changes in genes involved in glycolytic and lipid synthesis pathways	5
VLCD for 4 weeks or 2 days	Randomized, parallel-arm study	29 Obese subjects compared with 17 nonobese subjects *Subcutaneous adipose tissue*	100 Transcripts (including downregulation of proinflammatory factors and upregulation of anti-inflammatory molecules) regulated in obese individuals when eating a 28-day VLCD but not a 2-day VLCD Gene expression in obese subjects after 28-day VLCD closer to the profile of lean subjects than to the pattern of obese subjects before VLCD	6

LCD for 8 weeks	Free-living subjects with dietary advice	Nine obese men *PBMCs*	385 Differentially expressed transcripts Changes in pathways associated with carbohydrate, lipid, and protein metabolism, oxidative phosphorylation, immune response, and coagulation affected by intervention Downregulation of specific oxidative stress and inflammation genes	7
Energy-restriction phase with 4-week VLCD and weight stabilization period composed of 2-month LCD followed by 3–4 months of weight maintenance diet	Free-living subjects with dietary advice	22 Obese women *Subcutaneous adipose tissue*	464 Mostly adipocyte genes involved in metabolism downregulated during energy restriction, upregulated during weight stabilization, and not affected during the overall dietary intervention 511 Mainly macrophage genes involved in inflammatory pathways not affected or upregulated during energy restriction and downregulated during weight stabilization and the overall dietary intervention	8
Effects of energy restriction in conjunction with macronutrient changes on gene expression				
Moderate-fat, moderate-carbohydrate diet or LF, HC, hypoenergetic diet for 10 weeks	Randomized, parallel-arm, free-living study with dietary advice	Two sets of 47 obese women in each dietary arm matched for anthropometric and biological parameters *Subcutaneous adipose tissue*	1000 Genes regulated by energy restriction related to lipid metabolism, cellular assembly, and small molecule biochemistry	9
		40 Obese women *Subcutaneous adipose tissue*	52 Genes upregulated and 44 downregulated after both diets; no diet-specific effect Changes in the production of PUFAs including acetyl-CoA and malonyl-CoA downregulated No change in lipid-specific transcription factors, genes regulating signal transduction, lipolysis, or synthesis of acylglycerols	10
Basal diet for 1 week, one of four randomized diets for 3 weeks, acute weight loss with randomized diet for 5 weeks, and stabilization at reduced weight for 4 weeks	Randomized, four-arm study with semicontrolled diet (two out of three meals provided)	131 Moderately overweight men *Subcutaneous adipose tissue*	1473 (9.4%) Gene probes changed after acute weight loss, including lipogenic genes (in particular, stearoyl-CoA desaturase) 30 probes changed after isocaloric change in dietary composition No difference in response observed between diets	11
				12

(*Continues*)

TABLE I (*Continued*)

Intervention	Study design	Methodology (subjects and tissue type)	Outcome	Reference
After LCD for 8 weeks, randomly assigned to receive one of four diets differing in protein and glycemic index content for 6 weeks	Randomized, parallel-arm, free-living study with dietary advice	227 Obese subjects *Subcutaneous adipose tissue*	1338 Differentially expressed genes Cellular growth and proliferation, cell death, cellular function, and maintenance were the main biological processes represented in subcutaneous adipose tissue from subjects who regained weight Mitochondrial oxidative phosphorylation was the major pattern associated with continued weight loss	
Effects of macronutrient changes on gene expression				
Controlled diet for 4 days followed by isoenergetic high-fat/LC diet for 3 days	Controlled diet	10 Healthy young men *Muscle tissue*	369 Genes of 18,861 genes on the arrays differentially regulated Seven genes changed in the carbohydrate metabolism pathway	13
Protein intakes of 0.50 g (LPro), 0.75 g (MPro), and 1.00 g (HPro) of protein per kg body weight per day	Randomized, crossover, controlled feeding trials	12 Younger and 10 older men *Skeletal muscle tissue*	958 Transcripts differentially expressed by diet LPro associated with upregulation of transcripts related to ubiquitin-dependent protein catabolism and muscle contraction LPro and MPro resulted in upregulation of transcripts related to apoptosis and downregulation of transcripts related to cell differentiation, muscle and organ development, extracellular space, and responses to stimuli and stress 853 Transcripts had diet-by-age interaction: older males less responsive to anabolic stimuli and more responsive to catabolic state	14
HC or high-protein breakfast	Randomized, crossover, controlled diet	Eight healthy men *Leukocytes*	317 Genes differentially expressed for HC breakfast, in particular glycogen metabolism genes 919 Genes differentially expressed for high-protein breakfast, in particular genes involved in protein biosynthesis 141 Genes commonly differentially expressed in response to both breakfasts, including immune response and signal transduction, specifically T-cell receptor signaling and NF-κB signaling	15

Normocaloric diet intervention with 30 E%, 40 E%, and 30 E % from carbohydrates, fats, and proteins, respectively, compared to a prestudy diet with 41 E%, 40 E%, and 19 E %; each meal contained approximately equal caloric load of macronutrients	Free-living subjects with dietary advice	Five obese but otherwise healthy men *Blood and adipose tissue*	734 Downregulated genes, including genes that regulate immunological processes 299 Upregulated genes	16
Carbohydrate modification with 4-week baseline period and either oat–wheat–potato diet or rye–pasta diet for 12 weeks	Randomized, parallel-arm, supplemented diet	47 Subjects with the features of the metabolic syndrome *Subcutaneous adipose tissue*	71 Downregulated genes in rye–pasta group, including genes linked to insulin signaling and apoptosis 62 Upregulated genes in oat–wheat–potato group related to stress, cytokine–chemokine-mediated immunity, and the interleukin pathway	17
SFA-rich run-in diet for 2 weeks, followed by SFA-rich diet or MUFA-rich diet for 8 weeks	Randomized, parallel-arm, controlled diet	20 Abdominally overweight subjects *Subcutaneous adipose tissue*	1523 Genes differentially regulated after SFA-rich diet, mainly proinflammatory gene expression profile 592 Differentially regulated after MUFA-rich diet, mainly anti-inflammatory profile 76 Genes commonly differentially expressed on both diets	18
Consumed shakes enriched in PUFAs, MUFAs, or SFAs	Randomized, crossover study	21 Healthy male subjects *PBMCs*	437 Genes changed after PUFAs 297 Genes changed after SFAs 146 Commonly differentiated genes Genes linked to LXR signaling, oxidative stress, inflammation, carbohydrate metabolism, and a variety of other processes Opposite effects of PUFA and SFA intakes on expression of genes involved in LXR signaling MUFA intake had intermediate effect on several genes	19
Effects of the TMD on gene expression				
LF, carbohydrate-rich diet with VOO-based breakfast with either high or low content of phenolic compounds	Randomized, crossover design	20 Subjects suffering from metabolic syndrome *PBMCs*	79 Downregulated and 19 upregulated genes when comparing intake of phenol-rich olive oil with low-phenol olive oil Genes involved in inflammatory processes mediated by NF-κ B, activator protein 1 transcription factor complex, cytokines, mitogen-activated protein kinases, or arachidonic acid pathways	20

(*Continues*)

TABLE I (*Continued*)

Intervention	Study design	Methodology (subjects and tissue type)	Outcome	Reference
50 ml of olive oil at fasting state	Parallel-arm design, with a supplement	Six healthy male subjects *PBMCs*	259 Genes upregulated and 246 downregulated Genes related to metabolism, cellular processes, cancer, and atherosclerosis and associated processes such as inflammation and DNA damage	21
25 ml/day of olive oil for 3 weeks	Parallel-arm design, with a supplement	10 Healthy participants *PBMCs from pooled RNA samples*	1659 Probes, including 1034 upregulated and 628 downregulated Genes involved in atherosclerosis development and progression	22
Effects of $n-3$ PUFAs on gene expression				
Either 1.8 g EPA + DHA/day, 0.4 g EPA + DHA/day, or 4.0 g HOSF/day	Randomized, parallel design supplementation study	111 Healthy elderly subjects *PBMCs*	1040 Differentially expressed genes after EPA + DHA 298 Differentially expressed genes after HOSF 140 Commonly differentially expressed genes Changes in NF-kB signaling, eicosanoid synthesis, scavenger receptor activity, adipogenesis, and hypoxia signaling	23
3 g/day fish oil containing 26% EPA and 54% DHA for 2 months	Supplementation study	10 Male subjects *Lymphocytes*	588 Differentially expressed genes (including 6 upregulated and 71 downregulated) Changes in lymphocyte functions such as signaling, cell cycle, cytokine production, apoptosis, and stress response	24
1.8 g EPA + DHA/day alone or 1 8 g EPA + DHA/day + FG supplementation for 8 weeks	Randomized, crossover supplementation trial	16 Obese, insulin-resistant subjects *PBMCs*	805 Differentially expressed genes after EPA + DHA 184 Differentially expressed genes after EPA + DHA + FG Three commonly differentially expressed genes Changes in the PPARA pathway, oxidative stress response mediated by nuclear factor erythroid-derived 2-like 2, NF-kB, oxidative stress, and hypoxia-inducible factor signaling	25

Effects of antioxidants on gene expression				
Randomized to diet rich in various antioxidant-rich foods, kiwifruit diet, or control group for 8 weeks	Randomized, parallel design supplementation study	102 Healthy male smokers *Blood cells*	44 Gene transcripts differentially expressed in antioxidant-rich group compared to control Nine gene transcripts differentially expressed in kiwifruit diet group compared to control Genes involved in regulation of cellular stress defense, such as DNA repair, apoptosis, and hypoxia, upregulated by both diets compared to control group Genes with common regulatory motifs for AhR and AhR nuclear translocator upregulated by both interventions	26
Study 1: Quercetin either 50, 100, or 150 mg/day for 2 weeks *Study 2*: Randomized to receive 150 mg quercetin or placebo daily for 6 weeks each	*Study 1*: Supplementation study *Study 2*: Randomized, crossover supplementation study	*Study 1*: 10 Healthy subjects *Monocytes* *Study 2*: 20 Subjects exhibiting cardiovascular risk phenotype *Monocytes*	*Study 1*: 503 Genes upregulated and 788 genes downregulated Changes in the immune system, nucleic acid metabolism, apoptosis, and *O*-glycan biosynthesis *Study 2*: Four genes showed different expression changes (*O*-glycan biosynthesis, glycolipid catabolism, cell proliferation, and apoptosis) between quercetin and placebo but minimal fold change	27

AhR, aryl hydrocarbon receptor; DHA, docosahexaenoic acid; EPA, eicosapentaenoic acid; FG, fish gelatin; HC, high-carbohydrate; HOSF, high-oleic acid sunflower oil; HPro, higher protein; LC, low-carbohydrate; LF, low-fat; LCD, low-calorie diet; LPro, lower protein; LXR, liver X receptor; MPro, medium protein; MUFAs, monounsaturated fatty acids; NF-kB, nuclear transcription factor kappaB; E%, percentage of energy intake; PBMCs, peripheral blood mononuclear cells; PUFAs, polyunsaturated fatty acids; SFA, saturated fatty acid; VLCD, very low calorie diet; VOO, virgin olive oil.

Weight loss induced by caloric restriction is usually followed by a weight stabilization phase. In 2009, Capel and colleagues[8] demonstrated that metabolic pathways were downregulated during energy restriction (1-month VLCD), upregulated during weight stabilization (2-month low-calorie diet [LCD] and 3–4 months of weight maintenance diet), but unchanged over the entire dietary intervention. Secondly, inflammatory pathways were not changed or were upregulated during energy restriction and were downregulated during weight stabilization as well as over the entire dietary intervention.[8]

Overall, hypercaloric diets may produce negative effects on lipid metabolism. In contrast, there are benefits of both long-term moderate and short-term more severe dietary restrictions on gene expression levels, especially in angiogenesis, glycolysis, lipid synthesis, and inflammation pathways. Gene expression profiles can potentially reflect the different phases of a weight-loss program. Thus, understanding the molecular changes associated with weight gain, loss, or maintenance would be useful in the development of optimal strategies against obesity.

B. Energy Restriction in Conjunction with Changes in Dietary Composition

Although current weight-loss recommendations are to consume a moderately hypocaloric, high-carbohydrate (HC), low-fat (LF) diet, the obesity epidemic has led to a rise in the use of alternate dietary patterns, particularly very low carbohydrate (LC) diets. Favorable effects on serum triglyceride, high-density lipoprotein cholesterol, and low-density lipoprotein particle size have been shown with LC diets, while, in contrast, LF diets have favorable effects on blood total and low-density lipoprotein cholesterol. Thus, there are differences in the impact of each diet.

In 2005, Dahlman *et al.*[10] investigated the effects of a hypoenergetic diet with either an LF or a moderate-fat diet in obese subjects. They found that 96 genes were modified as a result of low-energy diets; however, there was no diet-specific effect observed. Further, no major effect on lipid-specific transcription factors or genes regulating signal transduction, lipolysis, or synthesis of acylglycerols was observed; yet genes regulating the formation of PUFAs were downregulated during the diets.[10] Another study showed that transcriptional expression of lipogenic genes is influenced by dietary macronutrient composition and energy restriction.[11] However, the gene expression responses to changes in dietary composition were minor in comparison with the energy restriction.[11] More recently, Capel and colleagues[9] found that two hypoenergetic diets (LF/HC and moderate-fat/LC) induced similar weight loss and similar gene expression changes except for components of the lipid profile. Finally, a study confirmed that differences in gene expression patterns are mainly due to weight variations rather than to differences in dietary macronutrient content.[12] Overall, these studies suggest that weight-reduction programs that involve restriction of specific macronutrients do not differentially affect transcriptional expression if they result in similar amounts of weight loss. However, restriction of specific macronutrients may contribute to altered systemic regulation of lipid metabolism genes.

C. Macronutrient Composition Changes

Studies examining the changes in gene expression after dietary macronutrient changes are reviewed in this section.

Sparks *et al.*[13] showed that carbohydrate metabolism and storage are under transcriptional control; thus, molecular pathways adapt to the intake of LC diets. More recently, Thalacker-Mercer and colleagues[14] assessed the effects of dietary protein on the skeletal muscle transcriptome. They showed an adaptive response to higher protein intake.[14] Further, van Erk *et al.* investigated the effects of both an HC and a high-protein meal on gene expression profiling.[15] Similar to previous studies, consumption of an HC meal resulted in differential expression of glycogen metabolism genes, and consumption of a high-protein meal resulted in differential expression of genes involved in protein biosynthesis.[15] After both meals, the immune response and signal transduction were the overrepresented functional groups.[15] Recently, Brattbakk *et al.*[16] conducted a normocaloric diet intervention with a higher protein/lower carbohydrate diet in obese men. In addition, each meal contained an approximately equal caloric load of macronutrients. The change from the prestudy diet and eating pattern to the diet intervention resulted in reduced low-grade systemic inflammation.[16] In sum, these studies have demonstrated an efficient transcriptional switch that influences substrate utilization in response to changes in macronutrient content in diets.

Diets rich in whole-grain cereals and foods with a low glycemic index may be beneficial for weight control and type 2 diabetes. Kallio and colleagues[17] examined two different carbohydrate modifications (a rye–pasta diet characterized by a low postprandial insulin response and an oat–wheat–potato diet characterized by a high postprandial insulin response) and their effects on gene expression in subjects with the metabolic syndrome. Results indicated that genes regulating insulin signaling and apoptosis were downregulated during the rye–pasta diet, and genes related mainly to metabolic stress were upregulated during the oat–wheat–potato diet.[17] These changes in gene expression appear to be driven by the nature of the dietary carbohydrates.

LF diets consist of a small amount of dietary fat ($\leq$25–35% of energy from fat), especially saturated fatty acids (SFAs) ($\leq$7–10% of energy from SFAs). There is a relationship between SFA intake, blood cholesterol levels, and the prevalence of cardiovascular disease (CVD). While many studies have found that including PUFAs in the diet to replace SFAs produces beneficial CVD outcomes, the effects of substituting monounsaturated fatty acids (MUFAs) are less clear. A study by van Dijk *et al.*[18] demonstrated that consumption of an SFA-rich diet resulted in increased expression of genes involved in inflammation processes. In contrast, an MUFA-rich diet led to a more anti-inflammatory gene expression profile.[18] Similarly, Bouwens and colleagues[19] showed that PUFA intake decreased the expression of genes in liver X receptor signaling, whereas SFA intake increased the expression of these genes. In addition, PUFA intake increased the expression of genes related to cellular stress responses, and MUFA intake had a slight effect on several of these genes.[19] In conclusion, these studies show that consumption of an SFA-rich diet, compared with a PUFA-rich or MUFA-rich diet, leads to a proinflammatory gene expression profile.

Overall, macronutrient composition can affect gene expression levels on a weight-stable background. Modifying carbohydrate and protein content changes the expression of molecular pathways in order to adjust to the nutrient mixture consumed. Increased intake of SFAs may have a detrimental role on the inflammation gene expression profile; in contrast, PUFAs and MUFAs may have a more beneficial effect on gene expression profiles. Finally, meal consumption patterns may also influence gene expression; for example, skipping meals may increase inflammation. Therefore, additional studies are needed to identify the exact molecular pathways induced or downregulated by various combinations of macronutrients, as well as optimal meal pattern profiles.

D. Functional Foods

Functional foods are foods claimed to have a health-promoting or disease-preventing property beyond the basic function of supplying nutrients. The patterns of gene expression associated with these functional foods are poorly understood. Here we examine the impact of olive oil and nuts in the traditional Mediterranean diet (TMD), $n-3$ PUFAs, and antioxidants on transcriptomic profiles, to infer on their mechanisms of action.

1. The Mediterranean Diet

Many experimental and epidemiological studies have shown the beneficial effects of the TMD on the incidence and progression of atherosclerosis. Virgin olive oil (VOO) and nuts are considered to be the main components responsible for the health benefits of the TMD; however, the molecular mechanisms of action are unclear. Several studies have demonstrated the effect of the TMD and VOO on transcriptomic profiles in humans, and these studies are summarized in Table I.

Camargo *et al.*[20] showed that several genes that seem to be involved in inflammatory processes had decreased expression after a single dose of VOO. Similarly, Konstantinidou and colleagues[21] demonstrated that genes related to metabolism, cellular processes, cancer, and atherosclerosis, and associated processes such as inflammation and DNA damage are modified after a single dose of VOO. Changes in the expression of seven insulin sensitivity-related genes also occurred.[29] Overall, these results suggest that a potentially short-term protective effect of VOO consumption could be mediated through gene expression changes, including modification in the expression of inflammatory- and insulin sensitivity-related genes.

However, one limitation of these short-term studies is that these effects on gene expression could be secondary to a time-course effect and to physiological changes following any fat meal intake. Longer studies on the TMD have also been conducted. Khymenets *et al.*[22] established that VOO supplementation for 3 weeks alters the expression of genes related to atherosclerosis development and progression. Konstantinidou *et al.*[30] recruited healthy volunteers who were randomized to the following intervention groups: TMD plus VOO, TMD with washed VOO (WOO, lower polyphenol content than VOO), and control diet for 3 months. Consumption of the TMD, either with VOO or WOO, decreased gene expression related to both inflammation and oxidative stress compared to the control diet.[30] In addition, VOO polyphenols in the TMD decreased expression of pro-atherogenic genes to a greater extent than consumption of the TMD with WOO.[30] Llorente-Cortes and colleagues[31] compared the effects on gene expression of TMD diets supplemented with either VOO or nuts versus a control diet for 3 months in asymptomatic participants with high CVD risk. These results suggest that the TMD with VOO or nuts influences to different extents the key genes involved in inflammation, vascular foam formation, and vascular modeling. Together, these clinical trials demonstrate that longer intake of the TMD with VOO, WOO, or nuts has advantageous effects on gene expression profiles.

Overall, these results suggest a molecular basis for the reduction in CVD via changes in expression of inflammatory, atherogenic, and insulin-sensitive genes after short-term and long-term consumption of the TMD, including VOO and nuts. However, changes in the expression of genes are modest because the bioactive components are part of a normal diet and were supplemented in nutritional doses. Moreover, these studies did not distinguish between the effects promoted by other bioactive components of the TMD, such as fish, fruits and vegetables, whole grains, and red wine. Thus, it is important to conduct further studies to decipher the exact mechanisms of action of the specific components of the TMD that may be beneficial for CVD prevention.

2. Omega-3 PUFAs

Omega-3 PUFAs, including eicosapentaenoic acid (EPA, 20:5, $n-3$) and docosahexaenoic acid (DHA, 22:6, $n-3$), are increasingly being used in the prevention and management of several CVD risk factors. The underlying beneficial mechanisms of $n-3$ PUFAs are still debated, and transcriptomics can perhaps add to our understanding of the mechanisms of action.

In 2009, Bouwens *et al.*[23] demonstrated that $n-3$ PUFA intake results in decreased expression of genes involved in inflammatory- and atherogenic-related pathways in healthy elderly subjects.[23] Earlier, Kabir and colleagues[32] showed that a subset of inflammation-related genes was reduced after $n-3$ PUFA supplementation in women with type 2 diabetes.[32] Recently, Rudkowska *et al.*[25] investigated gene expression changes following $n-3$ PUFA and $n-3$ PUFA plus fish protein supplementation in obese insulin-resistant subjects. Pathway analyses indicated changes in gene expression via the nuclear receptor peroxisome proliferator-activated receptor alpha and the inflammatory pathways after both supplementation periods.[25] Further, another study demonstrated that the proportion of DHA and EPA in an $n-3$ PUFA supplement may result in specific changes in gene expression.[24] Overall, these results show that intake of $n-3$ PUFAs can alter gene expression profiles to a more cardioprotective pattern in subjects with healthy and deteriorated metabolic profiles. Yet the impact on gene expression of individual PUFAs—EPA and DHA—is still unknown.

3. Antioxidant Consumption

Consuming a diet high in fruits and vegetables is associated with lower risks for numerous chronic diseases, including cancer and CVD. The majority of phytochemicals found in plants are antioxidants. Antioxidant-rich foods may limit oxidative damage caused by reactive oxygen species; however, the mechanisms behind the protective effect of antioxidant-rich foods are not fully elucidated.

Bohn *et al.*[26] showed that intake of antioxidant-rich foods can upregulate genes involved in cellular stress defense, such as DNA repair, apoptosis, and hypoxia. Boomgaarden *et al.*[27] investigated the molecular mechanisms behind the action of quercetin, a plant-derived flavonoid. They showed that functional groups of the immune system, nucleic acid metabolism, apoptosis, and *O*-glycan biosynthesis were modified. These studies suggest that the beneficial effects of an antioxidant-rich diet can be mediated through optimization of defense processes; however, further studies are needed to confirm this hypothesis.

The studies reviewed above focused on gene expression profile to describe the molecular basis for the effects of various dietary interventions for weight management. Briefly, both short-term and long-term energy restrictions with weight loss influence gene expression. Further, gene expression profiles can reflect appropriately the different phases of a weight-loss program. Research also suggests that weight-reduction programs involving restriction of specific macronutrients do not differentially affect transcriptional expression if they result in similar amounts of weight loss. However, restriction of specific macronutrients under weight-stable conditions may contribute to altered gene expression profiles. Further, gene expression patterns can be altered as a result of the distribution and timing of meals. In addition, potentially cardioprotective transcriptional changes are observed after intake of the TMD, $n-3$ PUFA supplementation, and antioxidants. Future clinical trials should incorporate gene expression patterns to shed light on the mechanism of action of novel diets or functional foods.

III. Nutrigenetics and Weight Management

Many popular diets (e.g., Atkins™, Zone, Weight Watchers, TMD, Ornish) are available for weight loss,[33] but there is considerable interindividual differences in responsiveness to these diets. This section examines the impact of nutrigenetics on body-weight management by providing an overview of the candidate gene polymorphisms that have been shown to influence body weight changes in response to various dietary interventions. Genes that have been shown to modulate the response of body weight to other types of interventions (e.g., bariatric surgery, drug treatment, caloric surplus) are not reviewed. Table II presents the list of candidate genes and a brief overview of the studies, with information on the intervention and the main outcome of each study.

A. Genes Related to Adipose Tissue or Lipid Turnover

Several genes related to adipogenesis or lipid turnover have been investigated for their role in mediating the response to weight-loss interventions. Adrenergic receptors play an important role in the regulation of energy balance through their effects on lipid metabolism and thermogenesis. The beta-3 adrenergic receptor (ADRB3) is involved in the regulation of catecholamine-induced lipolysis. One of the most widely investigated polymorphisms of the *ADRB3* gene is the Trp64Arg (rs4994) polymorphism. Several studies have investigated the impact of this polymorphism on the response to diet. Yoshida and colleagues[39] were the first to report that obese women carriers of the *ADRB3* Trp64Arg mutation were more resistant to weight loss in response to a combined LCD and exercise protocol. Other studies performed in obese[40] and nonobese[48] subjects also reported that the *ADRB3* Trp64Arg mutation was associated with resistance to weight loss in response to lifestyle intervention programs combining diet and exercise. A study of 36 Chinese obese children aged 8–11 years placed on a diet low in cholesterol and SFA found that children with the mutation ($n=13$) responded less to the diet as their increases in body weight and body mass index (BMI) were greater than those without the mutation.[44] Tchernof *et al.*[45] examined changes in body fat in relation to the *ADRB3* Trp64Arg variant in 34 obese postmenopausal women after a 12-month weight-loss program consisting of a 1200 kcal/day American Heart Association Step 2 diet. Changes in body weight and body fat were similar across genotypes, but the reduced visceral adipose tissue in response to the caloric

TABLE II

Summary of Candidate Gene Polymorphisms Influencing Weight-Related Phenotypes in Response to Nutritional Interventions

Gene (gene symbol)	Polymorphism (rs number)	Subjects	Intervention	Outcome	Reference
Acyl-CoA synthetase long-chain family member 5 (*ACSL5*)	rs2419621 (C>T)	141 Obese women	6-Week 900-kcal formula diet	Greater weight loss in T-allele carriers	34
Adiponectin, C12 and collagen domain containing (*ADIPOQ*)	-11391G/A (rs17300539)	180 Spanish overweight and obese subjects	8-Week LCD	Protection from weight regain in A-allele carriers	35
	G276T (rs1501299)	32 Japanese obese women	8-Week LCD	No reduction of waist circumference in T/T genotype compared to 5.9 cm reduction in carriers of G allele	36
Angiotensin I converting enzyme 1 (*ACE*)	Insertion/deletion	32 Japanese obese women	2-Month LCD	Smaller decreases of % body fat in subjects with D/D genotype	37
Apolipoprotein A-V (*APOA5*)	-1131T>C (rs662799)	606 Hyperlipidemic, overweight men	3-Month LF diet	Greater reduction of BMI in C-allele carriers	38
Beta-3 adrenergic receptor (*ADRB3*)	Trp64Arg (rs4994)	88 Japanese obese women	3-Month LCD and exercise	Smaller weight loss in carriers	39
		61 Japanese obese, type 2 diabetic women	3-Month LCD and exercise	Smaller weight loss and reduction of WHR in carriers	40
		113 Japanese obese women	3-Month LCD and exercise	Smaller weight loss in carriers of both *ADRB3* and *UCP1* -3826A>G polymorphisms	41
		85 Finnish obese women	12-Week VLCD	Smaller weight loss and weight regain in carriers of both *ADRB3* and *UCP1* -3826A>G polymorphisms	42
		210 Caucasian obese women	13-Week diet (Optifast®), exercise, and supportive group therapy	Smaller weight loss in carriers	43
		36 Chinese obese children	3-Month LF, low-cholesterol diet	Greater increases in body weight and BMI in carriers	44
		24 Obese, postmenopausal women	Caloric restriction for 13 months	Lower reduction of visceral adipose tissue in carriers	45
		90 Japanese obese women	3-Month diet and exercise	Smaller reduction of visceral to subcutaneous fat ratio in carriers	46

		224 Overweight and obese subjects	12-Week LCD (−300 kcal/day)	Smaller decreases of visceral fat in carriers of mutations in both *ADRB3* (rs4994) and *UCP3* (rs1800849) genes	47
		76 Japanese perimenopausal women	3-Month lifestyle intervention program combining diet and exercise	Changes in body weight, BMI, and waist circumference only in Trp64Trp women	48
Beta-2 adrenergic receptor (*ADRB2*)	Arg16Gly (rs1042713)	138 Japanese obese women	3-Month LCD and exercise	Greater weight loss in carriers	49
		154 Japanese overweight men	24-Month low-calorie and low-sodium diet and exercise	Resistance to weight loss and weight regain in carriers	50
Clock homolog (*CLOCK*)	g.3641252A/G (rs1801260)	500 Overweight and obese subjects	28-Week behavioral weight-reduction program based on the TMD	Smaller decreases of body weight in carriers of G allele	51
Cholesteryl ester transfer protein, plasma (*CETP*)	g.16519C>T (rs5883)	86 Subjects	4–12-Week LC diet	Greater weight loss in T-allele carriers	52
Fatty acid amide hydrolase (*FAAH*)	C385A (rs324420)	122 Obese subjects	3-Month LCD and exercise	Smaller decreases of body weight and waist circumference in carriers	53
Fatty acid-binding protein 2, intestinal (*FABP2*)	Ala54Thr (rs179883)	80 Japanese obese women	6-Month LCD with exercise	Resistance to loss of abdominal fat in Thr54-allele carriers	54
		204 Obese subjects	2-Month LF or LC diet with exercise	Reduced WHR only in Ala54/Ala54 subjects under LF diet	55
		69 Obese subjects	3-Month LCD and exercise	Resistance to loss of body fat in Thr54-allele carriers	56
Fat mass and obesity associated (*FTO*)	g.87653T>A (rs9939609)	280 Overweight children	1-Year intervention with diet, exercise, and behavior therapy	Smaller weight loss in children with A/A genotype	57
		771 European obese women and men	10-Week LF or high-fat LCD	Higher dropout rate in carriers of A allele No association with changes in body weight or body composition	58
Galanin prepropeptide (*GAL*)	rs694066G>A	86 Subjects	4–12-Week LC diet	Smaller weight loss in A-allele carriers	52
Glucocorticoid receptor (*GRL*; now known as nuclear receptor subfamily 3, group C, member 1, *NR3C1*)	BclI C>G (rs41423247)	120 Overweight and obese subjects	6-Week VCLD	Greater weight loss and loss of body fat and improved weight maintenance in subjects with G/G genotype	59

(*Continues*)

TABLE II (*Continued*)

Gene (gene symbol)	Polymorphism (rs number)	Subjects	Intervention	Outcome	Reference
Growth hormone secretagogue receptor (*GHSR*)	g.172175074G > C (rs490683)	507 Overweight subjects with impaired glucose tolerance	3-Year LF diet with exercise	Greater weight loss in subjects with C/C genotype	60
Glycogen synthase 2 (*GYS2*)	g.41149G > A (rs2306179)	86 Subjects	4–12-Week LC diet	Greater weight loss in A-allele carriers	52
Insulin induced gene 2 (*INSIG2*)	g.118836025C > G (rs7566605)	293 Obese children	1-Year intervention with diet, exercise, and behavior therapy	Smaller weight loss in children with C/C genotype	61
		280 Overweight children	1-Year intervention with diet, exercise, and behavior therapy	Smaller degree of overweight reduction in children with combination of C/C genotype and *FTO* rs9939609 A/A genotype	57
Insulin receptor substrate 1 (*IRS1*)	Gly971Arg (rs1801278)	210 Caucasian obese women	13-Week diet (Optifast®), exercise, and supportive group therapy	Smaller decreases of body weight in carriers	43
Interleukin 6 (*IL6*)	-174G > C (rs11800795)	67 Obese subjects	10-Week LCD	Improved weight maintenance in C-allele carriers	62
Leptin receptor (*LEPR*)	Ser (T) 343Ser (C) (rs1805134)	179 Overweight women	2–5-Month LCD	Greater weight loss in C-allele carriers	63
	3′UTR insertion/deletion	770 Subjects with impaired glucose tolerance	3-Year weight-reducing diet	Greater weight loss and reductions of BMI and waist circumference in I-allele carriers	64
	Lys109Arg (rs1137100)	170 Overweight and obese subjects	8-Week LCD	Smaller decreases of fat mass in Arg carriers	65
	Lys656Asn	67 Obese subjects	3-Month LCD with exercise	Resistance to loss of fat mass in Asn carriers	66
		78 Obese subjects	2-Month LF or LC diet with exercise	Resistance to reductions in waist circumference and WHR in Asn carriers on LC diet	67
Lipase, gastric (*LIPF*)	Ala161Thr (rs814628)	86 Subjects	4–12-Week LC diet	Greater weight loss in Thr carriers	52
Neuromedin B (*NMB*)	Pro73Thr (rs1051168)	292 Overweight and obese subjects	2.5-Year LCD with exercise and behavioral modification	Greater reduction of waist circumference in T-allele carriers for men only ($n = 37$)	68

Perilpin 1 (*PLIN1*)	11482G > A (rs894160)	48 Obese subjects	2-Week VLCD followed by 1-year LCD	Resistance to weight loss in A carriers	69
	14995A > T (rs1052700)	234 Obese children and adolescents	20-Week multidisciplinary behavioral and nutritional treatment	Greater weight loss and loss of BMI in T-allele carriers	70
	11482G > A 14995A > T	177 Overweigh/obese Koreans	12-Week LCD	Greater reduction in abdominal fat for subjects with GA/GA haplotype at SNPs 11482G > A and 14995A > T	71
	6209T > C (rs2289487)	118 Healthy overweight and obese subjects	6-Week VLCD	Greater weight loss and loss of fat mass in women ($n = 76$) with C alleles of 6209T > C and A alleles of 11482G > A	72
Period homolog 2 (*PER2*)		454 Overweight and obese subjects	1-Month behavioral weight-loss program based on the TMD	Higher frequency of minor T allele in withdrawers than in those who successfully completed treatment	73
Peroxisome proliferator-activated receptor gamma (*PPARG*)	Pro12Ala (rs1801282)	70 Postmenopausal women	6-Month LCD	Weight regain in carriers of Ala variant	74
		522 Subjects with impaired glucose tolerance	3-Year diet and exercise	Greater weight loss in subjects with Ala12Ala genotype	75
		120 Overweight and obese subjects	6-Week VCLD	Improved weight maintenance in subjects with Pro12Pro genotype	59
		67 Obese subjects	10-Week LCD	Improved weight maintenance in carriers of Ala allele and C allele of *IL6* -174G > C polymorphism	62
		141 Obese women	6-Week 900 kcal formula diet	Resistance to weight loss in Ala carriers	34
	rs2959272 rs1386835 rs709158 rs1175540 rs1175544 rs1797912	95 Japanese women	14-Week LCD	Greater weight reduction in carriers of variant	76
Transcription factor 7-like 2 (*TCF7L2*)	g.53341C > T (rs7903146)	309 Subjects at increased risk for type 2 diabetes	9-Month LF diet with exercise	Less favorable changes in BMI, body fat, and abdominal fat in T-allele carriers	77
		771 Obese subjects	10-Week LF or high-fat LCD	Smaller weight loss and waist circumference reduction in T-allele carriers	78

(*Continues*)

TABLE II (*Continued*)

Gene (gene symbol)	Polymorphism (rs number)	Subjects	Intervention	Outcome	Reference
Uncoupling protein 1 (*UCP1*)	-3826A > G	163 French obese subjects	2.5-Month LCD	Smaller weight loss in G-allele carriers	79
		113 Japanese obese women	3-Month LCD and exercise	Smaller weight loss in G-allele carriers and in carriers of mutations in both *UCP1* and *ADRB3* Trp64 Arg	41
		85 Finnish obese women	12-Week VLCD	Smaller weight loss and weight regain in carriers of mutations in both *UCP1* and *ADRB3* Trp64 Arg	42
		40 Korean obese women	6-Week meal replacement LCD	Smaller weight loss and BMI reduction in G-allele carriers	80
		17 Lean women	2-Week LCD and LF diet	Smaller weight loss and reductions of BMI and waist circumference in G-allele carriers	81
	-3826A > G -1766A > G +1068G > A	296 Korean overweight women	1-Month VCLD	Greater reductions of fat mass and WHR in carriers of GAG haplotype	82
Uncoupling protein 2 (*UCP2*)	-866G > A (rs659366)	301 Korean overweight women	1-Month VLCD	Smaller reductions of BMI and fat mass in A-allele carriers	83
Uncoupling protein 3 (*UCP3*)	-55C/T (rs1800849)	224 Overweight and obese subjects	12-Week LCD (-300 kcal/day)	Smaller decreases of visceral fat in carriers of mutations in both *UCP3* (rs1800849) and *ADRB3* (rs4994) genes	47
		131 Obese subjects	2-Month LF or LC diet with exercise	Decreased waist circumference in T-carriers under LF diet Decreased waist circumference in C/C subjects under LC diet	84
		107 Obese subjects	3-Month LCD with exercise	Decreases of fat mass, waist circumference, and WHR in C/C subjects	85
	rs1800840 rs2075576 rs1800006 rs1685325 rs2734827 rs2075577	214 Korean overweight women	1-Month VLCD	Two SNPs (rs2075577 and rs1685325) associated with weight loss One haplotype associated with increased weight loss and reductions in BMI and body fatness	86

BMI, body mass index; LCD, low-calorie diet; LC, low-carbohydrate; LF, low-fat; TMD, the Mediterranean diet; VLCD, very low calorie diet; WHR, waist-to-hip ratio.

restriction was 43% lower in women carrying the *ADRB3* Trp64Arg variant compared to noncarriers.[45] Another study in 90 Japanese obese women reported smaller reductions in the ratio of visceral to subcutaneous fat areas in carriers of the *ADRB3* Trp64Arg variant following a 3-month weight-loss program combining caloric restriction and exercise.[46]

The *ADRB3* Trp64Arg mutation was also investigated in combination with polymorphisms in other genes, including uncoupling protein (*UCP*) genes (see Section III.B) and the insulin receptor substrate 1 (*IRS1*) gene. Benecke and colleagues[43] examined associations between the *ADRB3* Trp64Arg mutation and the Gly971Arg polymorphism (rs1801278 G/A) of *IRS1* on body weight changes in 210 obese women who underwent a 13-week weight-loss program. Changes in body weight and BMI were significantly lower in subjects carrying mutations in both genes but were not significant in those carrying a mutation in only one of the two genes.

The results reviewed above suggest that subjects carrying the *ADRB3* Trp64Arg variant may have a reduced capacity to lose weight and/or body fat in response to diet. However, most of the studies reporting positive associations are based on small numbers of subjects, and some studies reported no evidence of association[79,87–90] (see also chapter "Nutrigenetics and Nutrigenomics of Caloric Restriction").

Two common polymorphisms in the beta-2 adrenergic receptor (*ADRB2*) gene (Gln27Glu and Arg16Gly) have been widely investigated for their role in obesity, but the number of studies that have examined their effects in response to dietary-induced weight loss is limited. One study found that Japanese obese women treated with a combined LCD and exercise program and carrying the *ADRB2* Arg16Gly variant (rs1042713) lost more weight (7.6 kg) than those without the mutation (5.5 kg).[49] Another study investigated the impact of the two *ADRB2* polymorphisms on weight management in 154 Japanese overweight men enrolled in a 24-month weight-loss program consisting of a low-calorie and low-sodium diet plus aerobic exercise.[50] Results showed that the frequency of the Gly16 allele was significantly higher in subjects who failed to lose significant weight during the 24-month duration of the program and in those who experienced weight regain after achieving significant weight loss at 6 months compared to subjects with successful weight-loss maintenance.

The adiponectin, C12 and collagen domain containing (*ADIPOQ*) gene, which encodes an adipose tissue-specific hormone commonly decreased in obese subjects and which has been associated with obesity, type 2 diabetes, and other features of metabolic syndrome,[91] has been investigated for its role in the response to LCDs. Goyenechea *et al.*[35] investigated the impact of the *ADIPOQ* promoter variant -11391G/A (rs17300539) on the risk of metabolic complications in 180 Spanish obese subjects at baseline and following an 8-week LCD. The G/G genotype was associated with an increased metabolic risk at baseline, but not following the diet. The diet induced significant weight losses that were not different between genotypes. However, carriers of the variant allele appeared to be protected from weight regain, as they were able to maintain their body weight and waist circumference 32 and 60 weeks postintervention.[35] Another study performed in Japanese obese women who underwent an 8-week LCD showed significantly decreased waist circumference in carriers of the G allele of the G276T (rs1501299) polymorphism, while no changes were noted in subjects with the T/T genotype.[36] In another study, the same polymorphism (G276T) was found to be associated with the response of circulating adiponectin levels and insulin resistance to a 12-week LCD, but not the changes in body weight.[71]

Peroxisome proliferator-activated receptor gamma (*PPARG*, also known as *PPARG2*) plays a role in the regulation of adipocyte differentiation and energy balance. One of the most studied variants of *PPARG*, the Pro12Ala variant (rs1801282), has been consistently associated with decreased risk of insulin resistance and type 2 diabetes. Several studies have examined the role of this polymorphism and other *PPARG* polymorphisms in weight management. A study of 70 postmenopausal obese women who completed a 6-month LCD showed no effect of the Pro12Ala variant on weight loss, but researchers found decreased fat oxidation and a greater weight regain during a 12-month follow-up in carriers of the Ala variant.[74] In the Finnish Diabetes Prevention Study, subjects with the Ala12Ala genotype lost more weight in response to a program aimed at reducing intake of dietary fat and increasing physical activity than subjects with the other genotypes.[75]

Vogels and colleagues[59] investigated the impact of the *PPARG* Pro12Ala polymorphism, as well as polymorphisms in the glucocorticoid receptor (*GRL*; now known as nuclear receptor subfamily 3, group C, member 1, *NR3C1*) and ciliary neurotrophic factor (*CNTF*) genes, for their associations with weight loss and weight maintenance. The results showed that subjects successful at maintaining weight loss (< 10% weight regain) had a different frequency distribution for the Pro12Ala *PPARG* and BclI *GRL* (rs41423247) polymorphisms than unsuccessful subjects.[59] In addition, subjects with the *PPARG* Pro12Pro genotype and the *GRL* G/G genotype appeared to lose more weight and body fat and showed better weight management.[59]

More recently, Matsuo *et al.*[76] examined the effects of eight SNPs in *PPARG* on weight reduction in response to a 14-week caloric restriction in 95 Japanese women. Although no evidence of association was found with the Pro12Ala polymorphism, they found that six *PPARG* SNPs were associated with weight reduction and that one of them (rs1175544) accounted for 7% of the variance in body weight changes. Goyenechea *et al.*[62] determined that carriers of the C allele of the interleukin 6 (*IL6*) gene -174G > C (rs11800795) polymorphism have protection against regain of weight lost. In addition, the presence of the Ala allele of *PPARG* (rs1801282) together with the C allele of the *IL6* -174G>C polymorphism further improved the weight maintenance.[62] Adamo *et al.*[34] investigated the impact of two polymorphisms in *PPARG* and eight polymorphisms in the acyl-CoA synthetase long-chain family member 5 (*ACSL5*) gene on weight loss in response to a 6-week caloric restriction in obese women. They found that the Pro12Ala polymorphism was associated with resistance to weight loss, while a polymorphism (rs2419621) located in the 5′UTR of *ACSL5* was associated with improved weight loss.

B. Genes Related to Regulation of Appetite or Energy Balance

The endocannabinoid system has emerged has an important factor in the regulation of feeding and energy balance.[92] The system comprises cannabinoid receptors and enzymes involved in the synthesis and degradation of endocannabinoids. A few studies have investigated the impact on weight loss of polymorphisms in candidate genes of this system. A missense polymorphism (G1359A) in the cannabinoid receptor 1 (*CNR1*) gene has been investigated for its role in weight loss in response to LF and LC hypocaloric diets[93] and in response to a combined LCD and exercise program.[94] The polymorphism was found to be associated with changes in adipokines and metabolic parameters, but not with changes in body weight or body fatness. The same research group also examined the effects of a missense polymorphism (C385A or rs324420) in the fatty acid amide hydrolase (*FAAH*) gene, which encodes the main inactivating enzyme of the endocannabinoid anandamide, on weight loss in response to a 3-month program consisting of a LCD and exercise.[53] Decreases in body weight and waist circumference were significantly greater in carriers of the A allele compared to wild-type homozygotes.[53] However, the same polymorphism (rs324420) was not associated with changes in body weight in response to an LF or LC diet.[95] Aberle and colleagues[96] examined the impact of both *CNR1* G1359A and *FAAH* C385A polymorphisms in response to a 6-week LF diet in 451 obese subjects but found no influence of these polymorphisms on changes in body weight, although carriers of the *FAAH* C385A mutation exhibited greater decreases in triglycerides and cholesterol in response to the diet. In sum, these polymorphisms seem to have more influence on metabolic parameters in response to weight loss than on the magnitude of weight loss.

Ghrelin, an orexigenic hormone produced by the stomach, is thought to play a role in the development of obesity through its role in the control of energy balance, food intake, and regulation of body weight. The effects of ghrelin are mediated via its receptor known as the growth hormone secretagogue receptor (GHSR). Seven polymorphisms in *GHSR* have been investigated for their role in obesity and body weight changes in individuals participating in the Finnish Diabetes Prevention Study, a study designed to assess the efficacy of an intensive diet and exercise program to prevent or delay the onset of type 2 diabetes in subjects with impaired glucose tolerance.[60] Results showed that individuals with the rs490683 C/C genotype exhibited greater weight loss than subjects with the other genotypes after a 3-year follow-up. Neuromedin-beta (NMB) is another peptide released from the gastrointestinal tract in response to food ingestion and inhibits food intake. The *NMB* Pro73Thr polymorphism (rs1051168), which has been associated with eating behaviors and increased risk of obesity,[97] has been tested for associations with anthropometric phenotypes in response to a 2.5-year weight-reduction program.[68] Results showed a greater reduction of waist circumference in carriers of the variant, but only in men.[68]

Polymorphisms in the leptin receptor (*LEPR*) gene have been investigated for their association with weight loss in response to diet, with various outcomes depending on the polymorphism examined. One study found that the *LEPR* Ser341Ser polymorphism was associated with greater weight loss,[63] while another found that I-allele carriers of an insertion/deletion polymorphism located in the 3′UTR of *LEPR* experienced greater reductions in body weight, BMI, and waist circumference after a 3-year diet program.[64] In a study of 170 overweight subjects who followed an 8-week LCD, the *LEPR* Lys109Arg polymorphism (rs1137100) was associated with a smaller decrease in fat mass.[65] Two studies examined the impact of the *LEPR* Lys656Asn polymorphism in response to an LCD[66] or to an LF or LC diet.[67] Results showed that the Asn variant was associated with resistance to the loss of body fat[66] and abdominal fat.[67] Results from the Finnish Diabetes Prevention study showed that two *LEPR* polymorphisms (Lys109Arg and Gln223Arg) were associated with an increased risk of type 2 diabetes in subjects with impaired glucose tolerance, but they were not associated with changes in body weight following a 3-year diet and exercise program.[98]

UCPs are a family of mitochondrial carrier proteins involved in the dissipation of the proton electrochemical gradient across the inner mitochondrial membrane, releasing the energy stored within the proton as heat. As such, they play an important role in the regulation of energy expenditure. Three forms of UCPs have been identified, and the genes encoding these various forms (*UCP1*, *UCP2*, and *UCP3*) have been investigated for their role in obesity. Several studies have investigated their role in modulating the response to weight loss, alone or in combination with the *ADRB3* Trp64Arg polymorphisms studies. Fumeron and colleagues[79] were the first to report that a polymorphisms in the promoter of *UCP1* (-3826A > G) was associated with resistance to weight loss. This was confirmed in a study in 113 Japanese obese women treated with a combined LCD and exercise program for 3 months; the resistance to weight loss was found to be more pronounced in subjects carrying variants in both *UCP1* and *ADRB3*.[41] Similar results were observed in 85 Finnish obese women who followed a 12-week VLCD, as women with both mutations had lower weight loss than those with no mutation.[42] Moreover, women with both mutations experienced faster weight gain during a 40-week postintervention period than those without mutation or mutation in one of the two genes.[42]

The effects of the same two polymorphisms were investigated in 40 Korean obese women randomly assigned to a low-calorie meal replacement diet (three meals/day) containing either white rice or mixed rice for a period of 6 weeks.[80] Results revealed that in the mixed rice group, women with the *UCP1* A/A genotype showed significant reductions in body weight compared to women carrying the G allele. No evidence of association was found with *ADRB3* or with the combination of both *ADRB3* and *UCP1* polymorphisms.[80] Recently, Nagai and colleagues[81] also found that the G allele of the *UCP1* -3826A>G polymorphism was associated with resistance to weight loss. Another study of Korean overweight women showed that two haplotypes based on three *UCP1* polymorphisms were associated with decreased body fatness and abdominal fat in response to a 1-month VLCD.[82]

Polymorphisms in *UCP2* and *UCP3* were also found to be associated with weight management. Yoon *et al.*[83] investigated the impact of 10 polymorphisms in *UCP2* and *UCP3* genes in overweight women after a 1-month VLCD (700 kcal/day). One polymorphism in *UCP2* (-866G>A), as well as one haplotype based on the 10 *UCP2* and *UCP3* polymorphisms, was found to be associated with changes in BMI and fat mass. Another study by the same research group[86] examined the effects of six polymorphisms in *UCP3* on changes in body weight and body fat following a 1-month VLCD. Two *UCP3* SNPs (rs2075577 and rs1685325) were associated with changes in body weight, while one of the three common haplotypes was associated with greater reductions in body weight and body fatness.[86] A polymorphism in the *UCP3* promoter (-55C>T) was tested for its effect on the response to either an LCD[85] or to an LF or LC diet,[84] and results showed that the variant was associated with changes in fat mass, waist circumference, and waist-to-hip ratio.

Another study by Kim *et al.*[47] investigated the combined effects of the *ADRB3* Trp64Arg and *UCP3* -55 C/T (rs1800849) polymorphisms on body fat distribution after a 12-week calorie-restricted diet in 224 overweight subjects. Subjects were subdivided in four groups based on the presence of mutation only in *ADRB3*, only in *UCP3*, in both genes, or in noncarriers of both mutations. Despite similar weight reductions in the four groups, subjects carrying both variants exhibited smaller reductions in visceral fat compared to the other groups.[47]

C. Genes Related to Lipid Metabolism

The fatty acid-binding protein 2, intestinal (*FABP2*) gene plays an important role in several steps of unsaturated and saturated long-chain fatty acids transport. The Ala54Thr (rs179883) polymorphism in *FABP2* has been associated with enhanced fat absorption in the intestine[99] and with obesity. Three studies have examined the effects of this polymorphism in response to dietary intervention. First, in a study of 69 obese subjects submitted to a 3-month LCD and exercise program, changes in body weight were similar between carriers and noncarriers of the Thr54 allele, but changes in body fatness were significant only in subjects with the Ala54/Ala54 genotype.[56] In a second study, 204 obese subjects submitted to either a 2-month LF ($n = 99$) or LC ($n = 105$) diet combined with aerobic exercise; weight loss and loss of body fat were similar between carriers and noncarriers of the variant under both diets, but under the LF diet, reduced waist-to-hip ratio was observed only in noncarriers.[55] Similar results were observed in a third study, which showed smaller reduction of waist circumference in Thr54-allele carriers, suggesting that the variant is associated with resistance to loss of abdominal fat.[54]

Several studies have examined the association between the perilipin 1 (*PLIN1*) gene, which encodes for a protein that coats lipid droplets in adipocytes and is involved in the regulation of triglyceride mobilization, and weight changes in response to an LCD. Corella and colleagues[69] found that carriers of the A allele of the *PLIN1* 11482G>A (rs894160) polymorphism were resistant to weight loss following a 1-year LCD. Seven *PLIN1* polymorphisms were investigated for associations with changes in abdominal fat and free fatty acids following a 12-week calorie-restriction program in 177 overweight and obese subjects.[99a] The *PLIN1* polymorphisms were associated with changes in free fatty acids, and greater reductions in waist circumference and total abdominal fat were observed for subjects with the nGA/nGA haplotype at SNPs 11482G/A and 14995 A/T (rs1052700). The *PLIN1* 14995 A/T polymorphism was also found to be associated with weight loss and decreases in BMI in obese children and adolescents who underwent a 20-week behavioral and nutritional intervention.[70] Finally, women with the C allele of the *PLIN1* 6209T > C (rs228487) polymorphism and A allele of the *PLIN1* 11482G > A polymorphism exhibited greater weight loss and loss of abdominal fat in response to a 6-week VLCD consisting of 500 kcal/day given in three sachets per day.[72]

D. Other Candidate Genes of Obesity

Genetic polymorphisms of the renin–angiotensin system have been implicated in CVD and obesity-related metabolic diseases. Recently, Hamada *et al.*[37] tested whether the insertion/deletion polymorphism of the angiotensin I converting enzyme 1 (*ACE*) gene and the 3123C/A polymorphism of the angiotensin II receptor, type 2 (*AGTR2*) gene were involved in modulating obesity-related metabolic changes in response to a 2-month LCD in 32 Japanese obese women. They observed that the reduction in percent body fat after the LCD was significantly less in D/D subjects than in carriers of the I allele. The *AGTR2* polymorphism was associated with improvements in some obesity-related metabolic parameters, but not with changes in body weight or body fatness.[37]

Evidence showing a relationship between chronobiology and obesity has raised interest for the investigation of genes of circadian rhythm regarding their role in obesity. The *clock* homolog (*CLOCK*) gene, which encodes a transcription factor essential for circadian rhythm, has been associated with energy intake and obesity.[100] Five genetic polymorphisms in *CLOCK* were investigated for their association with obesity and weight loss in response to a weight-reduction program based on the TMD.[51] Four of the five *CLOCK* SNPs were associated with obesity, including one (rs1801260) for which carriers of the variant allele exhibited smaller decreases in body weight in response to the intervention. The period homolog 2 (*PER2*) gene is another key component of the molecular mechanism that generates circadian rhythms. It has been shown that $mPer2^{-/-}$ mice display feeding abnormalities resembling that of the night-eating syndrome, which combines features of circadian rhythm disorder and an eating disorder.[101] A study demonstrated that two *PER2* polymorphisms (rs2304672 and rs4663302) were associated with abdominal obesity and that the minor allele of the rs4663302 was more frequent in withdrawers than in those who completed the weight-loss treatment.[73]

The fat mass and obesity associated (*FTO*) gene has been consistently associated with an increased risk of obesity.[102,103] A common *FTO* variant (rs9939609), which has been associated with this increased risk, has also been investigated for its role in the response to dietary interventions. However, most studies have found no evidence of associations between *FTO* polymorphisms and weight loss from dietary interventions.[104–107] A study of 771 obese individuals randomized to either an LF or high-fat LCD found that the A allele for the *FTO* rs9939609 polymorphism was associated with a higher dropout rate on both the LF (16.9% for AA vs. 6.7% for AT) and the high-fat (28.3% vs. 17.8%) diets.[58] But the *FTO* variant was not associated with changes in body weight or body composition. Another study investigated the impact of the *FTO* rs9939609 polymorphism in combination with the insulin-induced gene 2 (*INSIG2*) rs7566605 polymorphism on weight loss following an intervention based on nutrition education, physical activity, and behavioral therapy in 280 overweight children.[57] A trend toward lower weight loss was observed in children with the *FTO* A/A genotype, but the combination of the *INSIG2* C/C genotype and *FTO* A/A genotype was associated with the lowest degree of weight reduction in children.[57] In a previous study, the same authors found that the *INSIG2* C/C genotype alone was found to be associated with a smaller reduction of body weight in overweight individuals.[61]

The transcription factor 7-like 2 (*TCF7L2*) influences the transcription of several genes and is hypothesized to play a role in adipocyte differentiation. The gene has been associated with increased risk of diabetes, a risk that is modulated by obesity.[108,109] Two studies have investigated the impact of *TCF7L2* on weight management. Haupt and colleagues[77] examined the influence of four *TCF7L2* SNPs on weight loss in 309 subjects at risk of type 2 diabetes after a 9-month intervention program combining reduced caloric intake from fat and 3 h of moderate exercise per week. The type 2 diabetes risk alleles of two *TCF7L2* polymorphisms (rs7903146 and rs1255372) were associated with less favorable changes in BMI, body fatness, and abdominal fat response to the lifestyle intervention. Similarly, another study showed that the T-risk allele of the *TCF7L2* rs7903146 polymorphism was associated with smaller weight loss and a smaller reduction in waist circumference in response to a LF LCD.[78]

In a group of 606 hyperlipidemic men, a greater reduction in BMI after a 3-month LF diet was observed in carriers of the apolipoprotein A-V (*APOA5*) -1131T > C polymorphism.[38]

Two studies examined the impact of several candidate genes of obesity on weight loss in response to diet. The first study screened 27 SNPs in 15 candidate genes of obesity for an association with weight loss in 86 healthy adult subjects who were on an LC diet (carbohydrate intake accounting for about 10% of total energy intake) for a period ranging from 4 to 12 weeks.[52] The average weight loss was 6.4 kg, and the results showed that polymorphisms in the gastric lipase (*LIPF*), hepatic glycogen synthase 2 (*GYS2*), cholesteryl ester transfer protein, plasma (*CETP*), and galanin prepropeptide (*GAL*) genes were significantly associated with weight loss. A second larger study investigated the impact of 46 SNPs in 26 candidate genes of obesity on weight loss in response to either an LF or high-fat LCD aiming at reducing energy intake by 600 kcal in 771 obese subjects.[110] After the adjustment for multiple testing, the authors concluded that there was no evidence that the investigated polymorphisms influenced the clinical outcome of the intervention.

IV. Conclusions

The studies reviewed in this chapter provide suggestive evidence that variation in tissue-specific gene expression levels and DNA sequence variants influence weight management. Several studies show that gene expression profiles are influenced by nutritional intervention. However, even though microarray technology can detect small changes of expression in response to diet, gene expression changes do not necessarily reflect changes in protein concentrations or activity. Results from nutrigenomics studies should be reproduced and validated with established as well as novel biomarkers.[111] Our review of the literature also provides strong evidence for a role of common genetic polymorphisms in weight loss and weight-loss retention. Several candidate gene polymorphisms have been shown to influence weight-related phenotypes in response to various diets. However, many of the positive associations reported in the literature are based on a relatively small number of subjects, and not all findings have been replicated. For now, the evidence is incomplete and only suggestive. The need for more studies with large numbers of subjects and with specific dietary interventions to investigate the effects of polymorphisms in single genes, as well as multiple genes, is obvious. This is likely to require coordinated efforts from many laboratories.

Overall, the knowledge gained from the use of transcriptomics is setting the stage for a better understanding of the molecular impact of various dietary interventions on body-weight fluctuations. Moreover, there is now suggestive evidence that the success of obesity therapy is likely dependent on the genetic background of the patient and that multiple genes are probably involved. Nutrigenomics and nutrigenetics have the potential to help identify subjects who might profit the most from specific nutritional treatments. For this goal to become reality, much more research is needed.

References

1. Loos RJ. Recent progress in the genetics of common obesity. *Br J Clin Pharmacol* 2009; **68**:811–29.
2. Meugnier E, Bossu C, Oliel M, Jeanne S, Michaut A, Sothier M, et al. Changes in gene expression in skeletal muscle in response to fat overfeeding in lean men. *Obesity (Silver Spring)* 2007;**15**:2583–94.
3. Kolehmainen M, Salopuro T, Schwab US, Kekalainen J, Kallio P, Laaksonen DE, et al. Weight reduction modulates expression of genes involved in extracellular matrix and cell death: the GENOBIN study. *Int J Obes (Lond)* 2008;**32**:292–303.
4. Bouchard L, Rabasa-Lhoret R, Faraj M, Lavoie ME, Mill J, Perusse L, et al. Differential epigenomic and transcriptomic responses in subcutaneous adipose tissue between low and high responders to caloric restriction. *Am J Clin Nutr* 2010;**91**:309–20.
5. Ong KR, Sims AH, Harvie M, Chapman M, Dunn WB, Broadhurst D, et al. Biomarkers of dietary energy restriction in women at increased risk of breast cancer. *Cancer Prev Res (Phila)* 2009;**2**:720–31.
6. Clement K, Viguerie N, Poitou C, Carette C, Pelloux V, Curat CA, et al. Weight loss regulates inflammation-related genes in white adipose tissue of obese subjects. *FASEB J* 2004;**18**: 1657–69.
7. Crujeiras AB, Parra D, Milagro FI, Goyenechea E, Larrarte E, Margareto J, et al. Differential expression of oxidative stress and inflammation related genes in peripheral blood mononuclear cells in response to a low-calorie diet: a nutrigenomics study. *OMICS* 2008; **12**:251–61.
8. Capel F, Klimcakova E, Viguerie N, Roussel B, Vitkova M, Kovacikova M, et al. Macrophages and adipocytes in human obesity: adipose tissue gene expression and insulin sensitivity during calorie restriction and weight stabilization. *Diabetes* 2009;**58**:1558–67.
9. Capel F, Viguerie N, Vega N, Dejean S, Arner P, Klimcakova E, et al. Contribution of energy restriction and macronutrient composition to changes in adipose tissue gene expression

during dietary weight-loss programs in obese women. *J Clin Endocrinol Metab* 2008;**93**:4315–22.

10. Dahlman I, Linder K, Arvidsson NE, Andersson I, Liden J, Verdich C, et al. Changes in adipose tissue gene expression with energy-restricted diets in obese women. *Am J Clin Nutr* 2005;**81**:1275–85.
11. Mangravite LM, Dawson K, Davis RR, Gregg JP, Krauss RM. Fatty acid desaturase regulation in adipose tissue by dietary composition is independent of weight loss and is correlated with the plasma triacylglycerol response. *Am J Clin Nutr* 2007;**86**:759–67.
12. Marquez-Quinones A, Mutch DM, Debard C, Wang P, Combes M, Roussel B, et al. Adipose tissue transcriptome reflects variations between subjects with continued weight loss and subjects regaining weight 6 mo after caloric restriction independent of energy intake. *Am J Clin Nutr* 2010;**92**:975–84.
13. Sparks LM, Xie H, Koza RA, Mynatt R, Bray GA, Smith SR. High-fat/low-carbohydrate diets regulate glucose metabolism via a long-term transcriptional loop. *Metabolism* 2006;**55**:1457–63.
14. Thalacker-Mercer AE, Fleet JC, Craig BA, Campbell WW. The skeletal muscle transcript profile reflects accommodative responses to inadequate protein intake in younger and older males. *J Nutr Biochem* 2010;**21**:1076–82.
15. van Erk MJ, Blom WA, van OB, Hendriks HF. High-protein and high-carbohydrate breakfasts differentially change the transcriptome of human blood cells. *Am J Clin Nutr* 2006; **84**:1233–41.
16. Brattbakk HR, Arbo I, Aagaard S, Lindseth I, de Soysa AK, Langaas M, et al. Balanced caloric macronutrient composition downregulates immunological gene expression in human blood cells—adipose tissue diverges. *OMICS* 2011.
17. Kallio P, Kolehmainen M, Laaksonen DE, Kekalainen J, Salopuro T, Sivenius K, et al. Dietary carbohydrate modification induces alterations in gene expression in abdominal subcutaneous adipose tissue in persons with the metabolic syndrome: the FUNGENUT Study. *Am J Clin Nutr* 2007;**85**:1417–27.
18. van Dijk SJ, Feskens EJ, Bos MB, Hoelen DW, Heijligenberg R, Bromhaar MG, et al. A saturated fatty acid-rich diet induces an obesity-linked proinflammatory gene expression profile in adipose tissue of subjects at risk of metabolic syndrome. *Am J Clin Nutr* 2009; **90**:1656–64.
19. Bouwens M, Grootte BM, Jansen J, Muller M, Afman LA. Postprandial dietary lipid-specific effects on human peripheral blood mononuclear cell gene expression profiles. *Am J Clin Nutr* 2010;**91**:208–17.
20. Camargo A, Ruano J, Fernandez JM, Parnell LD, Jimenez A, Santos-Gonzalez M, et al. Gene expression changes in mononuclear cells in patients with metabolic syndrome after acute intake of phenol-rich virgin olive oil. *BMC Genomics* 2010;**11**:253.
21. Konstantinidou V, Khymenets O, Fito M, De La Torre R, Anglada R, Dopazo A, et al. Characterization of human gene expression changes after olive oil ingestion: an exploratory approach. *Folia Biol (Praha)* 2009;**55**:85–91.
22. Khymenets O, Fito M, Covas MI, Farre M, Pujadas MA, Munoz D, et al. Mononuclear cell transcriptome response after sustained virgin olive oil consumption in humans: an exploratory nutrigenomics study. *OMICS* 2009;**13**:7–19.
23. Bouwens M, van de Rest O, Dellschaft N, Bromhaar MG, de Groot LC, Geleijnse JM, et al. Fish-oil supplementation induces antiinflammatory gene expression profiles in human blood mononuclear cells. *Am J Clin Nutr* 2009;**90**:415–24.
24. Gorjao R, Verlengia R, Lima TM, Soriano FG, Boaventura MF, Kanunfre CC, et al. Effect of docosahexaenoic acid-rich fish oil supplementation on human leukocyte function. *Clin Nutr* 2006;**25**:923–38.

25. Rudkowska I, Ponton A, Jacques H, Lavigne C, Holub BJ, Marette A, et al. Effects of a supplementation of n-3 polyunsaturated fatty acids with or without fish gelatin on gene expression in peripheral blood mononuclear cells in obese, insulin-resistant subjects. *J Nutrigenet Nutrigenomics* 2011;**4**:192–202.
26. Bohn SK, Myhrstad MC, Thoresen M, Holden M, Karlsen A, Tunheim SH, et al. Blood cell gene expression associated with cellular stress defense is modulated by antioxidant-rich food in a randomised controlled clinical trial of male smokers. *BMC Med* 2010;**8**:54.
27. Boomgaarden I, Egert S, Rimbach G, Wolffram S, Muller MJ, Doring F. Quercetin supplementation and its effect on human monocyte gene expression profiles in vivo. *Br J Nutr* 2010; **104**:336–45.
28. de Mello VD, Kolehmainen M, Pulkkinen L, Schwab U, Mager U, Laaksonen DE, et al. Downregulation of genes involved in NFkappaB activation in peripheral blood mononuclear cells after weight loss is associated with the improvement of insulin sensitivity in individuals with the metabolic syndrome: the GENOBIN study. *Diabetologia* 2008;**51**:2060–7.
29. Konstantinidou V, Khymenets O, Covas MI, De La Torre R, Munoz-Aguayo D, Anglada R, et al. Time course of changes in the expression of insulin sensitivity-related genes after an acute load of virgin olive oil. *OMICS* 2009;**13**:431–8.
30. Konstantinidou V, Covas MI, Munoz-Aguayo D, Khymenets O, De La Torre R, Saez G, et al. In vivo nutrigenomic effects of virgin olive oil polyphenols within the frame of the Mediterranean diet: a randomized controlled trial. *FASEB J* 2010;**24**:2546–57.
31. Llorente-Cortes V, Estruch R, Mena MP, Ros E, Gonzalez MA, Fito M, et al. Effect of Mediterranean diet on the expression of pro-atherogenic genes in a population at high cardiovascular risk. *Atherosclerosis* 2010;**208**:442–50.
32. Kabir M, Skurnik G, Naour N, Pechtner V, Meugnier E, Rome S, et al. Treatment for 2 mo with n 3 polyunsaturated fatty acids reduces adiposity and some atherogenic factors but does not improve insulin sensitivity in women with type 2 diabetes: a randomized controlled study. *Am J Clin Nutr* 2007;**86**:1670–9.
33. Dansinger ML, Gleason JA, Griffith JL, Selker HP, Schaefer EJ. Comparison of the Atkins, Ornish, Weight Watchers, and Zone diets for weight loss and heart disease risk reduction: a randomized trial. *JAMA* 2005;**293**:43–53.
34. Adamo KB, Dent R, Langefeld CD, Cox M, Williams K, Carrick KM, et al. Peroxisome proliferator-activated receptor gamma 2 and acyl-CoA synthetase 5 polymorphisms influence diet response. *Obesity (Silver Spring)* 2007;**15**:1068–75.
35. Goyenechea E, Collins LJ, Parra D, Abete I, Crujeiras AB, O'Dell SD, et al. The - 11391 G/A polymorphism of the adiponectin gene promoter is associated with metabolic syndrome traits and the outcome of an energy-restricted diet in obese subjects. *Horm Metab Res* 2009; **41**:55–61.
36. Tsuzaki K, Kotani K, Nagai N, Saiga K, Sano Y, Hamada T, et al. Adiponectin gene single-nucleotide polymorphisms and treatment response to obesity. *J Endocrinol Invest* 2009; **32**:395–400.
37. Hamada T, Kotani K, Nagai N, Tsuzaki K, Sano Y, Matsuoka Y, et al. Genetic polymorphisms of the renin-angiotensin system and obesity-related metabolic changes in response to low-energy diets in obese women. *Nutrition* 2011;**27**:34–9.
38. Aberle J, Evans D, Beil FU, Seedorf U. A polymorphism in the apolipoprotein A5 gene is associated with weight loss after short-term diet. *Clin Genet* 2005;**68**:152–4.
39. Yoshida T, Sakane N, Umekawa T, Sakai M, Takahashi T, Kondo M. Mutation of beta 3-adrenergic-receptor gene and response to treatment of obesity. *Lancet* 1995;**346**:1433–4.
40. Sakane N, Yoshida T, Umekawa T, Kogure A, Takakura Y, Kondo M. Effects of Trp64Arg mutation in the beta 3-adrenergic receptor gene on weight loss, body fat distribution,

glycemic control, and insulin resistance in obese type 2 diabetic patients. *Diabetes Care* 1997;**20**:1887–90.

41. Kogure A, Yoshida T, Sakane N, Umekawa T, Takakura Y, Kondo M. Synergic effect of polymorphisms in uncoupling protein 1 and beta3-adrenergic receptor genes on weight loss in obese Japanese. *Diabetologia* 1998;**41**:1399.
42. Fogelholm M, Valve R, Kukkonen-Harjula K, Nenonen A, Hakkarainen V, Laakso M, et al. Additive effects of the mutations in the beta3-adrenergic receptor and uncoupling protein-1 genes on weight loss and weight maintenance in Finnish women. *J Clin Endocrinol Metab* 1998;**83**:4246–50.
43. Benecke H, Topak H, von zur Mühlen A, Schuppert F. A study on the genetics of obesity: influence of polymorphisms of the beta-3-adrenergic receptor and insulin receptor substrate 1 in relation to weight loss, waist to hip ratio and frequencies of common cardiovascular risk factors. *Exp Clin Endocrinol Diabetes* 2000;**108**:86–92.
44. Xinli W, Xiaomei T, Meihua P, Song L. Association of a mutation in the beta3-adrenergic receptor gene with obesity and response to dietary intervention in Chinese children. *Acta Paediatr* 2001;**90**:1233–7.
45. Tchernof A, Starling RD, Turner A, Shuldiner AR, Walston JD, Silver K, et al. Impaired capacity to lose visceral adipose tissue during weight reduction in obese postmenopausal women with the Trp64Arg beta3-adrenoceptor gene variant. *Diabetes* 2000;**49**:1709–13.
46. Nakamura M, Tanaka M, Abe S, Itoch K, Imai K, Masuda T, et al. Association between beta-3 adrenergenic receptor polymorphism and lower reduction in the ratio of visceral fat to subcutaneous fat area during weight loss in Japanese obese women. *Nutr Res* 2000;**20**:25–34.
47. Kim OY, Cho EY, Park HY, Jang Y, Lee JH. Additive effect of the mutations in the beta3-adrenoceptor gene and UCP3 gene promoter on body fat distribution and glycemic control after weight reduction in overweight subjects with CAD or metabolic syndrome. *Int J Obes Relat Metab Disord* 2004;**28**:434–41.
48. Shiwaku K, Nogi A, Anuurad E, Kitajima K, Enkhmaa B, Shimono K, et al. Difficulty in losing weight by behavioral intervention for women with Trp64Arg polymorphism of the beta3-adrenergic receptor gene. *Int J Obes Relat Metab Disord* 2003;**27**:1028–36.
49. Sakane N, Yoshida T, Umekawa T, Kogure A, Kondo M. Beta2-adrenoceptor gene polymorphism and obesity. *Lancet* 1999;**353**:1976.
50. Masuo K, Katsuya T, Kawaguchi H, Fu Y, Rakugi H, Ogihara T, et al. Rebound weight gain as associated with high plasma norepinephrine levels that are mediated through polymorphisms in the beta2-adrenoceptor. *Am J Hypertens* 2005;**18**:1508–16.
51. Garaulet M, Corbalan MD, Madrid JA, Morales E, Baraza JC, Lee YC, et al. CLOCK gene is implicated in weight reduction in obese patients participating in a dietary programme based on the Mediterranean diet. *Int J Obes (Lond)* 2010;**34**:516–23.
52. Ruano G, Windemuth A, Kocherla M, Holford T, Fernandez ML, Forsythe CE, et al. Physiogenomic analysis of weight loss induced by dietary carbohydrate restriction. *Nutr Metab (Lond)* 2006;**3**:20.
53. de Luis DA, Gonzalez SM, Aller R, Izaola O, Conde R. Effects of C358A missense polymorphism of the endocannabinoid degrading enzyme fatty acid amide hydrolase on weight loss after a hypocaloric diet. *Metabolism* 2011;**60**:730–4.
54. Takakura Y, Yoshioka K, Umekawa T, Kogure A, Toda H, Yoshikawa T, et al. Thr54 allele of the FABP2 gene affects resting metabolic rate and visceral obesity. *Diabetes Res Clin Pract* 2005;**67**:36–42.
55. de Luis DA, Aller R, Izaola O, Sagrado MG, Conde R. Influence of Ala54Thr polymorphism of fatty acid-binding protein 2 on weight loss and insulin levels secondary to two hypocaloric diets: a randomized clinical trial. *Diabetes Res Clin Pract* 2008;**82**:113–8.

56. de Luis DA, Aller R, Izaola O, Sagrado MG, Conde R. Influence of ALA54THR polymorphism of fatty acid binding protein 2 on lifestyle modification response in obese subjects. *Ann Nutr Metab* 2006;**50**:354–60.
57. Reinehr T, Hinney A, Toschke AM, Hebebrand J. Aggravating effect of INSIG2 and FTO on overweight reduction in a one-year lifestyle intervention. *Arch Dis Child* 2009;**94**:965–7.
58. Grau K, Hansen T, Holst C, Astrup A, Saris WH, Arner P, et al. Macronutrient-specific effect of FTO rs9939609 in response to a 10-week randomized hypo-energetic diet among obese Europeans. *Int J Obes (Lond)* 2009;**33**:1227–34.
59. Vogels N, Mariman EC, Bouwman FG, Kester AD, Diepvens K, Westerterp-Plantenga MS. Relation of weight maintenance and dietary restraint to peroxisome proliferator-activated receptor gamma2, glucocorticoid receptor, and ciliary neurotrophic factor polymorphisms. *Am J Clin Nutr* 2005;**82**:740–6.
60. Mager U, Degenhardt T, Pulkkinen L, Kolehmainen M, Tolppanen AM, Lindstrom J, et al. Variations in the ghrelin receptor gene associate with obesity and glucose metabolism in individuals with impaired glucose tolerance. *PLoS One* 2008;**3**:e2941.
61. Reinehr T, Hinney A, Nguyen TT, Hebebrand J. Evidence of an influence of a polymorphism near the INSIG2 on weight loss during a lifestyle intervention in obese children and adolescents. *Diabetes* 2008;**57**:623–6.
62. Goyenechea E, Dolores PM, Alfredo MJ. Weight regain after slimming induced by an energy-restricted diet depends on interleukin-6 and peroxisome-proliferator-activated-receptor-gamma2 gene polymorphisms. *Br J Nutr* 2006;**96**:965–72.
63. Mammes O, Aubert R, Betoulle D, Pean F, Herbeth B, Visvikis S, et al. LEPR gene polymorphisms: associations with overweight, fat mass and response to diet in women. *Eur J Clin Invest* 2001;**31**:398–404.
64. Zacharova J, Chiasson JL, Laakso M. Leptin receptor gene variation predicts weight change in subjects with impaired glucose tolerance. *Obes Res* 2005;**13**:501–6.
65. Abete I, Goyenechea E, Crujeiras AB, Martinez JA. Inflammatory state and stress condition in weight-lowering Lys109Arg LEPR gene polymorphism carriers. *Arch Med Res* 2009; **40**:306–10.
66. de Luis RD, de la Fuente RA, Sagrado MG, Izaola O, Vicente RC. Leptin receptor Lys656Asn polymorphism is associated with decreased leptin response and weight loss secondary to a lifestyle modification in obese patients. *Arch Med Res* 2006;**37**:854–9.
67. de Luis DA, Aller R, Izaola O, Sagrado MG, Conde R. Influence of Lys656Asn polymorphism of leptin receptor gene on leptin response secondary to two hypocaloric diets: a randomized clinical trial. *Ann Nutr Metab* 2008;**52**:209–14.
68. Spalova J, Zamrazilova H, Vcelak J, Vankova M, Lukasova P, Hill M, et al. Neuromedin beta: P73T polymorphism in overweight and obese subjects. *Physiol Res* 2008;**57**(Suppl. 1): S39–48.
69. Corella D, Qi L, Sorli JV, Godoy D, Portoles O, Coltell O, et al. Obese subjects carrying the 11482G>A polymorphism at the perilipin locus are resistant to weight loss after dietary energy restriction. *J Clin Endocrinol Metab* 2005;**90**:5121–6.
70. Deram S, Nicolau CY, Perez-Martinez P, Guazzelli I, Halpern A, Wajchenberg BL, et al. Effects of perilipin (PLIN) gene variation on metabolic syndrome risk and weight loss in obese children and adolescents. *J Clin Endocrinol Metab* 2008;**93**:4933–40.
71. Shin MJ, Jang Y, Koh SJ, Chae JS, Kim OY, Lee JE, et al. The association of SNP276G>T at adiponectin gene with circulating adiponectin and insulin resistance in response to mild weight loss. *Int J Obes (Lond)* 2006;**30**:1702–8.
72. Soenen S, Mariman EC, Vogels N, Bouwman FG, den Hoed M, Brown L, et al. Relationship between perilipin gene polymorphisms and body weight and body composition during weight loss and weight maintenance. *Physiol Behav* 2009;**96**:723–8.

73. Garaulet M, Corbalan-Tutau MD, Madrid JA, Baraza JC, Parnell LD, Lee YC, et al. PERIOD2 variants are associated with abdominal obesity, psycho-behavioral factors, and attrition in the dietary treatment of obesity. *J Am Diet Assoc* 2010;**110**:917–21.
74. Nicklas BJ, van Rossum EF, Berman DM, Ryan AS, Dennis KE, Shuldiner AR. Genetic variation in the peroxisome proliferator-activated receptor-gamma2 gene (Pro12Ala) affects metabolic responses to weight loss and subsequent weight regain. *Diabetes* 2001;**50**:2172–6.
75. Lindi VI, Uusitupa MI, Lindstrom J, Louheranta A, Eriksson JG, Valle TT, et al. Association of the Pro12Ala polymorphism in the PPAR-gamma2 gene with 3-year incidence of type 2 diabetes and body weight change in the Finnish Diabetes Prevention Study. *Diabetes* 2002; **51**:2581–6.
76. Matsuo T, Nakata Y, Katayama Y, Iemitsu M, Maeda S, Okura T, et al. PPARG genotype accounts for part of individual variation in body weight reduction in response to calorie restriction. *Obesity (Silver Spring)* 2009;**17**:1924–31.
77. Haupt A, Thamer C, Heni M, Ketterer C, Machann J, Schick F, et al. Gene variants of TCF7L2 influence weight loss and body composition during lifestyle intervention in a population at risk for type 2 diabetes. *Diabetes* 2010;**59**:747–50.
78. Grau K, Cauchi S, Holst C, Astrup A, Martinez JA, Saris WH, et al. TCF7L2 rs7903146-macronutrient interaction in obese individuals' responses to a 10-wk randomized hypoenergetic diet. *Am J Clin Nutr* 2010;**91**:472–9.
79. Fumeron F, Durack-Bown I, Betoulle D, Cassard-Doulcier AM, Tuzet S, Bouillaud F, et al. Polymorphisms of uncoupling protein (UCP) and beta 3 adrenoreceptor genes in obese people submitted to a low calorie diet. *Int J Obes Relat Metab Disord* 1996;**20**:1051–4.
80. Kim JY, Lee SS. The effects of uncoupling protein 1 and beta3-adrenergic receptor gene polymorphisms on weight loss and lipid profiles in obese women. *Int J Vitam Nutr Res* 2010; **80**:87–96.
81. Nagai N, Sakane N, Kotani K, Hamada T, Tsuzaki K, Moritani T. Uncoupling protein 1 gene -3826 A/G polymorphism is associated with weight loss on a short-term, controlled-energy diet in young women. *Nutr Res* 2011;**31**:255–61.
82. Shin HD, Kim KS, Cha MH, Yoon Y. The effects of UCP-1 polymorphisms on obesity phenotypes among Korean female subjects. *Biochem Biophys Res Commun* 2005; **335**:624–30.
83. Yoon Y, Park BL, Cha MH, Kim KS, Cheong HS, Choi YH, et al. Effects of genetic polymorphisms of UCP2 and UCP3 on very low calorie diet-induced body fat reduction in Korean female subjects. *Biochem Biophys Res Commun* 2007;**359**:451–6.
84. de Luis DA, Aller R, Izaola O, Gonzalez Sagrado M, Conde R. Modulation of insulin concentrations and metabolic parameters in obese patients by -55CT polymorphism of the UCP3 gene secondary to two hypocaloric diets. *Horm Metab Res* 2009;**41**:62–6.
85. de Luis DA, Aller R, Izaola O, Sagrado MG, Conde R. Modulation of adipocytokines response and weight loss secondary to a hypocaloric diet in obese patients by -55CT polymorphism of UCP3 gene. *Horm Metab Res* 2008;**40**:214–8.
86. Cha MH, Shin HD, Kim KS, Lee BH, Yoon Y. The effects of uncoupling protein 3 haplotypes on obesity phenotypes and very low-energy diet-induced changes among overweight Korean female subjects. *Metabolism* 2006;**55**:578–86.
87. de Luis DA, Gonzalez SM, Aller R, Izaola O, Conde R. Influence of the Trp64Arg polymorphism in the beta 3 adrenoreceptor gene on insulin resistance, adipocytokine response, and weight loss secondary to lifestyle modification in obese patients. *Eur J Intern Med* 2007; **18**:587–92.
88. de Luis DA, Gonzalez SM, Aller R, Izaola O, Conde R. Influence of Trp64Arg polymorphism of beta 3-adrenoreceptor gene on insulin resistance, adipocytokines and weight loss secondary to two hypocaloric diets. *Ann Nutr Metab* 2009;**54**:104–10.

89. Kim OY, Lee YA, Ryu HJ, Park HY, Jang Y, Lee JH. Effect of Trp64Arg mutation in the beta-3 adrenergic receptor gene on body fat distribution, glycemic control and lipids in response to hypocaloric diets in men with coronary artery disease. *Nutr Res* 2003;**23**:1013–25.
90. Rawson ES, Nolan A, Silver K, Shuldiner AR, Poehlman ET. No effect of the Trp64Arg beta (3)-adrenoceptor gene variant on weight loss, body composition, or energy expenditure in obese, caucasian postmenopausal women. *Metabolism* 2002;**51**:801–5.
91. Povel CM, Boer JM, Reiling E, Feskens EJ. Genetic variants and the metabolic syndrome: a systematic review. *Obes Rev* 2011;**12**:952–67.
92. Li C, Jones PM, Persaud SJ. Role of the endocannabinoid system in food intake, energy homeostasis and regulation of the endocrine pancreas. *Pharmacol Ther* 2011;**129**:307–20.
93. de Luis DA, Sagrado MG, Aller R, Conde R, Izaola O, de la Fuente B, et al. Role of G1359A polymorphism of the cannabinoid receptor gene on weight loss and adipocytokines levels after two different hypocaloric diets. *J Nutr Biochem* 2012;**23**:287–91.
94. de Luis DA, Gonzalez SM, Aller R, Conde R, Izaola O, de la Fuente B, et al. Roles of G1359A polymorphism of the cannabinoid receptor gene (CNR1) on weight loss and adipocytokines after a hypocaloric diet. *Nutr Hosp* 2011;**26**:317–22.
95. de Luis DA, Sagrado MG, Aller R, Izaola O, Conde R. Effects of C358A missense polymorphism of the degrading enzyme fatty acid amide hydrolase on weight loss, adipocytokines, and insulin resistance after 2 hypocaloric diets. *Metabolism* 2010;**59**:1387–92.
96. Aberle J, Fedderwitz I, Klages N, George E, Beil FU. Genetic variation in two proteins of the endocannabinoid system and their influence on body mass index and metabolism under low fat diet. *Horm Metab Res* 2007;**39**:395–7.
97. Bouchard L, Drapeau V, Provencher V, Lemieux S, Chagnon Y, Rice T, et al. Neuromedin beta: a strong candidate gene linking eating behaviors and susceptibility to obesity. *Am J Clin Nutr* 2004;**80**:1478–86.
98. Salopuro T, Pulkkinen L, Lindstrom J, Eriksson JG, Valle TT, Hamalainen H, et al. Genetic variation in leptin receptor gene is associated with type 2 diabetes and body weight: the Finnish Diabetes Prevention Study. *Int J Obes (Lond)* 2005;**29**:1245–51.
99. Levy E, Menard D, Delvin E, Stan S, Mitchell G, Lambert M, et al. The polymorphism at codon 54 of the FABP2 gene increases fat absorption in human intestinal explants. *J Biol Chem* 2001;**276**:39679–84.
99a. Jang Y, Kim OY, Lee JH, Koh SJ, Chae JS, Kim JY, et al. Genetic variation at the perilipin locus is associated with changes in serum free fatty acids and abdominal fat following mild weight loss. *Int J Obes (Lond)* 2006;**30**(11):1601–08.
100. Froy O. Metabolism and circadian rhythms—implications for obesity. *Endocr Rev* 2010;**31**:1–24.
101. Yang S, Liu A, Weidenhammer A, Cooksey RC, McClain D, Kim MK, et al. The role of mPer2 clock gene in glucocorticoid and feeding rhythms. *Endocrinology* 2009;**150**:2153–60.
102. Loos RJ, Bouchard C. FTO: the first gene contributing to common forms of human obesity. *Obes Rev* 2008;**9**:246–50.
103. Peng S, Zhu Y, Xu F, Ren X, Li X, Lai M. FTO gene polymorphisms and obesity risk: a meta-analysis. *BMC Med* 2011;**9**:71.
104. Dlouha D, Suchanek P, Lanskalanskalanska V, Hubacek JA. Body mass index change in females after short-time life style intervention is not dependent on the FTO polymorphisms. *Physiol Res* 2011;**60**:199–202.
105. Haupt A, Thamer C, Machann J, Kirchhoff K, Stefan N, Tschritter O, et al. Impact of variation in the FTO gene on whole body fat distribution, ectopic fat, and weight loss. *Obesity (Silver Spring)* 2008;**16**:1969–72.
106. Lappalainen TJ, Tolppanen AM, Kolehmainen M, Schwab U, Lindstrom J, Tuomilehto J, et al. The common variant in the FTO gene did not modify the effect of lifestyle changes on body weight: the Finnish Diabetes Prevention Study. *Obesity (Silver Spring)* 2009;**17**:832–6.

107. Muller TD, Hinney A, Scherag A, Nguyen TT, Schreiner F, Schafer H, et al. 'Fat mass and obesity associated' gene (FTO): no significant association of variant rs9939609 with weight loss in a lifestyle intervention and lipid metabolism markers in German obese children and adolescents. *BMC Med Genet* 2008;**9**:85.
108. Cauchi S, Choquet H, Gutierrez-Aguilar R, Capel F, Grau K, Proenca C, et al. Effects of TCF7L2 polymorphisms on obesity in European populations. *Obesity (Silver Spring)* 2008; **16**:476–82.
109. Grant SF, Thorleifsson G, Reynisdottir I, Benediktsson R, Manolescu A, Sainz J, et al. Variant of transcription factor 7-like 2 (TCF7L2) gene confers risk of type 2 diabetes. *Nat Genet* 2006;**38**:320–3.
110. Sorensen TI, Boutin P, Taylor MA, Larsen LH, Verdich C, Petersen L, et al. Genetic polymorphisms and weight loss in obesity: a randomised trial of hypo-energetic high- versus low-fat diets. *PLoS Clin Trials* 2006;**1**:e12.
111. Wittwer J, Rubio-Aliaga I, Hoeft B, Bendik I, Weber P, Daniel H. Nutrigenomics in human intervention studies: current status, lessons learned and future perspectives. *Mol Nutr Food Res* 2011;**55**:341–58.

Taste Preferences

María Mercedes Galindo,
Nanette Yvette Schneider,
Frauke Stähler, Jonas Töle,
and Wolfgang Meyerhof

*German Institute of Human Nutrition,
Potsdam-Rehbrücke, Nuthetal, Germany*

Personal experience, learned eating behaviors, hormones, neurotransmitters, and genetic variations affect food consumption. The decision of what to eat is modulated by taste, olfaction, and oral textural perception. Taste, in particular, has an important input into food preference, permitting individuals to differentiate nutritive and harmful substances and to select nutrients. To be perceived as taste, gustatory stimuli have to contact specialized receptors and channels expressed in taste buds in the oral cavity. Gustatory information is then conveyed via afferent nerves to the central nervous system, which processes the gustatory information at different levels, resulting in stimulus recognition, integration with metabolic needs, and control of ingestive reflexes. This review discusses physiological factors influencing the decision of what to eat, spanning the bow from the recognition of the nutritive value of food in the oral cavity, over the feedback received after ingestion, to processing of gustatory information to the central nervous system.

I. Introduction

Appetite is controlled at multiple levels, including variations of hormone and neurotransmitter levels that can promote or repress immediate food intake and long-term energy balance control (for review, see Refs. 1,2). Mutations in

Progress in Molecular Biology
and Translational Science, Vol. 108
DOI: 10.1016/B978-0-12-398397-8.00015-0

the genes encoding these hormones and neurotransmitters, their receptors, or modulators may be responsible for eating behavior disorders, eating-related diseases, and obesity (for review, see Ref. 3).

Individual diet is also modulated by cultural, religious, and ecological conditions, personal experiences, and learned eating behaviors, which influence individual taste preferences and limit or promote the consumption of food.[4–6] Moreover, taste, olfaction, and oral textural perception, which determine the flavor of a food item, influence its ingestion. Taste, in particular, has an important input into food preference, allowing individuals to differentiate between nutritive and harmful substances and to select nutrients fulfilling physiological requirements of caloric intake and body homeostasis.[7]

Taste preferences change during aging. Newborns are innately attracted to sweet taste and reject bitter-tasting substances. Adults, however, appreciate both sweet and bitter foods and beverages, with some people developing acceptance and even preference for bitter coffee, beer, and dark chocolate. On the other hand, taste sensitivity may be impaired in the elderly, affecting food selection and thereby health.[8–11]

To be perceived as taste, gustatory stimuli have to contact specialized receptors and ion channels expressed in chemosensory cells of taste buds in the oral cavity. Genetic variability in these taste receptors might be responsible for interindividual differences in taste perception, affecting taste and therefore food consumption (e.g., Refs. 12,13). Information from the peripheral taste sensors is conveyed via afferent gustatory nerves to the central nervous system, which processes the gustatory information at different levels, ruling stimulus recognition, integrating calorie intake with metabolic needs, and inducing oromotor and physiological reflexes.[14,15]

Food intake is controlled via multiple pathways involving hormones initiating or terminating a meal and hormones reflecting body adiposity and energy balance (for review, see Ref. 2). These hormones act on the vagus nerve, hypothalamus, and brain stem, regulating central neuropeptides that modulate feeding and energy expenditure (for review and examples, see Refs. 2,16–19).

Because of the broadness of the field of taste preference, we will not address all aspects in this chapter. While we will not further discuss cultural differences and learned eating behaviors and their critical roles in taste preferences (refer to Refs. 4–6,9), we will focus on how taste anatomy and physiology, acquisition of conditioned taste aversion/attraction, formation of taste memory, and the influence of genetic variability of components of the taste-signaling cascade may influence individual taste perception and taste preferences. We will start out by giving an overview of the cellular and molecular basis of taste perception at the level of the tongue leading over to its representation in the brain. Further, we will explain the differences between taste modalities. Then we will present how experience of taste perception in combination with an unconditioned stimulus (malaise or satiety) influences taste memory formation. We will consider as well how postingestive effects play an important role in this learning process. There is a connection between perception of nutrients in the gastrointestinal (GI) tract and expression of taste-related molecules; on the other hand, there are hormones involved in energy homeostasis that may also influence taste perception at the level of the tongue. Last but not least, we will discuss how genetic variability influences taste perception and may lead to variations in taste preference.

II. Biological Function of Taste Perception

A. Anatomy and Physiology of the Taste System

1. Taste Buds

To be perceived as taste, gustatory stimuli have to contact specialized cells in the oral cavity. In onion-shaped clusters called taste buds, 50–100 of those cells are grouped together. On the tip of the taste bud is the taste pore, where the apical poles of taste bud cells are in direct contact with the oral cavity, allowing them to detect tastants dissolved in saliva.[20–22] Taste buds are found either within the papillae on the tongue or within the soft palatal, pharyngeal, and laryngeal epithelium.[23] Taste-like cells also have been observed in extraoral tissues (for review, see Ref. 24).

Three types of lingual taste papillae are distributed on the dorsal surface of the tongue (Fig. 1). Close to 300 fungiform papillae are found on the anterior part of the tongue, containing approximately three taste buds in humans. Foliate papillae are observed arranged in groups of approximately 20 parallel ridges and valleys on the lateral-posterior margins of the tongue. On the posterior region, a reverted V-shape structure is found, formed by 4–18 circumvallate papillae in humans.[23,25–27] Taste buds contain distinct cell types with different morphology and functional characteristics, and they express specific cell markers.[21,28,29]

2. Type I Cells

Type I cells, representing the greatest proportion of cells found in taste buds, are elongated cells appearing dark on electron micrographs because of the electron-dense cytoplasm.[21,29] Proposed to have a glial-like function, they possess cytoplasmic protrusions wrapping around other taste cells[29] and express glial cell markers such as the glial glutamate transporter GLAST, which probably participates in the clearance of the transmitter glutamate.[30,31] The transmitter is secreted from the afferent gustatory nerves and might play a role in first-stage gustatory processing.[32] The ecto-ATPase NTPDase2 is expressed and hypothesized to be involved in the clearance of the neurotransmitter ATP in taste buds (see Section II.A.7).[33]

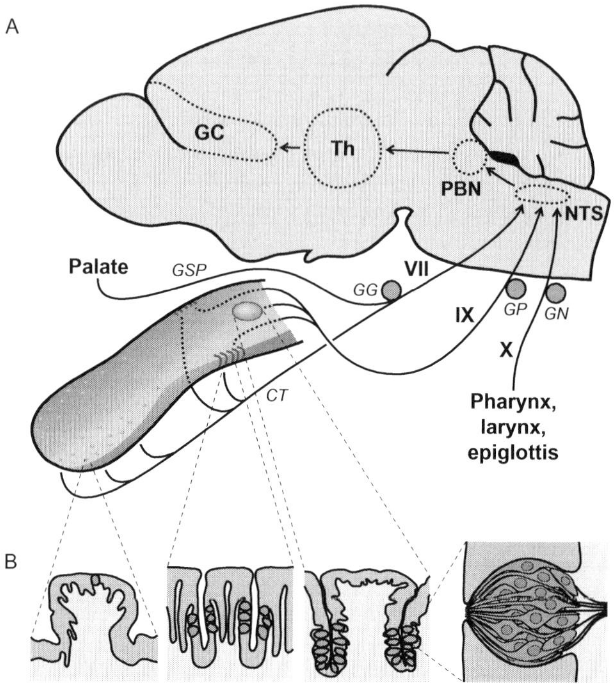

FIG. 1. Murine gustatory pathway. (A) Taste receptor cells in different oral areas are innervated by three cranial nerves. Two branches of the facial nerve (VII) convey the gustatory information from the anterior two-thirds of the tongue (chorda tympani, CT) and the palate (greater superficial petrosal nerve, GSP). The posterior third of the tongue is innervated by the glossopharyngeal nerve (IX). Pharynx, larynx, and epiglottis are innervated by gustatory fibers of the vagus nerve (X). The cell bodies of the gustatory fibers are located in the geniculate ganglion (GG), petrosal ganglion (GP), and nodosal ganglion (GN). The first station of central gustatory processing is the rostral part of the nucleus of the solitary tract (NTS), in which neurons project to the parabrachial nucleus of the pons (PBN). From there, the information is sent via the thalamus (Th) to the primary gustatory cortex (GC). (B) Taste receptor cells are organized in taste buds. On the tip is the taste pore, where the apical aspects of taste bud cells contact the oral cavity. On the bottom, nerve fibers enter the taste bud to form synapses with taste bud cells. Fungiform papillae are concentrated on the anterior tongue, foliate papillae can be found on the rim of the posterior tongue, and a single vallate papilla in rodents is situated close to the terminal sulcus of the tongue.

Type I cells do not have any voltage-gated Ca^{2+} currents and are unable to form synapses. They have small voltage-gated outward K^+ currents and, in some cases, small inward Na^+ currents.[34,35] The outward K^+ current is driven by the inwardly rectifying K^+ channel ROMK2 to excrete K^+ and maintain the membrane potential of taste cells.[36] Patch-clamp techniques suppose that type I cells participate in amiloride-sensitive salt taste sensitivity.[37]

3. Type II or Receptor Cells

Type II or receptor cells are electron-lucent light cells[29] expressing proteins for the transduction of sweet, umami, and bitter[30] but not sour taste stimuli.[38]

G-protein-coupled receptors (GPRs) are a family of cell membrane receptors with a characteristic structure of seven transmembrane domains and a large list of agonists from photons and anions to complex proteins (reviewed in Ref. 39). GPRs as taste receptors (taste receptor type 1 family, TAS1Rs and taste receptor type 2 family, TAS2Rs) are exclusively expressed in type II cells, giving them the name "receptor cells."[28,40,41]

Receptor cells are clearly dedicated to one taste modality expressing only one taste receptor type[42,43] and are tuned to the perception of a single taste quality.[44–46]

Receptor cells express proteins involved in the taste-signaling cascade such as a taste-specific isoform of phospholipase, phospholipase C-β2 (PLCβ2),[28,30,47,48] the taste-specific G protein gustducin,[49–51] the transient receptor potential cation channel subfamily M member 5 (TRPM5),[28,30,43] and the inositol 1,4,5-trisphosphate (IP_3) receptor (IP_3R3).[28,30,40] Type II cells express voltage-gated Na^+ and K^+ channels, which are necessary for generation of action potentials.[48,52,53] However, they do not possess synapses or conventional components of a neurotransmitter release machinery.[54]

4. Type III or Presynaptic Cells

Type III cells are named presynaptic cells because they form real synapses with the afferent taste fibers.[55] They express voltage-gated Na^+ and K^+ channels, which are important for action potential generation.[52,53] Presynaptic cells are sour-responsive cells,[38,56,57] and genetic ablation of these cells results in a specific loss of neural responses to acidic stimuli.[56] In addition, these cells respond to carbonated solutions.[58]

The identified transmitters of type III cells are serotonin (5-HT),[59,60] γ-aminobutyric acid (GABA),[28,61] and norepinephrine.[62] In accordance with their function as presynaptic cells, type III cells express synapse-related proteins, including synapsin-2 (neural cell adhesion molecule, NCAM) (reviewed in Refs. 28,63), SNAP25 (synaptosomal-associated protein 25),[28,64] the high-threshold-activated, voltage-dependent presynaptic P/Q-type calcium channel subunit alpha-1A,[28] and other neuronal-like genes, (e.g., *NCAM*[28,65]). A separate population from the 5-HT-positive cells coexpresses the neural markers ubiquitin carboxyl-terminal hydrolase isozyme L1 (protein gene product PGP 9.5) with the neuron-specific enolase.[66]

It is proposed that type III cells receive and integrate signals from type II cells, resulting in broadly tuned transmission of taste information across modalities.[67]

5. Type IV or Basal Cells

Type IV cells or basal cells are undifferentiated, nonpolarized basal cells, which have no contact to the oral cavity. Because the turnover of cells in taste buds is rapid, requiring approximately 250 h, a germinal epithelium warrants renewal of the taste cells. This role is undertaken by the basal cells.[68] The differentiation fate of new cells in taste buds must be carefully controlled. Many transcription factors have been described and are expressed in basal cells and governing the taste cell renewal and differentiation. Such transcription factors are SKN-1a,[69] sonic hedgehog protein, MASH1,[70,71] SOX2,[72] and HES6,[70] among others.

6. Signal Transduction in Receptor Cells

Taste cells are specialized epithelial cells that display apical and basal aspects.[73] The earliest transduction step, activation of ion channels or GPRs, occurs at the apical membrane of the taste cells that form the pore.[26] Activation of the taste GPRs triggers a stereotypic signaling cascade (Fig. 2) common to sweet, bitter, and umami cells, which is initiated by a guanine-nucleotide-binding protein (G protein). G proteins are heterotri meric proteins composed of α, β, and γ subunits localized at the cytoplasmic face of the plasma membrane bound to the GPR. Activation of a GPR causes the release of GDP and binding of GTP to the α subunit, as well as dissociation of the β/γ subunits. Gα and Gβγ subunits can trigger different signaling cascades to transmit the stimulus to cellular effector molecules (reviewed in Refs. 14,74).

The lingual G protein subunit α-gustducin is closely related to transducin, the retinal G protein that mediates the photoactivation in rod and cone cells.[75–77] This Gα subunit has been directly related to taste response to sweet,[49,50] bitter,[49] and umami taste qualities[51] but never to sour or salty stimuli.[49] Receptor cells express α-gustducin and other Gα proteins, mainly but not only from the Gi subfamily, which might couple to taste receptors.[78–83] *In vitro* studies have suggested that the active Gα subunit reduces intracellular cAMP levels, keeping the activity of cAMP-dependent protein kinase A (PKA) low, as well as maintaining the IP_3R3 in the endoplasmic reticulum in a dephosphorylated, sensitive state, which permits efficient release of Ca^{2+} from internal stores (Fig. 2). This process increases the taste cells' sensitivity to taste stimuli.[84] The main taste transduction pathway involves the GBγ subunit (Fig. 2), which activates PLCβ2.[85,86] PLCβ2 generates IP_3, which binds to its receptor IP_3R3, resulting in the release of Ca^{2+} from the endoplasmic reticulum to the cytosol. This, in turn, opens the nonselective cation channel TRPM5, leading to influx of Na^+ and plasma membrane depolarization of the type II cells.[87] Elevated intracellular Ca^{2+} levels and the

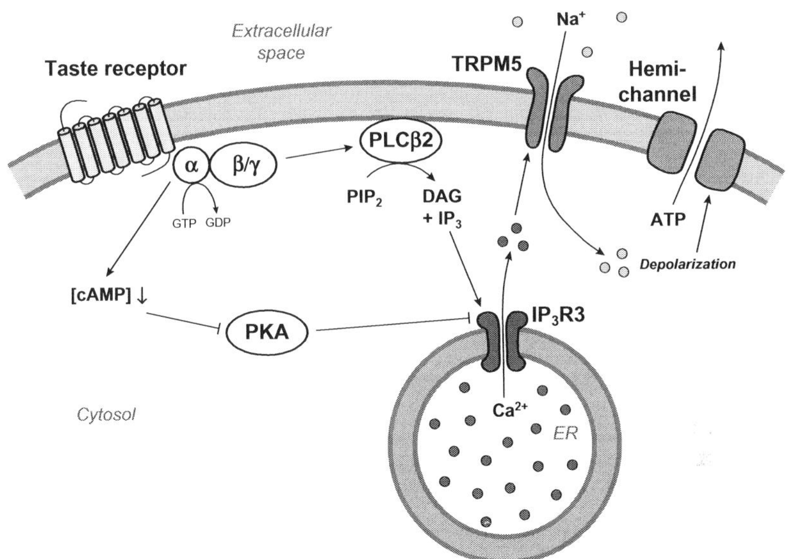

FIG. 2. Signaling pathway in type II receptor cells. Upon binding of tastants to the heptahelical taste receptors (TAS1Rs/TAS2Rs), the associated G protein gustducin is activated in a GTP-dependent manner. Alpha-gustducin leads to decreased intracellular cAMP concentration via a yet unknown mechanism, eventually reducing the inhibition of type 3 inositol 1,4,5-trisphosphate receptors (IP_3R3) by protein kinase A (PKA). The β/γ subunit of gustducin activates phospholipase C-β2 (PLCβ2), which cleaves phosphatidylinositol bisphosphate (PIP_2) into diacylglycerol (DAG) and inositol 1,4,5-trisphosphate (IP_3). IP_3, in turn, acts on IP_3R3 at the membrane of the endoplasmic reticulum (ER), releasing Ca^{2+} from intracellular stores. The rise in intracellular Ca^{2+} concentration leads to the opening of transient receptor potential cation channels (TRPM5), resulting in a Na^+ influx, membrane depolarization, and release of ATP via hemichannels, relaying the signal to other cells and nerve fibers.

depolarization result in the opening of hemichannels and the release of ATP.[14,60,88] This nonvesicular secretion of ATP occurs in receptor cells after intracellular calcium levels increase through pannexin-1[60] or connexin[89] hemichannels. ATP excites ionotropic P2X purinoceptors localized as homo- and heteromers of $P2X_2/P2X_3$ in taste afferent nerves.[88,90] Behavioral and taste nerve responses of genetically modified mice clearly demonstrate the role of both subunits, namely, $P2X_2$ and $P2X_3$, in taste perception, but not in the nontaste-related oral sensations of touch, temperature, or menthol.[88] However, it is unclear how the gustatory system distinguishes the ATP released from bitter-, sweet-, or umami-dedicated cells to correctly identify the quality of the stimulus.[14]

7. Intercellular Communication in Taste Buds

ATP not only transmits the signal from the receptor cells to the gustatory afferent nerves but also gives a positive autocrine feedback to the type II cells themselves, which express metabotropic P2Y purinoreceptors. Thereby, ATP secretion is increased, counteracting ATP degradation by ecto-ATPase.[91] Moreover, ATP also activates presynaptic type III cells via P2Y receptors to stimulate secretion of the neurotransmitters serotonin (5-HT) and norepinephrine.[14,60]

Sour stimulation of taste buds[59] and ATP released from receptor cells[60] result in the release of 5-HT. However, the role of 5-HT in taste signaling is not completely clear. It may excite the afferent taste fibers and exert a paracrine negative feedback on receptor cells, inhibiting Ca^{2+} mobilization and eventually ATP secretion.[14,60,91] A subset of presynaptic cells that secrete 5-HT also cosecrete norepinephrine.[62] Even though the target of norepinephrine remains to be established, both the hormone and receptors are present in taste cells and likely mediate paracrine regulatory effects.[92] Another taste bud neurotransmitter is GABA, which has a paracrine negative feedback on taste cells, directly inhibiting the secretion of ATP.[28,61] The synthesis occurs in type I and III cells, and the receptors are localized in both type II and type III cells.[61] In addition, glutamate functions as a neurotransmitter in taste. It is released by the gustatory fibers and modulates type III cells, which express specific glutamate receptors.[28,32] Thus, it is clear that taste buds are not only unidirectional devices that sense chemical cues at the apical aspect and release neurotransmitters at their basolateral aspects. They also appear to be a complex processing unit in which the incoming chemosensory signals induce chains of internal communication pathways. These include, as we have seen, feedforward and feedback loops, but most of the rules that govern taste bud communication remain to be uncovered.

8. Central Representation of Gustatory Information

Taste signals from taste buds are carried on by gustatory afferent fibers to the central nervous system. Taste buds of the posterior tongue, epiglottis, and esophagus are innervated by the lingual branch of the cranial nerve IX, the glossopharyngeal nerve, and the superior laryngeal branch of the vagus nerve. The anterior two-thirds of the tongue and palate are innervated by the chorda tympani and the greater superior petrosal branches of the facial nerve (cranial nerve VII). The cell bodies of these gustatory afferent fibers are located in three ganglia: the geniculate ganglion which contains the cell bodies of the facial nerve, the petrosal ganglion with cell bodies of the cranial nerve IX, and the nodose ganglion where the somata of the vagal fiber reside. From these ganglia, the nerves project to the gustatory nucleus of the medulla, that is, the rostral part of the nucleus tractus solitarius (rNTS).[15,93–95]

A group of gustatory ganglionic neurons engage primarily in local brain stem projections that participate in the digestive preparation and ingestive reflexes, including oromotor reflexes to swallow or spit food and salivation to prepare the organism for digestion. Other physiological reflexes from the cephalic phase of satiation are transmitted through the vagus nerve to the GI tract.[94–96] In addition, innate preference for or rejection of a tastant is mediated by the brain stem even in the absence of stimulus recognition. This has been observed in forebrain-lesioned experimental animals and humans.[97, 98, 236]

In rodents, the neurons from the rNTS project ipsilaterally to the posteromedial parabrachial nucleus of the pons (PBN). From there, the gustatory information is conveyed along the ventromedial forebrain axis to the central nucleus of the amygdala and lateral hypothalamus. Gustatory neurons in the PBN also project along the thalamocortical pathway to the parvocellular ventroposteromedial nucleus of the thalamus, a typical relay station. From there, gustatory information is conveyed to the gustatory cortex, an area of the agranular insular cortex (IC; Fig. 1). In primates, the rNTS projects directly to the thalamic relay nucleus and the primary gustatory cortex. This thalamocortical branch is involved in stimulus recognition and discrimination of taste qualities. Gustatory information descends from there to the central nucleus of the amygdala, the lateral hypothalamus, and midbrain dopaminergic regions and ascends to the secondary taste cortex. There, gustatory, olfactory, and visual signals are integrated to form flavors and processed together with information from the lateral hypothalamus in the context of the regulation of energy balance (for review, see Refs. 15,93–96).

B. Taste Modalities

1. Sweet and Umami

Sweet and umami taste are both rated as attractive taste modalities. Sweet taste is thought to help in detecting carbohydrate-rich food, thus ensuring the right level of energy intake. The word "umami" originates from the Japanese word for "savory." Umami taste is found in seaweed, tomatoes, meat, and cheese. This taste modality is important for the detection of L-amino acids and ribonucleotides,[99] which are metabolites of protein and nucleic acid degradation and indicate calories in the form of meat. Facial expression studies have revealed that attraction to sweet and umami is innate and universal. Newborns show preference for sweet-tasting substances and attraction to umami-tasting stimuli such as L-glutamate.[9,100] Breast milk contains high levels of L-glutamate, making it a likely candidate attracting newborns to mothers' milk.[101] While sweet taste has long been regarded as a taste modality, umami was accepted as such only after the cloning of an amino acid-specific taste receptor.[99]

These two taste modalities not only share the ability to detect the nutritional value of food, but they are both mediated by heteromeric receptors of the TAS1Rs.[42,102–107] In vertebrates, this receptor family comprises three members that form two heteromeric taste receptors. While TAS1R3 is found as a subunit in both receptors, the sweet taste receptor also includes the TAS1R2 subunit and the umami taste receptor TAS1R1 subunit.[42–44,99,108,109] Consistent with heteromer formation, the *Tas1r* gene was found to be expressed in distinct areas on the tongue in rodents. While *Tas1r1* was found expressed at low levels in rodent vallate and foliate papillae and highly expressed in fungiform taste buds, *Tas1r2* was mainly expressed in vallate and foliate taste buds but almost not detectable in fungiform papillae.[103] TAS1R3 was found to be mainly colocalized with TAS1R1 or TAS1R2 and expressed alone only in a small number of taste cells.[42,104] TAS1R3 and TAS1R2–TAS1R3 knockout mice have diminished if not abolished behavioral and nerve responses to sugars.[43,44,109] Moreover, all tested sweet substances were able to activate the sweet receptor heteromer in functional receptor assays, suggesting that TAS1R2–TAS1R3 is a universal sweet taste receptor (for review, see Ref. 93). Although TAS1R1–TAS1R3 plays a dominant role in umami taste,[99,108] other receptors have also been proposed to contribute.[110–112]

The structure of sweet and umami receptors can be separated into three parts: the heptahelical domain, the cysteine-rich region, and the N-terminal domain with a Venus flytrap motif. The Venus flytrap consists of two globular subdomains, the N-terminal upper lobe and the lower lobe, that are connected by a three-stranded flexible hinge closing over substances that bind its inside.[113] The great number of structurally diverse chemicals that elicit sweet taste and activate the receptor, such as mono- and disaccharides, amino acids, sweet proteins, plant secondary metabolites, various synthetic compounds, and guanidinoacetic acids,[93,114] suggests that they interact with different domains of the receptor complex. In fact, the sweet receptor interacts with sweet molecules using the N-terminal extracellular regions as well as the cysteine-rich and heptahelical domains of TAS1R3[108,115–128] (Fig. 3).

The umami receptor was found to interact with various L-amino acids in rodents, but not with their D-enantiomer counterparts; the human umami receptor binds specifically to L-glutamate.[42,108] In both species, the umami taste is further increased by 5′-ribonucleotides such as inosine-5′-monophosphate and guanosine-5′-monophosphate.[113,130,131] The binding sites of L-glutamate and inosine-5′-monophosphate are thought to be located at different positions of the Venus flytrap of TAS1R1.[108,113]

2. Salty and Sour

Salty and sour tastes are attractive at low and aversive at high concentrations.[132] Salty taste is necessary to drive intake of NaCl and other minerals in order to maintain electrolyte homeostasis; however, high salt concentrations

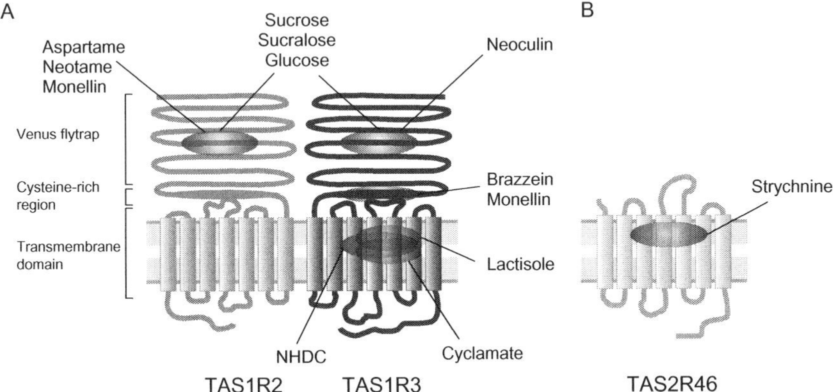

FIG. 3. Binding sites of sweet receptor and bitter receptor. (A) The heteromeric sweet receptor comprises the receptor subunits TAS1R2 and TAS1R3. Binding sites of the human sweet receptor are found in the Venus flytrap of TAS1R2 and TAS1R3, the cysteine-rich region, and the transmembrane domain of TAS1R3.[93] While aspartame, neotame, and monellin bind to the Venus flytrap of TAS1R2, neoculin binds to the Venus flytrap of TAS1R3. Some sweeteners such as sucrose, sucralose, and glucose interact with both Venus flytraps. Brazzein and monellin bind to the cysteine-rich region of TAS1R3. Neohesperidin dihydrochalcone (NHDC), cyclamate, and the sweet-inhibitor lactisole interact with different sites of the transmembrane domain of TAS1R3. (B) While the sweet receptor possesses several binding sites, bitter receptors seem to possess only one, here shown for the bitter receptor TAS2R46, in which the transmembrane domain interacts with strychnine.[129]

are rejected to avoid hypersalinity.[132–134] The main function of sour taste is to prevent ingestion of irritant concentrations of acids in spoiled food or unripe fruits, but it also promotes intake to balance concentrations of acid and bases in the body.[132,135,136] So far, the molecular identity of involved taste receptors is only partially uncovered, but it is likely that ion channel-mediated currents induce taste receptor cell depolarization.[14,137]

Salt taste in rodents is divided into two pathways: one is stimulated by sodium and sensitive to the diuretic drug amiloride and the other one is activated by various cations including Na^+ and modulated by cetylpyridiniumchloride.[138–141] On the tongue, these transduction pathways are segregated topographically. The anterior part of the tongue is amiloride-sensitive whereas the posterior part is not. For the latter, a transduction pathway has not been molecularly identified.[142–146] Therefore, further analyses are necessary to elucidate the mechanism of amiloride-insensitive salt taste transduction.

Amiloride is an effective inhibitor of the epithelial sodium channel (ENaC) composed of $\alpha_2/\beta/\gamma$ subunits. In rodents, ENaC was suggested to mediate the amiloride-sensitive salt taste.[140,147–149] This was recently confirmed by Chandrashekar and colleagues, who demonstrated that the α subunit is essential for amiloride-sensitive salt taste transduction.[150] Taste tissue-specific knockout animals lack attraction to NaCl solutions and amiloride-sensitive chorda tympani nerve responses. However, the precise subunit composition of the salt taste receptor and the host cell type has not been established.[150] An additional ENaC subunit, δ-ENaC, is present in humans and shows less pronounced amiloride sensitivity,[151,152] provoking research to confirm or reject its partial role as salt taste receptor.[153] Intriguingly, human salt taste is much less sensitive to amiloride compared to rodents.[154]

Even less is known about the molecular identity of sour taste receptors. However, several candidates have been proposed, including acid-sensing ion channels,[155] hyperpolarization-activated ion channels,[156] two-pore domain K^+ channels,[157,158] transient receptor potential cation channel subfamily P members PKD2L1/PKD1L3,[56,159,160] and apical proton conductance-mediating molecules.[161,162] Among these, an interesting molecule is the heteromeric ion channel PKD2L1/PKD1L3. Genetically engineered mice with ablated *Pkd2l1*-expressing cells were devoid of chorda tympani nerve responses to sour stimuli, while response to all other taste qualities remained unaffected. This reveals that *Pkd2l1*-expressing cells mediate sour taste.[56] The potential importance of PKDL1 is confirmed by a study that found no PKD sequences in sour-aguesic subjects.[163] Interestingly, loss of *Pkd2l1* or *Pkd1l3* genes in mice had little effect on acid-induced gustatory nerve responses,[164,165] suggesting that acid sensing is not dependent on PKD2L1 and that other transduction mechanisms exist.

Recently, Chang and colleagues identified an apical proton conductance that was activated by extracellular stimulation of vallate PKD2L1-positive taste cells with protons and led to membrane depolarization.[161] Another group also found evidence for participation of proton conductance in sour taste. DeSimone *et al.* observed that chorda tympani nerve responses to stimulation of oral taste buds with acids involve both NADPH-dependent and cAMP–PKA-sensitive proton channels.[162] Further investigations are needed to reveal the molecular identity of these acid sensors in taste tissue and confirm their role in sour taste.

3. Bitter

Bitter taste functions as a biological warning system leading to rejection of poisonous food. This hypothesis is based on the observation that toxic compounds often elicit bitter taste.[166–168] The aversion is innate and already evident in newborns, who demonstrate negative gustofacial reflexes.[8] However, there are also bitter-tasting foods that are accepted and even enjoyed by humans. Based on everyday experience, ingestion of bitter-tasting coffee, chocolate, or beer is harmless and sometimes preferred because of their waking-up, mood-lightening, and exhilarating effects, respectively. Thus, bitter taste acts as a defense system, but in low concentrations, bitter substances are sometimes sought and preferentially ingested.[167,168]

Bitter tastants include numerous structurally diverse substances, such as terpenoids, glycosides, alkaloids, amino acids, fatty acids, urea, phenols, amides, amines, and metal ions.[166,169,170] This multitude of substances is recognized by the TAS2Rs, with ~25 members in humans.[171–173] Cell-based receptor assays have revealed that the human TAS2Rs comprise three generalists that detect numerous compounds and possess extensively overlapping agonist spectra. Moreover, there are eight TAS2Rs with intermediate tuning breadth that still show overlap regarding their cognate bitter compounds. In addition, two TAS2Rs detect specific classes of bitter compounds with many representatives, and eight specialist receptors respond to very few or single chemicals. Bitter compounds for four TAS2Rs remain unknown.[137,173–187]

Given the fact that bitter receptors are combinatorially activated by various agonists, recent studies focused on the question of where the receptor–bitter compound interaction takes place. Computational receptor modeling[188,189] in combination with functional analyses revealed prominent ligand-binding sites in the upper half of transmembrane helices of the analyzed receptors.[129,190,191] However, bitter ligands have to access their binding sites through extracellular loops or lateral diffusion. Therefore, the extracellular loops may be indirectly involved in receptor activation.[129,178,186] This raises the question of whether bitter receptors possess several binding pockets, like the sweet taste receptor, or a single binding site.[93,137,167,185] Recently, Brockhoff *et al.* showed that TAS2R31, TAS2R43, and TAS2R46 accommodate their structurally different cognate compounds at one binding site contacting largely overlapping sets of amino acids of the receptors[129,186] (Fig. 3).

4. Fatty

On account of the importance of fats as energy sources, their presence in food needs to be detected to favor their ingestion. For years, this oral detection has been attributed to trigeminal,[192] olfactory,[193] and postingestive cues,[194] but, recently, evidence of gustatory cues has been uncovered.[195–200] Anosmic rats prefer free fatty acids over their triglyceride or vehicle.[197,200] Sham-feeding rats are able to recognize and prefer triglyceride-rich corn oil to nonnutritive mineral oil[201] in all cases despite the similarities in texture. In addition, bilateral sectioning of the gustatory nerves impairs the detection of fatty acid solutions.[202]

The main component of dietary lipids from vegetable and animal sources is triacylglycerides[203]; however, they do not seem to be the primary stimuli for fat taste. It has been reported that only free fatty acids and not triglycerides (or any derivative) activate the proposed fat taste candidates.[83,204] Furthermore, after inactivation of the lingual lipase, which is the enzyme responsible for hydrolyzing triglycerides to split them into their basic components, the oral fat perception in rats is blocked, indicating that this enzyme might play a role in providing free fatty acids to the oral fat receptors.[205]

Fat perception in humans occurs in a multimodal way, including gustatory mechanisms.[195,206] This perception occurs in the millimolar range of free fatty acids.[83,195,206–209] Detection of fatty acids seems not to depend on chain length,[207,209] degree of saturation,[195,206,209] or site of application on the tongue.[208] In contrast, the recognition of fatty acids is related to both chain length and degree of saturation, because the longer and the more unsaturated, the lower the recognition threshold concentration.[83]

Fat receptor candidates in rodents and in humans have been shown to be expressed in taste epithelium and to be responsible for oral fat taste detection. The fatty acid translocase CD36 is expressed in human[210] and rat apical membrane of the circumvallate and foliate papillae,[199,211] c with α-gustducin.[199] This transporter has been directly related to taste perception,[202] and its ablation in mice results in suppression of the spontaneous preference for long-chained fatty acids, as well as their postingestive effects.[199] Fatty acids applied on the tongue are able to inhibit delayed-rectifying potassium voltage-gated channels.[212] From this family, KCNA5 is the major channel expressed in rat lingual epithelium.[213] Two GPRs, GPR40[214] and GPR120,[215] have been deorphanized, binding lipids in the length range of dietary fatty acids. GPR120 is expressed in rodents[196,216,217] and human taste epithelium.[83] In mice, it is expressed mainly in type II cells of foliate and vallate papillae,[196,216,217] colocalizing with the specific markers PLCβ2, α-gustducin,[217] and TRPM5.[196] In humans, the expression of GPR120 is not restricted to taste cells but is extended to the surrounding nongustatory epithelium.[83] GPR40 has been colocalized in mice with the type I cell marker GLAST,[196] but it is not detected either in rats[216] or in humans.[83] Both GPRs have been directly related to oral fat taste perception, because their absence reduces both lipid preference and taste nerve response.[196]

The taste of fatty acids might represent the sixth taste quality,[195,197,198,200] and depending on the concentration, it can be an attractive or aversive taste stimulus or a mixture of both.[83,218] On the other hand, fatty acids might act as a modulator on other taste qualities.[212,219,220] More research is needed to determine the true role of fatty acids in taste and their putative receptors.

III. Development of Taste Preferences

A. Conditioned Taste Aversion and Preference

It is essential for all animals to discriminate between safe and toxic substances in order to survive.[221,222] While some taste preferences and aversions may be innate, such as the liking of sugars and disliking of bitter substances, others have to be obtained through the formation of taste memory.[223] Innate taste preference can be further modulated through exposure during peri- and postpartum development (for review, see Refs. 225,226). Memories of how previously encountered foods taste provide the means to distinguish food as "safe" or "dangerous or poisonous."

During the first encounter with a tastant, an animal consumes a small amount of the new substance (neophobic response), which reduces the risk of fatal effects if the meal is toxic. Each time a new substance is eaten, a memory of the tastant is formed. The taste becomes recognized as either safe if the consumption leads to satiety and lacks negative effect or as dangerous or poisonous if the consumption is followed by malaise.[224] These forms of recognition memory are referred to as conditioned taste attraction and conditioned taste aversion.[222,226] The neophobic response in mammals seems to be developed some time after birth. Rat pups, for example, show no neophobic behavior until 10 days after birth and are able to develop taste aversion only 11 days after birth.[225] Interestingly, this occurs approximately at the time they start eating solid food (around day 14) and are more likely to be exposed to toxic substances.

To test how conditioned taste aversion is induced, an unconditioned stimulus (e.g., an injected substance or exposure to radiation) causing illness is presented following the taste stimulus (conditioned stimulus) (e.g., Ref. 228). Thereby, a formally positive or neutral taste stimulus can be overwritten and is avoided in the future. Drugs used as unconditioned stimuli have included amphetamine, apomorphine, and nicotine, which directly act on the brain and lithium chloride which acts through activation of the vagal and splanchnic afferent nerves.[228–230] Thus, the formation of a conditioned taste aversion is induced only if the unconditioned stimulus directly affects the GI tract or its innervating fibers.[231–233]

Avoidance of a conditioned taste may manifest in mimetic signs by the animal following forced consumption.[226] These mimetic signs are caused by oralpharyngeal reflexes involved in motor responses such as feeding, chewing, swallowing, and respiration.[234] Oralpharyngeal reflexes are induced through changes in touch/pressure and temperature, water, and chemical stimuli.[234] They are partly evident in the mammalian neonate and gradually emerge as the animal or human matures.[235,236] While the tongue is important for mixing, holding, and transporting food to the pharynx, the pharynx transfers food to the stomach, as well as gulps of air to and from the lung, and the gag reflex is important to protect unwanted material from entering the pharynx.[237] Conditioned taste aversion caused by an unconditioned stimulus can also be found in humans and may affect diet as described by the American psychologist Martin Seligman (the Sauce-Béarnaise Syndrome). After eating Béarnaise sauce for the first time, Seligman suffered an intestinal infection.[238] This incidence conditioned him to avoid the sauce from then on, even though the infection had not been caused by the meal but by a virus.

While a new taste in combination with a negative unconditioned stimulus leads to the formation of a conditioned taste aversion, this effect is attenuated if a taste is familiar before being coupled with a malaise.[239–241] Another model frequently used in the study of safe taste memory is latent inhibition, defined as a delay of acquisition of a taste aversion to a conditioned stimulus that was previously known but not relevant.[242] Another phenomenon is the so-called blocking, in which a previously acquired aversion to a taste A hinders the formation of an aversion to taste B when both are presented together before the unconditioned stimulus.[243,244] In addition, the timing of the presentation of an unconditioned stimulus influences the likelihood of the acquisition of conditioned taste aversion. While memory modification through a new taste stimulus can take place a few minutes to a few hours after the presentation, the probability that new information will transform the taste representation from safe to aversive decreases as time passes after the presentation of the taste.[245–247] Thus, if there are 2–4 h of continuous time during which the noxious consequences of food ingestion are absent, a gustatory stimulus is classified as safe or neutral.[246–248]

Conditioned taste preference, the opposite of conditioned taste aversion, forms if the food is, for example, of high nutritional value.[246] This process can be differentiated into conditioned satiety, conditioned flavor preference, and conditioned flavor acceptance.[246] Conditioned satiety was first described by Le Magnen in 1955 and further defined by Booth as a change in the length of feeding phase and the strength of feeding inhibition dependent on the concentration of nutrients in a solution.[247] Meals with a higher nutritional value lead to a more pronounced inhibition, keeping the total level of nutrient intake more or less constant. Conditioned satiety is not attributed to the development of differences in the initial rate of feeding or preference for one diet over the other in two-stimulus tests, as found for conditioned taste aversion.[246] Only if fed with foods tasting like those they had been fed before did rats show feeding inhibition typical for conditioned satiety after sham feeding.[248] This finding indicates that the elevated consumption in sham feeding is due to the extinction of a learned inhibitory control of ingestion.[248]

Holman was the first to report nutrient-conditioned flavor preference.[249] Since then, various experiments have shown that flavor preferences can be induced by presentation of the flavor paired with intragastric infusions of complete diet or individual macronutrients in deprived or nondeprived animals.[250] Even normally avoided flavors can be conditioned to be preferred. Drucker *et al.* trained rats with bitter solutions paired with intragastric infusions, which led to the preference of the bitter solution over plain water.[251] Furthermore, if the absolute intake of a conditioned flavored solution is increased compared to another flavored solution in one-bottle tests, not showing a preference for one solution over the other, then this increased acceptance is described as conditioned flavor acceptance.[246] The pairing of a flavor with an intragastric infusion of carbohydrates in rats may lead to an increase in consumption of the flavored solution, but the strength of the effect seems to depend on at least three different factors: degree of aversiveness, stimulus intensity, and the innate predisposition to acquire certain stimulus–response relationships more readily than others.[252] Increased acceptance of the flavored solution is seen as conditioned, as it is still observed in the absence of the nutrient infusions.[252] While this observation can be made in some cases, the opposite, namely, a decrease in intake, may also be observed. It seems that nutrient concentration plays a role, but the underlying mechanisms are unclear.[246]

B. Formation of Taste Memory

1. Brain Areas Involved in Taste Memory Formation

The formation of taste memory starts with the detection of a taste on the tongue, which is the source of information delivered to brain areas involved in gustatory processing. Different parts of the taste neuronal network are important for different forms of taste memory. Lesion studies revealed that the rNTS is needed for the aversive or attractive response to sapid stimuli and for innate gustatory preference or aversion, but is not required for learned taste aversion.[253] On the contrary, the IC is important for taste memory formation but not necessary for taste responsiveness.[254,255] If food intake results in a positive effect such as satiety, then information is transferred from the GI tract via the vagus nerve (X) or via blood circulation to the taste system and from there to the neuronal pathway of the brain reward system, which drives the intake of food (for review, see Refs. 257–259) (Fig. 4). This pathway is mainly formed by the ventral tegmental area of the midbrain (the origin of the mesolimbic dopamine system), the nucleus accumbens of the ventral forebrain which is an essential interface between motivation and action, and the ventral pallidum from where information is sent to the lateral hypothalamus. If, on the contrary, ingestion is followed by malaise, the information leads to the activation of brain areas linked to anxiety and psychological stress, such as the supramammillary nucleus and thalamic paraventricular nucleus, as well as the circuit formed by the nucleus accumbens, the ventral pallidum, and the lateral hypothalamus (for review, see Ref. 260) (Fig. 4). These areas also seem to be involved in the acquisition and retrieval of a conditioned taste aversion.

2. Neurochemistry of Safe Taste Memory

The memory of attractive or safe taste is stored in the brain as safe taste memory trace. Release of the neurotransmitter acetylcholine seems to play a modulating role in the plastic modifications involved in the duration of stimulus representation.[261,262] It has been shown that the release of acetylcholine in the

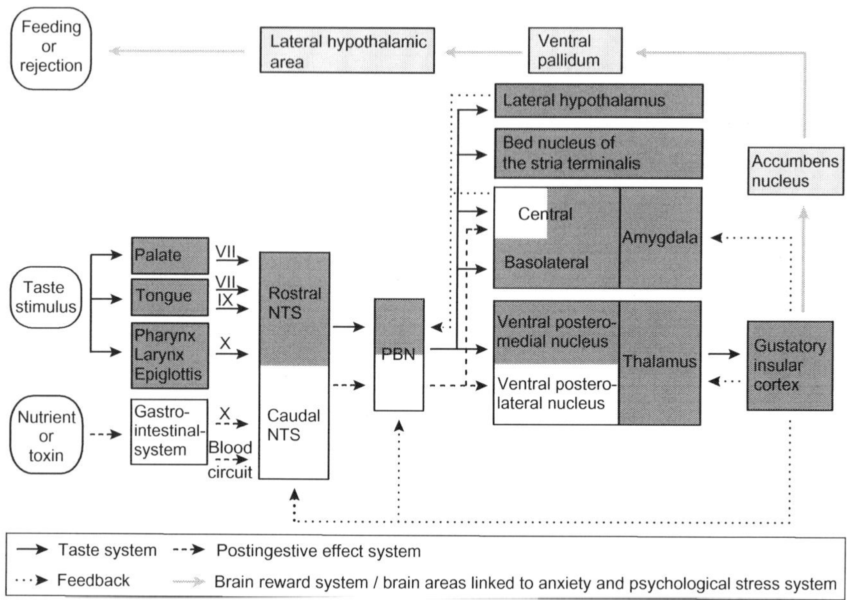

FIG. 4. Murine brain pathways of conditioned taste aversion/attraction. Taste information (conditioned stimulus) and unconditioned stimulus are both transferred from the GI tract to the taste-processing regions of the brain after food ingestion. Depending on the postingestive effect, the neuronal pathway of the brain reward system (safe), which drives food intake, or the areas of fear learning (dangerous or poisonous) will be activated.[224,259,260]

IC decreases following a stimulus if that stimulus is presented regularly. The release decreases with each presentation of the stimulus, finally reaching the level seen when presenting water.[263] Cholinergic activity seems to signal the novelty of a stimulus during the initial stages of memory formation.[264,265] Further, it has been shown that muscarinic acetylcholine receptors in the gustatory IC are important for taste memory formation.[266] The muscarinic pathway is further linked to plasticity stimulating activity of the extracellular signal-regulated kinase pathway.[267,268] Long-lasting activation of this pathway is seen for new but not for familiar tastes.[269–271]

Furthermore, it is well known that dopamine affects feeding behavior (for review, see Ref. 273). The dopamine[273] and the opioid systems[274] participate in flavor preference learning involving the PBN and the lateral hypothalamic area.[275,276] Opioid signaling in the nucleus accumbens strongly modulates flavor-based food choice.[277] Release of opioids in the nucleus accumbens during consumption of palatable foods produces a selective and transient increase in preference for a recently sampled flavor.[277]

The GABAergic system in the deep layers of the gustatory cortex also plays a role in the acquisition of a new taste and is involved in the offline processing and consolidation of taste information.[278] Still, much remains unknown regarding the formation of safe taste memory trace and the link between the taste and brain reward system, particularly at the cellular and molecular level.

3. Neurochemistry of Aversive Taste Memory

For the more extensively studied aversive taste memory trace, it is important that a taste is associated with the visceral consequences of an unconditioned stimulus (malaise).[224] In contrast to safe taste memory trace, acetylcholine does not seem to be involved in the association of taste with malaise or the consolidation of aversive taste memory trace.[224] Furthermore, disruption of the IC or nucleus basalis magnocellularis affects the acquisition but not the retrieval of the conditioned taste aversion.[279,280] In addition, the presentation of a conditioned aversive taste is followed by an increase of acetylcholine in the nucleus accumbens and the cortical gustatory area, which is thought to be related to behavioral expression to aversive taste stimuli.[281,282] The results indicate that muscarinic receptors of the gustatory IC may be involved in the early stages of taste memory trace formation. Cholinergic activity in the cortical gustatory area is supported by nerve growth factors.[279]

While cholinergic activity seems to be involved in taste memory trace formation, glutamatergic activity seems to play a role in the transformation of a taste memory trace from safe to aversive during memory consolidation.[224] Offline concomitant release of dopamine and glutamate within the IC is observed during consolidation of a conditioned taste aversion.[283] Blockade of dopamine D1 and/or *N*-methyl D-aspartate (NMDA) receptors before the offline activity impairs long- but not short-term memory.[283] In contrast, amphetamine facilitates the learning by strengthening the consolidation of short-term memory via D1 receptors in the shell region of the nucleus accumbens.[284] The formation of conditioned taste aversion further involves the sustained tyrosine phosphorylation of the NMDA receptor subunit NR2B in the IC,[285,286] while the presentation of a new taste is followed by tyrosine and serine phosphorylation of the NR2A and NR2B subunits.[287] After several presentations of the same taste stimulus, the serine phosphorylation of the NR2A and NR2B subunits is significantly reduced.[287] However, the exact role of the NR2A and NR2B subunit phosphorylation in the formation of conditioned taste aversion remains unclear.[224] Three types of glutamate receptors (NMDA, α-amino-3-hydroxy-5-methyl-4-isoxazolepropionic acid receptor [AMPA], and metabotropic glutamate receptors [mGluR]) are thought to play a role in taste aversion acquisition in the amygdala, with AMPA being the only one that is also important for recalling the conditioned taste aversion memory.[288,289]

Furthermore, protein synthesis of PKC, PKA, PKC isoform M zeta, and the cAMP response element-binding protein seems to be involved in long-term but not in short-term aversive taste memory trace.[290–295] While consolidation depends on protein synthesis in the central nucleus of the amygdala, reconsolidation does not seem to depend on protein synthesis in either the central nucleus or the basolateral nuclei of the amygdala.[296]

It is suggested that unconditioned stimulus actions induced by psychoactive drugs may be related to the central catecholamine system (e.g., Refs. 298,299). Deletion of noradrenaline in the amygdala abolishes a conditioned taste aversion acquisition while it has no effect on feeding preference of novel and familiar food.[299] On the contrary, 5-HT deletion in the amygdala neither changed food preferences nor impaired conditioned taste aversion formation.[299] This suggests that only noradrenaline is involved in conditioned taste aversion formation.

4. Taste Memory Extinction

Once an association between a conditioned stimulus and an unconditioned stimulus is developed, the conditioned response will diminish if it is present in the absence of the unconditioned stimulus.[224] This phenomenon is called extinction.[300] It is thought to be a form of learning in which the animal no longer associates the conditioned stimulus as being linked to the unconditioned stimulus, but instead to the absence of malaise.[301,302] As in the formation of other taste memory traces, protein synthesis in the basolateral nuclei of the amygdala, as well as the GABA type B receptor subunit 1 and β-adrenergic receptors, is important.[296,303–305] The process thus appears to be independent of muscarinic receptors and mitogen-activated protein kinase.[303] Further studies are needed to reveal how this form of memory formation equals or differs from the anatomical, cellular, and molecular basis of safe and aversive taste memory.

C. Postingestive Effects

The GI tract has to fulfill at least two major functions: protecting the body from harmful and toxic substances and digesting and absorbing nutrients. There are two ways a substance can still be rejected after entering the mouth and GI tract. First, the process of swallowing can be stopped through nausea and other oral and pharyngeal reflexes.[234,260] Second, nausea, diarrhea, or vomiting can occur after the substance reaches the GI tract (for review, see Ref. 307). The presence of nutrients in the GI tract, on the other hand, will lead to the secretion of peptide hormones that induce satiation or affect glycemic control, influencing motility, digestion, and absorption.[307] Furthermore, the vagus nerve pathways are involved in nutrient reflexes followed by control of glycemia, food intake, and satiety (for review, see Ref. 308). One response mediated by vagal motor neurons is the suppression of antral contractions and the stimulation of tonic and phasic pyloric pressures, which slows down the gastric emptying rate to match the absorption capacity of the small intestine.[308–310] Two classes of primary afferent neurons innervate the GI tract[311]: intrinsic primary afferent neurons innervate the mucosa and the muscular layers of the gut and are part of the enteric nervous system which induces reflexes affecting GI motility, blood flow, and water/electrolyte secretion[312,313]; extrinsic primary afferent neurons (vagal and spinal) innervate the entire length of the GI tract (for review, see Ref. 307). There are at least two possibilities for stimulus transfer from the GI tract to areas of taste memory in the brain: (1) direct irritation of the GI tract transmitted by the vagus nerve and (2) malaise-inducing substances transferred into the bloodstream and traveling to the area postrema, which is outside the blood–brain barrier.[314–316] Both pathways affect the caudal two-thirds of the NTS.[317,318]

D. Extraoral Taste Receptors and Metabolic Control

Taste receptors (sweet, bitter, and umami receptors) and other molecules of the taste-signaling cascade (α-gustucin, α-transducin, TRPM5) associated with taste perception on the tongue are also found in the GI tract, where they seem to take part in nutrient chemosensing (for review, see Refs. 24,320). Some are expressed in enteroendocrine cells which are found throughout the GI tract and which secrete hormones and signaling molecules involved in secretory production, motility, blood flow, and satiety.[320] These cells can detect the chemical contents of the GI tract.

The sweet taste receptor is found expressed in the duodenum and stomach.[321,322] It may be functionally involved in glucose-stimulated secretion of glucagon-like peptide 1 (GLP-1), gastric inhibitory polypeptide, and peptide YY,[182,323,324] and its expression is under regulation by dynamic metabolic and luminal control.[325] Current data suggest that there are additional nontaste chemosensory mechanisms involved in binding nutrient molecules and modulating the secretion of gut peptides.[326–330] Not all substances that evoke sweet taste also induce the release of gut peptides. Glucose seems to be the main agonist with this function. The sweet taste receptor may further play a role in mechanisms of intestinal glucose absorption into enterocytes.[324]

Bitter receptors have also been found expressed in the GI tract and are suggested to be involved in detection of toxic substances and in induction of behavioral and physiological changes leading to removal of the substance from the GI tract and prevention of its absorption.[330] This is underlined by the finding that bitter substances induce calcium signaling and cholecystokinin (CCK) release and that denatonium, a bitter substance, induces delay in gastric emptying.[331,332]

The umami receptor, as well as other L-glutamate-activated receptors, is found in GI tissue.[333] Gastric chief cells are found to express mGluR1, which could be responsible for the release of pepsinogen in the presence of

L-glutamate.[334,335] Secretion via goblet cells is thought to be regulated by mGluR4 and the extracellular calcium-sensing receptor (CasR).[336] Glutamate is a versatile amino acid, as it can evoke umami taste, regulate gastric acid, induce mucous and bicarbonate secretion, affect intercellular pH, and influence the speed of gastric emptying upon a rich protein meal, among other functions.[337]

Fat sensors are also found expressed in the GI tract (e.g., Refs. 215,338–344). GPR40 is thought to be involved in mediating incretin secretion through interaction with long-chain free fatty acids.[338] GPR41 and -43 were found expressed in endocrine cells, which produce peptide YY, and mucosal mast cells containing 5-HT.[341] Probably, the strongest evidence so far for lipid sensing in the intestine by GPRs is through GPR120. GPR120 is expressed throughout the intestine and in stanniocalcin-1 cells and is shown to promote the secretion of GLP-1 and CCK.[215,338,342,343,345]

G-cells that secrete the hormone gastrin also express CasR, which seems to be important in modulating taste perception in taste cells.[346–349] The release of gastrin, which induces acid secretion, is regulated through calcium and calcium receptor agonist.[350] In addition to neuroendocrine cells, gastric parietal cells are also involved in the detection of chemical components in the lumen. Gastric parietal cells can be activated by L-amino acids through CasR.[351]

Other components of the taste-signaling cascade found in the GI tract include α-gustducin, which was found in the duodenum, stomach, and colon[323,352] and expressed in enteroendocrine cells and intestinal brush cells.[353–356]

While most, if not all, taste-signaling components seem to be expressed in the GI tract, colocalization studies have shown that the components are rarely expressed in the same cells (for review, see Ref. 320). Still, much remains unknown about the role of taste receptors and other molecules of the taste-signaling cascade in postingestive processes.

Not only are components of the taste-signaling cascade found to be expressed in the GI tract, but hormones known to be involved in hunger and satiety seem to also play a role at the level of taste perception on the tongue. Receptors for several hormones, such as CCK, neuropeptide Y, vasoactive intestinal peptide, galanin, glucagon, and leptin, are found expressed in taste receptor cells.[357–362] Furthermore, GLP-1 receptor is expressed in intragemmal fibers of the afferent taste nerves.[363] All the hormones mentioned here besides leptin have been found to be produced in taste receptor cells.[359–363] These findings suggest that nutritional needs or metabolic status may modulate the mechanism of taste perception on the tongue to match the body's needs. It has been shown that glucagon, which is known to play a role in glucose homeostasis, may enhance responsiveness in sweet perception.[357] Similarly, it was found that endocannabinoids increase the response to sweeteners.[363]

In contrast, leptin, which signals satiety, suppresses the responsiveness to sweet stimuli.[363,364] This field of research in taste perception may turn out to be very important in furthering the understanding of links between taste perception, nutrition, and food preference.

IV. Genetic Variability of Taste Perception

It has been 80 years since we recognized that taste perception is influenced by heritable factors.[365,366] Based on this finding, it is conceivable that genetic variations in taste receptor genes are involved. Indeed, single-nucleotide polymorphisms (SNPs), copy number variations, and insertions/deletions have been observed.[181,367–370]

The best known genetically encoded taste difference relates to the human bitter taste receptor TAS2R38 with its agonists phenylthiocarbamide (PTC) and propylthiouracil (PROP). The ability to perceive PTC and PROP as bitter is linked to three SNPs resulting in different amino acid residues, namely, amino acids PAV at positions 49, 262, and 296, representing the taster allele present in ~65% of the population. The remaining 35% of subjects exhibit the AVI-nontaster variant.[175,371,372] These findings were confirmed by functional expression of the affected receptor, in which micromolar concentrations activated the PAV, but not the AVI variant of *TAS2R38*.[179] Besides PTC and PROP, other thiocyanate moieties (N—C═S) containing natural and synthetic compound activate the PAV-*TAS2R38* variant *in vitro*.[179,186,187] These isothiocyanates are naturally present in Brassicaceae vegetables such as broccoli or Brussels sprouts. PAV-taster individuals perceive these vegetables as more bitter than do nontasters.[12,187] Sensitivity to PROP and similar compounds has been associated with food choice and taste preferences.[373]

Interestingly, the AVI-nontaster variant of the *TAS2R38* gene persists in the human genome. This may be explained by balanced natural selection.[374] Another explanation would be that the AVI-variant detects other yet unknown bitter-tasting toxic compounds.[374,375] Moreover, ingestion of Brassicaceae plant-derived isothiocyanates, detected by PAV-*TAS2R38*, is associated with reduced cancer risk[376]; AVI carriers may demonstrate less rejection of these bitter foods, thereby promoting maintenance of the nonfunctional gene. On the other hand, excessive consumption of isothiocyanates in geographical regions with low iodine is associated with thyroid disease and goiter.[187,377] Thus *TAS2R38* variability is evolutionarily maintained in the human genome.[375] Intriguingly, chimps also possess functional and nonfunctional *TAS2R38* alleles. However, the underlying polymorphisms differ from those in humans, confirming the necessity of taking into account variability in tasting thioamides and isothiocyanates.[378]

Besides *TAS2R38*, other human bitter taste receptor genes contain variations that have been correlated with perceptual differences. In these genes, SNPs are mostly located within the coding region and are responsible for altered bitter taste sensitivity.[13,181,182,379] For example, one SNP affects the perception of ofloxacin in *TAS2R9*[182]; polymorphisms in *TAS2R43* alter the perception of aloin, aristolochic acid, saccharin, and acesulfame K.[177,178,181,186] The bitterness of saccharin and acesulfame K is linked not only to *TAS2R43* but also to its paralog *TAS2R31*.[181,370]

In addition, genetic variations are found in taste receptor genes of the *TAS1R* family, with most variations present in *TAS1R2*, fewer in *TAS1R1*, and least in *TAS1R3*.[380] These variants occurred under positive evolutionary selection and thus might account for variable taste perception.[380] Indeed, SNPs of *TAS1R1* are associated with impaired glutamate sensitivity in humans and in functional assays.[381,382] Nonsynonymous SNPs in *TAS1R3* transmembrane domains have been related to interindividual differences in glutamate perception.[381–385]

Furthermore, two independent allelic polymorphisms in the promoter region of *TAS1R3* are linked to human sucrose sensitivity.[386] Interestingly, the more sensitive alleles prevail in all but African populations. Researchers have speculated that the more sensitive version evolved to detect even small amounts of sugar in cold climates. This is less important in warm climates, where carbohydrate-rich plants are abundant.[386,387] Moreover, polymorphisms of α-gustducin, which are believed to affect α-gustducin transcript levels, may explain 13% of variation in sucrose perception.[388]

Another form of genetically determined taste variation is the loss of functional taste receptors. This is demonstrated in obligate carnivores of the Felidae family (domestic and predatory cats), which are indifferent to sweets.[389,390] Here, microdeletion and premature stop codons lead to pseudogenization of *TAS1R2* and subsequent loss of functional sweet taste receptors.[389] However, whether this molecular change is the cause or the consequence of the cats' obligate carnivore diet is not known.

A clear link between diet and genotype is also seen in bats. Because of mutations disrupting the open reading frame, vampire bats lack a functional *TAS1R2* subunit and are unable to recognize sweet tastants, while fruit-eating bats exhibit a functional sweet taste *TAS1R2* subunit.[391,392] Another interesting example is the giant panda, which is phylogenetically classified as carnivore, although its nutrition is 99% plant derived.[369] Sequencing of the giant panda genome revealed that the specific umami taste subunit *TAS1R1* is pseudogenized, which might explain pandas' preference for bamboo compared to meat.[393] Again, dietary reasons are probably not the general reason for the giant pandas' herbivorous way of life, as other herbivores like cows and horses posses an intact *TAS1R1* subunit.[369]

Altogether, this evidence indicates that taste complexity is ruled by interspecies and interindividual variations, and that it might involve more than a one-to-one association between phenotype and sequence variation within a single receptor gene. Whether the diet determines the evolution of the taste alleles or, on the contrary, whether the presence of an allele determines the diet is not known. More research is necessary to correlate taste variation and diet formation.

V. Conclusions and Outlook

Taste preferences influencing food choice vary among individuals, depending on many factors such as culture, learning experiences, and genetics. The first level of information determining whether to consume or avoid a food item is transmitted via sensory stimuli. Vision, olfaction, and taste, as well as perception of temperature or texture, provide this primary information.[7,9] Later on, post-ingestive effects like satiety and memory affect appetite and the craving for a certain tastant.[2,224] In this review, we discussed the role of taste in the development of taste preferences, which eventually determine food choice and diet.

Although much progress has been made in the past few years, many questions are still open. Further investigations are necessary to reveal the molecular identity of sour and fat sensors. In addition, the molecular characterization of taste receptors and taste transduction molecules can no longer focus on taste exclusively because they are also present in the GI tract.[24,323,352–356] However, the metabolic relevance of the GI tract in the taste-signaling cascade needs to be proven. Vice versa, metabolic responses appear to regulate the taste system at the level of the tongue via hunger and satiety hormones, which have receptors on taste cells. This cross talk could adjust taste sensitivity to metabolic needs,[357–362] but the underlying basic principles are still poorly understood.

The process of information transmission between taste bud cells and gustatory fibers has not been described in sufficient detail. It is necessary to elucidate the complete role of ATP and other neurotransmitters, as well as the identity of specific nerve fibers that contact the taste buds, to fully understand the mechanism of taste transduction. Such studies may result in the understanding of the broad tuning of the central gustatory neurons compared with the narrow tuning of the chemosensory cells.[14,67,103] Moreover, although the representation of taste has been examined at an anatomical level, the neuronal circuits that mediate the various functions of taste are only marginally described.

Genetic differences such as polymorphisms in taste receptors have been linked to differences in individual taste perception[12,13,175,181,182,371,372,386]; however, how exactly they influence taste preferences is not completely known. Therefore, the elucidation of the genetic influences on differences of alimentary preferences would have consequences for the understanding of interethnic and interindividual ingestive behavior and the correlation of taste variation and diet formation, as well as associated pathologies.

References

1. Small CJ, Bloom SR. Gut hormones and the control of appetite. *Trends Endocrinol Metab* 2004;**15**:259–63.
2. Wynne K, Stanley S, McGowan B, Bloom S. Appetite control. *J Endocrinol* 2005;**184**: 291–318.
3. Grimm ER, Steinle NI. Genetics of eating behavior: established and emerging concepts. *Nutr Rev* 2011;**69**:52–60.
4. Larson N, Story M. A review of environmental influences on food choices. *Ann Behav Med* 2009;**38**(Suppl. 1):S56–73.
5. Meyer-Rochow VB. Food taboos: their origins and purposes. *J Ethnobiol Ethnomed* 2009;**5**: 18.
6. Krebs JR. The gourmet ape: evolution and human food preferences. *Am J Clin Nutr* 2009;**90**: 707S–11S.
7. Mattes RD. Nutritional implications of taste and smell. In: Doty RL, editor. *Handbook of olfaction and gustation*. USA: Marcel Dekker, Inc.; 2003. pp. 881–903.
8. Steiner JE, Glaser D, Hawilo ME, Berridge KC. Comparative expression of hedonic impact: affective reactions to taste by human infants and other primates. *Neurosci Biobehav Rev* 2001;**25**:53–74.
9. Drewnowski A. Taste preferences and food intake. *Annu Rev Nutr* 1997;**17**:237–53.
10. Hetherington MM. Taste and appetite regulation in the elderly. *Proc Nutr Soc* 1998;**57**: 625–31.
11. Schiffman SS. The aging gustatory system. In: Firesten S, Beauchamp GK, editors. *The senses. Olfaction and taste*, vol. 4. Amsterdam: Elsevier; 2008. pp. 479–98.
12. Sandell MA, Breslin PA. Variability in a taste-receptor gene determines whether we taste toxins in food. *Curr Biol* 2006;**16**:R792–4.
13. Soranzo N, Bufe B, Sabeti PC, Wilson JF, Weale ME, Marguerie R, et al. Positive selection on a high-sensitivity allele of the human bitter-taste receptor TAS2R16. *Curr Biol* 2005;**15**: 1257–65.
14. Chaudhari N, Roper SD. The cell biology of taste. *J Cell Biol* 2010;**190**:285–96.
15. Simon SA, de Araujo IE, Gutierrez R, Nicolelis MA. The neural mechanisms of gustation: a distributed processing code. *Nat Rev Neurosci* 2006;**7**:890–901.
16. Ikeda H, West DB, Pustek JJ, Figlewicz DP, Greenwood MR, Porte Jr. D, et al. Intraventricular insulin reduces food intake and body weight of lean but not obese Zucker rats. *Appetite* 1986;**7**:381–6.
17. Wren AM, Small CJ, Abbott CR, Dhillo WS, Seal LJ, Cohen MA, et al. Ghrelin causes hyperphagia and obesity in rats. *Diabetes* 2001;**50**:2540–7.
18. Tschop M, Smiley DL, Heiman ML. Ghrelin induces adiposity in rodents. *Nature* 2000;**407**: 908–13.
19. Kirkham TC, Williams CM, Fezza F, Di Marzo V. Endocannabinoid levels in rat limbic forebrain and hypothalamus in relation to fasting, feeding and satiation: stimulation of eating by 2-arachidonoyl glycerol. *Br J Pharmacol* 2002;**136**:550–7.

20. Roper SD. The microphysiology of peripheral taste organs. *J Neurosci* 1992;**12**:1127–34.
21. Farbman AI, Hellekant G, Nelson A. Structure of taste buds in foliate papillae of the rhesus monkey, Macaca mulatta. *Am J Anat* 1985;**172**:41–56.
22. Jahnke K, Baur P. Freeze-fracture study of taste bud pores in the foliate papillae of the rabbit. *Cell Tissue Res* 1979;**200**:245–56.
23. Witt M, Reutter K, Miller Jr. IJ. Morphology of the peripheral taste system. In: Doty RL, editor. *Handbook of olfaction and gustation*. USA: Marcel Dekker, Inc.; 2003. pp. 1072–118.
24. Behrens M, Meyerhof W. Oral and extraoral bitter taste receptors. *Results Probl Cell Differ* 2010;**52**:87–99.
25. Cheng LH, Robinson PP. The distribution of fungiform papillae and taste buds on the human tongue. *Arch Oral Biol* 1991;**36**:583–9.
26. Lindemann B. Receptor seeks ligand: on the way to cloning the molecular receptors for sweet and bitter taste. *Nat Med* 1999;**5**:381–2.
27. Jung HS, Akita K, Kim JY. Spacing patterns on tongue surface-gustatory papilla. *Int J Dev Biol* 2004;**48**:157–61.
28. DeFazio RA, Dvoryanchikov G, Maruyama Y, Kim JW, Pereira E, Roper SD, et al. Separate populations of receptor cells and presynaptic cells in mouse taste buds. *J Neurosci* 2006;**26**: 3971–80.
29. Pumplin DW, Yu C, Smith DV. Light and dark cells of rat vallate taste buds are morphologically distinct cell types. *J Comp Neurol* 1997;**378**:389–410.
30. Clapp TR, Yang R, Stoick CL, Kinnamon SC, Kinnamon JC. Morphologic characterization of rat taste receptor cells that express components of the phospholipase C signaling pathway. *J Comp Neurol* 2004;**468**:311–21.
31. Lawton DM, Furness DN, Lindemann B, Hackney CM. Localization of the glutamate-aspartate transporter, GLAST, in rat taste buds. *Eur J Neurosci* 2000;**12**:3163–71.
32. Vandenbeuch A, Tizzano M, Anderson CB, Stone LM, Goldberg D, Kinnamon SC. Evidence for a role of glutamate as an efferent transmitter in taste buds. *BMC Neurosci* 2010;**11**:77.
33. Bartel DL, Sullivan SL, Lavoie EG, Sevigny J, Finger TE. Nucleoside triphosphate diphosphohydrolase-2 is the ecto-ATPase of type I cells in taste buds. *J Comp Neurol* 2006; **497**:1–12.
34. Romanov RA, Kolesnikov SS. Electrophysiologically identified subpopulations of taste bud cells. *Neurosci Lett* 2006;**395**:249–54.
35. Medler KF, Margolskee RF, Kinnamon SC. Electrophysiological characterization of voltage-gated currents in defined taste cell types of mice. *J Neurosci* 2003;**23**:2608–17.
36. Dvoryanchikov G, Sinclair MS, Perea-Martinez I, Wang T, Chaudhari N. Inward rectifier channel, ROMK, is localized to the apical tips of glial-like cells in mouse taste buds. *J Comp Neurol* 2009;**517**:1–14.
37. Vandenbeuch A, Clapp TR, Kinnamon SC. Amiloride-sensitive channels in type I fungiform taste cells in mouse. *BMC Neurosci* 2008;**9**:1.
38. Kataoka S, Yang R, Ishimaru Y, Matsunami H, Sevigny J, Kinnamon JC, et al. The candidate sour taste receptor, PKD2L1, is expressed by type III taste cells in the mouse. *Chem Senses* 2008;**33**:243–54.
39. Wellendorph P, Johansen LD, Brauner-Osborne H. Molecular pharmacology of promiscuous seven transmembrane receptors sensing organic nutrients. *Mol Pharmacol* 2009;**76**:453–65.
40. Clapp TR, Stone LM, Margolskee RF, Kinnamon SC. Immunocytochemical evidence for co-expression of Type III IP3 receptor with signaling components of bitter taste transduction. *BMC Neurosci* 2001;**2**:6.
41. Miyoshi MA, Abe K, Emori Y. IP(3) receptor type 3 and PLCbeta2 are co-expressed with taste receptors T1R and T2R in rat taste bud cells. *Chem Senses* 2001;**26**:259–65.

42. Nelson G, Hoon MA, Chandrashekar J, Zhang Y, Ryba NJ, Zuker CS. Mammalian sweet taste receptors. *Cell* 2001;**106**:381–90.
43. Zhang Y, Hoon MA, Chandrashekar J, Mueller KL, Cook B, Wu D, et al. Coding of sweet, bitter, and umami tastes: different receptor cells sharing similar signaling pathways. *Cell* 2003;**112**:293–301.
44. Zhao GQ, Zhang Y, Hoon MA, Chandrashekar J, Erlenbach I, Ryba NJ, et al. The receptors for mammalian sweet and umami taste. *Cell* 2003;**115**:255–66.
45. Mueller KL, Hoon MA, Erlenbach I, Chandrashekar J, Zuker CS, Ryba NJ. The receptors and coding logic for bitter taste. *Nature* 2005;**434**:225–9.
46. Chandrashekar J, Hoon MA, Ryba NJ, Zuker CS. The receptors and cells for mammalian taste. *Nature* 2006;**444**:288–94.
47. Rossler P, Kroner C, Freitag J, Noe J, Breer H. Identification of a phospholipase C beta subtype in rat taste cells. *Eur J Cell Biol* 1998;**77**:253–61.
48. Clapp TR, Medler KF, Damak S, Margolskee RF, Kinnamon SC. Mouse taste cells with G protein-coupled taste receptors lack voltage-gated calcium channels and SNAP-25. *BMC Biol* 2006;**4**:7.
49. Wong GT, Gannon KS, Margolskee RF. Transduction of bitter and sweet taste by gustducin. *Nature* 1996;**381**:796–800.
50. Danilova V, Damak S, Margolskee RF, Hellekant G. Taste responses to sweet stimuli in alpha-gustducin knockout and wild-type mice. *Chem Senses* 2006;**31**:573–80.
51. Ruiz CJ, Wray K, Delay E, Margolskee RF, Kinnamon SC. Behavioral evidence for a role of alpha-gustducin in glutamate taste. *Chem Senses* 2003;**28**:573–9.
52. Vandenbeuch A, Kinnamon SC. Why do taste cells generate action potentials? *J Biol* 2009; **8**:42.
53. Gao N, Lu M, Echeverri F, Laita B, Kalabat D, Williams ME, et al. Voltage-gated sodium channels in taste bud cells. *BMC Neurosci* 2009;**10**:20.
54. Yang R, Stoick CL, Kinnamon JC. Synaptobrevin-2-like immunoreactivity is associated with vesicles at synapses in rat circumvallate taste buds. *J Comp Neurol* 2004;**471**:59–71.
55. Vandenbeuch A, Zorec R, Kinnamon SC. Capacitance measurements of regulated exocytosis in mouse taste cells. *J Neurosci* 2010;**30**:14695–701.
56. Huang AL, Chen X, Hoon MA, Chandrashekar J, Guo W, Trankner D, et al. The cells and logic for mammalian sour taste detection. *Nature* 2006;**442**:934–8.
57. Huang YA, Maruyama Y, Stimac R, Roper SD. Presynaptic (Type III) cells in mouse taste buds sense sour (acid) taste. *J Physiol* 2008;**586**:2903–12.
58. Chandrashekar J, Yarmolinsky D, von Buchholtz L, Oka Y, Sly W, Ryba NJ, et al. The taste of carbonation. *Science* 2009;**326**:443–5.
59. Huang YJ, Maruyama Y, Lu KS, Pereira E, Plonsky I, Baur JE, et al. Mouse taste buds use serotonin as a neurotransmitter. *J Neurosci* 2005;**25**:843–7.
60. Huang YJ, Maruyama Y, Dvoryanchikov G, Pereira E, Chaudhari N, Roper SD. The role of pannexin 1 hemichannels in ATP release and cell-cell communication in mouse taste buds. *Proc Natl Acad Sci USA* 2007;**104**:6436–41.
61. Dvoryanchikov G, Huang YA, Barro-Soria R, Chaudhari N, Roper SD. GABA, its receptors, and GABAergic inhibition in mouse taste buds. *J Neurosci* 2011;**31**:5782–91.
62. Huang YA, Maruyama Y, Roper SD. Norepinephrine is coreleased with serotonin in mouse taste buds. *J Neurosci* 2008;**28**:13088–93.
63. De Camilli P, Benfenati F, Valtorta F, Greengard P. The synapsins. *Annu Rev Cell Biol* 1990; **6**:433–60.
64. Yang R, Crowley HH, Rock ME, Kinnamon JC. Taste cells with synapses in rat circumvallate papillae display SNAP-25-like immunoreactivity. *J Comp Neurol* 2000;**424**:205–15.

65. Nelson GM, Finger TE. Immunolocalization of different forms of neural cell adhesion molecule (NCAM) in rat taste buds. *J Comp Neurol* 1993;**336**:507–16.
66. Yee CL, Yang R, Bottger B, Finger TE, Kinnamon JC. "Type III" cells of rat taste buds: immunohistochemical and ultrastructural studies of neuron-specific enolase, protein gene product 9.5, and serotonin. *J Comp Neurol* 2001;**440**:97–108.
67. Tomchik SM, Berg S, Kim JW, Chaudhari N, Roper SD. Breadth of tuning and taste coding in mammalian taste buds. *J Neurosci* 2007;**27**:10840–8.
68. Beidler LM, Smallman RL. Renewal of cells within taste buds. *J Cell Biol* 1965;**27**:263–72.
69. Matsumoto I, Ohmoto M, Narukawa M, Yoshihara Y, Abe K. Skn-1a (Pou2f3) specifies taste receptor cell lineage. *Nat Neurosci* 2011;**14**:685–7.
70. Seta Y, Stoick-Cooper CL, Toyono T, Kataoka S, Toyoshima K, Barlow LA. The bHLH transcription factors, Hes6 and Mash1, are expressed in distinct subsets of cells within adult mouse taste buds. *Arch Histol Cytol* 2006;**69**:189–98.
71. Miura H, Kusakabe Y, Harada S. Cell lineage and differentiation in taste buds. *Arch Histol Cytol* 2006;**69**:209–25.
72. Okubo T, Pevny LH, Hogan BL. Sox2 is required for development of taste bud sensory cells. *Genes Dev* 2006;**20**:2654–9.
73. Rodriguez-Boulan E, Nelson WJ. Morphogenesis of the polarized epithelial cell phenotype. *Science* 1989;**245**:718–25.
74. Cabrera-Vera TM, Vanhauwe J, Thomas TO, Medkova M, Preininger A, Mazzoni MR, et al. Insights into G protein structure, function, and regulation. *Endocr Rev* 2003;**24**:765–81.
75. Yang R, Tabata S, Crowley HH, Margolskee RF, Kinnamon JC. Ultrastructural localization of gustducin immunoreactivity in microvilli of type II taste cells in the rat. *J Comp Neurol* 2000;**425**:139–51.
76. McLaughlin SK, McKinnon PJ, Margolskee RF. Gustducin is a taste-cell-specific G protein closely related to the transducins. *Nature* 1992;**357**:563–9.
77. Ruiz-Avila L, McLaughlin SK, Wildman D, McKinnon PJ, Robichon A, Spickofsky N, et al. Coupling of bitter receptor to phosphodiesterase through transducin in taste receptor cells. *Nature* 1995;**376**:80–5.
78. Kusakabe Y, Yamaguchi E, Tanemura K, Kameyama K, Chiba N, Arai S, et al. Identification of two alpha-subunit species of GTP-binding proteins, Galpha15 and Galphaq, expressed in rat taste buds. *Biochim Biophys Acta* 1998;**1403**:265–72.
79. Kusakabe Y, Yasuoka A, Asano-Miyoshi M, Iwabuchi K, Matsumoto I, Arai S, et al. Comprehensive study on G protein alpha-subunits in taste bud cells, with special reference to the occurrence of Galphai2 as a major Galpha species. *Chem Senses* 2000;**25**:525–31.
80. Tizzano M, Dvoryanchikov G, Barrows JK, Kim S, Chaudhari N, Finger TE. Expression of Galpha14 in sweet-transducing taste cells of the posterior tongue. *BMC Neurosci* 2008;**9**:110.
81. Ueda T, Ugawa S, Yamamura H, Imaizumi Y, Shimada S. Functional interaction between T2R taste receptors and G-protein alpha subunits expressed in taste receptor cells. *J Neurosci* 2003;**23**:7376–80.
82. McLaughlin SK, McKinnon PJ, Spickofsky N, Danho W, Margolskee RF. Molecular cloning of G proteins and phosphodiesterases from rat taste cells. *Physiol Behav* 1994;**56**:1157–64.
83. Galindo MM, Voigt N, Stein J, van Lengerich J, Raguse JD, Hofmann T. G Protein-coupled receptors in human fat taste perception. *Chem Senses* 2012;**37**:123–9. doi:10.1093/chemse/BJR069.
84. Clapp TR, Trubey KR, Vandenbeuch A, Stone LM, Margolskee RF, Chaudhari N, et al. Tonic activity of Galpha-gustducin regulates taste cell responsivity. *FEBS Lett* 2008;**582**:3783–7.
85. Rossler P, Boekhoff I, Tareilus E, Beck S, Breer H, Freitag J. G protein betagamma complexes in circumvallate taste cells involved in bitter transduction. *Chem Senses* 2000;**25**:413–21.

86. Yan W, Sunavala G, Rosenzweig S, Dasso M, Brand JG, Spielman AI. Bitter taste transduced by PLC-beta(2)-dependent rise in IP(3) and alpha-gustducin-dependent fall in cyclic nucleotides. *Am J Physiol Cell Physiol* 2001;**280**:C742–51.
87. Liu D, Liman ER. Intracellular Ca2+ and the phospholipid PIP2 regulate the taste transduction ion channel TRPM5. *Proc Natl Acad Sci USA* 2003;**100**:15160–5.
88. Finger TE, Danilova V, Barrows J, Bartel DL, Vigers AJ, Stone L, et al. ATP signaling is crucial for communication from taste buds to gustatory nerves. *Science* 2005;**310**:1495–9.
89. Romanov RA, Rogachevskaja OA, Bystrova MF, Jiang P, Margolskee RF, Kolesnikov SS. Afferent neurotransmission mediated by hemichannels in mammalian taste cells. *EMBO J* 2007;**26**:657–67.
90. Bo X, Alavi A, Xiang Z, Oglesby I, Ford A, Burnstock G. Localization of ATP-gated P2X2 and P2X3 receptor immunoreactive nerves in rat taste buds. *Neuroreport* 1999;**10**:1107–11.
91. Huang YA, Dando R, Roper SD. Autocrine and paracrine roles for ATP and serotonin in mouse taste buds. *J Neurosci* 2009;**29**:13909–18.
92. Herness S, Zhao FL, Kaya N, Lu SG, Shen T, Sun XD. Adrenergic signalling between rat taste receptor cells. *J Physiol* 2002;**543**:601–14.
93. Behrens M, Meyerhof W, Hellfritsch C, Hofmann T. Sweet and umami taste: natural products, their chemosensory targets, and beyond. *Angew Chem Int Ed Engl* 2011;**50**:2220–42.
94. Small DM. Central gustatory processing in humans. *Adv Otorhinolaryngol* 2006;**63**:191–220.
95. Rolls ET, Scott TR. Central taste anatomy and neurophysiology. In: Doty RL, editor. *Hand book of olfaction and gustation*. USA: Marcel Dekker, Inc.; 2003. pp. 679–706.
96. Spector AC, Glendinning JI. Linking peripheral taste processes to behavior. *Curr Opin Neurobiol* 2009;**19**:370–7.
97. Adolphs R, Tranel D, Koenigs M, Damasio AR. Preferring one taste over another without recognizing either. *Nat Neurosci* 2005;**8**:860–1.
98. Grill HJ, Norgren R. The taste reactivity test II. Mimetic responses to gustatory stimuli in chronic thalamic and chronic decerebrate rats. *Brain Res* 1978;**143**:281–97.
99. Nelson G, Chandrashekar J, Hoon MA, Feng L, Zhao G, Ryba NJ, et al. An amino-acid taste receptor. *Nature* 2002;**416**:199–202.
100. Steiner JE. What the neonate can tell us about umami. In: Kawamura Y, Kare MR, editors. *Umami: a basic taste*. New York and Basel: Marcel Dekker, Inc.; 1987. pp. 97–123.
101. Rassin DK, Sturman JA, Hayes KC, Gaull GE. Taurine deficiency in the kitten subcellular distribution of taurine and [35S]taurine in brain. *Neurochem Res* 1978;**3**:401–10.
102. Max M, Shanker YG, Huang L, Rong M, Liu Z, Campagne F, et al. Tas1r3, encoding a new candidate taste receptor, is allelic to the sweet responsiveness locus Sac. *Nat Genet* 2001;**28**: 58–63.
103. Hoon MA, Adler E, Lindemeier J, Battey JF, Ryba NJ, Zuker CS. Putative mammalian taste receptors: a class of taste-specific GPCRs with distinct topographic selectivity. *Cell* 1999;**96**: 541–51.
104. Montmayeur JP, Liberles SD, Matsunami H, Buck LB. A candidate taste receptor gene near a sweet taste locus. *Nat Neurosci* 2001;**4**:492–8.
105. Bachmanov AA, Li X, Reed DR, Ohmen JD, Li S, Chen Z, et al. Positional cloning of the mouse saccharin preference (Sac) locus. *Chem Senses* 2001;**26**:925–33.
106. Kitagawa M, Kusakabe Y, Miura H, Ninomiya Y, Hino A. Molecular genetic identification of a candidate receptor gene for sweet taste. *Biochem Biophys Res Commun* 2001;**283**:236–42.
107. Sainz E, Korley JN, Battey JF, Sullivan SL. Identification of a novel member of the T1R family of putative taste receptors. *J Neurochem* 2001;**77**:896–903.
108. Li X, Staszewski L, Xu H, Durick K, Zoller M, Adler E. Human receptors for sweet and umami taste. *Proc Natl Acad Sci USA* 2002;**99**:4692–6.

109. Damak S, Rong M, Yasumatsu K, Kokrashvili Z, Varadarajan V, Zou S, et al. Detection of sweet and umami taste in the absence of taste receptor T1r3. *Science* 2003;**301**:850–3.
110. Chaudhari N, Pereira E, Roper SD. Taste receptors for umami: the case for multiple receptors. *Am J Clin Nutr* 2009;**90**:738S–42S.
111. Faurion A. Are umami taste receptor sites structurally related to glutamate CNS receptor sites? *Physiol Behav* 1991;**49**:905–12.
112. Kurihara K, Kashiwayanagi M. Physiological studies on umami taste. *J Nutr* 2000;**130**: 931S–934S.
113. Zhang F, Klebansky B, Fine RM, Xu H, Pronin A, Liu H, et al. Molecular mechanism for the umami taste synergism. *Proc Natl Acad Sci USA* 2008;**105**:20930–4.
114. Schiffman SS, Cahn H, Lindley MG. Multiple receptor sites mediate sweetness: evidence from cross adaptation. *Pharmacol Biochem Behav* 1981;**15**:377–88.
115. Jiang P, Cui M, Ji Q, Snyder L, Liu Z, Benard L, et al. Molecular mechanisms of sweet receptor function. *Chem Senses* 2005;**30**(Suppl. 1):i17–8.
116. Temussi PA. Why are sweet proteins sweet? Interaction of brazzein, monellin and thaumatin with the T1R2-T1R3 receptor. *FEBS Lett* 2002;**526**:1–4.
117. Ohta K, Masuda T, Tani F, Kitabatake N. The cysteine-rich domain of human T1R3 is necessary for the interaction between human T1R2-T1R3 sweet receptors and a sweet-tasting protein, thaumatin. *Biochem Biophys Res Commun* 2011;**406**:435–8.
118. Winnig M, Bufe B, Meyerhof W. Valine 738 and lysine 735 in the fifth transmembrane domain of rTas1r3 mediate insensitivity towards lactisole of the rat sweet taste receptor. *BMC Neurosci* 2005;**6**:22.
119. Winnig M, Bufe B, Kratochwil NA, Slack JP, Meyerhof W. The binding site for neohesperidin dihydrochalcone at the human sweet taste receptor. *BMC Struct Biol* 2007;**7**:66.
120. Xu H, Staszewski L, Tang H, Adler E, Zoller M, Li X. Different functional roles of T1R subunits in the heteromeric taste receptors. *Proc Natl Acad Sci USA* 2004;**101**:14258–63.
121. Jiang P, Ji Q, Liu Z, Snyder LA, Benard LM, Margolskee RF, et al. The cysteine-rich region of T1R3 determines responses to intensely sweet proteins. *J Biol Chem* 2004;**279**:45068–75.
122. Jiang P, Cui M, Zhao B, Liu Z, Snyder LA, Benard LM, et al. Lactisole interacts with the transmembrane domains of human T1R3 to inhibit sweet taste. *J Biol Chem* 2005;**280**: 15238–46.
123. Jiang P, Cui M, Zhao B, Snyder LA, Benard LM, Osman R, et al. Identification of the cyclamate interaction site within the transmembrane domain of the human sweet taste receptor subunit T1R3. *J Biol Chem* 2005;**280**:34296–305.
124. Nie Y, Vigues S, Hobbs JR, Conn GL, Munger SD. Distinct contributions of T1R2 and T1R3 taste receptor subunits to the detection of sweet stimuli. *Curr Biol* 2005;**15**:1948–52.
125. Galindo-Cuspinera V, Winnig M, Bufe B, Meyerhof W, Breslin PA. A TAS1R receptor-based explanation of sweet 'water-taste'. *Nature* 2006;**441**:354–7.
126. Koizumi A, Nakajima K, Asakura T, Morita Y, Ito K, Shmizu-Ibuka A, et al. Taste-modifying sweet protein, neoculin, is received at human T1R3 amino terminal domain. *Biochem Biophys Res Commun* 2007;**358**:585–9.
127. Zhang F, Klebansky B, Fine RM, Liu H, Xu H, Servant G, et al. Molecular mechanism of the sweet taste enhancers. *Proc Natl Acad Sci USA* 2010;**107**:4752–7.
128. Morini G, Bassoli A, Temussi PA. From small sweeteners to sweet proteins: anatomy of the binding sites of the human T1R2_T1R3 receptor. *J Med Chem* 2005;**48**:5520–9.
129. Brockhoff A, Behrens M, Niv MY, Meyerhof W. Structural requirements of bitter taste receptor activation. *Proc Natl Acad Sci USA* 2010;**107**:11110–5.
130. Yamaguchi S, Yoshikawa T, Ikeda S, Ninomiya T. Measurement of the relative taste intensity of some L-α-amino acids and 5′-nucleotides. *J Food Sci* 1971;**36**:846–9.

131. Yamaguchi S. The synergistic taste effect of monosodium glutamate and disodium 5′-inosinate. *J Food Sci* 1967;**32**:473–8.
132. Lindemann B. Receptors and transduction in taste. *Nature* 2001;**413**:219–25.
133. Contreras RJ. Gustatory mechanisms of a specific appetite. In: Cagan RH, editor. *Neural mechanisms in taste*. Boca Raton, FL: CRC Press; 1989. pp. 119–45.
134. Duncan CJ. Salt preferences of birds and mammals. *Physiol Zool* 1962;**35**:120–32.
135. Ganchrow JR, Steiner JE, Daher M. Neonatal response to intensities facial expressions in different qualities and of gustatory stimuli. *Infant Behav Develop* 1983;**6**:189–200.
136. Blossfeld I, Collins A, Boland S, Baixauli R, Kiely M, Delahunty C. Relationships between acceptance of sour taste and fruit intakes in 18-month-old infants. *Br J Nutr* 2007;**98**:1084–91.
137. Max M, Meyerhof W. Taste receptors. In: Firesten S, Beauchamp GK, editors. *The senses. Olfaction and taste*, vol. 4. Amsterdam: Elsevier; 2008. pp. 197–217.
138. Lindemann B. Chemoreception: tasting the sweet and the bitter. *Curr Biol* 1996;**6**:1234–7.
139. Hettinger TP, Frank ME. Specificity of amiloride inhibition of hamster taste responses. *Brain Res* 1990;**513**:24–34.
140. Heck GL, Mierson S, DeSimone JA. Salt taste transduction occurs through an amiloride-sensitive sodium transport pathway. *Science* 1984;**223**:403–5.
141. DeSimone JA, Lyall V, Heck GL, Phan TH, Alam RI, Feldman GM, et al. A novel pharmacological probe links the amiloride-insensitive NaCl, KCl, and NH(4)Cl chorda tympani taste responses. *J Neurophysiol* 2001;**86**:2638–41.
142. Lyall V, Heck GL, Vinnikova AK, Ghosh S, Phan TH, Alam RI, et al. The mammalian amiloride-insensitive non-specific salt taste receptor is a vanilloid receptor-1 variant. *J Physiol* 2004;**558**:147–59.
143. Katsumata T, Nakakuki H, Tokunaga C, Fujii N, Egi M, Phan TH, et al. Effect of Maillard reacted peptides on human salt taste and the amiloride-insensitive salt taste receptor (TRPV1t). *Chem Senses* 2008;**33**:665–80.
144. Lyall V, Phan TH, Ren Z, Mummalaneni S, Melone P, Mahavadi S, et al. Regulation of the putative TRPV1t salt taste receptor by phosphatidylinositol 4,5-bisphosphate. *J Neurophysiol* 2010;**103**:1337–49.
145. Ruiz C, Gutknecht S, Delay E, Kinnamon S. Detection of NaCl and KCl in TRPV1 knockout mice. *Chem Senses* 2006;**31**:813–20.
146. Treesukosol Y, Lyall V, Heck GL, DeSimone JA, Spector AC. A psychophysical and electrophysiological analysis of salt taste in Trpv1 null mice. *Am J Physiol Regul Integr Comp Physiol* 2007;**292**:R1799–809.
147. Avenet P. Role of amiloride-sensitive sodium channels in taste. *Soc Gen Physiol Ser* 1992;**47**: 271–9.
148. Avenet P, Lindemann B. Amiloride-blockable sodium currents in isolated taste receptor cells. *J Membr Biol* 1988;**105**:245–55.
149. Doolin RE, Gilbertson TA. Distribution and characterization of functional amiloride-sensitive sodium channels in rat tongue. *J Gen Physiol* 1996;**107**:545–54.
150. Chandrashekar J, Kuhn C, Oka Y, Yarmolinsky DA, Hummler E, Ryba NJP, et al. The cells and peripheral representation of sodium taste in mice. *Nature* 2010;**464**:297–301.
151. Ossebaard CA, Polet IA, Smith DV. Amiloride effects on taste quality: comparison of single and multiple response category procedures. *Chem Senses* 1997;**22**:267–75.
152. Ossebaard CA, Smith DV. Effect of amiloride on the taste of NaCl, Na-gluconate and KCl in humans: implications for Na+ receptor mechanisms. *Chem Senses* 1995;**20**:37–46.
153. Stähler F, Riedel K, Demgensky S, Neumann K, Dunkel A, Täubert A, et al. A role of the epithelial sodium channel in human salt taste transduction? *Chem Percept* 2008;**1**:78–90.
154. Halpern BP. Amiloride and vertebrate gustatory responses to NaCl. *Neurosci Biobehav Rev* 1998;**23**:5–47.

155. Ugawa S, Minami Y, Guo W, Saishin Y, Takatsuji K, Yamamoto T, et al. Receptor that leaves a sour taste in the mouth. *Nature* 1998;**395**:555–6.
156. Stevens DR, Seifert R, Bufe B, Muller F, Kremmer E, Gauss R, et al. Hyperpolarization-activated channels HCN1 and HCN4 mediate responses to sour stimuli. *Nature* 2001;**413**:631–5.
157. Richter TA, Dvoryanchikov GA, Chaudhari N, Roper SD. Acid-sensitive two-pore domain potassium (K2P) channels in mouse taste buds. *J Neurophysiol* 2004;**92**:1928–36.
158. Lin W, Burks CA, Hansen DR, Kinnamon SC, Gilbertson TA. Taste receptor cells express pH-sensitive leak K+ channels. *J Neurophysiol* 2004;**92**:2909–19.
159. LopezJimenez ND, Cavenagh MM, Sainz E, Cruz-Ithier MA, Battey JF, Sullivan SL. Two members of the TRPP family of ion channels, Pkd1l3 and Pkd2l1, are co-expressed in a subset of taste receptor cells. *J Neurochem* 2006;**98**:68–77.
160. Ishimaru Y, Inada H, Kubota M, Zhuang H, Tominaga M, Matsunami H. Transient receptor potential family members PKD1L3 and PKD2L1 form a candidate sour taste receptor. *Proc Natl Acad Sci USA* 2006;**103**:12569–74.
161. Chang RB, Waters H, Liman ER. A proton current drives action potentials in genetically identified sour taste cells. *Proc Natl Acad Sci USA* 2010;**107**:22320–5.
162. DeSimone JA, Phan TH, Heck GL, Ren Z, Coleman J, Mummalaneni S, et al. Involvement of NADPH-dependent and cAMP-PKA sensitive H+ channels in the chorda tympani nerve responses to strong acids. *Chem Senses* 2011;**36**:389–403.
163. Huque T, Cowart BJ, Dankulich-Nagrudny L, Pribitkin EA, Bayley DL, Spielman AI, et al. Sour ageusia in two individuals implicates ion channels of the ASIC and PKD families in human sour taste perception at the anterior tongue. *PLoS One* 2009;**4**:e7347.
164. Horio N, Yoshida R, Yasumatsu K, Yanagawa Y, Ishimaru Y, Matsunami H, et al. Sour taste responses in mice lacking PKD channels. *PLoS One* 2011;**6**:e20007.
165. Nelson TM, Lopezjimenez ND, Tessarollo L, Inoue M, Bachmanov AA, Sullivan SL. Taste function in mice with a targeted mutation of the Pkd1l3 gene. *Chem Senses* 2010;**35**:565–77.
166. Meyerhof W. Elucidation of mammalian bitter taste. *Rev Physiol Biochem Pharmacol* 2005;**154**:37–72.
167. Roper SD. Signal transduction and information processing in mammalian taste buds. *Pflugers Arch* 2007;**454**:759–76.
168. Glendinning JI. Is the bitter rejection response always adaptive? *Physiol Behav* 1994;**56**:1217–27.
169. Belitz H, Wieser H. Bitter compounds: occurrence and structure-activity relationships. *Food Rev Int* 1985;**1**:271–354.
170. DuBois GE. Chemistry of gustatory stimuli. In: Firesten S, Beauchamp GK, editors. *The senses. Olfaction and taste*, vol. 4. Amsterdam: Elsevier; 2008. pp. 27–74.
171. Matsunami H, Montmayeur JP, Buck LB. A family of candidate taste receptors in human and mouse. *Nature* 2000;**404**:601–4.
172. Adler E, Hoon MA, Mueller KL, Chandrashekar J, Ryba NJ, Zuker CS. A novel family of mammalian taste receptors. *Cell* 2000;**100**:693–702.
173. Bufe B, Hofmann T, Krautwurst D, Raguse JD, Meyerhof W. The human TAS2R16 receptor mediates bitter taste in response to beta-glucopyranosides. *Nat Genet* 2002;**32**:397–401.
174. Chandrashekar J, Mueller KL, Hoon MA, Adler E, Feng L, Guo W, et al. T2Rs function as bitter taste receptors. *Cell* 2000;**100**:703–11.
175. Kim UK, Jorgenson E, Coon H, Leppert M, Risch N, Drayna D. Positional cloning of the human quantitative trait locus underlying taste sensitivity to phenylthiocarbamide. *Science* 2003;**299**:1221–5.

176. Behrens M, Brockhoff A, Kuhn C, Bufe B, Winnig M, Meyerhof W. The human taste receptor hTAS2R14 responds to a variety of different bitter compounds. *Biochem Biophys Res Commun* 2004;**319**:479–85.
177. Kuhn C, Bufe B, Winnig M, Hofmann T, Frank O, Behrens M, et al. Bitter taste receptors for saccharin and acesulfame K. *J Neurosci* 2004;**24**:10260–5.
178. Pronin AN, Tang H, Connor J, Keung W. Identification of ligands for two human bitter T2R receptors. *Chem Senses* 2004;**29**:583–93.
179. Bufe B, Breslin PA, Kuhn C, Reed DR, Tharp CD, Slack JP, et al. The molecular basis of individual differences in phenylthiocarbamide and propylthiouracil bitterness perception. *Curr Biol* 2005;**15**:322–7.
180. Brockhoff A, Behrens M, Massarotti A, Appendino G, Meyerhof W. Broad tuning of the human bitter taste receptor hTAS2R46 to various sesquiterpene lactones, clerodane and labdane diterpenoids, strychnine, and denatonium. *J Agric Food Chem* 2007;**55**:6236–43.
181. Pronin AN, Xu H, Tang H, Zhang L, Li Q, Li X. Specific alleles of bitter receptor genes influence human sensitivity to the bitterness of aloin and saccharin. *Curr Biol* 2007;**17**: 1403–8.
182. Dotson CD, Zhang L, Xu H, Shin YK, Vigues S, Ott SH, et al. Bitter taste receptors influence glucose homeostasis. *PLoS One* 2008;**3**:e3974.
183. Maehashi K, Matano M, Wang H, Vo LA, Yamamoto Y, Huang L. Bitter peptides activate hTAS2Rs, the human bitter receptors. *Biochem Biophys Res Commun* 2008;**365**:851–5.
184. Intelmann D, Batram C, Kuhn C, Haseleu G, Meyerhof W, Hofmann T. Three TAS2R bitter taste receptors mediate the psychophysical responses to bitter compounds of Hops (Humulus lupulus L.) and beer. *Chem Percept* 2009;**2**:118–32.
185. Behrens M, Meyerhof W. Mammalian bitter taste perception. *Results Probl Cell Differ* 2009; **47**:203–20.
186. Meyerhof W, Batram C, Kuhn C, Brockhoff A, Chudoba E, Bufe B, et al. The molecular receptive ranges of human TAS2R bitter taste receptors. *Chem Senses* 2010;**35**:157–70.
187. Wooding S, Gunn H, Ramos P, Thalmann S, Xing C, Meyerhof W. Genetics and bitter taste responses to goitrin, a plant toxin found in vegetables. *Chem Senses* 2010;**35**:685–92.
188. Floriano WB, Hall S, Vaidehi N, Kim U, Drayna D, Goddard 3rd WA. Modeling the human PTC bitter-taste receptor interactions with bitter tastants. *J Mol Model* 2006;**12**:931–41.
189. Miguet L, Zhang Z, Grigorov MG. Computational studies of ligand-receptor interactions in bitter taste receptors. *J Recept Signal Transduct Res* 2006;**26**:611–30.
190. Biarnes X, Marchiori A, Giorgetti A, Lanzara C, Gasparini P, Carloni P, et al. Insights into the binding of Phenyltiocarbamide (PTC) agonist to its target human TAS2R38 bitter receptor. *PLoS One* 2010;**5**:e12394.
191. Sakurai T, Misaka T, Ishiguro M, Masuda K, Sugawara T, Ito K, et al. Characterization of the beta-D-glucopyranoside binding site of the human bitter taste receptor hTAS2R16. *J Biol Chem* 2010;**285**:28373–8.
192. Tepper BJ, Nurse RJ. PROP taster status is related to fat perception and preference. *Ann N Y Acad Sci* 1998;**855**:802–4.
193. Ramirez I. Role of olfaction in starch and oil preference. *Am J Physiol* 1993;**265**:R1404–9.
194. Greenberg D, Smith GP. The controls of fat intake. *Psychosom Med* 1996;**58**:559–69.
195. Chale-Rush A, Burgess JR, Mattes RD. Multiple routes of chemosensitivity to free fatty acids in humans. *Am J Physiol Gastrointest Liver Physiol* 2007;**292**:G1206–12.
196. Cartoni C, Yasumatsu K, Ohkuri T, Shigemura N, Yoshida R, Godinot N, et al. Taste preference for fatty acids is mediated by GPR40 and GPR120. *J Neurosci* 2010;**30**: 8376–82.
197. Fukuwatari T, Shibata K, Iguchi K, Saeki T, Iwata A, Tani K, et al. Role of gustation in the recognition of oleate and triolein in anosmic rats. *Physiol Behav* 2003;**78**:579–83.

198. Hiraoka T, Fukuwatari T, Imaizumi M, Fushiki T. Effects of oral stimulation with fats on the cephalic phase of pancreatic enzyme secretion in esophagostomized rats. *Physiol Behav* 2003;**79**:713–7.
199. Laugerette F, Passilly-Degrace P, Patris B, Niot I, Febbraio M, Montmayeur JP, et al. CD36 involvement in orosensory detection of dietary lipids, spontaneous fat preference, and digestive secretions. *J Clin Invest* 2005;**115**:3177–84.
200. Takeda M, Sawano S, Imaizumi M, Fushiki T. Preference for corn oil in olfactory-blocked mice in the conditioned place preference test and the two-bottle choice test. *Life Sci* 2001;**69**:847–54.
201. Mindell S, Smith GP, Greenberg D. Corn oil and mineral oil stimulate sham feeding in rats. *Physiol Behav* 1990;**48**:283–7.
202. Gaillard D, Laugerette F, Darcel N, El-Yassimi A, Passilly-Degrace P, Hichami A, et al. The gustatory pathway is involved in CD36-mediated orosensory perception of long-chain fatty acids in the mouse. *FASEB J* 2008;**22**:1458–68.
203. Lawson H. *Food oils and fats: technology, utilization, and nutrition*. New York: Chapman & Hall; 1995.
204. Tsuruta M, Kawada T, Fukuwatari T, Fushiki T. The orosensory recognition of long-chain fatty acids in rats. *Physiol Behav* 1999;**66**:285–8.
205. Kawai T, Fushiki T. Importance of lipolysis in oral cavity for orosensory detection of fat. *Am J Physiol Regul Integr Comp Physiol* 2003;**285**:R447–54.
206. Chale-Rush A, Burgess JR, Mattes RD. Evidence for human orosensory (taste?) sensitivity to free fatty acids. *Chem Senses* 2007;**32**:423–31.
207. Mattes RD. Oral detection of short-, medium-, and long-chain free fatty acids in humans. *Chem Senses* 2009;**34**:145–50.
208. Mattes RD. Oral thresholds and suprathreshold intensity ratings for free fatty acids on 3 tongue sites in humans: implications for transduction mechanisms. *Chem Senses* 2009;**34**: 415–23.
209. Stewart JE, Feinle-Bisset C, Golding M, Delahunty C, Clifton PM, Keast RS. Oral sensitivity to fatty acids, food consumption and BMI in human subjects. *Br J Nutr* 2010;**104**:145–52.
210. Simons PJ, Kummer JA, Luiken JJ, Boon L. Apical CD36 immunolocalization in human and porcine taste buds from circumvallate and foliate papillae. *Acta Histochem* 2010;**113**:839–43.
211. Fukuwatari T, Kawada T, Tsuruta M, Hiraoka T, Iwanaga T, Sugimoto E, et al. Expression of the putative membrane fatty acid transporter (FAT) in taste buds of the circumvallate papillae in rats. *FEBS Lett* 1997;**414**:461–4.
212. Gilbertson TA, Fontenot DT, Liu L, Zhang H, Monroe WT. Fatty acid modulation of K+ channels in taste receptor cells: gustatory cues for dietary fat. *Am J Physiol* 1997;**272**:C1203–10.
213. Liu L, Hansen DR, Kim I, Gilbertson TA. Expression and characterization of delayed rectifying K+ channels in anterior rat taste buds. *Am J Physiol Cell Physiol* 2005;**289**: C868–80.
214. Briscoe CP, Tadayyon M, Andrews JL, Benson WG, Chambers JK, Eilert MM, et al. The orphan G protein-coupled receptor GPR40 is activated by medium and long chain fatty acids. *J Biol Chem* 2003;**278**:11303–11.
215. Hirasawa A, Tsumaya K, Awaji T, Katsuma S, Adachi T, Yamada M, et al. Free fatty acids regulate gut incretin glucagon-like peptide-1 secretion through GPR120. *Nat Med* 2005;**11**: 90–4.
216. Matsumura S, Mizushige T, Yoneda T, Iwanaga T, Tsuzuki S, Inoue K, et al. GPR expression in the rat taste bud relating to fatty acid sensing. *Biomed Res* 2007;**28**:49–55.
217. Matsumura S, Eguchi A, Mizushige T, Kitabayashi N, Tsuzuki S, Inoue K, et al. Colocalization of GPR120 with phospholipase-Cbeta2 and alpha-gustducin in the taste bud cells in mice. *Neurosci Lett* 2009;**450**:186–90.

218. Schiffman S, Dackis C. Taste of nutrients: amino acids, vitamins, and fatty acids. In: *Attention, perception, & psychophysics*, vol. 17. New York: Springer; 1975. pp. 140–6.
219. Gilbertson TA, Liu L, Kim I, Burks CA, Hansen DR. Fatty acid responses in taste cells from obesity-prone and -resistant rats. *Physiol Behav* 2005;**86**:681–90.
220. Pittman DW, Labban CE, Anderson AA, O'Connor HE. Linoleic and oleic acids alter the licking responses to sweet, salt, sour, and bitter tastants in rats. *Chem Senses* 2006;**31**:835–43.
221. Garcia J. The evolution of eating safely. *Bribulletin* 1982;**6**:3–5.
222. Garcia J. Learning without memory. *J Cogn Neurosci* 1990;**2**:287–305.
223. Bartoshuk LM, Beauchamp GK. Chemical senses. *Annu Rev Psychol* 1994;**45**:419–49.
224. Bermudez-Rattoni F. Molecular mechanisms of taste-recognition memory. *Nat Rev Neurosci* 2004;**5**:209–17.
225. Vogt MB, Rudy JW. Ontogenesis of learning: IV. Dissociation of memory and perceptual-altering processes mediating taste neophobia in the rat. *Dev Psychobiol* 1984;**17**:601–11.
226. Parker LA, Carvell T. Orofacial and somatic responses elicited by lithium-, nicotine- and amphetamine-paired sucrose solution. *Pharmacol Biochem Behav* 1986;**24**:883–7.
227. Garcia J, Kimmelfrof D, Koelling R. Conditioned taste aversion to saccharin resulting from exposure to gamma radiation. *Science* 1955;**122**:157–8.
228. Ivanova SF, Bures J. Acquisition of conditioned taste aversion in rats is prevented by tetrodotoxin blockade of a small midbrain region centered around the parabrachial nuclei. *Physiol Behav* 1990;**48**:543–9.
229. McAllister KH, Pratt JA. GR205171 blocks apomorphine and amphetamine-induced conditioned taste aversions. *Eur J Pharmacol* 1998;**353**:141–8.
230. Swank MW, Schafe GE, Bernstein IL. c-Fos induction in response to taste stimuli previously paired with amphetamine or LiCl during taste aversion learning. *Brain Res* 1995;**673**:251–61.
231. Ionescu E, Buresova O. Failure to elicit conditioned taste aversion by severe poisoning. *Pharmacol Biochem Behav* 1977;**6**:251–4.
232. Nachman M, Hartley PL. Role of illness in producing learned taste aversions in rats: a comparison of several rodenticides. *J Comp Physiol Psychol* 1975;**89**:1010–8.
233. Garcia J, Koelling R. Relation of cue to consequence in avoidance learning. *Psychon Sci* 1966;**4**:123–4.
234. Miller AJ. Oral and pharyngeal reflexes in the mammalian nervous system: their diverse range in complexity and the pivotal role of the tongue. *Crit Rev Oral Biol Med* 2002;**13**:409–25.
235. Steiner JE. The gustofacial response: observation on normal and anencephalic newborn infants. In: Bosma JF, editor. *Fourth symposium on oral sensation and perception*. Washington, DC: US Government Printing Office; 1973. pp. 254–78.
236. Weiffenbach JM, Thach BT. Elicited tongue movements: touch and taste in the mouth of the neonate. In: Bosma JF, editor. *Fourth symposium on oral sensation and perception*. Washington, DC: US Government Printing Office; 1973. pp. 232–43.
237. Leder SB. Gag reflex and dysphagia. *Head Neck* 1996;**18**:138–41.
238. Seligman M, Hager J. Biological boundaries of learning. The sauce-bearnaise syndrome. *Psychol Today* 1972;**6**:84–7.
239. Ackil JK, Carman HM, Bakner L, Riccio DC. Reinstatement of latent inhibition following a reminder treatment in a conditioned taste aversion paradigm. *Behav Neural Biol* 1992;**58**:232–5.
240. Fenwick S, Mikulka PJ, Klein SB. The effect of different levels of pre-exposure to sucrose on the acquisition and extinction of a conditioned aversion. *Behav Biol* 1975;**14**:231–5.
241. Revusky SH, Bedarf EW. Association of illness with prior ingestion of novel foods. *Science* 1967;**155**:219–20.
242. Lubow R. *Latent inhibition and conditioned attention theory*. New York: Cambridge University Press; 1989.

243. Gallo M, Candido A. Dorsal hippocampal lesions impair blocking but not latent inhibition of taste aversion learning in rats. *Behav Neurosci* 1995;**109**:413–25.
244. Gallo M, Candido A. Reversible inactivation of dorsal hippocampus by tetrodotoxin impairs blocking of taste aversion selectively during the acquisition but not the retrieval in rats. *Neurosci Lett* 1995;**186**:1–4.
245. Nunez-Jaramillo L, Ramirez-Lugo L, Herrera-Morales W, Miranda MI. Taste memory formation: latest advances and challenges. *Behav Brain Res* 2010;**207**:232–48.
246. Sclafani A. Learned controls of ingestive behaviour. *Appetite* 1997;**29**:153–8.
247. Booth DA. Conditioned satiety in the rat. *J Comp Physiol Psychol* 1972;**81**:457–71.
248. Weingarten HP, Kulikovsky OT. Taste-to-postingestive consequence conditioning: is the rise in sham feeding with repeated experience a learning phenomenon? *Physiol Behav* 1989;**45**: 471–6.
249. Holman GL. Intragastric reinforcement effect. *J Comp Physiol Psychol* 1968;**69**:432–41.
250. Sclafani A. How food preferences are learned—laboratory animal models. *Proc Nutr Soc* 1995;**54**:419–27.
251. Drucker DB, Ackroff K, Sclafani A. Nutrient-conditioned flavor preference and acceptance in rats: effects of deprivation state and nonreinforcement. *Physiol Behav* 1994;**56**:701–7.
252. Ramirez I. Stimulus specificity in flavor acceptance learning. *Physiol Behav* 1996;**60**:595–610.
253. Bures J, Bermudez-Rattoni F, Yamamoto T. *Conditioned taste aversion: memory of a special kind. Oxford psychology series no. 31*. New York: Oxford University Press; 1998.
254. Grill HJ, Norgren R. Chronically decerebrate rats demonstrate satiation but not bait shyness. *Science* 1978;**201**:267–9.
255. Kiefer SW, Braun JJ. Absence of differential associative responses to novel and familiar taste stimuli in rats lacking gustatory neocortex. *J Comp Physiol Psychol* 1977;**91**:498–507.
256. Berridge KC, Robinson TE. What is the role of dopamine in reward: hedonic impact, reward learning, or incentive salience? *Brain Res Brain Res Rev* 1998;**28**:309–69.
257. Wise RA. Brain reward circuitry: insights from unsensed incentives. *Neuron* 2002;**36**:229–40.
258. Maren S, Fanselow MS. The amygdala and fear conditioning: has the nut been cracked? *Neuron* 1996;**16**:237–40.
259. Yamamoto T. Neural substrates for the processing of cognitive and affective aspects of taste in the brain. *Arch Histol Cytol* 2006;**69**:243–55.
260. Welzl H, D'Adamo P, Lipp H-P. Conditioned taste aversion as a learining and memory paradigm. *Behav Brain Res* 2001;**125**:205–13.
261. Weinberger NM, Bakin JS. Learning-induced physiological memory in adult primary auditory cortex: receptive fields plasticity, model, and mechanisms. *Audiol Neurootol* 1998;**3**:145–67.
262. Fibiger HC, Damsma G, Day JC. Behavioral pharmacology and biochemistry of central cholinergic neurotransmission. *Adv Exp Med Biol* 1991;**295**:399–414.
263. Miranda MI, Ramirez-Lugo L, Bermudez-Rattoni F. Cortical cholinergic activity is related to the novelty of the stimulus. *Brain Res* 2000;**882**:230–5.
264. Miranda MI, Ferreira G, Ramirez-Lugo L, Bermudez-Rattoni F. Role of cholinergic system on the construction of memories: taste memory encoding. *Neurobiol Learn Mem* 2003;**80**: 211–22.
265. Ranganath C, Rainer G. Neural mechanisms for detecting and remembering novel events. *Nat Rev Neurosci* 2003;**4**:193–202.
266. Gutierrez R, Tellez LA, Bermudez-Rattoni F. Blockade of cortical muscarinic but not NMDA receptors prevents a novel taste from becoming familiar. *Eur J Neurosci* 2003;**17**: 1556–62.
267. Roberson ED, English JD, Adams JP, Selcher JC, Kondratick C, Sweatt JD. The mitogen-activated protein kinase cascade couples PKA and PKC to cAMP response element binding protein phosphorylation in area CA1 of hippocampus. *J Neurosci* 1999;**19**:4337–48.

268. Orban PC, Chapman PF, Brambilla R. Is the Ras-MAPK signalling pathway necessary for long-term memory formation? *Trends Neurosci* 1999;**22**:38–44.
269. Berman DE, Hazvi S, Neduva V, Dudai Y. The role of identified neurotransmitter systems in the response of insular cortex to unfamiliar taste: activation of ERK1-2 and formation of a memory trace. *J Neurosci* 2000;**20**:7017–23.
270. Swank MW, Sweatt JD. Increased histone acetyltransferase and lysine acetyltransferase activity and biphasic activation of the ERK/RSK cascade in insular cortex during novel taste learning. *J Neurosci* 2001;**21**:3383–91.
271. Berman DE, Hazvi S, Rosenblum K, Seger R, Dudai Y. Specific and differential activation of mitogen-activated protein kinase cascades by unfamiliar taste in the insular cortex of the behaving rat. *J Neurosci* 1998;**18**:10037–44.
272. Blackburn JR, Pfaus JG, Phillips AG. Dopamine functions in appetitive and defensive behaviours. *Prog Neurobiol* 1992;**39**:247–79.
273. Azzara AV, Bodnar RJ, Delamater AR, Sclafani A. D1 but not D2 dopamine receptor antagonism blocks the acquisition of a flavor preference conditioned by intragastric carbohydrate infusions. *Pharmacol Biochem Behav* 2001;**68**:709–20.
274. Azzara AV, Bodnar RJ, Delamater AR, Sclafani A. Naltrexone fails to block the acquisition or expression of a flavor preference conditioned by intragastric carbohydrate infusions. *Pharmacol Biochem Behav* 2000;**67**:545–57.
275. Sclafani A, Azzara AV, Touzani K, Grigson PS, Norgren R. Parabrachial nucleus lesions block taste and attenuate flavor preference and aversion conditioning in rats. *Behav Neurosci* 2001; **115**:920–33.
276. Touzani K, Sclafani A. Conditioned flavor preference and aversion: role of the lateral hypothalamus. *Behav Neurosci* 2001;**115**:84–93.
277. Woolley JD, Lee BS, Taha SA, Fields HL. Nucleus accumbens opioid signaling conditions short-term flavor preferences. *Neuroscience* 2007;**146**:19–30.
278. Doron G, Rosenblum K. c-Fos expression is elevated in GABAergic interneurons of the gustatory cortex following novel taste learning. *Neurobiol Learn Mem* 2010;**94**:21–9.
279. Gutierrez H, Miranda MI, Bermudez-Rattoni F. Learning impairment and cholinergic deafferentation after cortical nerve growth factor deprivation. *J Neurosci* 1997;**17**:3796–803.
280. Miranda MI, Bermudez-Rattoni F. Reversible inactivation of the nucleus basalis magnocellularis induces disruption of cortical acetylcholine release and acquisition, but not retrieval, of aversive memories. *Proc Natl Acad Sci USA* 1999;**96**:6478–82.
281. Mark GP, Weinberg JB, Rada PV, Hoebel BG. Extracellular acetylcholine is increased in the nucleus accumbens following the presentation of an aversively conditioned taste stimulus. *Brain Res* 1995;**688**:184–8.
282. Shimura T, Suzuki M, Yamamoto T. Aversive taste stimuli facilitate extracellular acetylcholine release in the insular gustatory cortex of the rat: a microdialysis study. *Brain Res* 1995;**679**:221–6.
283. Guzman-Ramos K, Osorio-Gomez D, Moreno-Castilla P, Bermudez-Rattoni F. Off-line concomitant release of dopamine and glutamate involvement in taste memory consolidation. *J Neurochem* 2010;**114**:226–36.
284. Fenu S, Di Chiara G. Facilitation of conditioned taste aversion learning by systemic amphetamine: role of nucleus accumbens shell dopamine D1 receptors. *Eur J Neurosci* 2003;**18**: 2025–30.
285. Rosenblum K, Berman DE, Hazvi S, Dudai Y. Carbachol mimics effects of sensory input on tyrosine phosphorylation in cortex. *Neuroreport* 1996;**7**:1401–4.
286. Yasoshima Y, Yamamoto T. Rat gustatory memory requires protein kinase C activity in the amygdala and cortical gustatory area. *Neuroreport* 1997;**8**:1363–7.
287. Nunez-Jaramillo L. Serine phosphorylation of the NMDA receptor subunits NR2A and NR2B in the insular cortex is related with taste recognition. *Soc Neurosci Abstr* 2003;**721**:7.

288. Yamamoto T, Fujimoto Y. Brain mechanisms of taste aversion learning in the rat. *Brain Res Bull* 1991;**27**:403–6.
289. Yasoshima Y, Morimoto T, Yamamoto T. Effects of functional blockers injection into the amygdala on conditioned taste aversion in the rat. In: Murphy C, editor. *International Symposium on Olfaction and Taste XII*, vol. 855. San Diego; 1997. p. 14.
290. Young E, Cesena T, Meiri KF, Perrone-Bizzozero NI. Changes in protein kinase C (PKC) activity, isozyme translocation, and GAP-43 phosphorylation in the rat hippocampal formation after a single-trial contextual fear conditioning paradigm. *Hippocampus* 2002; **12**:457–64.
291. Koh MT, Thiele TE, Bernstein IL. Inhibition of protein kinase A activity interferes with long-term, but not short-term, memory of conditioned taste aversions. *Behav Neurosci* 2002;**116**: 1070–4.
292. Lamprecht R, Hazvi S, Dudai Y. cAMP response element-binding protein in the amygdala is required for long- but not short-term conditioned taste aversion memory. *J Neurosci* 1997;**17**: 8443–50.
293. Swank MW. Phosphorylation of MAP kinase and CREB in mouse cortex and amygdala during taste aversion learning. *Neuroreport* 2000;**11**:1625–30.
294. Shema R, Sacktor TC, Dudai Y. Rapid erasure of long-term memory associations in the cortex by an inhibitor of PKM zeta. *Science* 2007;**317**:951–3.
295. Desmedt A, Hazvi S, Dudai Y. Differential pattern of cAMP response element-binding protein activation in the rat brain after conditioned aversion as a function of the associative process engaged: taste versus context association. *J Neurosci* 2003;**23**:6102–10.
296. Bahar A, Dorfman N, Dudai Y. Amygdalar circuits required for either consolidation or extinction of taste aversion memory are not required for reconsolidation. *Eur J Neurosci* 2004;**19**:1115–8.
297. Stricker EM, Zigmond MJ. Effects on homeostasis of intraventricular injections of 6-hydroxydopamine in rats. *J Comp Physiol Psychol* 1974;**86**:973–94.
298. Wagner GC, Foltin RW, Seiden LS, Schuster CR. Dopamine depletion by 6-hydroxydopamine prevents conditioned taste aversion induced by methylamphetamine but not lithium chloride. *Pharmacol Biochem Behav* 1981;**14**:85–8.
299. Borsini F, Rolls ET. Role of noradrenaline and serotonin in the basolateral region of the amygdala in food preferences and learned taste aversions in the rat. *Physiol Behav* 1984;**33**: 37–43.
300. Miyashita T, Williams CL. Glutamatergic transmission in the nucleus of the solitary tract modulates memory through influences on amygdala noradrenergic systems. *Behav Neurosci* 2002;**116**:13–21.
301. Rescorla RA. Preservation of Pavlovian association through extinction. *Q J Exp Psychol* 1996;**49B**:245–58.
302. Vianna MR, Igaz LM, Coitinho AS, Medina JH, Izquierdo I. Memory extinction requires gene expression in rat hippocampus. *Neurobiol Learn Mem* 2003;**79**:199–203.
303. Berman DE, Dudai Y. Memory extinction, learning anew, and learning the new: dissociations in the molecular machinery of learning in cortex. *Science* 2001;**291**:2417–9.
304. Berman DE, Hazvi S, Stehberg J, Bahar A, Dudai Y. Conflicting processes in the extinction of conditioned taste aversion: behavioral and molecular aspects of latency, apparent stagnation, and spontaneous recovery. *Learn Mem* 2003;**10**:16–25.
305. Jacobson LH, Kelly PH, Bettler B, Kaupmann K, Cryan JF. GABA(B(1)) receptor isoforms differentially mediate the acquisition and extinction of aversive taste memories. *J Neurosci* 2006;**26**:8800–3.
306. Blackshaw LA, Brookes SJ, Grundy D, Schemann M. Sensory transmission in the gastrointestinal tract. *Neurogastroenterol Motil* 2007;**19**:1–19.

307. Young RL. Sensing via intestinal sweet taste pathways. *Front Neurosci* 2011;**5**:23.
308. Azpiroz F, Malagelada JR. Gastric tone measured by an electronic barostat in health and postsurgical gastroparesis. *Gastroenterology* 1987;**92**:934–43.
309. Horowitz M, Dent J. Disordered gastric emptying: mechanical basis, assessment and treatment. *Baillieres Clin Gastroenterol* 1991;**5**:371–407.
310. Raybould HE. Does your gut taste? Sensory transduction in the gastrointestinal tract. *News Physiol Sci* 1998;**13**:275–80.
311. Furness JB, Kunze WA, Clerc N. Nutrient tasting and signaling mechanisms in the gut. II. The intestine as a sensory organ: neural, endocrine, and immune responses. *Am J Physiol* 1999;**277**:G922–8.
312. Furness JB, Jones C, Nurgali K, Clerc N. Intrinsic primary afferent neurons and nerve circuits within the intestine. *Prog Neurobiol* 2004;**72**:143–64.
313. Furness JB, Kunze WA, Bertrand PP, Clerc N, Bornstein JC. Intrinsic primary afferent neurons of the intestine. *Prog Neurobiol* 1998;**54**:1–18.
314. Bernstein IL, Chavez M, Allen D, Taylor EM. Area postrema mediation of physiological and behavioral effects of lithium chloride in the rat. *Brain Res* 1992;**575**:132–7.
315. Ritter S, McGlone JJ, Kelley KW. Absence of lithium-induced taste aversion after area postrema lesion. *Brain Res* 1980;**201**:501–6.
316. Ossenkopp KP, Sutherland C, Ladowsky RL. Motor activity changes and conditioned taste aversions induced by administration of scopolamine in rats: role of the area postrema. *Pharmacol Biochem Behav* 1986;**25**:269–76.
317. Yamamoto T. Electrophysiology of CTA. In: Bermúdez-Rattoni F, Yamamoto T, editors. *Memory of a special kind: conditioned taste aversion*. New York: Oxford University Press; 1998. pp. 76–91.
318. Tsukamoto G, Adachi A. Neural responses of rat area postrema to stimuli producing nausea. *J Auton Nerv Syst* 1994;**49**:55–60.
319. Behrens M, Meyerhof W. Gustatory and extragustatory functions of mammalian taste receptors. *Physiol Behav* 2011;**105**:4–13. doi:10.1016/j.physbeh.2011.02.010.
320. Hofer D, Asan E, Drenckhahn D. Chemosensory perception in the gut. *News Physiol Sci* 1999;**14**:18–23.
321. Hass N, Schwarzenbacher K, Breer H. T1R3 is expressed in brush cells and ghrelin-producing cells of murine stomach. *Cell Tissue Res* 2010;**339**:493–504.
322. Rozengurt E, Sternini C. Taste receptor signaling in the mammalian gut. *Curr Opin Pharmacol* 2007;**7**:557–62.
323. Jang HJ, Kokrashvili Z, Theodorakis MJ, Carlson OD, Kim BJ, Zhou J, et al. Gut-expressed gustducin and taste receptors regulate secretion of glucagon-like peptide-1. *Proc Natl Acad Sci USA* 2007;**104**:15069–74.
324. Margolskee RF, Dyer J, Kokrashvili Z, Salmon KS, Ilegems E, Daly K, et al. T1R3 and gustducin in gut sense sugars to regulate expression of Na+-glucose cotransporter 1. *Proc Natl Acad Sci USA* 2007;**104**:15075–80.
325. Young RL, Sutherland K, Pezos N, Brierley SM, Horowitz M, Rayner CK, et al. Expression of taste molecules in the upper gastrointestinal tract in humans with and without type 2 diabetes. *Gut* 2009;**58**:337–46.
326. Steinert RE, Frey F, Topfer A, Drewe J, Beglinger C. Effects of carbohydrate sugars and artificial sweeteners on appetite and the secretion of gastrointestinal satiety peptides. *Br J Nutr* 2011;**105**:1320–8.
327. Ma J, Bellon M, Wishart JM, Young R, Blackshaw LA, Jones KL, et al. Effect of the artificial sweetener, sucralose, on gastric emptying and incretin hormone release in healthy subjects. *Am J Physiol Gastrointest Liver Physiol* 2009;**296**:G735–9.

328. Little TJ, Gupta N, Case RM, Thompson DG, McLaughlin JT. Sweetness and bitterness taste of meals per se does not mediate gastric emptying in humans. *Am J Physiol Regul Integr Comp Physiol* 2009;**297**:R632–9.
329. Sirinek KR, Levine BA, O'Dorisio TM, Cataland S. Gastric inhibitory polypeptide (GIP) release by actively transported, structurally similar carbohydrates. *Proc Soc Exp Biol Med* 1983;**173**:379–85.
330. Shima K, Suda T, Nishimoto K, Yoshimoto S. Relationship between molecular structures of sugars and their ability to stimulate the release of glucagon-like peptide-1 from canine ileal loops. *Acta Endocrinol (Copenh)* 1990;**123**:464–70.
331. Glendinning JI, Yiin YM, Ackroff K, Sclafani A. Intragastric infusion of denatonium conditions flavor aversions and delays gastric emptying in rodents. *Physiol Behav* 2008;**93**: 757–65.
332. Chen MC, Wu SV, Reeve Jr. JR, Rozengurt E. Bitter stimuli induce Ca2+ signaling and CCK release in enteroendocrine STC-1 cells: role of L-type voltage-sensitive Ca2+ channels. *Am J Physiol Cell Physiol* 2006;**291**:C726–39.
333. Bezencon C, le Coutre J, Damak S. Taste-signaling proteins are coexpressed in solitary intestinal epithelial cells. *Chem Senses* 2007;**32**:41–9.
334. San Gabriel AM, Maekawa T, Uneyama H, Yoshie S, Torii K. mGluR1 in the fundic glands of rat stomach. *FEBS Lett* 2007;**581**:1119–23.
335. Zolotarev V, Khropycheva R, Uneyama H, Torii K. Effect of free dietary glutamate on gastric secretion in dogs. *Ann N Y Acad Sci* 2009;**1170**:87–90.
336. Akiba Y, Watanabe C, Mizumori M, Kaunitz JD. Luminal L-glutamate enhances duodenal mucosal defense mechanisms via multiple glutamate receptors in rats. *Am J Physiol Gastrointest Liver Physiol* 2009;**297**:G781–91.
337. Zai H, Kusano M, Hosaka H, Shimoyama Y, Nagoshi A, Maeda M, et al. Monosodium L-glutamate added to a high-energy, high-protein liquid diet promotes gastric emptying. *Am J Clin Nutr* 2009;**89**:431–5.
338. Edfalk S, Steneberg P, Edlund H. Gpr40 is expressed in enteroendocrine cells and mediates free fatty acid stimulation of incretin secretion. *Diabetes* 2008;**57**:2280–7.
339. Lauffer LM, Iakoubov R, Brubaker PL. GPR119 is essential for oleoylethanolamide-induced glucagon-like peptide-1 secretion from the intestinal enteroendocrine L-cell. *Diabetes* 2009; **58**:1058–66.
340. Chu ZL, Carroll C, Alfonso J, Gutierrez V, He H, Lucman A, et al. A role for intestinal endocrine cell-expressed g protein-coupled receptor 119 in glycemic control by enhancing glucagon-like Peptide-1 and glucose-dependent insulinotropic Peptide release. *Endocrinology* 2008;**149**:2038–47.
341. Cherbut C, Ferrier L, Roze C, Anini Y, Blottiere H, Lecannu G, et al. Short-chain fatty acids modify colonic motility through nerves and polypeptide YY release in the rat. *Am J Physiol* 1998;**275**:G1415–22.
342. Hirasawa A, Hara T, Katsuma S, Adachi T, Tsujimoto G. Free fatty acid receptors and drug discovery. *Biol Pharm Bull* 2008;**31**:1847–51.
343. Adachi T, Tanaka T, Takemoto K, Koshimizu TA, Hirasawa A, Tsujimoto G. Free fatty acids administered into the colon promote the secretion of glucagon-like peptide-1 and insulin. *Biochem Biophys Res Commun* 2006;**340**:332–7.
344. Abumrad NA. CD36 may determine our desire for dietary fats. *J Clin Invest* 2005;**115**: 2965–7.
345. Tanaka T, Katsuma S, Adachi T, Koshimizu TA, Hirasawa A, Tsujimoto G. Free fatty acids induce cholecystokinin secretion through GPR120. *Naunyn Schmiedebergs Arch Pharmacol* 2008;**377**:523–7.

346. Sternini C, Anselmi L, Rozengurt E. Enteroendocrine cells: a site of 'taste' in gastrointestinal chemosensing. *Curr Opin Endocrinol Diabetes Obes* 2008;**15**:73–8.
347. Ray JM, Squires PE, Curtis SB, Meloche MR, Buchan AM. Expression of the calcium-sensing receptor on human antral gastrin cells in culture. *J Clin Invest* 1997;**99**:2328–33.
348. San Gabriel A, Uneyama H, Maekawa T, Torii K. The calcium-sensing receptor in taste tissue. *Biochem Biophys Res Commun* 2009;**378**:414–8.
349. Tordoff MG, Shao H, Alarcon LK, Margolskee RF, Mosinger B, Bachmanov AA, et al. Involvement of T1R3 in calcium-magnesium taste. *Physiol Genomics* 2008;**34**:338–48.
350. San Gabriel A, Nakamura E, Uneyama H, Torii K. Taste, visceral information and exocrine reflexes with glutamate through umami receptors. *J Med Invest* 2009;**56**:209–17.
351. Busque M, Kerstetter JE, Geibel JP, Insogna K. L-type amino acids stimulate gastric acid secretion in parietal cells. *Am J Physiol Gastrointest Liver Physiol* 2005;**289**:G664–9.
352. Hass N, Schwarzenbacher K, Breer H. A cluster of gustducin-expressing cells in the mouse stomach associated with two distinct populations of enteroendocrine cells. *Histochem Cell Biol* 2007;**128**:457–71.
353. Rozengurt N, Wu SV, Chen MC, Huang C, Sternini C, Rozengurt E. Colocalization of the alpha-subunit of gustducin with PYY and GLP-1 in L cells of human colon. *Am J Physiol Gastrointest Liver Physiol* 2006;**291**:G792–802.
354. Hofer D, Drenckhahn D. Identification of the taste cell G-protein, alpha-gustducin, in brush cells of the rat pancreatic duct system. *Histochem Cell Biol* 1998;**110**:303–9.
355. Hofer D, Drenckhahn D. Identification of brush cells in the alimentary and respiratory system by antibodies to villin and fimbrin. *Histochemistry* 1992;**98**:237–42.
356. Gebhard A, Gebert A. Brush cells of the mouse intestine possess a specialized glycocalyx as revealed by quantitative lectin histochemistry. Further evidence for a sensory function. *J Histochem Cytochem* 1999;**47**:799–808.
357. Elson AE, Dotson CD, Egan JM, Munger SD. Glucagon signaling modulates sweet taste responsiveness. *FASEB J* 2010;**24**:3960–9.
358. Kawai K, Sugimoto K, Nakashima K, Miura H, Ninomiya Y. Leptin as a modulator of sweet taste sensitivities in mice. *Proc Natl Acad Sci USA* 2000;**97**:11044–9.
359. Herness S, Zhao FL, Lu SG, Kaya N, Shen T. Expression and physiological actions of cholecystokinin in rat taste receptor cells. *J Neurosci* 2002;**22**:10018–29.
360. Shen T, Kaya N, Zhao FL, Lu SG, Cao Y, Herness S. Co-expression patterns of the neuropeptides vasoactive intestinal peptide and cholecystokinin with the transduction molecules alpha-gustducin and T1R2 in rat taste receptor cells. *Neuroscience* 2005;**130**:229–38.
361. Zhao FL, Shen T, Kaya N, Lu SG, Cao Y, Herness S. Expression, physiological action, and coexpression patterns of neuropeptide Y in rat taste-bud cells. *Proc Natl Acad Sci USA* 2005;**102**:11100–5.
362. Martin B, Shin YK, White CM, Ji S, Kim W, Carlson OD, et al. Vasoactive intestinal peptide-null mice demonstrate enhanced sweet taste preference, dysglycemia, and reduced taste bud leptin receptor expression. *Diabetes* 2010;**59**:1143–52.
363. Shin YK, Martin B, Golden E, Dotson CD, Maudsley S, Kim W, et al. Modulation of taste sensitivity by GLP-1 signaling. *J Neurochem* 2008;**106**:455–63.
364. Horio N, Jyotaki M, Yoshida R, Sanematsu K, Shigemura N, Ninomiya Y. New frontiers in gut nutrient sensor research: nutrient sensors in the gastrointestinal tract: modulation of sweet taste sensitivity by leptin. *J Pharmacol Sci* 2010;**112**:8–12.
365. Blakeslee AF. Genetics of sensory thresholds: taste for phenyl thio carbamide. *Proc Natl Acad Sci USA* 1932;**18**:120–30.
366. Fox AL. The relationship between chemical constitution and taste. *Proc Natl Acad Sci USA* 1932;**18**:115–20.

367. Wong KK, deLeeuw RJ, Dosanjh NS, Kimm LR, Cheng Z, Horsman DE, et al. A comprehensive analysis of common copy-number variations in the human genome. *Am J Hum Genet* 2007;**80**:91–104.
368. Kim U, Wooding S, Ricci D, Jorde LB, Drayna D. Worldwide haplotype diversity and coding sequence variation at human bitter taste receptor loci. *Hum Mutat* 2005;**26**:199–204.
369. Zhao H, Yang JR, Xu H, Zhang J. Pseudogenization of the umami taste receptor gene Tas1r1 in the giant panda coincided with its dietary switch to bamboo. *Mol Biol Evol* 2010;**27**: 2669–73.
370. Roudnitzky N, Bufe B, Thalmann S, Kuhn C, Gunn HC, Xing C, et al. Genomic, genetic, and functional dissection of bitter taste responses to artificial sweeteners. *Hum Mol Genet* 2011; **20**:3437–49.
371. Kim UK, Breslin PA, Reed D, Drayna D. Genetics of human taste perception. *J Dent Res* 2004;**83**:448–53.
372. Guo SW, Reed DR. The genetics of phenylthiocarbamide perception. *Ann Hum Biol* 2001;**28**:111–42.
373. Tepper BJ. 6-n-Propylthiouracil: a genetic marker for taste, with implications for food preference and dietary habits. *Am J Hum Genet* 1998;**63**:1271–6.
374. Wooding S, Kim UK, Bamshad MJ, Larsen J, Jorde LB, Drayna D. Natural selection and molecular evolution in PTC, a bitter-taste receptor gene. *Am J Hum Genet* 2004;**74**: 637–46.
375. Meyerhof W, Born S, Brockhoff A, Behrens M. Molecular biology of mammalian bitter taste receptors. A review. *Flavour Frag J* 2011;**26**:260–8.
376. Fahey J, Stephensom K, Talalay P. Glucosinolates, myrosinase, and isothiocyanates: three reasons for eating brassica vegetables. In: Shibamoto editor, *Functional foods for disease prevention 1: Fruits, vegetables and teas* 1998;**701**:16–22.
377. Vanderpas J. Nutritional epidemiology and thyroid hormone metabolism. *Annu Rev Nutr* 2006;**26**:293–322.
378. Wooding S, Bufe B, Grassi C, Howard MT, Stone AC, Vazquez M, et al. Independent evolution of bitter-taste sensitivity in humans and chimpanzees. *Nature* 2006;**440**:930–4.
379. Hinrichs AL, Wang JC, Bufe B, Kwon JM, Budde J, Allen R, et al. Functional variant in a bitter-taste receptor (hTAS2R16) influences risk of alcohol dependence. *Am J Hum Genet* 2006;**78**:103–11.
380. Kim UK, Wooding S, Riaz N, Jorde LB, Drayna D. Variation in the human TAS1R taste receptor genes. *Chem Senses* 2006;**31**:599–611.
381. Raliou M, Grauso M, Hoffmann B, Schlegel-Le-Poupon C, Nespoulous C, Debat H. Human genetic polymorphisms in T1R1 and T1R3 taste receptor subunits affect their function. *Chem Senses* 2011;**36**:527–37.
382. Raliou M, Wiencis A, Pillias AM, Planchais A, Eloit C, Boucher Y, et al. Nonsynonymous single nucleotide polymorphisms in human tas1r1, tas1r3, and mGluR1 and individual taste sensitivity to glutamate. *Am J Clin Nutr* 2009;**90**:789S–99S.
383. Chen QY, Alarcon S, Tharp A, Ahmed OM, Estrella NL, Greene TA, et al. Perceptual variation in umami taste and polymorphisms in TAS1R taste receptor genes. *Am J Clin Nutr* 2009;**90**:770S–9S.
384. Shigemura N, Shirosaki S, Ohkuri T, Sanematsu K, Islam AS, Ogiwara Y. Variation in umami perception and in candidate genes for the umami receptor in mice and humans. *Am J Clin Nutr* 2009;**90**:764S–9S.
385. Shigemura N, Shirosaki S, Sanematsu K, Yoshida R, Ninomiya Y. Genetic and molecular basis of individual differences in human umami taste perception. *PLoS One* 2009;**4**: e6717.

386. Fushan AA, Simons CT, Slack JP, Manichaikul A, Drayna D. Allelic polymorphism within the TAS1R3 promoter is associated with human taste sensitivity to sucrose. *Curr Biol* 2009;**19**: 1288–93.
387. Mainland JD, Matsunami H. Taste perception: how sweet it is (to be transcribed by you). *Curr Biol* 2009;**19**:R655–6.
388. Fushan AA, Simons CT, Slack JP, Drayna D. Association between common variation in genes encoding sweet taste signaling components and human sucrose perception. *Chem Senses* 2010;**35**:579–92.
389. Li X, Li W, Wang H, Cao J, Maehashi K, Huang L, et al. Pseudogenization of a sweet-receptor gene accounts for cats' indifference toward sugar. *PLoS Genet* 2005;**1**:27–35.
390. Li X, Li W, Wang H, Bayley DL, Cao J, Reed DR, et al. Cats lack a sweet taste receptor. *J Nutr* 2006;**136**:1932S–4S.
391. Zhao H, Zhou Y, Pinto CM, Charles-Dominique P, Galindo-Gonzalez J, Zhang S, et al. Evolution of the sweet taste receptor gene Tas1r2 in bats. *Mol Biol Evol* 2010;**27**:2642–50.
392. Thompson RD, Elias DJ, Shumake SA, Gaddis SE. Taste preferences of the common vampire (Desmodus rotundus). *J Chem Ecol* 1982;**8**:715–21.
393. Li R, Fan W, Tian G, Zhu H, He L, Cai J, et al. The sequence and de novo assembly of the giant panda genome. *Nature* 2010;**463**:311–7.

Nutrition and the Epigenome

Paul Haggarty

Lifelong Health, Rowett Institute of Nutrition and Health, University of Aberdeen, Aberdeen, UK

Epigenetic regulation is central to genome structure and function. Epigenetic status varies between individuals, and there is increasing awareness of the importance of this variation in health and disease. Epigenetic mechanisms include DNA methylation, histone modification, and regulation by noncoding RNAs. Epigenetic control is central to the way in which the genome interacts with, and responds to, the environment and even potentially the way in which the genome can influence its own environment via effects on behavior. The substrates for epigenetic reactions (acetyl and methyl groups) are central to nutritional metabolism, and there is ample evidence for nutritional effects on the epigenome. Challenges in human nutritional epigenetics research include the problem of tissue-specific epigenomes and heterogeneity of response by epigenetic loci. The promise of nutritional epigenetics is that it will help elucidate the way in which nutrition can influence health through direct effects on the genome.

I. Epigenetics

Epigenetics has been variously defined as "heritable changes in gene function that cannot be explained by changes in DNA sequence,"[1] a collection of mechanisms that define the phenotype of a cell without affecting the genotype,[2] and "the structural adaptation of chromosomal regions so as to register, signal or perpetuate altered activity states."[3] The latter definition in particular captures the central importance of chromatin structure in epigenetics. At its most fundamental level, epigenetics is about information and specifically the information present in the genome over and above that coded in the DNA sequence. This epigenetic information determines how, when, and where the

Progress in Molecular Biology
and Translational Science, Vol. 108
DOI: 10.1016/B978-0-12-398397-8.00016-2

1877-1173/12 $35.00

sequence information is used. Epigenetics lies at the heart of a series of feedback loops involving genome structure and genetic variation (Fig. 1), and there is an increasing awareness of the importance of epigenetics in determining biological function, including susceptibility to disease,[4–9] cognitive function,[10,11] and reproduction.[12] Epigenetics is also central to the way in which the genome interacts with and responds to the environment and even potentially the way in which the genome can influence its own environment via effects on behavior (Fig. 1).

Genome organization has been the subject of a number of recent excellent reviews.[13–17] The simplest and most basic level of genome organization is the linear sequence of bases in DNA. Above that is the nucleosome which consists of DNA wrapped around an octamer core of histone proteins. Multiple nucleosomes form fibers of around 10 nm in diameter, and these fibers can further compact to form more complex higher order fibers, which in turn give rise to subchromosomal domains. These domains, which span around 1 million bases, undergo further folding to produce chromosomes. Chromosomes are themselves ordered within the nucleus, and all of the above, with the exception of the DNA sequence, may differ between different cell types. Epigenetic processes operate at all of these levels of organization higher order chromatin organization, nucleosome modeling, gene expression—through a range of mechanisms that include DNA methylation, histone modification, and regulation by noncoding RNAs.[2]

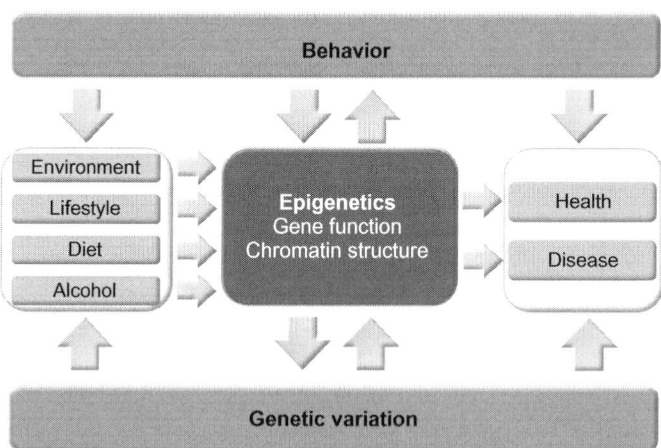

FIG. 1. Epigenetics lies at the heart of multiple processes linking environmental exposures, health, behavior, and genotype.

Chromatin is organized into accessible regions of euchromatin and poorly accessible regions of heterochromatin, and epigenetic control is fundamental to the transition between these states.[18,19] The N-terminal tails of histones may be modified by acetylation or methylation, and changes within the chromatin structure brought about by these modifications (sometimes called the histone code) influence many biological processes.[18,20,21] Histone acetyltransferases add acetyl groups to the amino groups of several lysine residues in the exposed tails of the histone octamer, a modification associated with transcription[18] (Fig. 2). Histone deacetylases (HDACs) give rise to transcriptional repression by removing acetyl groups from the histone tails.[18] Transcription may also be affected by histone methylation at specific arginine and lysine residues. This process is catalyzed by the histone methyltransferases.

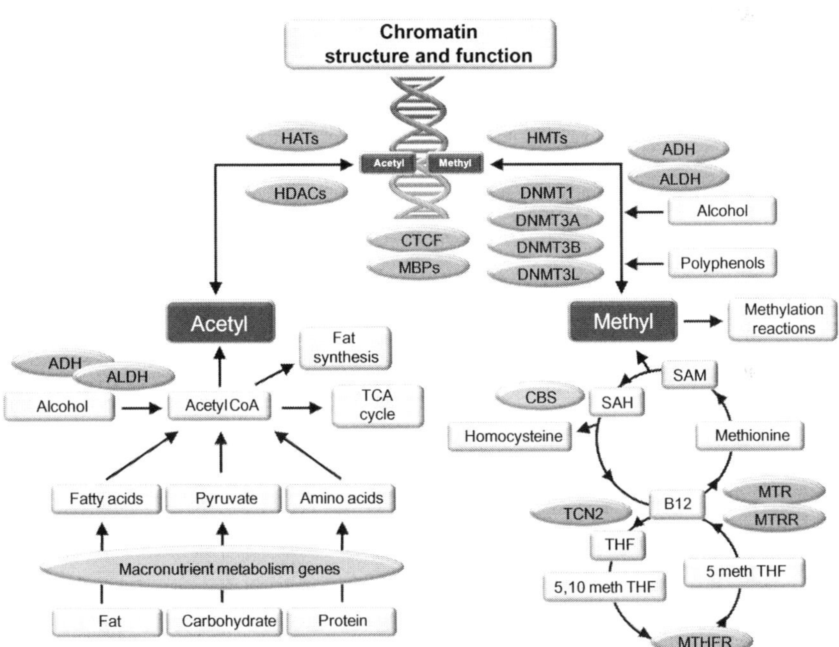

Fig. 2. Potential links between epigenetic processes, nutrients, nutrient metabolism, and diet. Acetyl, acetyl groups; Methyl, methyl groups; SAH, *S*-adenosylhomocysteine; SAM, *S*-adenosylmethionine; THF, tetrahydrofolate; HDACs, histone deacetylases; HATs, histone acetyltransferases; HMTs, histone methyltransferases; DNMTs, DNA methyltransferases; ALDH, acetaldehyde dehydrogenase; ADH, alcohol dehydrogenase; CBS, cystathionine beta synthase; TCN2, transcobalamin; MTR, methionine synthase; MTRR, methionine synthase reductase; MTHFR, methylenetetrahydrofolate reductase; CTCF, transcriptional repressor (CCCTC-binding factor); MBPs, methyl binding proteins; TCA, tricarboxylic acid.

Mammalian DNA methylation occurs on cytosine located 5′ to guanosine (CpG site).[18,22] Regions rich in CpG sites are often found in gene bodies, endogenous repeats, and transposable elements.[23] These CpG "islands" are thought to be important in transcriptional repression.[23] A significant component of the global methylation signature (average level of methylation across the entire genome) is accounted for by the transposons. Classes of transposons include the long interspersed nuclear elements (LINE1), the intracisternal A particle, the short interspersed transposable nuclear elements (SINE), and the *Alu* family of human *SINE* elements.[24,25] Some transposon classes are able to move around the genome and have the potential to cause abnormal function and disease if inserted into areas of the genome where the sequence is important for function.[19,24,25] These are often heavily methylated (~90%), and this has the effect of repressing transposition and protecting the early embryo in particular from potentially damaging genome rearrangement during critical periods of development.

Most autosomal genes are expressed equally from both parental alleles, but imprinted genes are an exception. Imprinted genes are known to be important in determining in prenatal growth, placental function, and brain function and behavior.[11,26,27] Imprinting occurs when genes are epigenetically marked within the germ cells in a parent-of-origin-specific manner such that the subsequent expression pattern depends on the parent the allele was derived from.[2,7,18,19] The germ line imprint may be distinct from the promoter region, but it is necessary for subsequent promoter methylation.[19] Around 80% of imprinted genes are found in clusters with other imprinted genes, suggesting coordinated regulation of the genes within a chromosomal domain, often downstream of regions of DNA that have a high density of CpG sites.[19] Imprinting centers may occur in these clusters, and these are thought to exert regional control of imprinted expression and methylation. An example of such a cluster contains the maternally expressed transcript (*H19*) gene and the paternally expressed insulin-like growth factor 2 (*IGF2*) gene. This imprinting center can also function as a methylation-sensitive insulator that binds the transcriptional repressor CTCF (CCCTC-binding factor) and controls the interaction of enhancers with maternal *IGF2* promoters.[28] Repetitive regions flanking imprinted genes are thought to be important in setting and maintaining the imprint.[29–31] Some imprinted regions can acquire tissue-specific expression and may vary with stage of development,[11,19,32] but, in general, the imprint is relatively stable over decades and is retained in multiple tissues throughout the life span.[33] Although imprinting methylation is reasonably stable within human populations, there is variation around the mean value of 50% characteristic of imprinted genes, with considerable interest in the biological significance of this variation.[9,33,34] Epigenetic status varies between individuals[34–36] and even between genetically identical monozygotic twins.[37]

Leukocyte LINE1 methylation has been reported to differ by gender and race/ethnicity,[38] but other reports suggest no differences between Caucasian and African-American women.[39]

DNA methylation is carried out by the DNA methyltransferases (DNMTs) (Fig. 2). Humans have four main variants: *DNMT1*, *DNMT3A*, *DNMT3B*, and *DNMT3L*. There is some overlap in function between these variants, but the main function of DNMT1 is to facilitate the propagation of existing methylation patterns, while DNMT3A, 3B, and 3L are primarily required for *de novo* methylation.[22] Although the primary role of DNMT1 is maintenance methylation, there are germ-cell-specific isoforms, suggesting that it may also have some role in supporting *de novo* methylation.[19,40] DNMT3L does not itself exhibit methyltransferase activity, but it is required for the establishment of parent-of-origin-specific methylation through a direct interaction with DNMT3A and 3B.[40–42] DNMT3A, 3B, and 3L are expressed in oocytes at the same time as imprinting methylation occurs,[31] and DNMT3A and DNMT3L in particular are necessary for the imprint to be set.[43,44] The molecular interaction between DNMT3A and 3L, which occurs during DNA methylation, has been well characterized,[41,42] and it has been proposed that DNMT3A activity may compensate to some extent for the loss of DNMT3L, resulting in imperfect stochastic imprinting.[45] In sperm, *Dnmt1*, *Dnmt3a*, *Dnmt3b*, and *Dnmt3l* expression is linked to stage of development in the testis after birth, and the absence of 3a or 3l results in failure of spermatogenesis and infertility.[40] Loss of *Dnmt3l* function in animal knockouts results in loss of imprinting, biallelic expression of imprinted genes, altered methylation at nonimprinted loci, and impaired reproductive function.[45] The critical importance of the DNMTs is reflected in the fact that the functional domains are highly conserved, and in a 2009 study, no nonsynonymous polymorphisms were detected in the catalytic domains of DNMT3A, DNMT3B, or DNMT1 in a European population.[46]

There is overlap and interaction between different types of epigenetic regulation. DNA methylation affects histone acetylation and histone methylation,[3] and numerous mechanisms have been identified by which this might occur. DNMT3L promotes *de novo* DNA methylation by recruitment or activation of DNMT3A in response to histone H3 tails that are unmethylated at lysine 4.[47] Altered DNA methylation can correspond to changes in the histones (e.g., trimethylation of the lysine 9 residue of H3); however, the mapping between these two epigenetic processes is not absolute, and the nature of the interactions between DNA methylation and histone acetylation and methylation are not fully understood. The processes controlling propagation of the epigenetic mark across regions of the genome, as well as how the epigenetic signal in one gene/region may influence another, is also the subject of intense research activity. It has been proposed that the protein CTCF may be involved in this process[48,49]; CTCF is a zinc finger protein insulator that mediates intra- and interchromosomal contacts and coordinates DNA methylation and higher order chromatin structure.[48,49]

DNA demethylation is as important as methylation in determining epigenetic status, but this process has not been fully elucidated. Demethylation can occur passively or actively. Passive demethylation may occur when DNMT1 fails to reestablish the fully methylated state following DNA replication and cell division. However, this cannot explain the demethylation that occurs in cells not undergoing division.[50] In the mismatch repair pathway, 5-methyl cytosine is deaminated to thymine, and the T-G mismatch repair mechanism replaces the thymine with an unmethylated cytosine, resulting in excision of the entire methylated cytosine from DNA.[5,51] However, the exact mechanism underpinning the process of active demethylation is still controversial, and other mechanisms have also been proposed.[50]

II. Nutritional Effects

There is a growing body of empirical evidence demonstrating nutritional influences on epigenetic status and a growing understanding of the mechanisms by which this can occur. This is perhaps not surprising because the methyl and acetyl groups that constitute the key epigenetic marks are at the heart of nutritional metabolism (Fig. 2). Nutrition can influence epigenetic status through the availability of substrate in epigenetic reactions, direct effects on the proteins involved in epigenetic marking, direct effects on the genome, and the selection and propagation of cells with particular epigenetic profiles.

The ultimate methyl donor for epigenetic-methylation reactions is *S*-adenosylmethionine (SAM), which is produced as part of the folate–methylation cycle (Fig. 2). Nutritional and genetic factors that affect the activity of this cycle also influence epigenetic marking, and Stover recently reviewed the effect of the B vitamins, which are an intrinsic part of the cycle.[52,53] Human DNA lymphocyte hypomethylation is associated with low folate status and elevated homocysteine.[54,55] There is also evidence of gene–nutrient interaction in respect of the genes of the folate–methylation cycle (Fig. 2). Polymorphisms in the methylenetetrahydrofolate reductase (*MTHFR*) gene interact with folate status to influence DNA methylation.[56,57] In addition, folate has direct effects on genome structure and function, perhaps also operating through epigenetic mechanisms (see chapter "Genetic and Epigenomic Footprints of Folate"). Folate-sensitive fragile sites are regions of chromatin that fail to compact normally during mitosis when folate and thymidine are deficient,[8] and over 20 sites have been observed in the human genome. Other effects of the B vitamins on epigenetic regulation have been reported. Examples include the effect of niacin on chromatin structure and function,[58] as well as biotin binding to histones and its influence on the retrotransposons.[59]

Acetyl groups provide the substrate for acetylation of histones. While methylation is likely to be particularly sensitive to the B vitamins, by virtue of their direct involvement in the methylation cycle, acetyl metabolism is probably more important to a greater range of nutrients (Fig. 2). It is the final common intermediate in the catabolism of protein, fat, and carbohydrate; it is the substrate for many reactions, including fatty acid and cholesterol synthesis; and it is the primary substrate for energy production within the tricarboxylic acid cycle. However, less work has been done on the effects of nutritional metabolism relevant to acetylation than methylation (see chapter "Genetic and Epigenomic Footprints of Folate").

Alcohol may influence epigenetic processes through both the methyl and acetyl pathways. Alcohol is known to interact with methyl group metabolism, and it also has the potential to influence the epigenetic pathway through acetylation as it is metabolized to acetate and acetyl-CoA. DNA methylation is altered in animal models of chronic alcohol exposure,[60,61] and alcohol exposure during early neurulation results in changes in DNA methylation patterns and gene expression.[62] Ethanol consumption inhibits fetal DNA methylation in mice, and even relatively low concentrations of acetaldehyde inhibit DNMT activity in animal models *in vitro*.[61] Chronic administration of alcohol to rats results in myelocytomatosis oncogene (*Myc*) hypomethylation and altered expression of the methionine adenosyltransferases that catalyze the formation of SAM.[63] Paternal alcohol exposure affects sperm cytosine methyltransferase messenger RNA levels.[64] DNA methylation has also been shown to vary with alcohol exposure[65,66] in humans.

In addition, there is evidence of direct nutritional effects on the enzymes involved in the epigenetic pathway. HDAC is inhibited by sulforaphane, which is a compound found in cruciferous vegetables,[67] while polyphenols in green tea, coffee, and soybean influence DNA methylation and DNMT activity.[68–70]

The timing of nutritional exposures is important, as many epigenetic events are restricted to specific phases of development and processes such as cellular differentiation and cell division. The period before birth in particular is marked by intense epigenetic activity,[18] and the transgenerational nature of imprinting provides a mechanism whereby epigenetic risk may be passed across the generations. Studies in rodent models have demonstrated that epigenetic status in the offspring is influenced by the maternal intake during pregnancy of nutrients that influence the availability of methyl groups. Examples include folic acid, choline, betaine,[7,9] low-protein diets,[71] and phytoestrogens.[7,72] Cord blood DNA in babies of mothers who took folic acid supplements during pregnancy had higher levels of *IGF2* methylation.[73] The level of *IGF2* methylation also was related to birth weight,[73] which is a predictor of disease risk in later life.[74] Changes in *IGF2* methylation have been observed in women decades after prenatal exposure to famine,[75] and these changes appear to be associated with increased breast cancer risk.[75]

III. Behavior and Epigenetic Feedback

Behaviors related to appetite and lifestyle have been linked to epigenetic processes. DNA methylation and epigenetic processes may be involved in the regulation of appetite[76–79] and alcohol intake.[80] Epigenetic status has been linked to body fatness, and there is a report that a polymorphism within the fat mass and obesity-associated (*FTO*) gene interacts with *DNMT3B* genetic variants to influence postprandial levels of hunger and satiety.[81] DNA methylation in the ATPase, class V, type 10A (*ATP10A*) and CD44 molecule (*CD44*) genes were associated with the level of weight loss achieved.[82] These authors concluded that DNA methylation patterns may be used as epigenetic markers that could help to predict weight loss.[82]

Interactions between epigenetics, nutrition, and behavior could have important consequences for our understanding of both the way in which diet influences health and how health-related behaviors may be improved. There is currently a great deal of interest in the importance of epigenetic factors in brain function, cognition, and cognitive aging.[83–86] A number of lines of evidence point in particular to the importance of imprinted genes[11,32] and repeat elements[87–92] in determining brain development and cognitive ability. Imprinted genes make up a small proportion of all genes, but they are primarily thought to affect brain function and behavior and prenatal growth.[11,26,27] These two key effects of the imprinted genes are consistent with the epidemiological link between intelligence and birth weight.[93–95] The differential maternal and paternal inheritance patterns of many mental disorders are also consistent with the imprinting control of brain function.[10,86,96]

Syndromes in which the imprint is disrupted, such as Angelman and Prader–Willi, are associated with low IQ. Prader–Willi is associated with an approximately normal distribution of IQ but with a mean 40 points below that of the general population, suggesting a global effect on IQ of the imprinted genes.[97] Angelman is similarly linked to low IQ, and other mental impairment syndromes have been associated with imprinting changes or polymorphisms in genes essential for imprinting.[98,99] The region 15q11–q13 is associated with Prader–Willi and Angelman syndromes and includes the imprinted genes ubiquitin protein ligase E3A (*UBE3A*) and SNRPN upstream reading frame (*SNURF*)/small nuclear ribonucleoprotein polypeptide N (*SNRPN*). The Beckwith–Wiedemann and Prader–Willi epimutations are thought to arise in the early embryo or the germ line, either by failure to erase the grandparental imprint or by establishing the wrong imprint.[19] Imprinting syndromes may result from serious chromosomal abnormalities such as uniparental disomy, but they can also arise from more subtle effects such as altered methylation and loss of imprinting, making these syndromes potentially relevant to normal epigenetic variation in the general population.[7]

Indirect evidence for the importance of imprinting in cognition comes from an association between a polymorphism in the gene *DNMT3L*—critical to the process of imprinting—and childhood intelligence.[100] Human cognitive ability, and its decline with age, have a high level of heritability,[101,102] and imprinting and epigenetic control of repetitive elements have the potential to account for some of this heritability.

Early experience is known to influence neural function and behavior, and it has been proposed that this occurs through epigenetic mechanisms (Fig. 3).[103,104] There is a growing body of evidence from animal studies demonstrating that factors such as maternal care can influence the

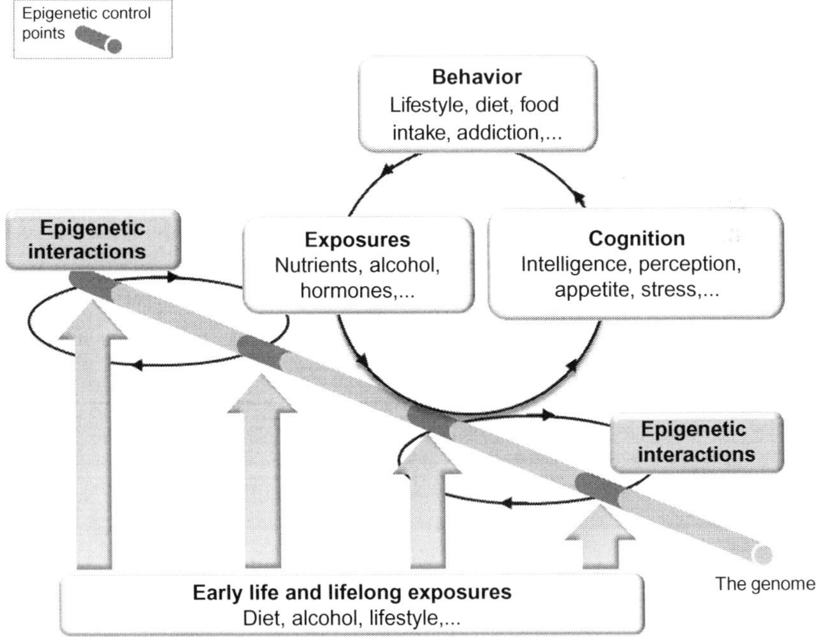

Fig. 3. Schematic illustration of the way in which exposures in early life and through the life span may affect multiple epigenetic control points in the genome and how these can influence behaviors and new environmental exposures in a way that can ultimately feed back to reinforce or modify epigenetic status.

hypothalamic–pituitary–adrenal (HPA) function in the offspring through epigenetic programming of glucocorticoid receptor expression and altered histone acetylation and transcription factor (nerve growth factor-inducible protein-A) binding to the glucocorticoid receptor promoter.[103–105] Abuse in children is known to alter HPA stress responses and increase the risk of suicide in later life, and studies of the hippocampus in suicide victims have shown epigenetic changes in the glucocorticoid receptor associated with a history of childhood abuse.[106]

High levels of maternal stress are associated with low birth weight,[107,108] and plausible biological mechanisms, involving the programming of HPA function, have been proposed to explain this link.[109] In animals, prenatal stress and altered glucocorticoid exposure not only alter the behavior of the offspring after birth, but also influence birth weight, risk of hypertension and hyperglycemia, and HPA axis activity in the offspring.[110] Cortisol levels of low-birth-weight babies are higher throughout life, and neuropsychiatric disorders in later life have been linked to maternal stress in pregnancy, with some of these effects being passed to subsequent generations.[110] Interesting from the point of view of nutrition is the observation that central infusion of methionine (a precursor of SAM; Fig. 2) into adult animals appears to reverse the effect of maternal behavior on responses to stress in the offspring and the epigenetic changes that result in altered glucocorticoid receptor expression and HPA function.[111]

IV. The Future of Nutritional Epigenetics Research

The quality, reliability, and coverage of epigenetics technologies are now reaching a stage where it is possible to study epigenetic changes in fine detail while also carrying out epigenome scanning equivalent to genome-wide association studies.[112,113] This holds out the promise of identifying predictive epigenetic markers of disease and, more usefully, epigenetic changes that may be causal in the development of disease in response to dietary and other environmental influences. The disadvantages inherent in genome-wide scanning technologies are the same regardless of whether they are applied to genetic or epigenetic variants; these include the problem of multiple testing, the generation of false positive results, and the need to adjust the level of significance for the number of tests performed.

There is a more fundamental problem in human epigenetics studies that does not apply to genetics studies and that is that the epigenome is specific to each tissue type and stage of development.[114] Indeed, epigenetic change is central to the process of differentiation. In animal models, it is possible to study epigenetic changes in response to diet in all tissues, but in most human nutrition studies involving epigenetic outcomes, the samples measured are usually blood or buccal cells, as these are easily accessible. It may be valid only in certain circumstances to extrapolate from these peripheral tissues to epigenetic status in tissues likely to be linked directly to health. These include studies of epigenetic marks (e.g., imprinting), which are set very early in development and are retained in multiple tissues; studies in which the peripheral epigenetic mark is known to change in concert with the tissue of interest; and studies in which the epigenetic mark in different tissues may respond in the same way to environmental influences under certain circumstances (e.g., global DNA methylation).

Other experimental designs that do not require direct measurement of epigenetic status in the tissues of interest can also be used to look for evidence of epigenetic involvement in health and disease and the response to diet. Although the evidence is indirect, the problem of tissue-specific epigenomes can be overcome in standard genetic association studies. Association between genetic variants in the genes involved in epigenetic processes, epigenetic status, and health has been used in this way.[100,115–120]

A further barrier to understanding nutritional effects on epigenetics is the fact that we are only just beginning to understand epigenetic processes themselves. Uncertainty about the precise way in which epigenetics influences disease risk gives rise to uncertainty as to what parameters should be measured in human studies and how the results should be interpreted. Genetic variation in an individual can be considered to be binary (based on two alleles), and for most loci, the genotype is typically the same in all normal haploid cells regardless of the tissue they are derived from. However, epigenetic variation is stochastic. For any given sample of cells, there is a range of epigenetic status at each loci, even for a single cell type (Fig. 4), with the spread of values depending on the loci measured, cell type, physiological status, and a range of other factors. This variation may be as important as the mean value, as many human diseases have the potential to develop from a subset of cells or even one cell as in cancer. The traditional view of cancer etiology is that it arises from a single cell that undergoes a series of genetic mutations (e.g., *HER2/NEU* in breast cancer and *ERBB2* in lung cancer) or even epigenetic alterations.[5]

Whether the studies are dietary interventions or observational as in "epigenetic epidemiology," almost universally it is the population median or mean epigenetic value (e.g., methylation percent) that is reported (Fig. 4). However, it may be that critical information is contained in the variance or spread of epigenetic status within an individual. A higher proportion of cells with epigenetic states in the higher risk extreme values could occur in an individual on the basis of a change in the variance with little change in the median/mean value for epigenetic status or even no change at all. Such a mechanism could explain reported links between disease outcomes and relatively minor changes in mean

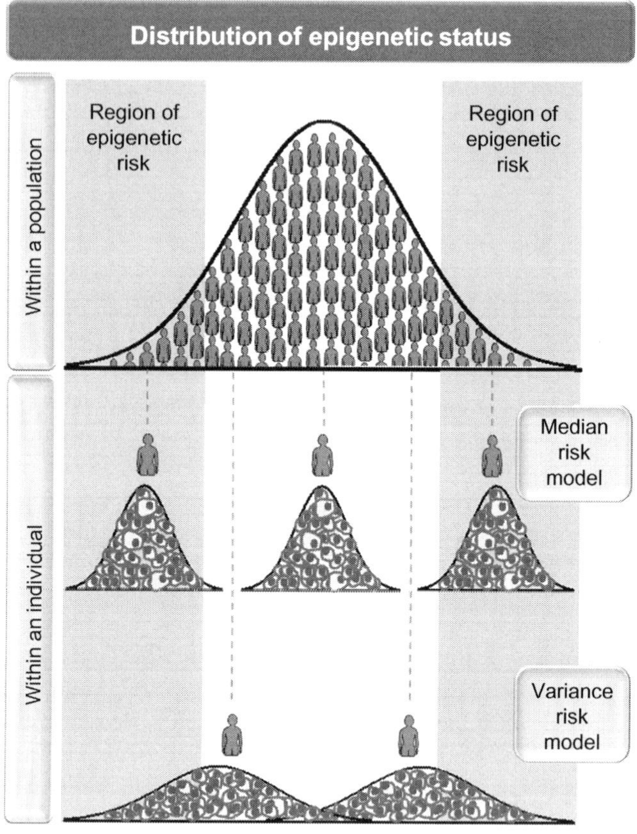

Fig. 4. Representation of the median epigenetic status at a given locus within a population and how this may relate to epigenetic variation within individuals and epigenetic risk of disease.

epigenetic status (see above). It may be that such small changes in the mean status reflect more marked changes in the distributions, putting a greater proportion of cells into the high-risk category.

V. Implications

There is a considerable body of evidence to suggest that diet can influence epigenetic status in ways that are likely to have important effects on biology and health. However, the translation of that knowledge into strategies designed to influence health is more problematic. Nutritional strategies designed to alter global epigenetic status or the epigenetic control of a gene or region of the genome in one tissue or cell type may have unintended consequences because of the tissue-specific nature of the epigenome and differential epigenetic regulation of genes/regions. This is illustrated by the example of cancer in which hypermethylation of specific genes thought to be involved in carcinogenesis and disease progression (e.g., tumor suppressor genes) occurs within the context of global hypomethylation.[8] The development of any nutritional strategy designed to protect against cancer via an epigenetic mechanism would require knowledge of the causal epigenetic changes, how the desirable epigenetic status within the relevant target tissue could be achieved, and whether there may be unintended consequences for other epigenetic processes from the nutritional intervention.

Our emerging understanding of the role of epigenetics at the heart of a feedback loop with the environment, and the way in which nutrition can directly influence the epigenome, could have important implications for individuals and society as a whole. One of the most intriguing of these implications is in the area of behavior. Many health problems related to diet in developed societies center around behaviors—consuming too much energy, consuming the wrong types of foods, lack of exercise. Epigenetics has the potential to explain some aspects of the persistence of poor health behaviors and even the transmission of disadvantage across generations.[121]

The concept of transmission of epigenetic risk from one generation to the next is also of considerable interest as a scientific concept. This process is sometimes referred to as "Lamarkian" in the sense that it appears to make possible the inheritance of acquired characteristics. However, the evidence for such epigenetic processes resulting in "Lamarkian evolution" is weak. The most commonly cited example of epigenetic inheritance is the effect of maternal diet during pregnancy on coat color in the Agouti mouse,[7,9] but this is not truly Lamarkian in the sense that a change in maternal coat color acquired during the lifetime of the mother is somehow encoded within the germ cells and passed on to the offspring. Also, the likely duration of these effects is inconsistent with evolutionary timescales. The primary explanation for the apparent ability of the offspring genome to be programmed in response to the maternal environment before birth is that it may be advantageous to the offspring to be born with a metabolic response profile most appropriate for the environment where it will live. However, the logic of this is that the genome should be reprogrammed in each generation to optimize the response to the environment, something which would preclude a multigeneration effect operating over evolutionary time scales. An alternative to the Lamarkian interpretation is that this type of flexible epigenetic response to the environment may confer a fitness advantage that is genetically coded[122] and selected for in a truly Darwinian manner. An emerging area of interest in epigenetics is the linkage between genetic and epigenetic variation and how sequence variation in the genes involved in epigenetic processes may alter epigenetic status.[119]

Epigenetics provides a powerful new paradigm for the way in which diet, lifestyle, and the environment can directly influence the human genome and human health. It can operate over a wide range of timescales, from minutes to lifetimes, and it can even span more than one generation. Practical measurement tools are now available to investigate epigenetic mechanisms, provided that the problems of tissue-specific genomes and tissue-specific epigenetic responses can be overcome in human studies.

Acknowledgment

The author is grateful to the Scottish Government for support.

References

[1]. Russo EA, Martienssen RA, Riggs AD. *Epigenetic mechanisms of gene regulation.* Woodbury: Cold Spring Harbor Laboratory Press; 1996.
[2]. Sasaki H, Matsui Y. Epigenetic events in mammalian germ-cell development: reprogramming and beyond. *Nat Rev Genet* 2008;**9**:129–40.
[3]. Bird A. Perceptions of epigenetics. *Nature* 2007;**447**:396–8.
[4]. Burdge GC, Lillycrop KA. Nutrition, epigenetics, and developmental plasticity: implications for understanding human disease. *Annu Rev Nutr* 2010;**30**:315–39.
[5]. Feinberg AP, Ohlsson R, Henikoff S. The epigenetic progenitor origin of human cancer. *Nat Rev Genet* 2006;**7**:21–33.
[6]. Issa JP. Epigenetic variation and human disease. *J Nutr* 2002;**132**:2388S–92S.
[7]. Jirtle RL, Skinner MK. Environmental epigenomics and disease susceptibility. *Nat Rev Genet* 2007;**8**:253–62.
[8]. Robertson KD. DNA methylation and human disease. *Nat Rev Genet* 2005;**6**:597–610.
[9]. Waterland RA, Jirtle RL. Early nutrition, epigenetic changes at transposons and imprinted genes, and enhanced susceptibility to adult chronic diseases. *Nutrition* 2004;**20**:63–8.
[10]. Davies W, Isles AR, Wilkinson LS. Imprinted genes and mental dysfunction. *Ann Med* 2001;**33**:428–36.
[11]. Wilkinson LS, Davies W, Isles AR. Genomic imprinting effects on brain development and function. *Nat Rev Neurosci* 2007;**8**:832–43.
[12]. Horsthemke B, Ludwig M. Assisted reproduction: the epigenetic perspective. *Hum Reprod Update* 2005;**11**:473–82.
[13]. Belmont AS. Mitotic chromosome structure and condensation. *Curr Opin Cell Biol* 2006;**18**:632–8.
[14]. Bickmore WA, Teague P. Influences of chromosome size, gene density and nuclear position on the frequency of constitutional translocations in the human population. *Chromosome Res* 2002;**10**:707–15.
[15]. Felsenfeld G, Groudine M. Controlling the double helix. *Nature* 2003;**421**:448–53.

[16]. Misteli T. Beyond the sequence: cellular organization of genome function. *Cell* 2007; **128**:787–800.
[17]. Misteli T. Higher-order genome organization in human disease. *Cold Spring Harb Perspect Biol* 2010;**2**:a000794.
[18]. Strachan T, Read AP. *Human molecular genetics*. 3rd ed. New York: Garland Science; 2004.
[19]. Reik W, Walter J. Genomic imprinting: parental influence on the genome. *Nat Rev Genet* 2001;**2**:21–32.
[20]. Margueron R, Trojer P, Reinberg D. The key to development: interpreting the histone code? *Curr Opin Genet Dev* 2005;**15**:163–76.
[21]. Hake SB. Chromatin modifications. In: Niculescu MD, Haggarty P, editors. *Nutrition in epigenetics*. Ames, USA: Wiley-Blackwel; 2011. pp. 47–71.
[22]. Bestor TH. The DNA methyltransferases of mammals. *Hum Mol Genet* 2000;**9**:2395–402.
[23]. Illingworth RS, Bird AP. CpG islands—'a rough guide'. *FEBS Lett* 2009;**583**:1713–20.
[24]. Walter J, Hutter B, Khare T, Paulsen M. Repetitive elements in imprinted genes. *Cytogenet Genome Res* 2006;**113**:109–15.
[25]. Waterland RA, Jirtle RL. Transposable elements: targets for early nutritional effects on epigenetic gene regulation. *Mol Cell Biol* 2003;**23**:5293–300.
[26]. Reik W, Davies K, Dean W, Kelsey G, Constancia M. Imprinted genes and the coordination of fetal and postnatal growth in mammals. *Novartis Found Symp* 2001;**237**:19–31.
[27]. Tycko B, Morison IM. Physiological functions of imprinted genes. *J Cell Physiol* 2002; **192**:245–58.
[28]. Ferguson-Smith AC, Surani MA. Imprinting and the epigenetic asymmetry between parental genomes. *Science* 2001;**293**:1086–9.
[29]. Allegrucci C, Thurston A, Lucas E, Young L. Epigenetics and the germline. *Reproduction* 2005;**129**:137–49.
[30]. Constancia M, Pickard B, Kelsey G, Reik W. Imprinting mechanisms. *Genome Res* 1998; **8**:881–900.
[31]. Lucifero D, Mann MRW, Bartolomei MS, Trasler JM. Gene-specific timing and epigenetic memory in oocyte imprinting. *Hum Mol Genet* 2004;**13**:839–49.
[32]. Kopsida E, Mikaelsson MA, Davies W. The role of imprinted genes in mediating susceptibility to neuropsychiatric disorders. *Horm Behav* 2011;**59**:375–82.
[33]. Sandovici I, Leppert M, Hawk PR, Suarez A, Linares Y, Sapienza C. Familial aggregation of abnormal methylation of parental alleles at the IGF2/H19 and IGF2R differentially methylated regions. *Hum Mol Genet* 2003;**12**:1569–78.
[34]. Sakatani T, Wei M, Katoh M, Okita C, Wada D, Mitsuya K, et al. Epigenetic heterogeneity at imprinted loci in normal populations. *Biochem Biophys Res Commun* 2001;**283**:1124–30.
[35]. Bjornsson HT, Sigurdsson MI, Fallin MD, Irizarry RA, Aspelund T, Cui H, et al. Intra-individual change over time in DNA methylation with familial clustering. *JAMA* 2008;**299**:2877–83.
[36]. Sandovici I, Kassovska-Bratinova S, Loredo-Osti JC, Leppert M, Suarez A, Stewart R, et al. Interindividual variability and parent of origin DNA methylation differences at specific human Alu elements. *Hum Mol Genet* 2005;**14**:2135–43.
[37]. Fraga MF, Ballestar E, Paz MF, Ropero S, Setien F, Ballestar ML, et al. Epigenetic differences arise during the lifetime of monozygotic twins. *Proc Natl Acad Sci USA* 2005;**102**:10604–9.
[38]. Zhang FF, Cardarelli R, Carroll J, Fulda KG, Kaur M, Gonzalez K, et al. Significant differences in global genomic DNA methylation by gender and race/ethnicity in peripheral blood. *Epigenetics* 2011;**6**:623–9.
[39]. Axume J, Smith SS, Pogribny IP, Moriarty DJ, Caudill MA. Global leukocyte DNA methylation is similar in African American and Caucasian women under conditions of controlled folate intake. *Epigenetics* 2007;**2**:66–8.

[40]. Trasler JM. Gamete imprinting: setting epigenetic patterns for the next generation. *Reprod Fertil Dev* 2006;**18**:63–9.
[41]. Chedin F, Lieber MR, Hsieh CL. The DNA methyltransferase-like protein DNMT3L stimulates de novo methylation by Dnmt3a. *Proc Natl Acad Sci USA* 2002; **99**:16916–21.
[42]. Jia D, Jurkowska RZ, Zhang X, Jeltsch A, Cheng X. Structure of Dnmt3a bound to Dnmt3L suggests a model for de novo DNA methylation. *Nature* 2007;**449**:248–51.
[43]. Bourc'his D, Xu GL, Lin CS, Bollman B, Bestor TH. Dnmt3L and the establishment of maternal genomic imprints. *Science* 2001;**294**:2536–9.
[44]. Kaneda M, Okano M, Hata K, Sado T, Tsujimoto N, Li E, et al. Essential role for de novo DNA methyltransferase Dnmt3a in paternal and maternal imprinting. *Nature* 2004; **429**:900–3.
[45]. Arnaud P, Hata K, Kaneda M, Li E, Sasaki H, Feil R, et al. Stochastic imprinting in the progeny of Dnmt3L(−/−) females. *Hum Mol Genet* 2006;**15**:589–98.
[46]. El-Maarri O, Kareta MS, Mikeska T, Becker T, Diaz-Lacava A, Junen J, et al. A systematic search for DNA methyltransferase polymorphisms reveals a rare DNMT3L variant associated with subtelomeric hypomethylation. *Hum Mol Genet* 2009;**18**:1755–68.
[47]. Ooi SK, Qiu C, Bernstein E, Li K, Jia D, Yang Z, et al. DNMT3L connects unmethylated lysine 4 of histone H3 to de novo methylation of DNA. *Nature* 2007;**448**:714–7.
[48]. Filippova GN. Genetics and epigenetics of the multifunctional protein CTCF. *Curr Top Dev Biol* 2008;**80**:337–60.
[49]. Phillips JE, Corces VG. CTCF: master weaver of the genome. *Cell* 2009;**137**:1194–211.
[50]. Carey N, Marques CJ, Reik W. DNA demethylases: a new epigenetic frontier in drug discovery. *Drug Discov Today* 2011;**16**:683–90.
[51]. Reik W. Stability and flexibility of epigenetic gene regulation in mammalian development. *Nature* 2007;**447**:425–32.
[52]. Stover PJ. Folate, vitamin B_{12}, and vitamin B_6. In: Niculescu MD, Haggarty P, editors. *Nutrition in epigenetics*. Ames, USA: Wiley-Blackwell; 2011. pp. 211–23.
[53]. Stover PJ. One-carbon metabolism-genome interactions in folate-associated pathologies. *J Nutr* 2009;**139**:2402–5.
[54]. Jacob RA, Gretz DM, Taylor PC, James SJ, Pogribny IP, Miller BJ, et al. Moderate folate depletion increases plasma homocysteine and decreases lymphocyte DNA methylation in postmenopausal women. *J Nutr* 1998;**128**:1204–12.
[55]. Yi P, Melnyk S, Pogribna M, Pogribny IP, Hine RJ, James SJ. Increase in plasma homocysteine associated with parallel increases in plasma S-adenosylhomocysteine and lymphocyte DNA hypomethylation. *J Biol Chem* 2000;**275**:29318–23.
[56]. Friso S, Choi SW, Girelli D, Mason JB, Dolnikowski GG, Bagley PJ, et al. A common mutation in the 5,10-methylenetetrahydrofolate reductase gene affects genomic DNA methylation through an interaction with folate status. *Proc Natl Acad Sci USA* 2002; **99**:5606–11.
[57]. Stern LL, Mason JB, Selhub J, Choi SW. Genomic DNA hypomethylation, a characteristic of most cancers, is present in peripheral leukocytes of individuals who are homozygous for the C677T polymorphism in the methylenetetrahydrofolate reductase gene. *Cancer Epidemiol Biomarkers Prev* 2000;**9**:849–53.
[58]. Kirkland JB. Niacin status impacts chromatin structure. *J Nutr* 2009;**139**:2397–401.
[59]. Zempleni J, Chew YC, Bao B, Pestinger V, Wijeratne SS. Repression of transposable elements by histone biotinylation. *J Nutr* 2009;**139**:2389–92.
[60]. Choi SW, Stickel F, Baik HW, Kim YI, Seitz HK, Mason JB. Chronic alcohol consumption induces genomic but not p53-specific DNA hypomethylation in rat colon. *J Nutr* 1999;**129**:1945–50.

[61]. Garro AJ, McBeth DL, Lima V, Lieber CS. Ethanol consumption inhibits fetal DNA methylation in mice: implications for the fetal alcohol syndrome. *Alcohol Clin Exp Res* 1991;**15**:395–8.

[62]. Liu Y, Balaraman Y, Wang G, Nephew KP, Zhou FC. Alcohol exposure alters DNA methylation profiles in mouse embryos at early neurulation. *Epigenetics* 2009;**4**:500–11.

[63]. Lu SC, Huang ZZ, Yang H, Mato JM, Avila MA, Tsukamoto H. Changes in methionine adenosyltransferase and S-adenosylmethionine homeostasis in alcoholic rat liver. *Am J Physiol Gastrointest Liver Physiol* 2000;**279**:G178–85.

[64]. Bielawski DM, Zaher FM, Svinarich DM, Abel EL. Paternal alcohol exposure affects sperm cytosine methyltransferase messenger RNA levels. *Alcohol Clin Exp Res* 2002;**26**:347–51.

[65]. Bonsch D, Lenz B, Reulbach U, Kornhuber J, Bleich S. Homocysteine associated genomic DNA hypermethylation in patients with chronic alcoholism. *J Neural Transm* 2004; **111**:1611–6.

[66]. Bonsch D, Lenz B, Kornhuber J, Bleich S. DNA hypermethylation of the alpha synuclein promoter in patients with alcoholism. *Neuroreport* 2005;**16**:167–70.

[67]. Ho E, Clarke JD, Dashwood RH. Dietary sulforaphane, a histone deacetylase inhibitor for cancer prevention. *J Nutr* 2009;**139**:2393–6.

[68]. Lee WJ, Zhu BT. Inhibition of DNA methylation by caffeic acid and chlorogenic acid, two common catechol-containing coffee polyphenols. *Carcinogenesis* 2006;**27**:269–77.

[69]. Fang M, Chen D, Yang CS. Dietary polyphenols may affect DNA methylation. *J Nutr* 2007;**137**:223S–8S.

[70]. Fang MZ, Wang Y, Ai N, Hou Z, Sun Y, Lu H, et al. Tea polyphenol (−)-epigallocatechin-3-gallate inhibits DNA methyltransferase and reactivates methylation-silenced genes in cancer cell lines. *Cancer Res* 2003;**63**:7563–70.

[71]. Lillycrop KA, Phillips ES, Jackson AA, Hanson MA, Burdge GC. Dietary protein restriction of pregnant rats induces and folic Acid supplementation prevents epigenetic modification of hepatic gene expression in the offspring. *J Nutr* 2005;**135**:1382–6.

[72]. Dolinoy DC, Weidman JR, Waterland RA, Jirtle RL. Maternal genistein alters coat color and protects Avy mouse offspring from obesity by modifying the fetal epigenome. *Environ Health Perspect* 2006;**114**:567–72.

[73]. Steegers-Theunissen RP, Obermann-Borst SA, Kremer D, Lindemans J, Siebel C, Steegers EA, et al. Periconceptional maternal folic acid use of 400 microg per day is related to increased methylation of the IGF2 gene in the very young child. *PLoS One* 2009;**4**:e7845.

[74]. Scientific Advisory Committee on Nutrition. *The influence of maternal fetal and child nutrition on the development of chronic disease in later life*. A report of the SACN subgroup on Maternal and Child Nutrition. London: TSO; 2007..

[75]. Heijmans BT, Tobi EW, Stein AD, Putter H, Blauw GJ, Susser ES, et al. Persistent epigenetic differences associated with prenatal exposure to famine in humans. *Proc Natl Acad Sci USA* 2008;**105**:17046–9.

[76]. Martin-Gronert MS, Ozanne SE. Programming of appetite and type 2 diabetes. *Early Hum Dev* 2005;**81**:981–8.

[77]. Levin BE. Epigenetic influences on food intake and physical activity level: review of animal studies. *Obesity (Silver Spring)* 2008;**16**(Suppl. 3):S51–4.

[78]. Hillemacher T, Kahl KG, Heberlein A, Muschler MA, Eberlein C, Frieling H, et al. Appetite- and volume-regulating neuropeptides: role in treating alcohol dependence. *Curr Opin Investig Drugs* 2010;**11**:1097–106.

[79]. Ostan I, Poljsak B, Simcic M, Tijskens LM. Appetite for the selfish gene. *Appetite* 2010; **54**:442–9.

[80]. Hillemacher T. Biological mechanisms in alcohol dependence—new perspectives. *Alcohol Alcohol* 2011;**46**:224–30.

[81]. den Hoed M, Westerterp-Plantenga MS, Bouwman FG, Mariman EC, Westerterp KR. Postprandial responses in hunger and satiety are associated with the rs9939609 single nucleotide polymorphism in FTO. *Am J Clin Nutr* 2009;**90**:1426–32.
[82]. Milagro FI, Campion J, Cordero P, Goyenechea E, Gomez-Uriz AM, Abete I, et al. A dual epigenomic approach for the search of obesity biomarkers: DNA methylation in relation to diet-induced weight loss. *FASEB J* 2011;**25**:1378–89.
[83]. Levenson JM, Sweatt JD. Epigenetic mechanisms in memory formation. *Nat Rev Neurosci* 2005;**6**:108–18.
[84]. Mattson MP. Methylation and acetylation in nervous system development and neurodegenerative disorders. *Ageing Res Rev* 2003;**2**:329–42.
[85]. Tremolizzo L, Rodriguez-Menendez V, Brighina L, Ferrarese C. Is the inverse association between Alzheimer's disease and cancer the result of a different propensity to methylate DNA? *Med Hypotheses* 2006;**66**:1251–2.
[86]. Badcock C, Crespi B. Imbalanced genomic imprinting in brain development: an evolutionary basis for the aetiology of autism. *J Evol Biol* 2006;**19**:1007–32.
[87]. Coufal NG, Garcia-Perez JL, Peng GE, Yeo GW, Mu Y, Lovci MT, et al. L1 retrotransposition in human neural progenitor cells. *Nature* 2009;**460**:1127–31.
[88]. Kan PX, Popendikyte V, Kaminsky ZA, Yolken RH, Petronis A. Epigenetic studies of genomic retroelements in major psychosis. *Schizophr Res* 2004;**67**:95–106.
[89]. Muotri AR, Marchetto MC, Coufal NG, Oefner R, Yeo G, Nakashima K, et al. L1 retrotransposition in neurons is modulated by MeCP2. *Nature* 2010;**468**:443–6.
[90]. Sasaki T, Nishihara H, Hirakawa M, Fujimura K, Tanaka M, Kokubo N, et al. Possible involvement of SINEs in mammalian-specific brain formation. *Proc Natl Acad Sci USA* 2008;**105**:4220–5.
[91]. Singer T, McConnell MJ, Marchetto MC, Coufal NG, Gage FH. LINE-1 retrotransposons: mediators of somatic variation in neuronal genomes? *Trends Neurosci* 2010;**33**:345–54.
[92]. Temtamy SA, Aglan MS, Valencia M, Cocchi G, Pacheco M, Ashour AM, et al. Long interspersed nuclear element-1 (LINE1)-mediated deletion of EVC, EVC2, C4orf6, and STK32B in Ellis-van Creveld syndrome with borderline intelligence. *Hum Mutat* 2008;**29**:931–8.
[93]. Shenkin SD, Starr JM, Deary IJ. Birth weight and cognitive ability in childhood: a systematic review. *Psychol Bull* 2004;**130**:989–1013.
[94]. Matte TD, Bresnahan M, Begg MD, Susser E. Influence of variation in birth weight within normal range and within sibships on IQ at age 7 years: cohort study. *BMJ* 2001;**323**:310–4.
[95]. Richards M, Hardy R, Kuh D, Wadsworth MEJ. Birth weight and cognitive function in the British 1946 birth cohort: longitudinal population based study. *BMJ* 2001;**322**:199–203.
[96]. Curley JP, Mashoodh R. Parent-of-origin and trans-generational germline influences on behavioral development: the interacting roles of mothers, fathers, and grandparents. *Dev Psychobiol* 2010;**52**:312–30.
[97]. Whittington J, Holland A, Webb T, Butler J, Clarke D, Boer H. Cognitive abilities and genotype in a population-based sample of people with Prader-Willi syndrome. *J Intellect Disabil Res* 2004;**48**:172–87.
[98]. Amir RE, Van den Veyver IB, Wan M, Tran CQ, Francke U, Zoghbi HY. Rett syndrome is caused by mutations in X-linked MECP2, encoding methyl-CpG-binding protein 2. *Nat Genet* 1999;**23**:185–8.
[99]. Samaco RC, Hogart A, LaSalle JM. Epigenetic overlap in autism-spectrum neurodevelopmental disorders: MECP2 deficiency causes reduced expression of UBE3A and GABRB3. *Hum Mol Genet* 2005;**14**:483–92.
[100]. Haggarty P, Hoad G, Harris SE, Starr JM, Fox HC, Deary IJ, et al. Human intelligence and polymorphisms in the DNA methyltransferase genes involved in epigenetic marking. *PLoS One* 2010;**5**:e11329.

[101]. Bouchard Jr. TJ. Genetic and environmental influences on adult intelligence and special mental abilities. *Hum Biol* 1998;**70**:257–79.
[102]. Deary IJ, Spinath FM, Bates TC. Genetics of intelligence. *Eur J Hum Genet* 2006; **14**:690–700.
[103]. Meaney MJ, Szyf M. Environmental programming of stress responses through DNA methylation: life at the interface between a dynamic environment and a fixed genome. *Dialogues Clin Neurosci* 2005;**7**:103–23.
[104]. Zhang TY, Meaney MJ. Epigenetics and the environmental regulation of the genome and its function. *Annu Rev Psychol* 2010;**61**:439–66.
[105]. Weaver IC, D'Alessio AC, Brown SE, Hellstrom IC, Dymov S, Sharma S, et al. The transcription factor nerve growth factor-inducible protein a mediates epigenetic programming: altering epigenetic marks by immediate-early genes. *J Neurosci* 2007;**27**:1756–68.
[106]. McGowan PO, Sasaki A, D'Alessio AC, Dymov S, Labonte B, Szyf M, et al. Epigenetic regulation of the glucocorticoid receptor in human brain associates with childhood abuse. *Nat Neurosci* 2009;**12**:342–8.
[107]. Borders AE, Grobman WA, Amsden LB, Holl JL. Chronic stress and low birth weight neonates in a low-income population of women. *Obstet Gynecol* 2007;**109**:331–8.
[108]. Paarlberg KM, Vingerhoets AJ, Passchier J, Dekker GA, Heinen AG, van Geijn HP. Psychosocial predictors of low birthweight: a prospective study. *Br J Obstet Gynaecol* 1999;**106**:834–41.
[109]. Meaney MJ, Szyf M, Seckl JR. Epigenetic mechanisms of perinatal programming of hypothalamic-pituitary-adrenal function and health. *Trends Mol Med* 2007;**13**:269–77.
[110]. Seckl JR, Meaney MJ. Glucocorticoid "programming" and PTSD risk. *Ann N Y Acad Sci* 2006;**1071**:351–78.
[111]. Weaver IC, Champagne FA, Brown SE, Dymov S, Sharma S, Meaney MJ, et al. Reversal of maternal programming of stress responses in adult offspring through methyl supplementation: altering epigenetic marking later in life. *J Neurosci* 2005;**25**:11045–54.
[112]. Laird PW. Principles and challenges of genomewide DNA methylation analysis. *Nat Rev Genet* 2010;**11**:191–203.
[113]. Rakyan VK, Down TA, Balding DJ, Beck S. Epigenome-wide association studies for common human diseases. *Nat Rev Genet* 2011;**12**:529–41.
[114]. Schones DE, Zhao K. Genome-wide approaches to studying chromatin modifications. *Nat Rev Genet* 2008;**9**:179–91.
[115]. Cebrian A, Pharoah PD, Ahmed S, Ropero S, Fraga MF, Smith PL, et al. Genetic variants in epigenetic genes and breast cancer risk. *Carcinogenesis* 2006;**27**:1661–9.
[116]. Liu Z, Wang L, Wang LE, Sturgis EE, Wei Q. Polymorphisms of the DNMT3B gene and risk of squamous cell carcinoma of the head and neck: a case–control study. *Cancer Lett* 2008;**268**:158–65.
[117]. Park BL, Kim LH, Shin HD, Park YW, Uhm WS, Bae SC. Association analyses of DNA methyltransferase-1 (DNMT1) polymorphisms with systemic lupus erythematosus. *J Hum Genet* 2004;**49**:642–6.
[118]. Shen H, Wang L, Spitz MR, Hong WK, Mao L, Wei Q. A novel polymorphism in human cytosine DNA-methyltransferase-3B promoter is associated with an increased risk of lung cancer. *Cancer Res* 2002;**62**:4992–5.
[119]. Wang J, Bhutani M, Pathak AK, Lang W, Ren H, Jelinek J, et al. Delta DNMT3B variants regulate DNA methylation in a promoter-specific manner. *Cancer Res* 2007;**67**: 10647–52.
[120]. Zhu Y, Brown HN, Zhang Y, Holford TR, Zheng T. Genotypes and haplotypes of the methyl-CpG-binding domain 2 modify breast cancer risk dependent upon menopausal status. *Breast Cancer Res* 2005;**7**:R745–52.

[121]. Haggarty P, Campbell DM, Duthie S, Andrews K, Hoad G, Piyathilake C, et al. Diet and deprivation in pregnancy. *Br J Nutr* 2009;**102**:1487–97.
[122]. Feinberg AP, Irizarry RA. Evolution in health and medicine Sackler colloquium: stochastic epigenetic variation as a driving force of development, evolutionary adaptation, and disease. *Proc Natl Acad Sci USA* 2010;**107**(Suppl. 1):1757–64.

Gene–Exercise Interactions

Tuomo Rankinen and
Claude Bouchard

Human Genomics Laboratory, Pennington Biomedical Research Center, Baton Rouge, Louisiana, USA

A growing body of evidence indicates that genotype-by-physical activity interactions on various health-related outcomes do exist. Observational studies have shown that relationships between DNA sequence variants and risk factors are significantly different between sedentary and physically active individuals, while exercise intervention studies have demonstrated that genetic variation contributes significantly to interindividual variation in responsiveness to exercise training. The knowledge base on gene–activity interactions will grow considerably within a few years when large observational genome-wide association study (GWAS) consortia will report their findings. Progress with exercise intervention studies will be slower because of resource requirements. However, such studies are desperately needed to fully understand the genetics as well as the exercise biology of complex traits and to confirm the gene–exercise interactions derived from observational studies. Furthermore, development of personalized exercise medicine applications will be difficult or even impossible without a proper understanding of gene–exercise interactions.

I. Introduction

Pharmacogenomics, nutrigenetics, and exercise genomics represent the major anchoring pillars of personalized preventive and therapeutic medicine. It appears at this time that pharmacogenomics is more advanced and that nutrigenetics has a larger body of data than exercise genomics. However, the pace of growth in exercise genomics research has increased substantially in the past several years, and more and better data are being generated. Nutrigenetics and nutrigenomics having much in common with exercise genomics, it was deemed appropriate to include a chapter on the latter in this volume.

Progress in Molecular Biology
and Translational Science, Vol. 108
DOI: 10.1016/B978-0-12-398397-8.00017-4

The past two decades have been an exciting time for exercise geneticists. Since the early 1990s, numerous epidemiological studies and several exercise intervention trials have demonstrated the wide spectrum of health benefits associated with a physically active lifestyle. As evidenced by several consensus statements and guidelines promulgated by national health organizations as well as the U.S. government, regular exercise is recognized as a central component of primary and secondary prevention of chronic diseases. During the same time period, advances in DNA genotyping and sequencing technologies have revolutionized human genetics and genomics research. High-density single-nucleotide polymorphism (SNP) genotyping arrays and massively parallel next-generation sequencing applications have made it possible to study millions of DNA sequence variants in large numbers of subjects fairly rapidly and affordably. Such approaches have identified hundreds of novel genes and DNA sequence variants that contribute to the pathogenesis of rare and common diseases as well as to interindividual variation in complex traits such as risk factors for chronic diseases.

Despite these major advances, both exercise science and genetics still wrestle with quite similar questions: exercise scientists try to understand why health benefits of standardized regular physical activity programs vary from person to person, while geneticists try to figure out why key DNA markers from genome-wide association studies (GWASs) do not fully explain the heritability of a given trait. Both of these issues may be related, at least to some extent, by a common phenomenon best defined as "genotype-by-physical activity interactions." The classic definition of an interaction effect states that an association between two factors is modified by a third factor. For example, the inverse relationship between physical activity level and risk of hypertension may vary as a function of a genotype at a key gene locus, or as a geneticist would formulate it, the effect of genotype on the risk of hypertension is modified by the physical activity level. Without proper modeling of such an interaction effect, there is a risk that the contributions of both physical activity and genotype to hypertension risk will be underestimated.

Two different study designs are used to investigate such interactions. The most common is an observational, usually cross-sectional, cohort study design in which genotype and physical activity main effects, as well as their interactions on dependent variables, are modeled statistically: a significant interaction term is taken as evidence supporting the interaction effect. The other, physiologically more powerful, approach is to conduct a standardized exercise intervention study and to test for associations between genetic variants and exercise-induced changes in the phenotype of interest. Since the phenotype response is driven by a standardized change in physical activity behavior (exercise intervention), the association between the sequence variant and training response constitutes a formal test of a genotype-by-exercise interaction. However, the exercise intervention-based design is more powerful, because the physical activity behavior is documented and controlled far much more accurately than in observational studies. In contrast, observational studies are potentially affected by uncontrolled cofounders.

The purpose of this review is to summarize some of the latest developments in gene–exercise interaction studies based on both study designs, to illustrate the differences between the designs in terms of information gained, and finally to discuss opportunities and challenges for future gene–exercise interaction studies.

II. Exercise Intervention Studies

The fundamental justification for the critical importance of genetic differences on exercise training responsiveness was established through a series of twin studies in the 1980s[1] and confirmed later on in the HERITAGE Family Study.[2] The initial attempts to identify genetic variants affecting training responsiveness were based on a candidate gene approach, but the results were mainly negative.[3] The main weakness of the candidate gene strategy is that the gene selection is limited by our understanding of the physiology regulating the trait of interest. It is safe to say that such an approach is less than optimal to identify all relevant genes. Rapid technical improvements in microarray-based high-throughput methods have made it possible to genotype hundreds of thousands of SNPs in a single assay and to measure the expression levels of thousands of gene transcripts simultaneously. These advances in technologies have made it possible to undertake detailed genome-wide screening explorations focused on DNA sequence variants (mainly SNPs) and gene transcripts abundance, both of which are leading us to objective, largely unbiased, and hypothesis-free association studies. Because the candidate gene-based exercise training studies have been reviewed in detail elsewhere,[3] we focus here on studies utilizing genome-wide screening approaches.

The first exercise training response-related study using a high-density genome-wide screening approach was published in 2010 and it targeted maximal oxygen uptake (VO_{2max}) as a response phenotype.[4] A study by Timmons and colleagues used a combination of global skeletal muscle gene expression profiling and DNA sequence variation screening to identify genes associated with VO_{2max} training response status.[4] RNA expression profiling of skeletal muscle samples prior to training was carried out in two independent exercise training trials. The first stage of the study ($n=24$) identified a panel of 29 transcripts that were associated with VO_{2max} training response, which were subsequently confirmed in the second stage ($n=17$). Next, haplotype tagging SNPs in the 29 predictor genes were identified and genotyped in HERITAGE.

A multivariable regression analysis using the transcriptome-derived predictor gene SNPs and a set of SNPs from positional cloning studies of HERITAGE identified a set of 11 SNPs that explained about 23% of the variance in VO_{2max} training response, which corresponded to about 50% of the maximal heritability estimate. Seven of the SNPs were from the RNA predictor gene set, and four were from the HERITAGE QTL projects. Interestingly, when incorporated into the original RNA transcript prediction model, three of the four QTL-derived genes improved the performance of the model.[4]

While transcriptomics-based genomic screening provides a comprehensive view of the gene expression profile in the targeted tissue(s), it cannot detect those genes that exert their effects on the phenotype of interest in other tissues. In contrast, a GWAS based on dense screening of nuclear DNA sequence variation is not restricted by tissue specificity. The first full GWASs applied to regular exercise-induced changes in VO_{2max} and submaximal exercise heart rate (HR50) were published in 2011.[5,6] In a single SNP analysis, a total of 39 SNPs were associated with VO_{2max} training response at $p < 1.5 \times 10^{-4}$; five SNPs showed associations at $p < 1 \times 10^{-5}$, while another 20 SNPs showed significance levels at $1.5 \times 10^{-5} < p < 9.3 \times 10^{-4}$. The strongest evidence of association ($p = 1.3 \times 10^{-6}$) was observed with an SNP located in the first intron of the acyl-CoA synthetase long-chain family member 1 (*ACSL1*) gene, located on chromosome 4q35. When all 39 SNPs were analyzed simultaneously in multivariate regression models, 21 SNPs were retained in the final model. Of these, nine SNPs explained at least 2% (range 2.2–7.0%) of the trait variance ($p < 0.0001$ for all), while seven markers contributed between 1% and 2% each (Table I). Collectively, the 16 SNPs explained 45% of the variance in VO_{2max} training response, which is very close to the maximal heritability estimate of 47% reported previously in HERITAGE.[7]

The strongest associations for HR50 training response (adjusted for age, sex, body mass index [BMI], and baseline HR50) were detected with SNPs at the tyrosine 3-monooxygenase/tryptophan 5-monooxygenase activation protein, theta polypeptide (*YWHAQ*) locus on chromosome 2p25 ($p = 8.1 \times 10^{-7}$), the RNA-binding protein with multiple splicing (*RBPMS*) locus on 8p12 ($p = 3.8 \times 10^{-6}$), and the cAMP responsive element-binding protein 1 (*CREB1*) locus on 2q34 ($p = 1.6 \times 10^{-5}$). In addition, 37 other SNPs showed p-values $< 9.9 \times 10^{-5}$. After removing redundant SNPs, the 10 most significant SNPs explained 35.9% of the ΔHR50 variance in a multivariate regression model (Table I). Furthermore, conditional heritability tests showed that nine of these SNPs (all located within a gene locus) accounted for 100% of the ΔHR50 heritability.[6]

In both GWAS reports, a predisposition score was constructed using the most informative SNPs from the final regression models. Each SNP was recoded on the basis of the number of positive training response (increase in VO_{2max},

TABLE I

SUMMARY OF THE MULTIVARIATE REGRESSION MODELS FOR THE VO_{2MAX} AND HR50 TRAINING RESPONSES USING THE MOST SIGNIFICANT SNPS FROM GENOME-WIDE ASSOCIATION SCANS IN HERITAGE WHITES

SNP	Chr	Map	Freq	Gene[a]	Regression model: Partial R^2	Model R^2	p-Value
ΔVO_{2max}							
rs10499043	6	106,353,830	0.13	*PRDM1* (287 kb)	0.070	0.070	<0001
rs1535628	9	104,056,570	0.09	*GRIN3A* (516 kb)	0.052	0.122	<0001
rs4973706	3	18,896,776	0.24	*KCNH8* (268 kb)	0.045	0.167	<0001
rs12115454	9	117,759,871	0.11	*C9orf27* (33 kb)	0.041	0.208	<0001
rs6552828	4	185,962,410	0.37	*ACSL1*	0.035	0.243	<0001
rs11715829	3	148,439,856	0.08	*ZIC4* (146 kb)	0.032	0.275	<0001
rs884736	1	6,937,692	0.41	*CAMTA1*	0.029	0.304	<0001
rs10921078	1	190,325,645	0.15	*RGS18* (70 kb)	0.025	0.329	<0001
rs6090314	20	61,327,997	0.16	*BIRC7* (10 kb)	0.022	0.351	<0001
rs10500872	11	20,202,299	0.15	*DBX1* (65 kb)	0.018	0.369	0.0003
rs1956197	14	58,547,167	0.16	*DAAM1* (174 kb)	0.017	0.386	0.0004
rs824205	15	21,559,164	0.15	*NDN* (75 kb)	0.014	0.401	0.0013
rs7933007	11	118,235,879	0.23	*CXCR5* (24 kb)	0.013	0.414	0.0015
rs12896790	14	37,343,673	0.09	*TTC6*	0.013	0.427	0.0014
rs4952535	2	41,985,027	0.41	*LOC400950* (11 kb)	0.012	0.439	0.0024
rs2053896	4	137,374,246	0.10	*LOC100289626* (134 kb)	0.012	0.451	0.0022
ΔHR50							
rs2979481	8	30,382,328	0.645	*RBPMS*	0.0605	0.0605	8.1×10^{-8}
rs6432018	2	9,639,347	0.524	*YWHAQ*	0.0457	0.1062	5.0×10^{-7}
rs2253206	2	208,100,223	0.522	*CREB1*	0.0447	0.1509	2.2×10^{-6}
rs1560488	4	90,444,858	0.791	*GPRIN3*	0.0423	0.1932	2.4×10^{-6}
rs10248479	7	115,395,591	0.876	*TFEC*	0.0333	0.2264	1.1×10^{-5}
rs857838	1	157,017,174	0.60	*OR6N2*	0.0302	0.2566	1.8×10^{-5}
rs909562	6	16,238,312	0.875	*MYLIP*	0.0296	0.2861	1.7×10^{-5}
rs4759659	12	129,403,241	0.538	*PIWIL1*	0.0276	0.3138	2.3×10^{-5}
rs2057368	14	54,373,759	0.804	*GCH1*	0.0238	0.3375	6.3×10^{-5}
rs4498613	9	95,545,460	0.808	*PHF2* (60 kb)	0.0218	0.3593	0.0001

[a]The gene located nearest to the SNP. Distance to the gene in kilo bases (1000 bp) is shown in parentheses. If no distance is shown, the SNP is located within the gene locus.

Modified from Bouchard *et al.*[5] and Rankinen *et al.*[6]

decrease in HR50) alleles: a low-response allele homozygote was assigned 0, a heterozygote received 1, and a homozygote for the high-response allele was assigned 2. The sum of the recoded SNPs was used as the genomic predictor score; the theoretical range of the score values was from 0 (no beneficial alleles) to 42 (two copies of the beneficial alleles in all 21 loci) for VO_{2max} response and 0 to

20 for HR50 response. The differences in VO_{2max} and HR50 training responses between those with the lowest and the highest predisposition score categories were 383 mL/min and 21 beats per minute (bpm), respectively (Fig. 1).

An obvious question with the HR50 training response predictor SNPs is whether they have any impact on the resting heart rate. While the individual SNPs were associated with resting heart rate phenotypes only modestly, if at all,

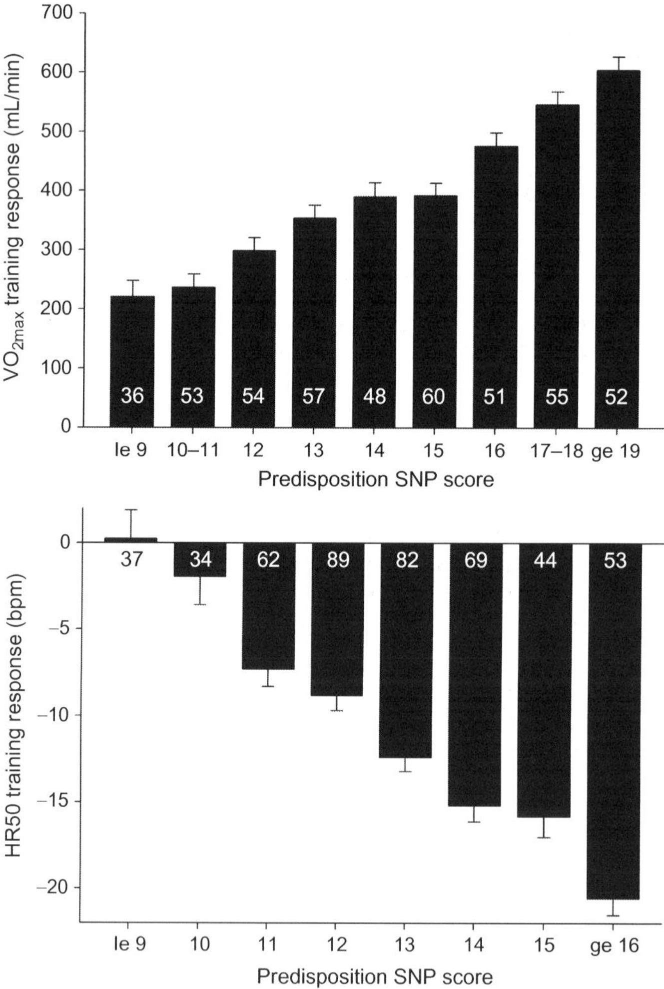

FIG. 1. VO_{2max} (top) and HR50 (bottom) training responses across GWAS predisposition SNP score categories in HERITAGE Whites. Number of subjects within each SNP score category is indicated inside each histogram bar. Modified from Bouchard *et al.*[5] and Rankinen *et al.*[6]

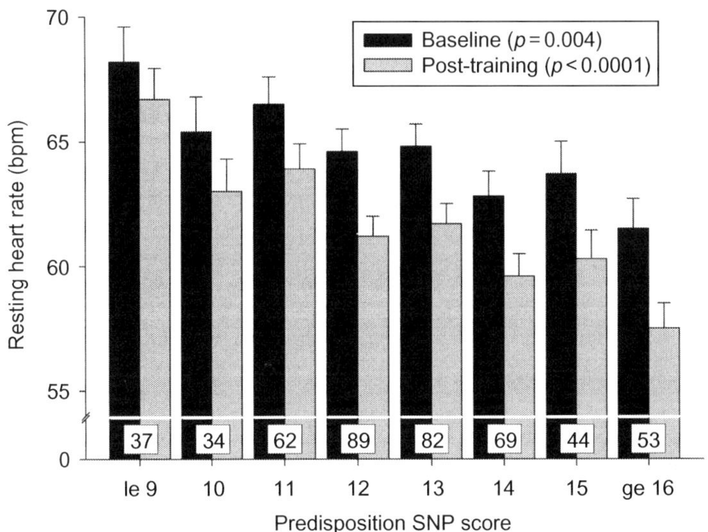

FIG. 2. Relationship between ΔHR50 predisposition SNP scores and resting heart rates at baseline (black bars) and post training (gray bars) in the HERITAGE Family Study.

the HR50 training response SNP predisposition score showed significant associations. In the sedentary state (baseline), subjects in the lowest predisposition score category had a 6.5 bpm higher ($p = 0.004$) resting heart rate than those in the highest category. Moreover, the association was even stronger in the physically active state (post training); the difference in resting heart rate between the lowest and highest predisposition score groups was more than 9 bpm, and the proportion of resting heart rate variance explained by the score increased from 3.9% at baseline to 7.4% post training ($p = 0.0001$) (Fig. 2), which is another classic example of a genotype–exercise interaction. Thus, it seems that the SNPs and genes that modify submaximal exercise heart rate adaptation to regular physical activity may also be involved in the regulation of resting heart rate.

III. Observational Studies

A major limitation with exercise training-related genetics studies is that few studies with large sample sizes and carefully monitored exercise programs are available. Therefore, the evaluation of genotype-by-physical activity interaction effects using large epidemiological cohorts has been proposed as a less expensive alternative to exercise intervention trials. While interest in gene–physical activity interactions in observational studies has increased recently, the topic is by no means a new one. Such analyses have been an integral part of candidate gene studies for a few decades, when appropriate physical activity measures were available. For example, the latest comprehensive human gene map for physical activity traits published in 2009, covering the pre-GWAS-era literature, listed about 30 studies that reported positive genotype-by-activity interaction effects on blood pressure and hypertension, body size and composition, plasma lipid levels, glucose and insulin metabolism, etc.[3]

The recent popularity of interaction studies has been fueled by large GWAS consortia. A good example comes from recent studies dealing with physical activity–*FTO* (fat mass and obesity associated) gene interaction effects on body weight and body composition. *FTO* was the first gene identified, using high-density GWAS, as a strong candidate for obesity-related phenotypes in Caucasians.[8–10] Minor alleles of obesity risk SNPs located in the first intron of the gene were associated with about a 65% greater risk of obesity in the homozygous state; the homozygotes of the minor allele were 3–4 kg heavier, and heterozygotes had 1–2 kg more body weight than the common allele homozygotes. The population-attributable risk of *FTO* for obesity has been estimated to be as high as 20%. The initial finding has been replicated in several large cohorts, and the association holds both in children and in adults.[11]

Subsequently, two observational studies reported that the *FTO* gene-related risk of obesity was modified by the physical activity level of the subjects.[12,13] The greater BMI level found in the *FTO* risk allele homozygotes was particularly evident in those who were sedentary, while the difference in BMI between the risk allele and common allele homozygotes was weaker among physically active individuals.[12,13] These studies were followed by several other reports, some confirming the interaction effect, some not. Consequently, Kilpelainen *et al.* carried out a comprehensive large-scale meta-analysis in which *FTO*-by-physical activity results from 45 independent studies comprising 218,166 adult subjects were combined.[14] The random-effects meta-analysis of the studies confirmed the *FTO*-by-physical activity interaction effects on BMI ($p_{\text{interaction}} = 0.005$) and obesity ($p_{\text{interaction}} = 0.001$). The risk of obesity associated with the *FTO* SNP minor allele was greater in sedentary adults than in their physically active counterparts; the odds of obesity were 1.30-fold greater (95% confidence interval [CI]:1.24–1.36) per *FTO* allele in the sedentary subjects, while the corresponding odds ratio was 1.22 per allele (95% CI: 1.19–1.25) in those classified as physically active.[14]

The results of this large meta-analysis raise several interesting points. First, they demonstrate that the *FTO*-by-activity interaction effect on obesity is a consistent finding. However, a closer inspection of the results shows that it may actually be physical inactivity that amplifies the *FTO* genotype-related obesity risk rather than physical activity attenuating it. The odds of obesity in physically active subjects were almost identical to the odds ratio observed in the full cohort (1.23/allele [95% CI: 1.19–1.25]), while the odds in the sedentary subjects were clearly higher. Thus, the combination of *FTO* risk genotype and lack of physical activity seems to be particularly detrimental for body weight regulation, whereas the effect of regular physical activity on *FTO*-related obesity risk remains unclear and deserves to be further investigated.

A second interesting observation is the marked heterogeneity across the individual studies of the meta-analysis. For example, when meta-analyses were stratified by geographical location of the cohort, only studies from North America ($n=47{,}938$) showed significant interactions ($p_{\text{interaction}}=1.6\times10^{-9}$), while in European ($n=164{,}307$) and Asian ($n=5921$) studies, the interactions did not reach statistical significance ($p=0.18$ and 0.15, respectively).[14] Also, about 44% of the European studies reported interaction beta coefficients that were in the opposite direction to the final meta-analysis estimates. Finally, while the *FTO* genotype has been shown to be associated with body size and risk of obesity in children, a meta-analysis of 19,268 children from nine studies found no evidence of an FTO-by-physical activity interaction effect on BMI ($p=0.98$). Thus, despite the overall statistical significance of the meta-analysis interaction term, it remains to be clarified how much additional information the *FTO*-by-activity interaction term really provides and what the explanation is for the heterogeneity of outcomes between individual studies.

Although *FTO* is so far the strongest genetic predictor of common obesity, many more genes and sequence variants have been identified through massive GWAS meta-analyses relying cumulatively on hundreds of thousands of subjects. Using data from the European Prospective Investigation of Cancer (EPIC)-Norfolk cohort, Li *et al.* tested in 20,430 individuals whether physical activity level modifies the association between BMI and a composite score of 12 obesity risk SNPs.[15] Total physical activity (both occupational and leisure-time) was quantified with a self-administered questionnaire and expressed as average daily hours of activity. For the statistical analyses, subjects were classified in four groups on the basis of their level of activity: inactive, moderately inactive, moderately active, and active. The genetic predisposition score for obesity was derived from 12 SNPs that had been identified in previously published GWAS for obesity. Each SNP was coded to reflect the number of obesity risk alleles (0, 1, 2), and the score was calculated by adding up the 12 recoded SNPs. Thus, the theoretical predisposition score values ranged from 0 to 24.[15]

In the full EPIC-Norfolk cohort, both the genetic predisposition score and physical activity level were associated with BMI. There was no direct association between the predisposition score and physical activity level. A statistically significant interaction effect between genetic predisposition score and physical activity level was observed both on BMI ($p_{\text{interaction}}=0.016$) and risk of obesity ($p_{\text{interaction}}=0.038$).[15] The association of the genetic score with BMI and risk of obesity was considerably stronger in inactive subjects, while in moderately inactive, moderately active, and active individuals, the genetic score effect was slightly smaller than in the total cohort (Fig. 3). However, the genetic associations with BMI level and obesity risk did not differ across the moderately inactive, moderately active, and active groups.[15]

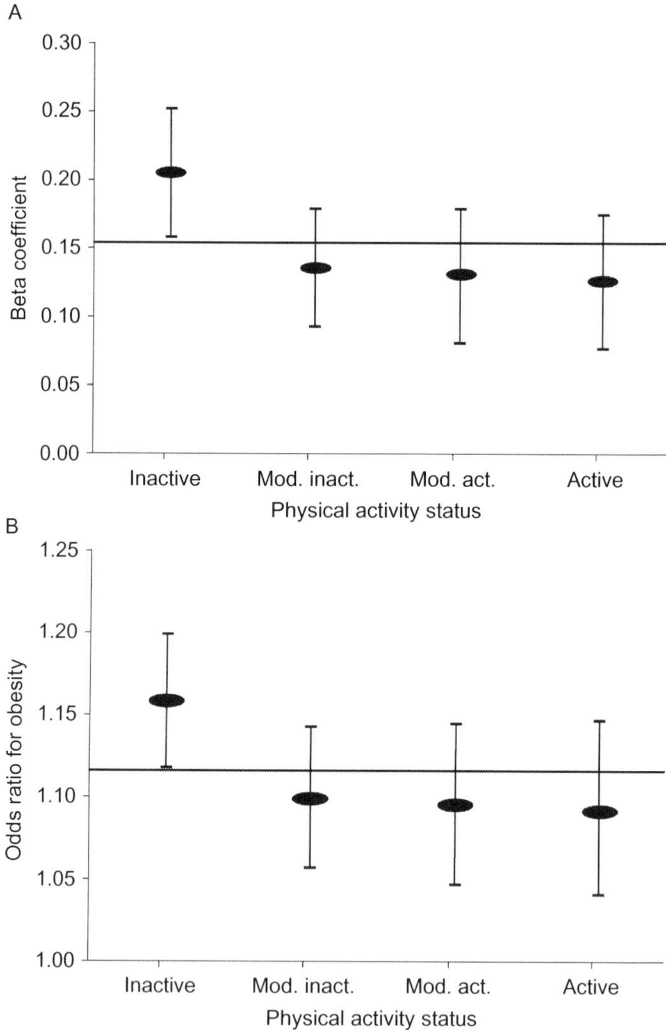

FIG. 3. Associations of the genetic predisposition score with BMI (A) and risk of obesity (B) stratified by physical activity level. Reference lines indicate the beta coefficient (A) and odds ratio (B) observed in the total EPIC-Norfolk cohort. Data derived from Table 2 of Li *et al.*[15]

IV. Interpretations, Future Directions, and Challenges

As summarized above, we already have a body of evidence indicating that genotype-by-physical activity interaction effects are ubiquitous and that they can be detected with adequate sample sizes and study designs. Given the first wave of GWAS reports from observational population studies, it is likely that many more genome-wide SNP-by-environment (including physical activity) interaction effect studies will appear soon. While these studies will generate impressive levels of significance given their sample sizes, it would be desirable for the investigators to quantify the true level and the nature of additional information provided by these interactions. Although the general expectation is that these interaction studies will uncover at least some of the "missing heritability," we still have no clear idea how much additional information is really provided by, for example, the *FTO*-by-activity interaction effects on obesity risk. The best-case scenario is one in which such studies can generate valuable information that could be utilized in preventive exercise medicine or to provide a foundation for hypothesis-driven controlled exercise interaction studies.

We also need to be cautious in the interpretation of the results of interaction studies. The basic rule of epidemiological studies applies to gene-by-physical activity interaction studies: never make claims of causality based on observational cross-sectional studies. For example, the genotype-by-physical activity interaction effects on BMI and obesity risk have been widely assumed to indicate that regular physical activity attenuates the genotype effect. However, the results in support of such a statement remain ambiguous. Another interpretation could be that physical inactivity amplifies the obesity risk associated with the risk genotype, because the greatest deviation from the obesity risk observed in the full study population was observed in the inactive subjects, not in those who were physically active.[14] Moreover, in the EPIC-Norfolk cohort, the genetic predisposition score-related obesity risk did not really decrease when physical activity level increased from less than 30 min per day to more than 60 min per day.[15] Thus, the current data likely could be interpreted to suggest that a complete lack of physical activity is particularly harmful in those individuals who have a genetic predisposition to gain weight and to become obese. Given the study designs, the only justified statement we can make at the moment is that the strength of association between genetic markers and the risk of obesity is significantly different between active and inactive individuals. But whether increasing the amount of physical activity truly protects against weight gain in those who have a genetic predisposition for obesity is still unclear, and the interaction effect size with inactivity or an increasing amount of activity remains to be quantified.

We need to keep in mind the kind of information a given study design provides and its limitations. Gene–activity interactions from observational studies tell us nothing about responsiveness to regular exercise. On the other hand, observational studies provide valuable clues regarding genotype and physical activity in the primary prevention of adverse health outcomes, a question that would be very difficult to address with exercise intervention design. However, a controlled exercise trial is the only way to investigate the genetics of interindividual differences in responsiveness to regular physical activity, and such data are absolutely necessary for clinical exercise applications, especially in the context of personalized exercise medicine.

The *FTO*-by-activity interaction effect on body weight and body composition provides a good example of the limitations associated with observational data. When the first interaction studies were published, the lay media, in particular, interpreted the results to indicate that *FTO* risk allele homozygotes could eliminate their genotype-related obesity risk and lose weight more efficiently by exercising. However, data from the HERITAGE Family Study suggested that the situation in terms of body fat loss may be exactly the opposite.[16] The *FTO* risk allele homozygotes showed no change in fat mass during 20 weeks of carefully controlled and supervised exercise training, while subjects who did not carry the risk allele lost on average almost 1 kg of body fat. One possible interpretation of these findings is that physical activity may prevent *FTO*-related weight gain in the risk allele homozygotes, but that exercise is not associated with greater adiposity loss in the same genotype compared to others. Obviously, this is an interesting question that requires further studies in order to be resolved.

While the future of gene–physical activity interaction studies based on observational studies seems promising, the situation with exercise intervention trials is more complex. As reviewed earlier in this chapter, exercise training studies provide information about the genetics of adaptation to regular physical activity that is impossible to obtain from other types of study designs. Furthermore, intervention studies allow much more precise assessment of and control for physical activity levels of the volunteers than observational studies. This is probably one of the main reasons why the SNP effect sizes observed in exercise training studies are considerably larger than those reported in cross-sectional epidemiological studies. For example, data from HERITAGE suggest that the SNPs identified in the GWAS analyses fully account for the heritability of VO_{2max} and HR50 training responses observed in the cohort.[5,6] Exercise training studies also provide clinically important information about the effectiveness of physical activity on various health outcomes and risk factor levels and how that effectiveness varies as a function of genotype. These types of data will be absolutely essential for the development of personalized exercise medicine approaches.

The main obstacle to new exercise intervention trials is the cost. In order to address genetic hypotheses in such studies, sample size has to be sufficiently large; at least several hundreds, preferably thousands, of participants need to be recruited. Furthermore, the exercise training program has to be fully standardized and controlled. The adherence rate of the volunteers has to be as close to 100% as possible in order to minimize random trait variation due to shifts in intervention protocol. These factors, together with greater phenotyping needs (both pre- and posttrial measurements) than in observational studies, impose major labor, material, and infrastructure requirements. Considering the current research funding environment, conducting appropriately powered exercise intervention studies is a major challenge. An alternative, although not quite as powerful, strategy is to combine results or even raw data from multiple smaller existing studies using meta-analytic approaches. However, even with meta-analysis, the individual studies should have at least 300–400 subjects each, with comparable and standardized exercise training protocols and appropriate end-point measurements. Unfortunately, such studies are still very rare, which seriously limits the usefulness of the meta-analytic approach. Therefore, it is vital that the exercise medicine community makes every possible effort to conduct sufficiently large exercise intervention trials.

V. Summary

A growing body of evidence shows that gene-by-exercise interaction effects on several health-related outcomes exist and can be detected with appropriate study designs. Such interactions derived from observational population studies and experimental exercise intervention studies have great potential to further our understanding of how genetic factors modify the effects of regular physical activity and sedentary behavior on health status and on the risk of morbidity and even mortality. These data would be particularly important for the development of more personalized strategies to use physical activity in primary and secondary prevention of chronic diseases, as well as in quality of life improvements for the general population. We expect to gain much new information regarding gene–physical activity interactions from observational studies within the next few years. However, in order to generate the necessary data to develop strategies and methods for personalized exercise medicine applications, new exercise intervention trials are needed with appropriately large sample sizes and well-standardized exercise programs as well as end-point measurement protocols. Only then can we take a full advantage of the vast information potential embedded in gene–exercise interactions.

References

1. Bouchard C, Dionne FT, Simoneau JA, Boulay MR. Genetics of aerobic and anaerobic performances. *Exerc Sport Sci Rev* 1992;**20**:27–58.

2. Bouchard C, Rankinen T. Individual differences in response to regular physical activity. *Med Sci Sports Exerc* 2001;**33**:S446–51.
3. Bray MS, Hagberg JM, Perusse L, Rankinen T, Roth SM, Wolfarth B, et al. The human gene map for performance and health-related fitness phenotypes: the 2006-2007 update. *Med Sci Sports Exerc* 2009;**41**:37–72.
4. Timmons JA, Knudsen S, Rankinen T, Koch LG, Sarzynski M, Jensen T, et al. Using molecular classification to predict gains in maximal aerobic capacity following endurance exercise training in humans. *J Appl Physiol* 2010;**108**:1487–96.
5. Bouchard C, Sarzynski MA, Rice TK, Kraus WE, Church TS, Sung YJ, et al. Genomic predictors of maximal O2 uptake response to standardized exercise training programs. *J Appl Physiol* 2011;**110**:1160–70.
6. Rankinen T, Sung YJ, Sarzynski MA, Rice TK, Rao DC, Bouchard C. The heritability of submaximal exercise heart rate response to exercise training is accounted for by nine SNPs. *J Appl Physiol* 2012;**112**:892–7.
7. Bouchard C, An P, Rice T, Skinner JS, Wilmore JH, Gagnon J, et al. Familial aggregation of VO2max response to exercise training: results from the HERITAGE Family Study. *J Appl Physiol* 1999;**87**:1003–8.
8. Frayling TM, Timpson NJ, Weedon MN, Zeggini E, Freathy RM, Lindgren CM, et al. A common variant in the FTO gene is associated with body mass index and predisposes to childhood and adult obesity. *Science* 2007;**316**:889–94.
9. Zeggini E, Weedon MN, Lindgren CM, Frayling TM, Elliott KS, Lango H, et al. Replication of genome-wide association signals in UK samples reveals risk loci for type 2 diabetes. *Science* 2007;**316**:1336–41.
10. Scott LJ, Mohlke KL, Bonnycastle LL, Willer CJ, Li Y, Duren WL, et al. A genome-wide association study of type 2 diabetes in Finns detects multiple susceptibility variants. *Science* 2007;**316**:1341–5.
11. Loos RJ, Bouchard C. FTO: the first gene contributing to common forms of human obesity. *Obes Rev* 2008;**9**:246–50.
12. Andreasen CH, Stender-Petersen KL, Mogensen MS, Torekov SS, Wegner L, Andersen G, et al. Low physical activity accentuates the effect of the FTO rs9939609 polymorphism on body fat accumulation. *Diabetes* 2008;**57**:95–101.
13. Rampersaud E, Mitchell BD, Pollin TI, Fu M, Shen H, O'Connell JR, et al. Physical activity and the association of common FTO gene variants with body mass index and obesity. *Arch Intern Med* 2008;**168**:1791–7.
14. Kilpelainen TO, Qi L, Brage S, Sharp SJ, Sonestedt E, Demerath E, et al. Physical activity attenuates the influence of FTO variants on obesity risk: a meta-analysis of 218,166 adults and 19,268 children. *PLoS Med* 2011;**8**:e1001116.
15. Li S, Zhao JH, Luan J, Ekelund U, Luben RN, Khaw KT, et al. Physical activity attenuates the genetic predisposition to obesity in 20,000 men and women from EPIC-Norfolk prospective population study. *PLoS Med* 2010;**7**:e1000332.
16. Rankinen T, Rice T, Teran-Garcia M, Rao DC, Bouchard C. FTO genotype is associated with exercise training-induced changes in body composition. *Obesity (Silver Spring)* 2010;**18**:322–6.

Adaptive Genetic Variation and Population Differences

CHAO-QIANG LAI

Nutrition and Genomics Laboratory, Jean Meyer USDA Human Nutrition Research Center on Aging, Tufts University, Boston, Massachusetts, USA

Since the expansion of modern humans (*Homo sapiens*) from Africa to the rest of the world between 50,000 and 100,000 years ago, the human genome has been shaped not only by demographic history but also by adaptation to local environments, including regional climate, landscape, food sources, culture, and pathogens. Genetic differences among populations interact with environmental factors, such as diet and lifestyle, leading to differences in nutrient metabolism, which translate into differences in susceptibility to a variety of diseases.

Progress in Molecular Biology and Translational Science, Vol. 108
DOI: 10.1016/B978-0-12-398397-8.00018-6

1877-1173/12 $35.00

Individuals from different populations sharing the same environments can exhibit differences in disease risk, as do individuals from the same population living in various regions of the globe. Therefore, it is important to understand how adaptive genetic variations interact with environments to influence health. This knowledge will provide a broad foundation for designing experiments and approaches in nutrigenomics research and strengthening the knowledge base for dietary recommendations for disease prevention. The objectives of this chapter are to (1) understand the methodology employed in examining adaptive genetic variation across populations, (2) establish the importance of adaptive genetic variation to human health, and (3) discuss the implications for nutrigenomics research and disease prevention.

I. Brief History of Human Migration

We are physiologically and morphologically diverse, but our genomes are 99% identical. This puzzling fact relates to the origin of modern humans (*Homo sapiens*). The combination of earliest existing archeological evidence in East Africa and genetic evidence[1–3] from the human genomes of populations from across the world strongly supports the fact that modern humans lived in Africa over 160,000 years ago.[4,5] About 90,000 years ago, the first group of *H. sapiens* to successfully move out of Africa migrated along the southern coast of the Arabian Peninsula toward India, eventually reaching South China around 85,000–75,000 years ago. A subgroup of the same population (perhaps traveling with a new group from Africa) moved into Europe around 50,000 years ago after recovering from the 10,000-year ice age. Around 16,000 years ago, the ancestors of Native Americans from Asia voyaged across the Bering land bridge linking Siberia and Alaska.[6–8] A subgroup of this population migrated to South America around 15,000 years ago.[7,9]

Although the timing for specific human migration patterns is imprecise, molecular evidence has provided crucial supportive evidence. Because there is little recombination between genetic markers on the mitochondrial DNA and Y chromosome, both act as molecular clocks for the history of modern humans.[3] Based on the relationship between haplogroups of mitochondrial DNA derived from populations worldwide, all of the haplogroups can be traced to origins in Africa, and haplogroup distributions are very consistent with migration patterns of human populations as already summarized.[7,9–11] In addition, the suggested times of migration are well supported by the estimated divergence times of the DNA sequences of mitochondria[7,9–11] and Y chromosomes[12] among worldwide populations.

In the history of their expansion within and out of Africa, the habitats and lifestyles of modern humans have evolved greatly in accordance with the passage of time and changes of the environments.[13,14] It is believed that most modern humans lived on forest-based diets or the so-called hunter-gatherer lifestyle from 50,000 years ago until about 10,000 years ago.[15] Such lifestyles are believed to have shaped the genome and to favor the development of physiological strategies to achieve effective energy storage and resistance to starvation, the so-called thrifty genotypes.[15,16] Farming began about 10,000 years ago in the Middle East,[17] which marked the starting point of modern civilization, and spread to Europe, Asia, and the rest of the world.[17] Advances in science and technology that accompanied and followed the development of agriculture substantially altered lifestyles, particularly over the past 100 years.[13,17] As a result, humans in many developed and developing countries now embrace modern civilization and its consequences, which include energy-dense processed foods and a sedentary lifestyle.

II. Human Population Differences Defined by Major Geographic Regions

Genetic diversity among populations is of great interest to geneticists. Based on a genome survey of 377 small microsatellite repeats (or simple sequence repeats) in 1058 individuals from 52 populations around the world, and without using prior information about the origins of the individuals, Rosenberg *et al.*[2] identified six major genetic clusters. Five of these clusters each represents a major geographic region: Africa, America, East Asia, Europe/West Asia, and Oceania. A similar pattern was demonstrated by Wang *et al.*[11] based on a larger number of microsatellite repeats (678 small microsatellite repeats) and a larger sample size of Native Americans. Li *et al.*[18] also came to a similar conclusion using 650,000 single-nucleotide polymorphisms (SNPs) in 938 subjects from 51 populations. The diversities across groups are further proved by the phylogenic analyses of human mitochondrial DNA haploid groups[6–8,19] and Y chromosomes.[12]

The International HapMap project[20] that started in 2002 has devoted much effort to examining genetic variation among humans, initially working with only three major groups: African (YRI: Yoruban individuals in Nigeria), Asian (ASN = CHB: Han Chinese individuals in Beijing + JPT: Japanese individuals in Tokyo), and European (CEU: individuals of European ancestry in Utah); later, the project expanded to 11 groups. Based on the Phase I data, allele frequency distributions of many SNPs are distinctly different between populations.[21] Among the 900,000 SNPs genotyped in the YRI, ASN, and CEU groups by Perlegen and the HapMap project, Coop and colleagues[22] showed that 6955 SNPs between ASN and YRI, 3500 SNPs between CEU and YRI, and 792 SNPs between ASN and CEU displayed an absolute allele frequency difference of greater than 70%. The numbers of genetic variants that are fixed (unique) for specific populations, however, are small. Genetic diversity can be also measured using Wright's F_{st} (see Section III), which is the fraction of the genetic difference between populations normalized to the total genetic variation. The average F_{st} based on SNPs between CEU and ASN is 0.111, whereas that for YRI versus ASN is 0.185 and CEU versus YRI is 0.157.[23] These measures are further supported by other studies.[18,22,24,25] The results suggest that, when compared genetically, ASN and YRI are more dissimilar than CEU and ASN. Pairwise F_{st} across the world has been estimated.[18] Overall, human genetic differences are substantial and mainly determined by geographic location, suggesting that both demographic history (genetic drift and geographic isolation) and adaptation to local environments play crucial roles in the genetic divergence between populations.

III. Detection of Positive Selection in Human Populations

Why do we care about adaptive genetic variation? As human lifestyles have shifted because of modern civilization, the prevalence of cardiovascular disease (CVD), diabetes mellitus, metabolic syndrome, and cancer has increased dramatically worldwide, both in developed and in developing countries. This increase in disease despite improvements in medical knowledge and technologies has been the focus of public health for more than 50 years. For example, the thrifty genotype hypothesis was first proposed by Neel[16] when he postulated that the genetic variants that increase disease risk in our modern age are likely to be the same variants that conferred survival advantage to our ancestors in harsh environments. Over the course of human evolution, these variants have been subject to positive selection which altered their frequencies.

The great interest in selection stems from the recognition that genes targeted by selection are more likely to be associated with protection from or predisposition to diseases.[26–28] A selection signal is usually a functional variant that is responsible for a particular phenotype. Through such phenotypes, natural selection acting on the human genome can translate into negative or positive selection. Negative selection that removes deleterious mutations can be observed in conserved exons in protein-coding genes,[29] and in conserved microRNA sequences and microRNA-binding sites in noncoding regions.[30] An estimated 38% of the mutations that change amino acids in genomes are washed out by negative selection.[29] In contrast, positive selection drives new mutations to higher frequencies, conferring survival advantage to individuals in a population (see Fig. 1). Major effort has focused on positive selection, which can leave a "footprint" in the genome and be detected at the molecular level. Over the past 20 years, many methods in population genetics have been developed to detect selection based on model organisms, as well as in human populations.[31] Excellent reviews have discussed the pros and cons of each method.[32–38] Described below is a list of common methods that can be applied to the human genome.

A. Tests Based on Substitution Rate Within and Among Species

The first method for detecting positive selection was proposed by Kimura[39] and is based on the examination of protein-coding DNA sequence differences between related species. Assuming that synonymous mutation is neutral, a test statistic quantified the ratio of the nonsynonymous substitution rate (K_a) and the synonymous substitution rate (K_s) in the same protein, with $K_a/K_s > 1$ suggesting that selection exists in a gene.[31] This method was first formulized by Hudson, Kreitman, and Aguadé (HKA test)[40] and was further extended by McDonald and Kreitman (MK test).[41] In contrast to previous tests, the primacy of the MK test is that the ratio K_a/K_s between species is the same as the ratio within species if the observed substitutions are neutral. Otherwise, a significant deviation from the null model will be evident in the selection. Other tests

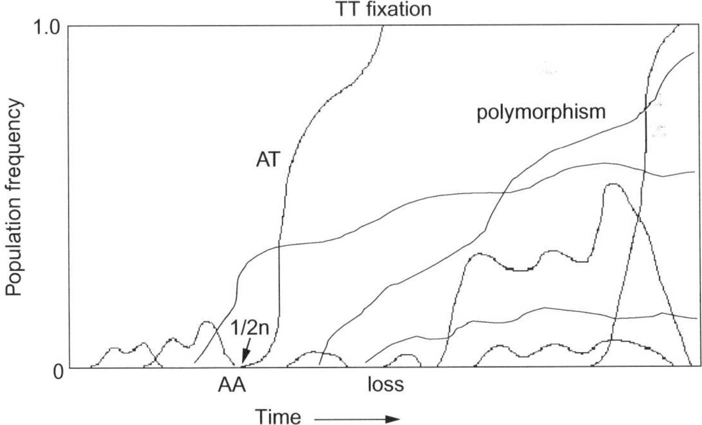

FIG. 1. The fate of new mutations. Mutations (i.e., A>T, frequency [T] = 1/2n, n = population size), which can have deleterious, neutral, or favorable effects on a phenotype, occur spontaneously. If a mutation has a favorable effect, which gives an advantage to an organism to survive in a certain environment (e.g., resistance to fatal infection), the frequency of such a mutation would increase rapidly in the population (fixation when frequency [T] = 1). This process is also called positive selection. On the contrary, negative selection eliminates mutations (loss when frequency [T] = 0) that give a deleterious effect to an organism. New mutations can also be maintained as polymorphisms by selection or drift or can be lost by negative selection or drift.

similar to the MK test have been developed,[42,43] but they are limited to protein-coding sequences, have little power in detecting recent selection, and cannot distinguish between past and present selection.[36]

B. Selection by Population Subdivisions

After migration to new environments, populations often reproduce in isolation from other groups because of geographical or cultural obstacles. This eventually leads to genetic divergence between groups. Such divergence can be measured by *F*st, an *F*-statistic measuring the variance in allele frequency between subpopulations relative to the total variance of the whole population.[44] F_{st}, however, can be confounded by the demographic history of groups and the bottleneck effect.[24] Thus, one should be cautious in relying solely on F_{st} criteria as evidence of selection.[24] Weir *et al.*[24] suggested that the population-specific F_{st} value is the best indicator of selection. When this F_{st} value is more than three standard deviations higher than the chromosome mean, such regions are considered under selection. Akey *et al.*[45] derived genome-wide criteria under a neutral model to be $F_{st} \geq 0.45$ for autosomes and $F_{st} \geq 0.59$ for the X chromosome. Myles *et al.*[46] proposed more extreme F_{st} values (within the top 1% of the F_{st} distribution) as indicators of positive selection.

C. Tests Based on Allele Frequency Spectrum

When a mutation is neutral, the expected frequency of the allele is proportional to $1/2n$, where n is the effective population size, that is, the number of independent breeding individuals. When the selection removes deleterious mutations (i.e., negative selection), it leads to a large number of mutations existing at low frequencies. In contrast, positive selection leads to a large number of mutations segregating at high frequencies. To detect such selection signals, many methods have been developed to measure allele frequency spectrums in a population. One of the most popular methods is Tajima's D,[47] which tests the difference between the number of segregating sites and the average number of nucleotide differences between pairs of DNA sequences among individuals within a population. Positive values of the test suggest deviation from neutrality with many variants at intermediate frequencies, whereas negative values indicate deviation from neutrality with many variants at low frequencies. Fu and Li[48] expanded this method to consider an outgroup (a closely related species) in order to infer the ancestral state of the mutations, which allows weighting allele frequencies differentially. This method was further refined by Fay and Wu's *H*-test,[49] which is more sensitive to detect positive selection when an excess of derived variants (derived refers to the new mutation as determined from an outgroup) are at high frequencies.

D. Linkage Disequilibrium and Haplotypes

Linkage disequilibrium (LD) between two loci decays gradually in proportion to the recombination rate and time as measured in numbers of generations. When mutations are under positive selection, the LD surrounding the mutations is maintained because of the hitchhiking effect (see Fig. 2); thus, longer haplotypes at high frequencies can be maintained within the population. Many LD-based methods have been developed to detect positive selection. Hudson *et al.*[51] proposed the first method to detect positive selection by measuring haplotype patterns. Using the extended haplotype homozygosity (EHH) test, Sabeti *et al.*[52] developed a more robust method to detect positive selection by measuring longer haplotypes at high frequencies. This method was further refined by Voight *et al.*,[53] who standardized the EHH test using the genome-wide empirical distributions of EHH. Based on a similar rationale, Wang *et al.*[54] developed a new version of the LD-based method called LD decay. One major advantage of the LD-based method is that it can detect recent positive selection, including partial sweep (selective alleles are not fixed), more effectively than earlier approaches.

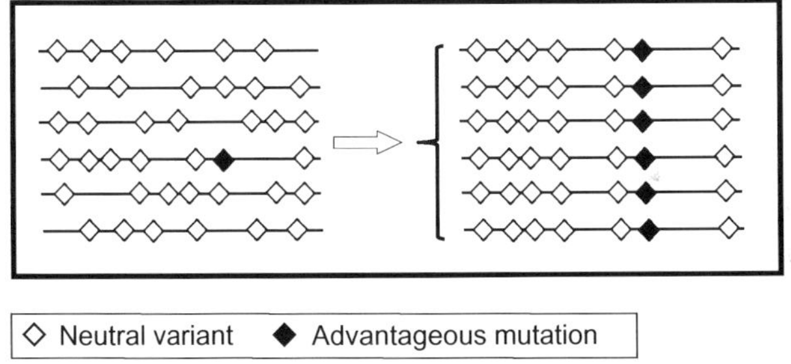

◇ Neutral variant ◆ Advantageous mutation

FIG. 2. Positive selection drives an advantageous mutation to a high frequency in a population. This is also called selective sweep.[50] Neutral variants that are in strong linkage disequilibria with the advantageous mutation will hitch-hike to high frequencies similar to that of the mutation. Based on this phenomenon, extended haplotype methods were developed to measure positive selection.[51,52]

E. Detecting Selection Due to Climatic Pressure

Adaptation occurs when populations move to a new location and survive in an unfamiliar environment, which imposes new challenges for survival. Climate is considered the most important environmental factor for adaptation because it largely determines other environmental factors.[55] Therefore, if climate plays a crucial role in shaping the process of adaptation in humans, the allele frequency distribution should be in accordance with climate patterns. Hence, to detect climatic selection, one can measure the correlation between allele frequencies at candidate genes and climate parameters of local regions, such as temperature, humidity, rainfall, and solar radiation. The strength of such correlations may reflect the pressure of selection. Hancock and colleagues[55] employed two methods: the Spearman rank correlation coefficient and Bayesian geographic analysis, to measure the correlations between potential adaptive variants and the climatic patterns. This approach was later expanded to the whole genome.[56,57]

IV. Genome-Wide Detection of Positive Selection Among Human Populations

The complete sequencing of multiple human genomes and the advances in the HapMap project[20,58] have laid the foundation for identifying selection signatures on a genome-wide scale. In addition, the application of early theories in population genetics[31] combined with new analytical methods of detecting natural selection enabled genome-wide screens for genetic signatures of early and recent positive selection.[52,53,59] Twenty-five genome-wide surveys have been reported to identify selection signals in the human genome, and most of these have been discussed extensively in previous reviews.[36–38] Here we focus on some of the key surveys that have increased our understanding of the major differences in genetic adaptation among populations.

A. Recent Adaptation Based on Extended Haplotype Scan

The idea that haplotype patterns can detect recent adaptation led to a comprehensive survey of the human genome for positive selection that was based on SNP genotype data from the HapMap project. Extending the EHH test[52] to the integrated haplotype score (iHS) method, Voight and coworkers[53] examined 800,000 SNPs in 89 Asians (Chinese and Japanese, ASN), 60 Europeans (CEU), and 60 Yorubans (YRI) and identified 250 loci in each population. Most of the identified adaptation signals were specific to one of the three populations, with only 10 loci shared among all three. Enrichment gene categories include carbohydrate metabolism, lipid, and fatty acid binding proteins, and vitamin/cofactor transport in ASN; olfaction, electron transport, MHC class 1-mediated immunity, and phosphate metabolism in CEU; and steroid metabolism and peroxisome transport in YRI. Using the 3.8 million SNPs from Phase II of the HapMap project, 961 (ASN), 1013 (CEU), and 1000 (YRI) regions at the top 5% of the iHS selection signal were identified. Among them, 12.8% of the selective signals are shared between ASN and CEU; 6.4% between CEU and YRI; and 5.9% between ASN and YRI. Only 20 regions (2%) are shared between the three populations.

Similarly, using a different haplotype-based method, namely, the LD decay method, Wang *et al.*[54] identified 1800 genes that are likely to be under positive selection based on 1.6 million SNPs from Perlegen Sciences[60] and 1 million SNPs from Phase I of the HapMap project. The overrepresented genes include those involved in host–pathogen interactions, reproduction, DNA metabolism, cell cycle, protein metabolism, and neuronal function. Among these potential selective signals, 112 genes were common among Africans, Asians, and Europeans. Compared to the result provided by Voight *et al.*,[53] 27% of the genes from Wang *et al.*[54] are shared between the two studies.[34]

Based on 3.1 million SNPs from Phase II of the HapMap project, Grossman and colleagues[61] conducted a genome survey using the composite of multiple signal approach by combining five tests: long-range haplotype, iHS, cross-population EHH, F_{st}, and the test for derived alleles at high frequency relative to other populations. Using this approach, 178 regions of positive selection containing about 1500 genes were mapped to regions less than 55 kb. Among them, 53 (ASN), 69 (CEU), and 49 (YRI) regions are unique to one of three populations, with none being shared by all three populations.

B. Older Adaptive Variants Detected by Allele Frequency Spectrum

Using the human genome data generated by Perlegen,[60] Carlson *et al.*[62] identified a small set of candidate regions that displayed selection sweep based on Tajima's *D*-test and Fay and Wu's *H*-test. These included 7, 23, and 29 regions of African, European, and Chinese populations, respectively, with no overlap. Using the same dataset but a composite likelihood ratio method that accounts for population demography and recombination rate variation, Williamson *et al.*[63] identified 101 genes that have been subject to positive selection. Among them, 10 (Africans), 26 (Europeans), and 57 (Asians) are unique to each of the three populations, respectively. These genes are involved in nervous system development and function, the immune system, and heat shock biology. In particular, 56 unlinked genes are related to heat shock, with 28 genes having likely experienced recent selection sweep.

C. Population Subdivision

Akey *et al.*[45] reported the first genome-wide study to detect positive selection based on population subdivisions, using allele frequency panels consisting of 42 African-Americans, 42 European-Americans, and 42 East Asians. From 26,530 SNPs, 174 genes were identified as candidate genes of positive selection. Based on 1 million SNPs from Perlegen and 600,000 SNPs from Phase I of the HapMap project, Weir *et al.*[24] calculated F_{st} between three populations, as well as population-specific F_{st}, and identified a set of regions with extreme values of F_{st} (three standard deviations from the mean). Using 1.5 million SNPs from three populations (24 European-American, 23 African-American, 24 Han Chinese) from Perlegen, Myles *et al.*[46] identified 1074 genome regions within the top 1% of the F_{st} distribution. Among them, there were 354 regions for Europeans versus African-Americans, 357 regions for Chinese versus African-Americans, and 363 regions for Chinese versus Europeans. This suggests that the identified genetic variation of positive selection by the F_{st} method is diverse among groups. One shortcoming of the F_{st} method of detecting positive selection is that it is often confounded by the demographic history (migration and isolation) of groups.

D. Climate Adaptation at Genome-Wide Level

There are two genome-wide studies using climatic measures to detect positive selection. Hancock *et al.*[38] categorized the global climates into four ecologic regions (dry, polar, humid temperate, humid tropical); classified subsistence into four types (agriculture, foraging, horticulture, pastoralism); and examined the correlation between such climatic measures and allele frequencies on a genome-wide scale for 61 populations from around the globe. The study identified the polar domain as the strongest selective factor and a small number of potential adaptive variants that exhibit consistent allele frequency differences in relation to the climatic measures. In addition, genic and nonsynonymous SNPs showed stronger associations compared to non-genic SNPs with environment measures among the identified SNPs. By defining specific climatic factors for local environments (sunlight, rain precipitation, temperature, humidity, latitude) over worldwide geographic locations, Hancock *et al.*[56] examined more than 500,000 SNPs in 61 global populations for correlations with defined climatic variables using the Bayesian linear model method.[64] Among 604 identified SNPs of potential climate adaptation, a small set of SNPs overlapped with the SNPs identified on the basis of genome-wide association studies (GWASs), especially for pigmentation, immune and autoimmune function, and CVD. Again, genic and nonsynonymous SNPs exhibited the strongest correlations between SNPs and environmental variables. Importantly, most adaptive variants of strong signals are unique to each population, suggesting that selection pressures in climates are usually local or, alternatively, that the identified variants experienced convergent evolution.

In summary, so far, genome-wide scans have detected significant differences in adaptive variation among populations. The results of multiple studies overlapped when similar methods were used. The differences between studies are mainly due to differences in methodology, which is not surprising, as the methods are designed to detect specific types of positive selection. The results of several genome-wide scans of positive selection are now available in searchable databases, including Haplotter (http://haplotter.uchicago.edu/),[53] SNP@Evolution (http://bighapmap.big.ac.cn/),[65] and dbCLINE (http://genapps2.uchicago.edu:8081/dbcline/main.jsp).[56]

V. Population Differences Due to Climatic Adaptation

Environmental changes are the primary driver of adaptation, and climate is the major component of regional environments. After moving out of Africa, where an equatorial climate is the norm, and spreading to the Middle East, Asia, and Europe, our ancestors were subject to the challenges of high latitudes, hot and cold temperatures, and changes in humidity and sunlight radiation. These climatic pressures have changed human bodies in shape and size and in skin pigmentation, energy and nutrient metabolism, and regulation of fluid and electrolyte balance. These phenotypic changes are recorded as variations in the genes involved in these biological processes. Thus, population differences in morphology and physiology are determined to a great extent by the climatic differences around the globe.

A. Skin Pigmentation Among Populations

The major morphological difference among populations around the world is in skin pigmentation. This human phenotype is highly correlated with the latitude of the location where the humans inhabited. Darker-skinned groups lived in or near the tropics, whereas those with lighter skin were distributed in northern regions of the globe. Adaptation of skin pigmentation has been subject to two types of climatic selection.[66] First, strong UV radiation near the equator favored the development of a darker skin with a large amount of melanin in the skin cells, providing better protection against UV light from the sun. Several genetic loci associated with skin pigmentation show evidence of strong natural selection. For example, melanocortin 1 receptor (*MC1R*), which regulates melanin production, experienced intensive negative selection.[67,68] Second, weak sunlight in high latitudes leads to lighter, less pigmented skin so that effective synthesis of vitamin D3 in a low-UVB (290–310 nm) environment is possible. In populations of European ancestry, strong positive selection was detected in oculocutaneous albinism II (*OCA2*), or the blue eye gene, as its mutation reduces the production of melanin in the iris (so that brown eyes become blue eyes), hair, and skin.[69,70] The vitamin D receptor (*VDR*) gene has been subject to positive selection in European and Asian populations.[71] In addition, sexual selection could be another factor contributing to differences in skin and eye color.[72]

B. Adaptations to Climate and Common Metabolic Disorders

Temperature and humidity are the key factors in the climate of regional environments. To survive in harsh environments, the human body requires robust metabolic systems to tolerate hot or cold temperatures and wet or dry surroundings. Therefore, genes involved in energy metabolism are likely to be subject to climatic adaptation. Hancock and coworkers[55] examined 873 tag SNPs of 82 candidate genes encoding metabolic regulators for evidence of climatic selection. They analyzed correlations between allele frequencies in 54 populations worldwide and climate variables (temperature, humidity, precipitation, solar radiation) measured at the regions where the populations live. Seventy-two SNPs of 32 genes were identified as potential targets of climate selection. Among them, 33 SNPs in 19 genes, including cholesteryl ester transfer protein, plasma (*CETP*); paraoxonase 1 (*PON1*); transcription factor 7-like 2 (*TCF7L2*); superoxide dismutase 1, soluble (*SOD1*); clock homolog (mouse) (*CLOCK*); epidermal growth factor receptor (*EGFR*); leptin receptor (*LEPR*); and regulatory-associated protein of MTOR, complex 1 (*RPTOR*), were associated with summer climate; whereas 32 SNPs in 17 genes, including angiotensin I converting enzyme 1 (*ACE*); *LEPR*; mitogen-activated protein kinase 1 (*MAPK1*); uncoupling protein 3 (*UCP3*); and *TCF1* (now known as HNF1 homeobox A [*HNF1A*]), were correlated with winter climate. In addition, 14 SNPs in 7 genes were associated with latitude parameters. In particular, several SNPs in the *RPTOR* gene, which is responsible for nutrient signaling, mitochondrial oxygen consumption, and oxidative capacity, exhibited significant signals associated with summer and winter climate-related variables, as well as latitude. *TCF7L2* variants consistently associated with type 2 diabetes showed significant associations with summer climatic variables. This study demonstrated that genes under climate selection are also likely to be associated with risk factors for common metabolic disorders.

C. Climatic Difference and Hypertension Prevalence

The prevalence of high blood pressure or hypertension differs significantly among populations. In the United States, African-Americans experience a higher rate of hypertension than other ethnic groups.[73] Genetic susceptibility has been considered one of the major contributors to this disease.[74] By examining the worldwide allelic distribution of seven functional variants from five genes responsible for salt and water balance and vascular reactivity (angiotensingen [*AGT*]; guanine nucleotide binding protein, beta polypeptide 3 [*GNB3*]; beta-2 adrenergic receptor, surface [*ADRB2*]; non-voltage-gated sodium channel 1 alpha [*ENaC*α, now known as *SCNN1A*]; and gamma [*ENaC*γ, now known as *SCNN1G*]), researchers found that heat-adapted allele frequencies were highly correlated with low latitude or hot and wet climates.[75] The results further demonstrated that the heat-adapted and derived allele (a non-ancestral allele) *GNB3*-825T was associated with hypertension in the African populations. The results strongly suggest that high-risk alleles for hypertension were subject to selection during the out-of-African expansion. Thompson and colleagues[76] showed that the frequency of the non-expressor allele of the cytochrome P450, family 3, subfamily A, polypeptide 5 gene (*CYP3A5**3) is strongly correlated with distance from the equator by displaying differential minor allele frequencies across populations and latitudes (Nigeria [Yoruba], 0.06, 8°N; China [Han], 0.75, 32°N; France [Basque], 0.96, 43°N). Haplotype pattern analysis further indicated that this variant was selected. This variant is also associated with blood pressure when salt intake is considered.[77]

VI. Dietary Adaptation and Population Differences

During the history of modern humans living in and expanding out of Africa, as well as in recent civilization, major shifts in dietary habits have led to dietary diversity across populations. Modern dietary habits can be traced back to the hunter-gatherer lifestyle of feast and famine, which was followed by the domestication of animals and plants, with heavy dependence on cereals, roots, and tubers. In contrast, modern diets are characterized by high energy (foods high in fat, protein, and carbohydrate). Dietary changes are considered one of the major environmental determinants of human genome adaptation. The hunter-gatherer lifestyle led to evolution of the so-called thrifty genotype,[16] whereas farming overcame the unpredictable patterns of feast and starvation and freed more hands for the development of advanced technology. Advanced technology created modern civilization, which in turn changed the lifestyle of humans dramatically. The combination of a genome adapted to periods of feast and starvation and the current environment with its energy-dense diets and low physical activity levels has led to increased prevalence of obesity and related diseases in many populations.

A. Adaptation to Dairy Farming

The development of agriculture is hypothesized to have radically altered the dietary habits of modern humans. Cattle raising and dairy farming are important examples of agricultural development, which dates back to Europe about 9000 years ago.[78] Lifelong consumption of dairy products requires the ability to digest lactose in milk. This capability varies dramatically, with most Europeans demonstrating lactase persistence (see also Chapter 5). The distribution of lactase persistence is consistent with the dairy farming history.[79,80] Using the long-range haplotype method,[52] Bersaglieri *et al.*[81] showed that the lactase (*LCT*) gene has been subject to strong positive selection in European ancestral populations. This is a classic example of strong selection that facilitated dietary adaptation in humans. Other studies[53,82] further confirmed that *LCT* was subject to positive selection in European groups. In contrast, the *LCT* gene was not under positive selection in Asian populations, in which the lactase-persistence 13910T allele is absent or present at low frequencies.[81] In addition, three additional *LCT* variants detected in east Africans confer lactose tolerance independently of those variants occurring in Europeans. These promoter variants emerged in response to selection sweep (see Fig. 2; an extreme selection leads to 100% of the relevant allele) and represent an example of convergent adaptation.[83] On the other hand, these *LCT* variants also are present at very low frequencies or do not exist at all in East Asian groups. Nucleotide-binding oligomerization domain containing 2 (*NOD2*) gene variants that are associated with Crohn's disease are present at high frequencies in populations of European ancestry and are likely the result of positive selection that confers resistance against bacteria that may be present in cow's milk.[84]

Using the iHS method, Voight *et al.*[53] identified a small set of genes (including *LCT*) that are under positive selection and likely involved in dietary adaptation in different populations. Hancock *et al.*[57] identified two genes (tenascin XB [*TNXB*] and activating transcription factor 6 beta [*ATF6B*, also known as *CREBL1*]) that are subject to climatic selection and are correlated with consumption of fat, meat, or milk.

B. Adaptation in Carbohydrate Metabolism

Using the iHS method, Voight *et al.*[53] identified several genes associated with carbohydrate metabolism (*MAN2A1*, *SI*, *LCT*) in Africans and East Asians. Taking a different approach to detect recent dietary adaptation, Hancock *et al.*[57] correlated allele frequency differences at the genome-wide level with three main food components (cereals; fat, meat, and milk; roots and tubers) in diets of 61 populations across the world. Five genes with strong selection signals were detected: *TNXB* and *CREBL1* for fat, meat, and milk; potassium voltage-gated channel, KQT-like subfamily, member 1 (*KCNQ1*) for cereals; and *KCNJ2* and zinc finger, matrin-type 4 (*ZMAT4*) for roots and tubers. The strongest correlation was observed in populations that consume roots and tubers as their staple food. Importantly, these genes have been shown to be associated with disease risk in GWAS of different populations.[57]

A genome-wide targeted scan based on SNPs is a powerful approach to detect selection signals, but other genome footprints such as copy number variation of positive selections are often overlooked. Although not detected on the basis of genome-wide scans, copy number variants of the amylase (*AMY1*) gene were identified as being under strong selection.[85] Starch is the main nutrient of farming and foraging societies with corn, rice, and potatoes as the staple foods. Populations with high-starch diets tend to have, on average, more copies of *AMY1* than those with low-starch diets. Furthermore, salivary amylase activity is highly correlated with the *AMY1* copy number,[85] with higher *AMY1* copy numbers and protein levels likely giving advantages to individuals who most effectively digest starchy foods. This is another classic example of differences in dietary adaptations among human populations.

C. Culture Adaption and Alcohol Use

Culture can be defined biologically as knowledge, beliefs, values, and skills that shape an individual's behavior, which are learned from others and transferred from one generation to the next.[86] Culture coexists alongside human evolution; thus, it plays a key role in human adaptation to environments. Alcohol use is one example of a cultural adaptation. Its consumption has a long history, likely starting with the practice of agriculture. Certain religions restrict alcohol use, while customs or traditions of some ethnic groups promote alcohol consumption. Thus, such cultural factors can exert pressure on alleles that confer alcohol intolerance or tolerance resulting in increased frequencies of either trait (positive selection).

Strong signals of selection were identified in the alcohol dehydrogenase genes (*ADH1A*, *ADH1B*, *ADH1C*, *ADH4*, *ADH5*, *ADH6*, *ADH7*) in East Asians based on the iHS method.[53] Using F_{st} and long-range haplotype methods, Han *et al.*[87] showed that a functional variant of *ADH1B*, Arg47His, experienced strong positive selection in East Asians, with the 47His allele conferring protection against alcoholism. After extending the original study to 24 populations from a wide selection of geographic areas and diverse ethnic groups in China and Laos, Li *et al.*[88] determined that the *ADH1B* genomic regions are subject to positive selection in different ethnic groups, and that genotype distributions are strongly correlated with ethnicity. The populations (Korean-Japanese, Han Chinese, Hmong-Mien, Daic, and Austronesian) with high frequency of *ADH1B*-47His allele have practiced agriculture for a long time. On the other hand, the populations with a low frequency of this allele tend to practice pastoral lifestyles or hunting or began farming relatively recently. This observation suggests that cultural adaptation could have driven the *ADH1B*-47His to high frequencies in some of East Asian groups.

Another well-known nonsynonymous mutation (Glu487Lys or rs671) is at the aldehyde dehydrogenase 2 family (*ALDH2*) gene, which encodes an enzyme that metabolizes acetaldehyde (a metabolite of alcohol). This gene gives carriers of a mutant allele the "flushing" face with alcohol ingestion but also protects them against alcoholism.[89] In addition, carriers of the mutation have a higher risk of developing alcohol-related diseases, such as liver and esophageal cancers and other digestive tract cancers, when being a heavy drinker.[90] This flushing face null allele was driven to a high frequency by positive selection in the East Asian population.[91,92] However, the force of selection remains to be determined. It is unlikely that the high frequency of these risk alleles was under recent selection by alcoholism. Goldman and Enoch[93] proposed two possible types of selection. First, mycotoxins (aflatoxins, in particular), which can cause hepatic disease and cancer, are commonly present in moldy rice, and ALDH enzymes can convert fungal protoxins found in moldy rice to toxins. Thus, heterozygotes of the mutation at *ALDH* produce less toxins and have an advantage over those who are homozygous for the major allele of *ALDH*. Second, many infectious bacteria and protozoa are very susceptible to acetaldehyde. Thus, carriers of *ALDH2*2*, who have high plasma concentrations of acetaldehyde, may better resist microbial infections and thus possess a selective survival advantage. While this hypothesis remains to be proven, this speculation suggests that adaptation based on one selective force may result in higher risk for other diseases.

D. Adaptation on Taste

Humans can recognize five tastes: sweet, bitter, sour, salty, and umami (taste like glutamate). Taste has a great influence on food preference and intake, as it is an important indicator of the toxicity and nutrient qualities of a food. Bitter tastes usually indicate poison, whereas sweet is linked to fruit and other often favored foods (see also Chapter 15). Natural selection has been driving the evolution of human tastes, likely in progress for more than 1.5 million years.[94] Cultural differences could favor genetic variation for a particular taste or food.[94,95] About 33 bitter taste genes are known in humans[95,96]; at least 25 of them are functional genes, whereas the other 8 are likely pseudogenes. Using the F_{st} method, Kim and colleagues[97] showed that 25 taste receptor, type 2 (*TAS2R*) bitter taste genes, on average, had significantly higher F_{st} than the estimated average F_{st} for genome-wide SNPs (0.22 vs. 0.123)[45] among five continental populations (21 Cameroonians, 10 Native Americans, 10 Japanese, 9 Hungarians, 5 Mbuti Pygmies from the Ituri Forest in Northeast Democratic Republic of Congo). In addition, the ratio of synonymous and nonsynonymous nucleotide substitution rates K_a/K_s in these taste genes was significantly greater than that of most other genes (152 random genes) (0.94 vs. 0.11).[98] This observation strongly supports the idea that taste genes are under positive selection across these five groups. In particular, strong positive selection has been detected in the *TAS2R16* gene in 19 populations worldwide using Fay and Wu's *H*-test[99]; a nonsynonymous variant K172N is a likely target of positive selection. The age of the taster allele N172 is estimated to be at least 78,700 years old and allele frequencies vary greatly across the globe, suggesting that selection occurred before the expansion out of Africa (50,000–100,000 years ago) and continued after humans spread to the rest of the world. Bitter tasters tend to avoid beer and alcohol, both of which are bitter.[100] Interestingly, the nontaster allele K172 associated with alcoholism is at a higher frequency (26%) in Africans but at a very low frequency (0.0–0.6%) in Europeans and Asians.[101] The *TAS2R38* gene is shown to be under selection in all three major populations (Asians, Africans, and Europeans) using Tajima's *D*- and Fu and Li's *D*- and *F*-statistics.[94]

VII. Adaptation to Pathogens

Pathogens have played a crucial role in adaptation. As infection could be fatal, pathogens can serve as strong selection factors to eliminate vulnerable individuals from the population. Thus, carriers of resistance alleles would survive and prevail while noncarriers would be more vulnerable to pathogens. This presents a typical selection sweep as first described by Maynard-Smith and Haigh (Fig. 2).[50] For instance, malaria has driven the O allele of the *FY* (now known as Duffy blood group, chemokine receptor, *DARC*) gene to fixation in the population of sub-Saharan Africa by conferring resistance to *Plasmodium vivax* infection in its carriers.[102,103] Other variants have been similarly selected for into high frequencies.[104,105] In European populations, smallpox epidemics are believed to be the drivers of the high frequency of the chemokine receptor 5 (*CCR5*)-*Δ32* variant (i.e., 32-bp deletion), which confers resistance to the human immunodeficiency virus (HIV).[106]

Human migration and colonization of new environments introduced new pathogens that were often detrimental and fatal. After farming started about 10,000 years ago, agriculture facilitated the transfer of pathogens from animals to humans,[107] while people living in larger and more stable groups facilitated the transmission from humans to humans. The diversity of pathogens at the global level and temporal spectrum has shaped human genome adaptation to pathogens. Furthermore, genetic variants selected for resistance to one disease can become risk factors for other diseases. For example, variants in the interferon induced with helicase C domain 1 (*IFIH1*) gene that confer resistance to diverse viral infections and have been driven to high frequencies in South Americans also pose a high risk for type 1 diabetes.[108] In genome-wide scans for positive selection, many genetic variants were associated with risk of multiple diseases. As discussed in the previous section, the high frequencies of the *ALDH2*2* null allele might have been driven by selection favoring heterozygotes against infectious disease in East Asian populations.[93]

VIII. Population Admixture

Populations gradually become homogenous and adapted to local environments after isolation from other groups for thousands of generations. However, recent migrations and intermarriage between formerly isolated groups could lead to population admixture, as recombination history between chromosomes from ancestral populations has been short. Thus, ancestral haplotypes with long-range LD blocks segregate in the population.

Admixture can result in genetic subgroups within a population, and such subgroups or substructures may cause false positive or negative associations between genetic markers and disease risk if the subgroups are not equally represented in cases and controls.[109,110] To overcome this problem, ancestral index or population admixture can be estimated using ancestral informative markers (i.e., genetic loci with alleles displaying significant frequency differences between ancestral populations). Three approaches have been used: genomic control,[111] ancestry estimated using STRUCTURE[112,113] and principal component analysis (PCA).[114]

Estimation of admixture has become the general practice in association studies, particularly in subjects of African and Hispanic ancestry. The African-American population admixture varies greatly.[115] In Hispanics, Mexican populations consist of European and Native American ancestries.[116] In Puerto Ricans, we have estimated admixture of 57% European, 27% African, and 15% Native American.[117] Even within populations of exclusively European ancestry, stratification has been detected.[118]

Population ancestry has been associated with increased risk of specific diseases and with anthropometric traits. In Puerto Rican populations, in the presence of a Western lifestyle, we have shown that African ancestry is associated with hypertension, whereas Native American ancestry is correlated with CVD and European ancestry with type 2 diabetes.[117] Insulin-like growth factor 1 that is responsible for early growth has been correlated with African and European ancestry.[119] Children of African ancestry tend to have a higher level of insulin-like growth factor 1 than those of European descent. Recently, by examining children from three ethnic groups (African, European, and Hispanic-American), Casazza and coworkers[120] showed that European and Native American ancestries are positively associated with all adiposity traits (total body fat, intra-abdominal fat, subcutaneous fat), whereas African ancestry is negatively associated with all measures of fat deposition. Furthermore, strong interaction between European ancestry and energy intake was observed in boys. Individuals with high European ancestry tend to have high fat deposition when their energy intake is high. In contrast, girls with high African ancestry tend to have low-fat deposition when they consume a carbohydrate-rich and low-fat diet. We showed that when ancestry is associated with a disease, adjustment for ancestry has substantial effect on tests of association between genetic variants and disease traits.[117] In contrast, if ancestry is not associated with the disease, adjustment for stratification has little impact on the association test. In summary, the fact that different ancestries are associated with differential risk of human diseases strongly reflects genetic divergence among populations.

IX. Adaptive Genetic Variation and Common Diseases

As discussed earlier, adaptive genetic variation contributes significantly to the divergence between populations. A remaining question is whether genetic variants that have experienced positive selection contribute disproportionately to the risk of common diseases, such as CVD, type 2 diabetes, metabolic diseases, and cancer. To address this question, Myles *et al.*[46] examined 25 SNPs that showed strong associations with six complex diseases (Crohn's disease, type 1 diabetes, type 2 diabetes, rheumatoid arthritis, coronary artery disease, and obesity) for differences in allele frequencies over worldwide populations. While the global F_{st} was not significantly different from the expected value of a random set of SNPs, the allele frequencies exhibited substantial differences and were associated with disease prevalence rates. The main caveat with the study is that these 25 disease-associated SNPs are mainly based on Europeans, not worldwide populations, and global F_{st} is not the best measure for positive selection that occurs only in specific groups.[24]

Using a similar approach, Ding and Kullo[121] examined 158 SNPs associated with CVD or related traits based on a GWAS database[122] for population-specific selection. Eight SNPs exhibited significantly higher global F_{st} ($P < 0.05$). In addition, for CVD and obesity, we[123] and others[124] summarized potential adaptive genetic variants that have been identified mainly in populations of European ancestry. Based on 151 published GWAS, Hindorff *et al.*[122] determined that about 10% of genetic variants identified by GWAS are mapped to the regions of positive selections based on iHS.[53]

Taking a different approach, Casto and Feldman[125] correlated the distribution of 1300 GWAS SNPs allele frequencies and the pattern of iHS scores and found that GWAS SNPs, on average, have elevated scores. In particular, SNPs associated with skin pigmentation, blood pressure, infectious disease, and autoimmune disease traits show diverse distribution in specific regions of the world and have higher iHS scores. In general, GWAS SNPs exhibit diverse allele frequencies and tend to be located within the genomic regions of positive selections. Furthermore, disease-associated SNPs identified in one population may not be present in another population, which could be because of differences in LD structure or demographic history. This suggests that GWAS results are population specific and therefore that GWASs should be performed separately for each ethnic group. As more GWASs are being performed in ethnic groups worldwide, it will become more apparent how much adaptive genetic variation contributes to common diseases in humans. Overall, one can clearly conclude that adaptive genetic variants contribute substantially to the risk of common diseases, at least in the population of European ancestry.

Thus far, the majority of studies examining population differences have focused on three major groups—African, Asian, and European—but as the HapMap project expands, it will reveal worldwide genetic diversity. For the three major groups, most studies have consistently shown that selection is stronger in non-African than in African populations, with Asian groups experiencing the strongest selection.[18,22,63] Furthermore, most adaptive variants with strong signals are unique to a single population, suggesting that selection pressures are usually local or the identified variants experienced convergent evolution.[22] However, one caveat applicable to most genome scans reported for selection signals is that they use SNP genotypes, and that may lead to ascertainment bias.[126] The completion of the 1000 Genomes Project[127] combined with more robust methods of detecting positive selection may overcome this problem.

Another challenging question is the understanding of the links between adaptation and the risk of modern-day diseases, which are often indirect. For example, adaptive *OCA2* variants give rise to the blue eye and lighter skin phenotype,[68,70] but it can also increase the risk of skin cancer.[128] However, the selection force did not directly modify skin cancer risk in the populations of European ancestry, but rather arose to favor the development of lighter skin color so that adequate vitamin D synthesis could occur in sunlight-limited regions. Evidently, the relationship between selection forces and disease risk is complex.

Another complexity of adaptive variation and an important source that contributes to divergence among populations is genetic drift, which can be defined as the change in allele frequencies of genetic variants due to chance. It can occur when a population splits into multiple groups or when subpopulations migrate. Allele frequency change due to genetic drift is highly correlated with divergence times and the effective population size. Genetic drift is usually considered in concert with selection, mutation, and migration to provide a more complete understanding of the changes in allele frequency.

X. Population Differences and Dietary Guidelines

There is no doubt that a significant amount of diversity among populations worldwide is caused by adaptation. Adaptive genetic variation contributes significantly to the risk of common diseases, including metabolic diseases,[108] obesity,[123,129] CVD,[121,123,124] and cancer.[122,125] It is conceivable that adaptive genetic variants tend to interact with environmental factors ($G \times E$), which are not usually evaluated in most GWAS; this possibility merits more study. Furthermore, gene–gene interactions ($G \times G$) are usually ignored in GWAS due to the complex relationships among loci. Both types of interactions ($G \times G$ and $G \times E$) are likely population- and region-specific. In addition, the fact that specific ancestries are associated with different disease risks further supports the hypothesis that genetically based population differences present varying levels of risk for disease. Furthermore, certain nutrient deficiencies differ among populations because of varied prevalence of disease-associated alleles across populations, as well as local climates, and geographical environments. Therefore, understanding the differences among populations helps us develop effective dietary guidelines for specific groups.[130] For example, the supplementation of folate for the prevention of neural tube defects lifted the selection pressure on methylenetetrahydrofolate reductase (*MTHFR*)-667T, thus affecting the distribution of the *MTHFR* 677C > T genotype and allowing for the increase of the 667T frequency in the population.[131] In addition to a greater need for folate, 667T carriers have higher homocysteine, which is associated with increased risk of CVD. While supplementation prevents one disease (neural tube defects), it could also increase risk of another disease (CVD).[132,133]

The fact that obesity and metabolic syndrome have become worldwide epidemics and that adaptive genetic variation detected at molecular levels (DNA sequence variation) contributes to common diseases provides strong evidence for the "thrifty" genotype hypothesis.[16] Namely, the hypothesis suggests that the genotypes we carry, which adapted to the diet and lifestyle of our ancestors, are in conflict with modern civilization. Therefore, while prevention of nutrient deficiency is the primary goal for all populations, dietary guidelines for specific populations should also advocate more traditional diets and lifestyle. For example, the prevalence of type 2 diabetes in Pima Indians of Arizona in the United States is more than five times higher (38% vs. 6.9%) than that of the Pima Indians in Mexico.[134] While the two groups share the same genetic background, as they separated only about 700 years ago, they lived in very different environments and with different lifestyles. Mexican Pima Indians lived in a remote mountainous area and practiced a "traditional" lifestyle; their diets were low in animal fat (15% of total energy intake) and high in complex carbohydrates, and they maintained a high intensity of physical labor. Conversely, Pima Indians in Arizona resided in a suburban area and lived in an "affluent" environment; their diets were high in animal fat (40% of total energy intake) and simple carbohydrates, and they lived a less labor-intensive lifestyle.[135] A functional variant (R230C or rs9282541) at the ATP-binding cassette, subfamily A, member 1 (*ABCA1*) gene that exists only in the Native Americans and Mexicans experienced positive selection and is associated with metabolic syndrome in these populations.[136] Mexican subjects with metabolic syndrome who carry the risk allele 230C showed significant reductions in body weight and increases in serum adiponectin concentration when compared to the noncarriers in response to a 2-month intervention, during which a daily supplement of a mixture of dehydrated nopal, chia seeds, soybean, and oats was provided.[137] Both nopal (*Opuntia* spp.) and chia (*Salvia hispanica*) seeds, rich in vitamin C, beta-carotene, proteins, fiber, and antioxidants, are traditional foods of Native Americans and Mexicans.[138,139] These observations suggest that continuing use of traditional diets and lifestyles of specific populations is an important strategy for prevention of metabolic syndrome.

Dietary guidelines in Western countries recommend drinking three cups of milk per day. However, a significant proportion of populations worldwide suffer from lactose intolerance, characterized by stomachache, bloating, gas, and diarrhea when a large quantity of milk is consumed. In the United States, about 90% of Asian-Americans and Native Americans, 75% of African-Americans, and 20% of European-Americans experience such symptoms because of their intolerance.[140] As described earlier (Section VI.A), the difference in prevalence of lactose intolerance among ethnic groups is likely due to the divergence in allele frequencies of the *LCT* variant (C-13910 > T) that are responsible for lactose intolerance.[83] Obviously, to circumvent the deficiency of calcium and proteins in the populations with high prevalence of lactose intolerance, dietary guidelines should be modified to suit the group. Lactose-free milk, cheese, and/or yogurt; calcium-fortified tofu; and calcium-fortified orange juice may be alternatives for whole milk to compensate for the deficiency of nutrients for those with lactose intolerance.[141]

While drinking a small amount of alcohol daily is recommended in Western countries to reduce CVD risk, approximately 36% of East Asians display alcohol flush, nausea, and/or tachycardia after consuming alcohol.[142] This "Asian flush" symptom is associated with a functional variant (Glu487Lys or rs671) at *ALDH2*, which is known to be subject to positive selection in East Asian populations.[91,92] Heavy alcohol drinkers with this *ALDH2* mutation are at higher risk of esophageal cancer than those with the normal *ALDH2* allele.[143] About 540 million individuals are ALDH2-deficient, and many now live in Western societies. Thus, drinking alcohol should not be advocated for carriers of the ALDH2-deficient allele. It has been predicted that about 53% of squamous cell esophageal cancers can be prevented in Japanese men by reducing alcohol consumption.[143] Therefore, restriction of alcohol use should be recommended for East Asian populations, particularly for carriers of the ALDH2-deficient allele, for the prevention of esophageal cancer and possibly liver cancer.

XI. Conclusion

In summary, it has become evident that a substantial amount of adaptive and drift variation exists among populations and that such genetic variation interacts with modern environmental changes, including alterations in climate, dietary intake, and lifestyle. All of these differences contribute to variation in susceptibility to diseases. In addition, certain nutrient deficiencies differ because of the variable prevalence of disease-associated alleles, local climates, and geographical environments. Thus, in light of these differences among populations, dietary guidelines developed for one population may not be appropriate for another. Understanding the differences between populations helps us develop more effective dietary guidelines for specific groups.[130] Obviously, individualized dietary recommendations are the ultimate goal of nutrigenomics and nutrigenetics. But identifying and understanding adaptive and nonadaptive genetic variation and environmental factors, their interactions (G × G), and interactions with environments (G × E) should facilitate optimization of dietary recommendations for specific groups.

Acknowledgments

The author wishes to thank Caren Smith and reviewers for their valuable suggestions and comments on the contents of this chapter. Additional note: Mention of trade names or commercial products in this publication is solely for the purpose of providing specific information and does not imply recommendation or endorsement by the U.S. Department of Agriculture. The USDA is an equal opportunity provider and employer.

References

1. Bowcock AM, Ruiz-Linares A, Tomfohrde J, Minch E, Kidd JR, Cavalli-Sforza LL. High resolution of human evolutionary trees with polymorphic microsatellites. *Nature* 1994;**368**:455–7.
2. Rosenberg NA, Pritchard JK, Weber JL, Cann HM, Kidd KK, Zhivotovsky LA, et al. Genetic structure of human populations. *Science* 2002;**298**:2381–5.
3. Cavalli-Sforza LL, Feldman MW. The application of molecular genetic approaches to the study of human evolution. *Nat Genet* 2003;**33**(Suppl.):266–75.
4. Stringer CB, Andrews P. Genetic and fossil evidence for the origin of modern humans. *Science* 1988;**239**:1263–8.
5. Tattersall I. Out of Africa: modern human origins special feature: human origins: out of Africa. *Proc Natl Acad Sci USA* 2009;**106**:16018–21.
6. Mulligan CJ, Hunley K, Cole S, Long JC. Population genetics, history, and health patterns in native Americans. *Annu Rev Genomics Hum Genet* 2004;**5**:295–315.
7. Tamm E, Kivisild T, Reidla M, Metspalu M, Smith DG, Mulligan CJ, et al. Beringian standstill and spread of native American founders. *PLoS One* 2007;**2**:e829.
8. O'Rourke DH, Raff JA. The human genetic history of the Americas: the final frontier. *Curr Biol* 2010;**20**:R202–7.
9. Achilli A, Perego UA, Bravi CM, Coble MD, Kong QP, Woodward SR, et al. The phylogeny of the four pan-American MtDNA haplogroups: implications for evolutionary and disease studies. *PLoS One* 2008;**3**:e1764.
10. Ingman M, Kaessmann H, Paabo S, Gyllensten U. Mitochondrial genome variation and the origin of modern humans. *Nature* 2000;**408**:708–13.

11. Wang S, Lewis CM, Jakobsson M, Ramachandran S, Ray N, Bedoya G, et al. Genetic variation and population structure in native Americans. *PLoS Genet* 2007;**3**:e185.
12. Thomson R, Pritchard JK, Shen P, Oefner PJ, Feldman MW. Recent common ancestry of human Y chromosomes: evidence from DNA sequence data. *Proc Natl Acad Sci USA* 2000; **97**:7360–5.
13. Leonard WR. Food for thought. Dietary change was a driving force in human evolution. *Sci Am* 2002;**287**:106–15.
14. Rotilio G, Marchese E. Nutritional factors in human dispersals. *Ann Hum Biol* 2010; **37**:312–24.
15. Chakravarthy MV, Booth FW. Eating, exercise, and "thrifty" genotypes: connecting the dots toward an evolutionary understanding of modern chronic diseases. *J Appl Physiol* 2004; **96**:3–10.
16. Neel JV. Diabetes mellitus: a "thrifty" genotype rendered detrimental by "progress"? *Am J Hum Genet* 1962;**14**:353–62.
17. Pinhasi R, Fort J, Ammerman AJ. Tracing the origin and spread of agriculture in Europe. *PLoS Biol* 2005;**3**:e410.
18. Li JZ, Absher DM, Tang H, Southwick AM, Casto AM, Ramachandran S, et al. Worldwide human relationships inferred from genome-wide patterns of variation. *Science* 2008;**319**: 1100–4.
19. van Oven M, Kayser M. Updated comprehensive phylogenetic tree of global human mitochondrial DNA variation. *Hum Mutat* 2009;**30**:E386–94.
20. The International HapMap Consortium. The International HapMap Project. *Nature* 2003; **426**:789–96.
21. Olivier M. A haplotype map of the human genome. *Physiol Genomics* 2003;**13**:3–9.
22. Coop G, Pickrell JK, Novembre J, Kudaravalli S, Li J, Absher D, et al. The role of geography in human adaptation. *PLoS Genet* 2009;**5**:e1000500.
23. Altshuler DM, Gibbs RA, Peltonen L, Altshuler DM, Gibbs RA, Peltonen L, et al. Integrating common and rare genetic variation in diverse human populations. *Nature* 2010;**467**:52–8.
24. Weir BS, Cardon LR, Anderson AD, Nielsen DM, Hill WG. Measures of human population structure show heterogeneity among genomic regions. *Genome Res* 2005;**15**:1468–76.
25. Myles S, Hradetzky E, Engelken J, Lao O, Nurnberg P, Trent RJ, et al. Identification of a candidate genetic variant for the high prevalence of type II diabetes in Polynesians. *Eur J Hum Genet* 2007;**15**:584–9.
26. Clark AG, Glanowski S, Nielsen R, Thomas PD, Kejariwal A, Todd MA, et al. Inferring nonneutral evolution from human-chimp-mouse orthologous gene trios. *Science* 2003;**302**: 1960–3.
27. Bustamante CD, Fledel-Alon A, Williamson S, Nielsen R, Hubisz MT, Glanowski S, et al. Natural selection on protein-coding genes in the human genome. *Nature* 2005;**437**:1153–7.
28. Nielsen R, Bustamante C, Clark AG, Glanowski S, Sackton TB, Hubisz MJ, et al. A scan for positively selected genes in the genomes of humans and chimpanzees. *PLoS Biol* 2005;**3**:e170.
29. Eyre-Walker A, Keightley PD. High genomic deleterious mutation rates in hominids. *Nature* 1999;**397**:344–7.
30. Friedman RC, Farh KK, Burge CB, Bartel DP. Most mammalian mRNAs are conserved targets of microRNAs. *Genome Res* 2009;**19**:92–105.
31. Kreitman M. Methods to detect selection in populations with applications to the human. *Annu Rev Genomics Hum Genet* 2000;**1**:539–59.
32. Bamshad M, Wooding SP. Signatures of natural selection in the human genome. *Nat Rev Genet* 2003;**4**:99–111.
33. Nielsen R. Molecular signatures of natural selection. *Annu Rev Genet* 2005;**39**:197–218.
34. Biswas S, Akey JM. Genomic insights into positive selection. *Trends Genet* 2006;**22**:437–46.

35. Harris EE, Meyer D. The molecular signature of selection underlying human adaptations. *Am J Phys Anthropol* 2006;**Suppl. 43**:89–130.
36. Nielsen R, Hellmann I, Hubisz M, Bustamante C, Clark AG. Recent and ongoing selection in the human genome. *Nat Rev Genet* 2007;**8**:857–68.
37. Akey JM. Constructing genomic maps of positive selection in humans: where do we go from here? *Genome Res* 2009;**19**:711–22.
38. Hancock AM, Alkorta-Aranburu G, Witonsky DB, Di Rienzo A. Adaptations to new environments in humans: the role of subtle allele frequency shifts. *Philos Trans R Soc Lond B Biol Sci* 2010;**365**:2459–68.
39. Kimura M. How genes evolve; a population geneticist's view. *Ann Genet* 1976;**19**:153–68.
40. Hudson RR, Kreitman M, Aguade M. A test of neutral molecular evolution based on nucleotide data. *Genetics* 1987;**116**:153–9.
41. McDonald JH, Kreitman M. Adaptive protein evolution at the Adh locus in Drosophila. *Nature* 1991;**351**:652–4.
42. McDonald JH. Detecting non-neutral heterogeneity across a region of DNA sequence in the ratio of polymorphism to divergence. *Mol Biol Evol* 1996;**13**:253–60.
43. McDonald JH. Improved tests for heterogeneity across a region of DNA sequence in the ratio of polymorphism to divergence. *Mol Biol Evol* 1998;**15**:377–84.
44. Weir BS, Hill WG. Estimating F-statistics. *Annu Rev Genet* 2002;**36**:721–50.
45. Akey JM, Zhang G, Zhang K, Jin L, Shriver MD. Interrogating a high-density SNP map for signatures of natural selection. *Genome Res* 2002;**12**:1805–14.
46. Myles S, Tang K, Somel M, Green RE, Kelso J, Stoneking M. Identification and analysis of genomic regions with large between-population differentiation in humans. *Ann Hum Genet* 2008;**72**:99–110.
47. Tajima F. Statistical method for testing the neutral mutation hypothesis by DNA polymorphism. *Genetics* 1989;**123**:585–95.
48. Fu YX, Li WH. Statistical tests of neutrality of mutations. *Genetics* 1993;**133**:693–709.
49. Fay JC, Wu CI. Hitchhiking under positive Darwinian selection. *Genetics* 2000;**155**:1405–13.
50. Smith JM, Haigh J. The hitch-hiking effect of a favourable gene. *Genet Res* 1974;**23**:23–35.
51. Hudson RR, Bailey K, Skarecky D, Kwiatowski J, Ayala FJ. Evidence for positive selection in the superoxide dismutase (Sod) region of Drosophila melanogaster. *Genetics* 1994;**136**:1329–40.
52. Sabeti PC, Reich DE, Higgins JM, Levine HZ, Richter DJ, Schaffner SF, et al. Detecting recent positive selection in the human genome from haplotype structure. *Nature* 2002;**419**:832–7.
53. Voight BF, Kudaravalli S, Wen X, Pritchard JK. A map of recent positive selection in the human genome. *PLoS Biol* 2006;**4**:e72.
54. Wang ET, Kodama G, Baldi P, Moyzis RK. Global landscape of recent inferred Darwinian selection for Homo sapiens. *Proc Natl Acad Sci USA* 2006;**103**:135–40.
55. Hancock AM, Witonsky DB, Gordon AS, Eshel G, Pritchard JK, Coop G, et al. Adaptations to climate in candidate genes for common metabolic disorders. *PLoS Genet* 2008;**4**:e32.
56. Hancock AM, Witonsky DB, Alkorta-Aranburu G, Beall CM, Gebremedhin A, Sukernik R, et al. Adaptations to climate-mediated selective pressures in humans. *PLoS Genet* 2011;**7**: e1001375.
57. Hancock AM, Witonsky DB, Ehler E, Alkorta-Aranburu G, Beall C, Gebremedhin A, et al. Colloquium paper: human adaptations to diet, subsistence, and ecoregion are due to subtle shifts in allele frequency. *Proc Natl Acad Sci USA* 2010;**107**(Suppl. 2):8924–30.
58. The International HapMap Consortium. A haplotype map of the human genome. *Nature* 2005;**437**:1299–320.
59. Sabeti PC, Varilly P, Fry B, Lohmueller J, Hostetter E, Cotsapas C, et al. Genome-wide detection and characterization of positive selection in human populations. *Nature* 2007; **449**:913–8.

60. Hinds DA, Stuve LL, Nilsen GB, Halperin E, Eskin E, Ballinger DG, et al. Whole-genome patterns of common DNA variation in three human populations. *Science* 2005;**307**:1072–9.
61. Grossman SR, Shylakhter I, Karlsson EK, Byrne EH, Morales S, Frieden G, et al. A composite of multiple signals distinguishes causal variants in regions of positive selection. *Science* 2010;**327**:883–6.
62. Carlson CS, Thomas DJ, Eberle MA, Swanson JE, Livingston RJ, Rieder MJ, et al. Genomic regions exhibiting positive selection identified from dense genotype data. *Genome Res* 2005;**15**:1553–65.
63. Williamson SH, Hubisz MJ, Clark AG, Payseur BA, Bustamante CD, Nielsen R. Localizing recent adaptive evolution in the human genome. *PLoS Genet* 2007;**3**:e90.
64. Coop G, Witonsky D, Di Rienzo A, Pritchard JK. Using environmental correlations to identify loci underlying local adaptation. *Genetics* 2010;**185**:1411–23.
65. Cheng F, Chen W, Richards E, Deng L, Zeng C. SNP@Evolution: a hierarchical database of positive selection on the human genome. *BMC Evol Biol* 2009;**9**:221.
66. Jablonski NG, Chaplin G. Colloquium paper: human skin pigmentation as an adaptation to UV radiation. *Proc Natl Acad Sci USA* 2010;**107**(Suppl. 2):8962–8.
67. John PR, Makova K, Li WH, Jenkins T, Ramsay M. DNA polymorphism and selection at the melanocortin-1 receptor gene in normally pigmented southern African individuals. *Ann N Y Acad Sci* 2003;**994**:299–306.
68. Sturm RA. Molecular genetics of human pigmentation diversity. *Hum Mol Genet* 2009;**18**: R9–R17.
69. Sturm RA, Frudakis TN. Eye colour: portals into pigmentation genes and ancestry. *Trends Genet* 2004;**20**:327–32.
70. Duffy DL, Montgomery GW, Chen W, Zhao ZZ, Le L, James MR, et al. A three-single-nucleotide polymorphism haplotype in intron 1 of OCA2 explains most human eye-color variation. *Am J Hum Genet* 2007;**80**:241–52.
71. Ramagopalan SV, Heger A, Berlanga AJ, Maugeri NJ, Lincoln MR, Burrell A, et al. A ChIP-seq defined genome-wide map of vitamin D receptor binding: associations with disease and evolution. *Genome Res* 2010;**20**:1352–60.
72. Aoki K. Sexual selection as a cause of human skin colour variation: Darwin's hypothesis revisited. *Ann Hum Biol* 2002;**29**:589–608.
73. Burt VL, Whelton P, Roccella EJ, Brown C, Cutler JA, Higgins M, et al. Prevalence of hypertension in the US adult population. Results from the Third National Health and Nutrition Examination Survey, 1988–1991. *Hypertension* 1995;**25**:305–13.
74. Cooper RS, Zhu X. Racial differences and the genetics of hypertension. *Curr Hypertens Rep* 2001;**3**:19–24.
75. Young JH, Chang YP, Kim JD, Chretien JP, Klag MJ, Levine MA, et al. Differential susceptibility to hypertension is due to selection during the out-of-Africa expansion. *PLoS Genet* 2005;**1**:e82.
76. Thompson EE, Kuttab-Boulos H, Witonsky D, Yang L, Roe BA, Di Rienzo A. CYP3A variation and the evolution of salt-sensitivity variants. *Am J Hum Genet* 2004;**75**:1059–69.
77. Zhang L, Miyaki K, Wang W, Muramatsu M. CYP3A5 polymorphism and sensitivity of blood pressure to dietary salt in Japanese men. *J Hum Hypertens* 2010;**24**:345–50.
78. Simoons FJ. Primary adult lactose intolerance and the milking habit: a problem in biologic and cultural interrelations. II. A culture historical hypothesis. *Am J Dig Dis* 1970;**15**:695–710.
79. Simoons FJ. Primary adult lactose intolerance and the milking habit: a problem in biological and cultural interrelations. I. Review of the medical research. *Am J Dig Dis* 1969;**14**:819–36.
80. Kretchmer N. Lactose and lactase—a historical perspective. *Gastroenterology* 1971; **61**:805–13.

81. Bersaglieri T, Sabeti PC, Patterson N, Vanderploeg T, Schaffner SF, Drake JA, et al. Genetic signatures of strong recent positive selection at the lactase gene. *Am J Hum Genet* 2004;**74**:1111–20.
82. Ingram CJ, Mulcare CA, Itan Y, Thomas MG, Swallow DM. Lactose digestion and the evolutionary genetics of lactase persistence. *Hum Genet* 2009;**124**:579–91.
83. Tishkoff SA, Reed FA, Ranciaro A, Voight BF, Babbitt CC, Silverman JS, et al. Convergent adaptation of human lactase persistence in Africa and Europe. *Nat Genet* 2007;**39**:31–40.
84. Gasche C, Nemeth M, Grundtner P, Willheim-Polli C, Ferenci P, Schwarzenbacher R. Evolution of Crohn's disease-associated Nod2 mutations. *Immunogenetics* 2008;**60**:115–20.
85. Perry GH, Dominy NJ, Claw KG, Lee AS, Fiegler H, Redon R, et al. Diet and the evolution of human amylase gene copy number variation. *Nat Genet* 2007;**39**:1256–60.
86. Laland KN, Odling-Smee J, Myles S. How culture shaped the human genome: bringing genetics and the human sciences together. *Nat Rev Genet* 2010;**11**:137–48.
87. Han Y, Gu S, Oota H, Osier MV, Pakstis AJ, Speed WC, et al. Evidence of positive selection on a class I ADH locus. *Am J Hum Genet* 2007;**80**:441–56.
88. Li H, Gu S, Cai X, Speed WC, Pakstis AJ, Golub EI, et al. Ethnic related selection for an ADH Class I variant within East Asia. *PLoS One* 2008;**3**:e1881.
89. Chao YC, Liou SR, Chung YY, Tang HS, Hsu CT, Li TK, et al. Polymorphism of alcohol and aldehyde dehydrogenase genes and alcoholic cirrhosis in Chinese patients. *Hepatology* 1994; **19**:360–6.
90. Ding JH, Li SP, Cao HX, Wu JZ, Gao CM, Liu YT, et al. Alcohol dehydrogenase-2 and aldehyde dehydrogenase-2 genotypes, alcohol drinking and the risk for esophageal cancer in a Chinese population. *J Hum Genet* 2010;**55**:97–102.
91. Lin YP, Cheng TJ. Why can't Chinese Han drink alcohol? Hepatitis B virus infection and the evolution of acetaldehyde dehydrogenase deficiency. *Med Hypotheses* 2002;**59**:204–7.
92. Oota H, Pakstis AJ, Bonne-Tamir B, Goldman D, Grigorenko E, Kajuna SL, et al. The evolution and population genetics of the ALDH2 locus: random genetic drift, selection, and low levels of recombination. *Ann Hum Genet* 2004;**68**:93–109.
93. Goldman D, Enoch MA. Genetic epidemiology of ethanol metabolic enzymes: a role for selection. *World Rev Nutr Diet* 1990;**63**:143–60.
94. Wooding S, Kim UK, Bamshad MJ, Larsen J, Jorde LB, Drayna D. Natural selection and molecular evolution in PTC, a bitter-taste receptor gene. *Am J Hum Genet* 2004;**74**:637–46.
95. Kim UK, Breslin PA, Reed D, Drayna D. Genetics of human taste perception. *J Dent Res* 2004;**83**:448–53.
96. Shi P, Zhang J, Yang H, Zhang YP. Adaptive diversification of bitter taste receptor genes in Mammalian evolution. *Mol Biol Evol* 2003;**20**:805–14.
97. Kim U, Wooding S, Ricci D, Jorde LB, Drayna D. Worldwide haplotype diversity and coding sequence variation at human bitter taste receptor loci. *Hum Mutat* 2005;**26**:199–204.
98. Nekrutenko A, Makova KD, Li WH. The K(A)/K(S) ratio test for assessing the protein-coding potential of genomic regions: an empirical and simulation study. *Genome Res* 2002;**12**: 198–202.
99. Soranzo N, Bufe B, Sabeti PC, Wilson JF, Weale ME, Marguerie R, et al. Positive selection on a high-sensitivity allele of the human bitter-taste receptor TAS2R16. *Curr Biol* 2005; **15**:1257–65.
100. Duffy VB, Peterson JM, Bartoshuk LM. Associations between taste genetics, oral sensation and alcohol intake. *Physiol Behav* 2004;**82**:435–45.
101. Hinrichs AL, Wang JC, Bufe B, Kwon JM, Budde J, Allen R, et al. Functional variant in a bitter-taste receptor (hTAS2R16) influences risk of alcohol dependence. *Am J Hum Genet* 2006;**78**:103–11.

102. Miller LH, Mason SJ, Clyde DF, McGinniss MH. The resistance factor to Plasmodium vivax in blacks. The Duffy-blood-group genotype, FyFy. *N Engl J Med* 1976;**295**:302–4.
103. Tournamille C, Le Van Kim C, Gane P, Cartron JP, Colin Y. Molecular basis and PCR-DNA typing of the Fya/fyb blood group polymorphism. *Hum Genet* 1995;**95**:407–10.
104. Kwiatkowski DP. How malaria has affected the human genome and what human genetics can teach us about malaria. *Am J Hum Genet* 2005;**77**:171–92.
105. Maestre A, Muskus C, Duque V, Agudelo O, Liu P, Takagi A, et al. Acquired antibody responses against Plasmodium vivax infection vary with host genotype for duffy antigen receptor for chemokines (DARC). *PLoS One* 2010;**5**:e11437.
106. Galvani AP, Slatkin M. Evaluating plague and smallpox as historical selective pressures for the CCR5-Delta 32 HIV-resistance allele. *Proc Natl Acad Sci USA* 2003;**100**:15276–9.
107. Pearce-Duvet JM. The origin of human pathogens: evaluating the role of agriculture and domestic animals in the evolution of human disease. *Biol Rev Camb Philos Soc* 2006;**81**:369–82.
108. Fumagalli M, Cagliani R, Riva S, Pozzoli U, Biasin M, Piacentini L, et al. Population genetics of IFIH1: ancient population structure, local selection, and implications for susceptibility to type 1 diabetes. *Mol Biol Evol* 2010;**27**:2555–66.
109. Li CC. Population subdivision with respect to multiple alleles. *Ann Hum Genet* 1969;**33**:23–9.
110. Ziv E, Burchard EG. Human population structure and genetic association studies. *Pharmacogenomics* 2003;**4**:431–41.
111. Devlin B, Roeder K. Genomic control for association studies. *Biometrics* 1999;**55**:997–1004.
112. Pritchard JK, Stephens M, Donnelly P. Inference of population structure using multilocus genotype data. *Genetics* 2000;**155**:945–59.
113. Falush D, Stephens M, Pritchard JK. Inference of population structure using multilocus genotype data: linked loci and correlated allele frequencies. *Genetics* 2003;**164**:1567–87.
114. Price AL, Patterson NJ, Plenge RM, Weinblatt ME, Shadick NA, Reich D. Principal components analysis corrects for stratification in genome-wide association studies. *Nat Genet* 2006;**38**:904–9.
115. Stefflova K, Dulik MC, Barnholtz-Sloan JS, Pai AA, Walker AH, Rebbeck TR. Dissecting the within-Africa ancestry of populations of African descent in the Americas. *PLoS One* 2011;**6**:e14495.
116. Gonzalez Burchard E, Borrell LN, Choudhry S, Naqvi M, Tsai HJ, Rodriguez-Santana JR, et al. Latino populations: a unique opportunity for the study of race, genetics, and social environment in epidemiological research. *Am J Publ. Health* 2005;**95**:2161–8.
117. Lai CQ, Tucker KL, Choudhry S, Parnell LD, Mattei J, Garcia-Bailo B, et al. Population admixture associated with disease prevalence in the Boston Puerto Rican health study. *Hum Genet* 2009;**125**:199–209.
118. Seldin MF, Shigeta R, Villoslada P, Selmi C, Tuomilehto J, Silva G, et al. European population substructure: clustering of northern and southern populations. *PLoS Genet* 2006;**2**:e143.
119. Higgins PB, Fernandez JR, Goran MI, Gower BA. Early ethnic difference in insulin-like growth factor-1 is associated with African genetic admixture. *Pediatr Res* 2005;**58**:850–4.
120. Casazza K, Willig AL, Gower BA, Nagy TR, Hunter GR, Wallace S, et al. The role of European genetic admixture in the etiology of the insulin resistance syndrome in children: are the effects mediated by fat accumulation? *J Pediatr* 2010;**157**:50–56.e1.
121. Ding K, Kullo IJ. Geographic differences in allele frequencies of susceptibility SNPs for cardiovascular disease. *BMC Med Genet* 2011;**12**:55.
122. Hindorff LA, Sethupathy P, Junkins HA, Ramos EM, Mehta JP, Collins FS, et al. Potential etiologic and functional implications of genome-wide association loci for human diseases and traits. *Proc Natl Acad Sci USA* 2009;**106**:9362–7.
123. Parnell LD, Lee YC, Lai CQ. Adaptive genetic variation and heart disease risk. *Curr Opin Lipidol* 2010;**21**:116–22.

124. Ding K, Kullo IJ. Evolutionary genetics of coronary heart disease. *Circulation* 2009;**119**:459–67.
125. Casto AM, Feldman MW. Genome-wide association study SNPs in the human genome diversity project populations: does selection affect unlinked SNPs with shared trait associations? *PLoS Genet* 2011;**7**:e1001266.
126. Clark AG, Hubisz MJ, Bustamante CD, Williamson SH, Nielsen R. Ascertainment bias in studies of human genome-wide polymorphism. *Genome Res* 2005;**15**:1496–502.
127. Wise J. Consortium hopes to sequence genome of 1000 volunteers. *BMJ* 2008;**336**:237.
128. Jannot AS, Meziani R, Bertrand G, Gerard B, Descamps V, Archimbaud A, et al. Allele variations in the OCA2 gene (pink-eyed-dilution locus) are associated with genetic susceptibility to melanoma. *Eur J Hum Genet* 2005;**13**:913–20.
129. Lai CQ, Parnell LD, Arnett DK, Garcia-Bailo B, Tsai MY, Kabagambe EK, et al. WDTC1, the ortholog of Drosophila adipose gene, associates with human obesity, modulated by MUFA intake. *Obesity (Silver Spring)* 2009;**17**:593–600.
130. Stover PJ. Influence of human genetic variation on nutritional requirements. *Am J Clin Nutr* 2006;**83**:436S–42S.
131. Soloway PD. Gene nutrient interactions and evolution. *Nutr Rev* 2006;**64**:S52–4 discussion S72–S91.
132. Munoz-Moran E, Dieguez-Lucena JL, Fernandez-Arcas N, Peran-Mesa S, Reyes-Engel A. Genetic selection and folate intake during pregnancy. *Lancet* 1998;**352**:1120–1.
133. Whitehead AS. Changes in MTHFR genotype frequencies over time. *Lancet* 1998;**352**:1784–5.
134. Schulz LO, Bennett PH, Ravussin E, Kidd JR, Kidd KK, Esparza J, et al. Effects of traditional and western environments on prevalence of type 2 diabetes in Pima Indians in Mexico and the U.S. *Diabetes Care* 2006;**29**:1866–71.
135. Ravussin E, Valencia ME, Esparza J, Bennett PH, Schulz LO. Effects of a traditional lifestyle on obesity in Pima Indians. *Diabetes Care* 1994;**17**:1067–74.
136. Acuna-Alonzo V, Flores-Dorantes T, Kruit JK, Villarreal-Molina T, Arellano-Campos O, Hunemeier T, et al. A functional ABCA1 gene variant is associated with low HDL-cholesterol levels and shows evidence of positive selection in Native Americans. *Hum Mol Genet* 2010;**19**:2877–85.
137. Guevara-Cruz M, Tovar AR, Aguilar-Salinas CA, Medina-Vera I, Gil-Zenteno L, Hernandez-Viveros I, et al. A dietary pattern including nopal, chia seed, soy protein, and oat reduces serum triglycerides and glucose intolerance in patients with metabolic syndrome. *J Nutr* 2012;**142**(1):64–9.
138. Stintzing FC, Carle R. Cactus stems (Opuntia spp.): a review on their chemistry, technology, and uses. *Mol Nutr Food Res* 2005;**49**:175–94.
139. Chicco AG, D'Alessandro ME, Hein GJ, Oliva ME, Lombardo YB. Dietary chia seed (Salvia hispanica L.) rich in alpha-linolenic acid improves adiposity and normalises hypertriacylglycerolaemia and insulin resistance in dyslipaemic rats. *Br J Nutr* 2009;**101**:41–50.
140. Scrimshaw NS, Murray EB. The acceptability of milk and milk products in populations with a high prevalence of lactose intolerance. *Am J Clin Nutr* 1988;**48**:1079–159.
141. Jarvis JK, Miller GD. Overcoming the barrier of lactose intolerance to reduce health disparities. *J Natl Med Assoc* 2002;**94**:55–66.
142. Eng MY, Luczak SE, Wall TL. ALDH2, ADH1B, and ADH1C genotypes in Asians: a literature review. *Alcohol Res Health* 2007;**30**:22–7.
143. Brooks PJ, Enoch MA, Goldman D, Li TK, Yokoyama A. The alcohol flushing response: an unrecognized risk factor for esophageal cancer from alcohol consumption. *PLoS Med* 2009;**6**:e50.

Index

Note: Page numbers followed by "*f*" indicate figures, and "*t*" indicate tables.

B

C

D

E

F

G

H

I

K

L

M

N

O

P

T

U

V

W

Z

Hierarchical clustering
Method = complete
Dendrogram

TBC1D30
CALR
YWHAE
EIF2S3
EIF5A
NUTF2
HNRNPL
C9orf38
ZNF117
DHX9
PLK4
ABCB1
ABCB4
COPA
TPR
AXL
GPR137B
METTL7A
RASA4
CCNG2
ATP1B1
SETD2
GPNMB
PNPLA3
CG012
ACAT2
PPARD
SAT1
LSS
HMGCR
ACSL3
IDI1
DKK2
SQLE
CCL3
GPR183
ENPP2
BHLHE40
TIMP3
PTX3
VEGFC
KLF2
FADS2
AQP1
TNF
C3AR1
ALOX5
DHCR7
TUBA1A
TMEM135
CYP51A1
HMGCS1
LDLR
FADS1
SC4MOL
INSIG1
SCD
ABCA1

ALA DHA EPA

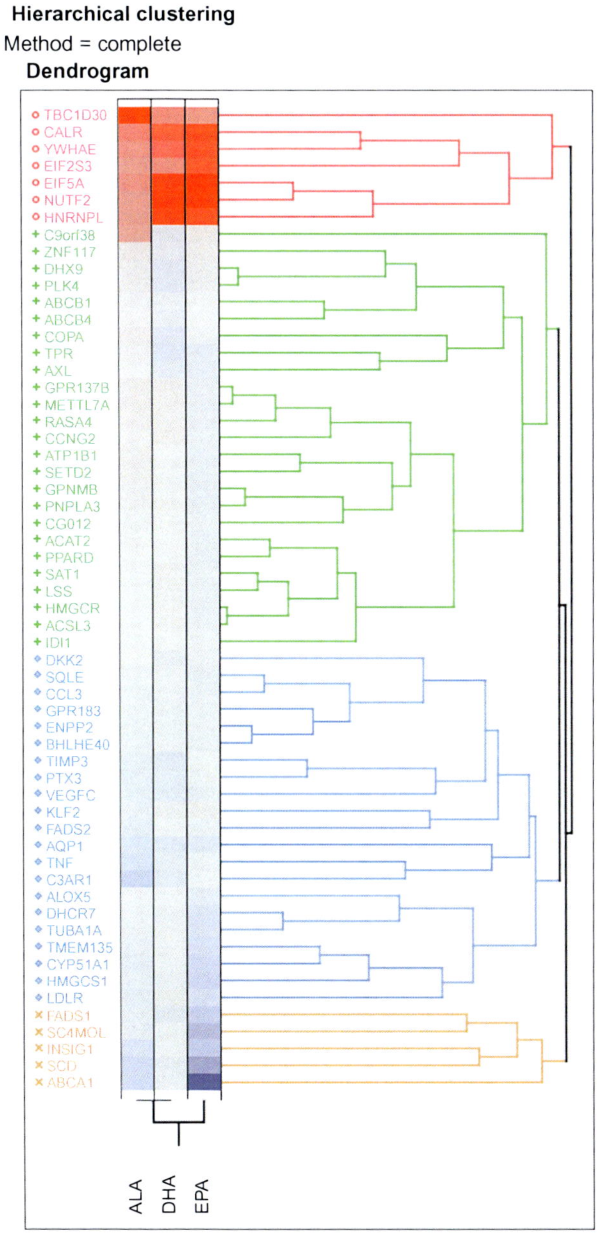

JOHN P. VANDEN HEUVEL, FIG. 3.